Foundations and Industrial Applications of Microwave and Radio Frequency Fields

Foundations and Industrial Applications of Microwave and Radio Frequency Fields

Physical and Chemical Processes

G. Roussy
Université de Nancy 1, France

and

J. A. Pearce
University of Texas at Austin, USA

JOHN WILEY & SONS
Chichester • New York • Brisbane • Toronto • Singapore

National 01243 779777
International (+44) 1243 779777

Other Wiley Editorial Offices

John Wiley & Sons, Inc., 605 Third Avenue,
New York, NY 10158-0012, USA

Jacaranda Wiley Ltd, G.P.O. Box 859, Brisbane,
Queensland 4001, Australia

John Wiley & Sons (Canada) Ltd, 22 Worcester Road,
Rexdale, Ontario M9W 1L1, Canada

John Wiley & Sons (SEA) Pte Ltd, 37 Jalan Pemimpin #05-04,
Block B, Union Industrial Building, Singapore 2057

Library of Congress Cataloging-in-Publication Data

Roussy, G. (Georges)
Foundations and industrial applications of microwave and radio frequency fields: physical and chemical processes / G. Roussy, J.A. Pearce.
p. cm.
Includes bibliographical references and index.
ISBN 0 471 93849 1
1. Microwave devices. 2. Microwaves — Industrial applications. 3. Radio frequency. I. Pearce, John A., 1946- . II. Title.
TK7876.R68 1995
621.381′3 — dc20 94-36676
CIP

British Library Cataloguing in Publication Data

A catalogue record for this book is available from the British Library

ISBN 0 471 93849 1

Typeset in 10/12pt Times by Laser Words, Madras, India
Printed and bound in Great Britain by Bookcraft (Bath) Ltd

Contents

Foreword

Why write a book on microwave and radio frequency fields? At first glance, it is a familiar topic and rather straight forward. I mean, every engineer and physicist knows what an electromagnetic field is, what the electric permittivity is, and that dielectric losses can heat a material. Furthermore, energy efficiency concerns have made this topic both popular and attractive. Most every paper on the efficient use of electricity includes at least one statement regarding the high value of dielectric drying, because the required energy can be directly delivered to the water without wasting energy heating the substrate or surroundings. Anyone who has been surprised by boiling ketchup sauce in a nearly-cold cheeseburger has an intuitive perception of what this means. Also, microwave cooking appliances do not seem at all tricky: a few pounds of power source, waveguide segment, cavity and packaging are all that is required. As a technical topic it seems not to be of much interest except to a few tens of specialized engineers and technicians. Unfortunately, such a view is highly misleading.

At second glance, things are seen to be, in fact, far more complicated. In truth, the complexities of the material interactions are often beyond the multidisciplinary capabilities of most of the designers involved. Those familiar with heat transfer may not know much more than the telegrapher's equations in electromagnetic theory. Specialists in electrical insulation or waveguide design may have no knowledge at all of heat transfer. Also, scholars well-versed in theoretical considerations are consistently disappointed by the real world: real materials are, unfortunately, hardly ever linear, homogeneous or isotropic. Scientists, engineers and manufacturers are often tempted to give up at this point for different reasons. Depending on their point of view they may restrict the consideration to over-simplified analysis, never really grappling with the fundamental problem. They may focus on understanding the scientific details and develop a solution much too sophisticated to be practically marketable. Or, they may emphasize design over understanding and design a device which heats or dries but is far too crude to become cost-effective. I would insist that the above categories are not purely imaginary. I have studied several examples of each instance and analyzed the market quite carefully: my assessment is that microwave and radio frequency applications are definitely underdeveloped.

Although I really feel that society is losing something by not coming to a complete understanding of how dielectric heating works, I leave it to others to comment on the issue; it is not my main point. I would suggest that the major loss is economic, which results in a more generalized loss of public welfare. There is consensus on the prerequisites for sustainable growth and energy supply. Energy efficiency at an affordable cost is a key consideration. It has often been observed that the lag between best-available technologies and most-often-used technologies is a good measure on which to base energy and economic policies.

We must not be idle. The fact is that, in the radio frequency and microwave case, the gap is substantial. We need efficient equipment and design tools for industrial applications. Cost-effective design depends critically upon successful initial designs which minimize re-design costs; these, in turn, depend on a clear and complete understanding of the material-field interaction. Only a few teams worldwide have the extensive experience needed; two examples are those in Austin, Texas and Nancy, France.

The authors of this book are filling the gaps: this is the major added value of this book, and, by the same token, the key for readers. Most people involved in this type of work will browse through the pages which they find straightforward or even common knowledge, and spend some considerable time exploring aspects that they are not familiar with and should not skip. The basic truth is that very few people already know all things presented in this book, very few know none of it, and most need insights and an update. However, all may understand the entire work. This is the other added value of this book: it is within reach. No specialized prerequisites are required; every concept is clearly and carefully explained and references for further reading are given. Examples of applications are provided, as well as summaries of the main subtopics.

It will be a pleasure for scholars and scientists to read of industrial applications: they will develop a feel for the marketplace. Even more importantly, it will provide a lift for engineers and technicians on the market side to understand the rationale behind their practice and thus improve their products. Again this is, in the long run, the most important gap that this book bridges. I am confidant that the best available technologies in this field will become more cost-effective, and that the most-often-used technologies will become more efficient. It is a unique contribution to sustainable growth.

Antoine Bastin
Sous Directeur
Direction de Etudes et Recherches, EdF
Paris, March 1994

Preface

The aim of our book is to review the basic knowledge upon which industrial applications of microwave and radio frequency energy are evolving. These applications are developing in many different domains, so it immediately becomes apparent that we must restrict our discussion to general physical principles. We have attempted to cover physics, engineering, chemistry and process engineering.

An important part of the book is allocated to a scientific approach to the construction of a model of a physical system. The model is a non-unique simplified description of the physical system. We have taken some care to present different points of view of the same problem to illustrate the diversity of approach, perhaps at the risk of disturbing the reader. We do hope that the few redundancies included will stimulate new visions of familiar problems.

The origin of the book was a lecture presented by GR at the Université de Nancy I. It has been fully reviewed and amplified by JP with an effort to match the text to readers of both the French and English languages. We refer to a limited list of basic but comprehensive texts in the hope that our work will be self-explanatory. We have had many discussions to homogenize the contents. The pleasure of our collaboration was limited only by the attempt to do our best. We sincerely hope that the reader will find the contents useful and applicable and enjoyable. We apologize for any puns which might be embedded in the work: one of the authors is under suspicion (JP) but there is insufficient evidence to prosecute.

We would like to gratefully acknowledge the kind tolerance of our families, especially our wives Ann and Jeannine, and the many and significant contributions of our colleagues and students. Specifically, J. M. Thiebaut and P. S. Schmidt have been exceptionally helpful. We also thank the Université de Nancy I, The University of Texas at Austin, EPRI and Electricité de France for their support, without which the research could not have been accomplished.

Georges Roussy
Directeur de Recherche, CNRS
Université Nancy I

John A. Pearce
B. N. Gafford Professor of
Electrical Engineering
University of Texas at Austin

Introduction

High power microwave (MW) and high frequency — i.e. radio frequency (RF) — techniques were originally developed during World War II for use in navigation, radar target detection, and to glue plywood sections of the "Mosquito" bomber (in the UK). The field has developed enormously since then. Microwave and RF methods have been successfully applied in industry to dry and/or thermally transform products ranging from macaroni to airplanes. In medicine MW and RF diathermy is used for the treatment of muscle pain (in physical and sports medicine) and tumors (as one the methods of inducing hyperthermia). The remarkable success of home microwave ovens in the United States and Japan has also been realized in the French market, in spite of the traditionalism of French cooking. The recent increased interest in electromagnetic heating has contributed to research, in general, and to increased public awareness; yet only a very small fraction of the public (and, ironically, only a small fraction of the technical community) are familiar with the fundamental principles of electromagnetic heating.

Presently, most of the industrial applications of electromagnetic heating are found where a change of state of non-conductive matter is involved. For example, defrosting, and dehydration involve the change of state of water, and are easily done using microwave and radio frequency techniques. Another important domain of application is where the transformation of a product requires thermal energy, such as polymerization, fusion of solids, and sintering processes. The technical advantages of microwave techniques are: (1) rapid heating is often easily obtained; (2) volumetric deposition of energy within the material avoids surface limitations; (3) economy of energy can be realized in that it is not necessary to heat the environment as well as the product (although in many important applications, such as drying, the environment must be at the product temperature so no energy advantage may be realized); (4) electromagnetic heating is non-polluting — at least it removes the source of pollution from the processing plant to the electric generating station — and non-contact, so food processing may be simplified; (5) in many important cases electromagnetic heating is easy to apply; (6) electromagnetic heating can be automated.

One very important aspect of the development of microwave techniques has been the cost of the technology. Studies show that MW and RF installations cost around 1 to 3 k$ (US) per kW with operating costs slightly higher than those of conventional heating processes. In many cases, a correct prediction of the value of the product versus the cost of the necessary equipment is not easily done. Nevertheless, in applications where this analysis is tractable, it has been shown that most often the value of electromagnetic heating lies either in increased production rates, improved quality, and reduced loss of high value products, or in accomplishing heating in applications for which conventional methods fail, two clear examples of which are the treatment of large volumes of toxic waste and devolatilization of contaminated soil *in situ*. There is also apparently a correlation between the "information content" of a product and the attractiveness of electromagnetic heating. For example, sand (SiO_2) is of low information content (low value) and not attractive for EM heating; however, fine crystal ware or a boule of silicon for integrated circuit fabrication, which have much higher information contents (and thus high value), might be very attractive candidates. The same may be said of molded zeolite catalysts which are extremely difficult to dry by conventional methods but very receptive to RF heating.

Although the economic problem is an important one from an engineering point of view, this text addresses electromagnetic heating from a scientific viewpoint, only. Economic considerations often vary with extremely short time constants and frequently depend as much on public policy as on technical considerations.

Our goal is to create a text which is equally useful for classroom or self-study by electrical engineers, physicists, chemists and materials scientists. This text will present microwave and radio frequency techniques from the point of view of industrial applications with special attention to electromagnetic energy and material interaction at the microscopic level. The physical principles and engineering methods presented are common to the other applications as well. We begin with a brief description of the complete set of macroscopic governing equations, including conduction processes, which will form a concise review for physicists and electrical engineers and may be instructive for materials scientists. We then describe in detail the microscopic interaction effects including many useful results obtained from spectroscopic studies, which are not usually treated in books of this type. The purpose of this section is to describe in some detail inhomogeneous but approximately isotropic composite materials, a recently significant class of materials. The text concludes with a discussion of representative industrial applications including the emerging new field of microwave catalysis. The primary concern in this text will be the description of systems which are known to be well treated by electromagnetic heating. A second aspect is how to treat an application which has been discovered, and how to interpret the results.

PART 1

Electrical Aspects

The first part of the book, Chapters 1 through 5, contains a concise summary of the governing electrical relationships and their use in designing and constructing practical radio frequency and microwave systems. Chapter 1 presents the notation and overall approach which we will use in analyzing the electromagnetic fields. The formulation of a complete system of field equations — Maxwell's equations, boundary conditions and constitutive relations — is described. The first chapter also contains a brief description of the inter-relationship of the electrical, thermal and thermodynamic principles which are explored in more detail in the second part of the book. Chapter 2 treats the problem of transmitting the electromagnetic energy from the generator to the load — the transmission line — in both radio frequency and microwave realizations. We place the discussion of transmission lines before that of generators and loads in order to motivate those discussions and to introduce Smith chart analysis methods in an easily understood context. Chapter 3 presents radio frequency and microwave circuit design principles. We pay special attention to the non-ideal nature of ordinary circuit elements. We also include a detailed discussion of vacuum tubes and radio frequency generator design since these are described only extremely rarely, and not at all in recent publications. We leave detailed treatment of magnetron tubes to the other authors, who have done an excellent job. Special attention is given to both radio frequency and microwave impedance matching networks because of their importance in 50 Ω RF technology and in high power microwave systems where reflections cannot be tolerated. Chapter 4 contains a complete survey of methods which may be used to model field applicators and load materials. We discuss several alternative analytical and numerical methods for creating models of the applicator–load combinations. We also include discussion of the heating characteristics of various applicators in both radio frequency and microwave use. The chapter is quite long as a consequence, and we beg the reader's pardon if the particular applicator which they are most interested in is not completely described. Finally, Chapter 5 describes radio frequency and microwave instrumentation techniques, particularly those for estimating applied field strengths in the load. The special problem of temperature measurement in an EM field is described: fortunately, there are several forms of commercial optical measurement systems with which these measurements can be made.

1 Governing Electromagnetic and Thermal Field Relations

In this chapter, we will review the physical aspects of microwave (MW) and radio frequency (RF) fields. These have been well described in the literature, especially in books concerned with telecommunication or radar techniques. The chapter contains a brief review of the field relations in order to establish the working notation. Special attention is paid to the transition between quasi-static and time-varying analysis. Additional fundamental relations necessary for the understanding of material interactions at high power density will also be presented here.

All electromagnetic radiation comes from the acceleration of charge. The topic at hand is one of radiative energy transfer to achieve heating or chemical transformation. Therefore, we begin with the time-varying governing equations for electromagnetic fields. We will treat non-radiating static (i.e. DC) electric and magnetic fields as a special case of the more general Maxwell equations. A discussion of the static case is of considerable use since many practical geometries can be treated approximately as static or quasi-static problems, especially when the maximum dimension of the problem is small. In general, sinusoidally varying fields are used to obtain electromagnetic heating — in part due to the relative ease of analysis and in part due to the achievable coupling efficiency. Consequently, the majority of the analysis will be done assuming sinusoidal source fields. The boundary conditions for the electromagnetic fields are of controlling importance in many applications, so a careful development is included in this chapter with special attention to the case of semiconductor–dielectric interfaces in sinusoidally varying (time harmonic) fields. We conclude the chapter with a brief discussion of the governing transient thermal equations including interfacial heat transfer, surface convection and radiation boundary conditions.

1.1 OVERVIEW OF THE ELECTROMAGNETIC SPECTRUM

Industrial microwave and radio frequencies are limited by international convention. The allowed ISM (Industrial, Scientific and Medical) frequencies differ slightly from country to country, however. A representative list is:

6.78 MHz	(US)
13.56 MHz	
27.12 MHz($\pm$ 0.05 %)	
40.56 MHz	(US)
434 MHz	(Germany, Italy)

894 MHz	(UK)
915 MHz	(US)
2450 MHz (± 25 MHz)	

By comparison, the standard frequency bands used for telecommunications are:

175–280 kHz	Long wave AM radio broadcast (France)
500–1600 kHz	AM radio (US)
1–25 MHz	Shortwave and amateur communications (US)
54–88 MHz	Television broadcast channels 2–6 (US)
88–106 MHz	FM radio
174–216 MHz	Television broadcast channels 7–13 (US)
2–2.7 GHz	Wireless cable systems, consumer (US)
2.3–2.45 GHz	E-band radar (US, formerly S-band)
3.3–3.5 GHz	F-band radar (US, formerly S-band)
3.7–4.2 GHz	C-band satellite communications for television
6 GHz	Postal telecommunications (Francc)
11.2–12.7 GHz	K-band satellite communications for television

The ISM frequencies are surrounded by bands allocated for communication purposes. Consequently, ISM apparatus must be carefully designed to prevent radiated signals from interfering with other uses of the EM spectrum.

1.2 CHARGE AND FIELD QUANTITIES

The original measurements of "electric flux" were made by Benjamin Franklin in his kite experiment of 1752. The first recorded accident in electrical science followed soon after when Professor Georg Richmann (of St. Petersburg) was killed while attempting to duplicate Franklin's experiment (Stillings, 1973). In a classic series of physiologic experiments Volta (*ca.* 1775) demonstrated the electric potentials generated by living organisms, thus facilitating both electro-medical quackery and modern prosthetic devices such as pacemakers. Colonel Charles de Coulomb invented a very sensitive torsion balance (in 1785) for measuring small forces with which he established an empirical relationship between two discrete electrostatic point charges, q_1 and q_2 (C), and the force of attraction for charges of opposite sign (or repulsion for charges of like sign):

$$\mathbf{F}_{12} = -\mathbf{F}_{21} = \frac{q_1 q_2}{4\pi\varepsilon_0 R^2}\mathbf{a}_R. \tag{1.1}$$

In our standard notation boldface type will be used to represent a vector quantity. Somewhat ironically, the inverse square law had first been studied by Cavendish in secret experiments which were not published until the manuscripts were discovered much later, in 1879. In this expression the constant of proportionality, $\varepsilon_0 = 8.85 \times 10^{-12}$ F/m, is the electric permittivity of free space (i.e. a vacuum). The vector direction of the force is along the separation vector between the charges, $\mathbf{R}$, which has the normalized unit vector $\mathbf{a}_R$ (see Figure 1.1a).

The total electrostatic force on a point charge is the vector sum of all forces from neighboring point charges (Figure 1.1b):

$$\mathbf{F}_1 = \frac{q_1}{4\pi\varepsilon_0}\sum_{i=2}^{N}\frac{q_i \mathbf{a}_{R_i}}{R_i^2}. \tag{1.2}$$

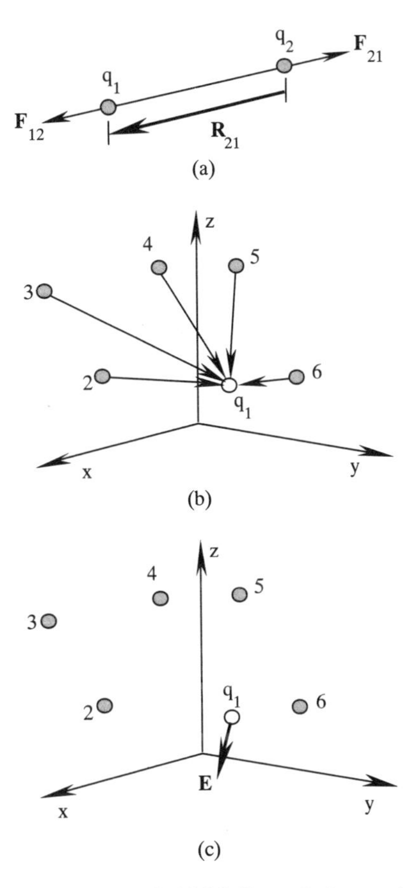

Fig. 1.1 *Coulomb force law and electric field (a) Force between two point charges. (b) Force exerted by multiple point charges. (c) Electric field due to surrounding charges.*

Expressing the force as a discrete vector sum is usually inconvenient. So, the net force of other charges on, for example, q_1 is often more conveniently expressed in terms of a vector field, the electric field strength, $\mathbf{E}(V/m)$, a function of position and time (Figure 1.1c). The electric field "felt" by charge q_1 is, then, the sum on the right-hand side of equation (1.2). So, the electric field created by a single discrete point charge is:

$$\mathbf{E}_{\text{point charge}} = \frac{q}{4\pi\varepsilon_0 R^2}\mathbf{a}_R \tag{1.3}$$

where:

$$\mathbf{R} = (x - x')\mathbf{a}_x + (y - y')\mathbf{a}_y + (z - z')\mathbf{a}_z.$$

If q is located at the origin then the position vector, $\mathbf{R}$, becomes the usual spherical coordinate system radius and the vector direction of $\mathbf{E}$ is also radial, as is expected.

Charges, (electrons, protons and positrons) are discrete by definition. However, it is often difficult to calculate the electric field of even simple charge systems with discrete models. For the class of problems in which the minimum dimension of interest is large compared to inter-atomic distances or molecular dimensions, it is much simpler to define an approximately continuous scalar volume charge distribution, ρ_v (C/m^3), so that we may apply vector calculus techniques to find the **E**-field:

$$\rho_v = \lim_{\Delta v \to 0} \sum_{i=1}^{N} q_i \Big/ \Delta v. \tag{1.4}$$

In this formulation the sum is over all discrete charges enclosed within the control volume, Δv. Consequently, when the continuous volume charge is used the results are inherently valid only for the macroscopic field case. We may use this notation to conveniently describe volume charge distributions, ρ_v (the general case) or sheet charge, ρ_s (C/m^2), or line charges, ρ_L (C/m), as well: the differential charge $dq(C) = \rho_v\,dv = \rho_s\,dS = \rho_L\,dL$. Using this definition the calculation of the static vector electric field at point $P = (x, y, z)$ (see Figure 1.2) from a charge at (x', y', z') may be formalized:

$$\mathbf{E}(x, y, z) = \iiint_{-\infty}^{\infty} \frac{\rho_v(x', y', z')\,dx'dy'dz'[\mathbf{r} - \mathbf{r}']}{4\pi\varepsilon_0\left[(x - x')^2 + (y - y')^2 + (z - z')^2\right]^{3/2}} \tag{1.5}$$

where:

$$\mathbf{r} - \mathbf{r}' = (x - x')\mathbf{a}_x + (y - y')\mathbf{a}_y + (z - z')\mathbf{a}_z.$$

Of course, the charges around the reference charge q_1 may be in motion. If so their net motion generates a vector magnetic flux density field, $\mathbf{B}(T = \text{Wb/m}^2)$, which also exerts a force on q_1 if it is in motion as well (i.e. with velocity $\mathbf{u}_1$). In that case the net force on point charge q_1 has two contributions:

$$\mathbf{F}_1 = q_1(\mathbf{E} + \mathbf{u}_1 \times \mathbf{B}) \tag{1.6}$$

where the $\times$ symbol denotes the vector cross product and the **E** and **B** fields are due to all other charges except q_1. We note here (without proof) that the electric field strength at a point is determined by the relative positions of the external charges (i.e. not q_1) and the magnetic field strength by their relative velocities.

The field strength vector fields (both electric and magnetic) have flux density vector fields associated with them. The electric field has an electric flux density, **D** (C/m^2), associated with it. The source of the electric **E** and **D** fields is the volume charge distribution, ρ_v. The physical interpretation usually assigned to **D** is that it is an indication of the "lines of force" of the electric field, a somewhat nebulous but useful description. In use, the closed surface

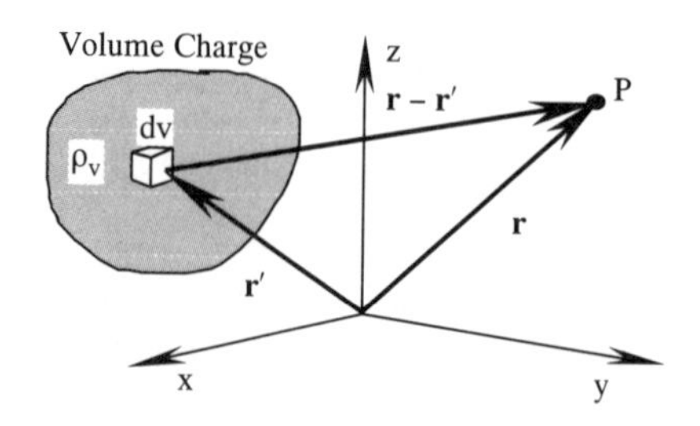

Fig. 1.2 *Geometry for calculation of the electric field at point P from a volume distribution of charge.*

integral of **D** is the total electric flux for the volume which is equal to the charge enclosed; so **D** (or, more rigorously, its divergence) is more clearly described as a measure of the local density of charge. The relationship between **E** and **D** will be described in some detail in Chapter 7.

In conductive media the current density, **J** (A/m^2), also has the characteristics of a flux density in the sense that it represents a charge flux associated with **E** and/or **B**. Finally, the magnetic field strength, **H** (A/m) is associated with the magnetic flux density, **B**. The source of magnetic **B** and **H** vector fields is current density, **J**. The magnetic field strength may be determined from the Biot-Savart law:

$$\begin{aligned}\mathbf{H}(x, y, z) &= \iiint_{-\infty}^{\infty} \frac{\mathbf{J}(x', y', z')\,\mathrm{d}v' \times \mathbf{a}_R}{4\pi R^2} \\ &= \iiint_{-\infty}^{\infty} \frac{\mathbf{J}(x', y', z')\,\mathrm{d}x'\mathrm{d}y'\mathrm{d}z' \times [\mathbf{r} - \mathbf{r}']}{4\pi \left[(x - x')^2 + (y - y')^2 + (z - z')^2\right]^{3/2}}\end{aligned} \tag{1.7}$$

where:

$$\mathbf{r} - \mathbf{r}' = (x - x')\mathbf{a}_x + (y - y')\mathbf{a}_y + (z - z')\mathbf{a}_z.$$

The relationships between the flux density fields and the field strengths are described by constitutive relations, which will be discussed in Section 1.4 of this chapter.

1.3 ELECTROMAGNETIC FIELD EQUATIONS, MAXWELL RELATIONS

By about 1860 four fundamental empirical/theoretical laws of electric and magnetic fields had been established. Based on the above electric field relations and vector mathematics which he developed, Gauss established that the net electric flux density through a closed surface is equal to the net charge enclosed (Gauss' Electric Law):

$$\oint_{\Sigma} \mathbf{D} \cdot \mathrm{d}\mathbf{S} = Q_{\text{enclosed}} = \iiint_{\text{vol}} \rho_v \,\mathrm{d}x\mathrm{d}y\mathrm{d}z \tag{1.8}$$

and that the net magnetic flux density through a closed surface is zero (Gauss' Magnetic Law):

$$\oint_{\Sigma} \mathbf{B} \cdot \mathrm{d}\mathbf{S} = 0. \tag{1.9}$$

In these expressions Σ denotes a closed surface and the vector sense of d**S** is always out of a closed surface (Figure 1.3). The integration volume on the right hand side of (1.8) is the volume defined by Σ, and $\cdot$ denotes the vector dot product. These expressions are members of the class of continuity relations. We may conclude from them that, in a macroscopic sense at least, the electric field lines have point sources (discrete origins and terminations) and the magnetic field lines close on themselves (magnetic fields are "solenoidal") and have no discrete sources.

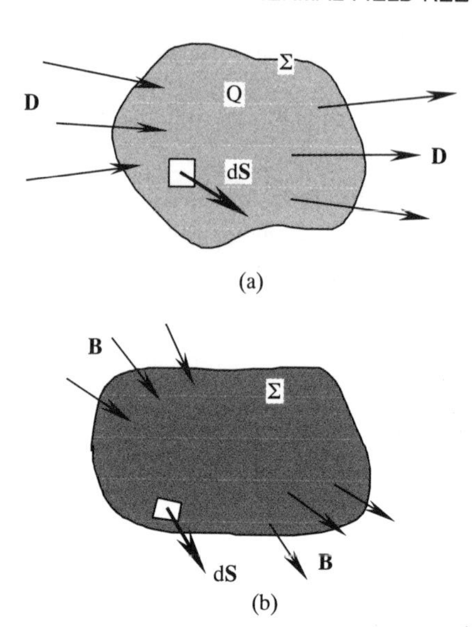

Fig. 1.3 *Geometry for Gauss' Law calculations. (a) Gauss' electric law geometry. (b) Gauss' magnetic law geometry.*

Also by Maxwell's time, the relationship between time-varying magnetic flux density and induced electric field (Faraday's Law) had been established:

$$\oint_C \mathbf{E} \cdot d\mathbf{L} = -\frac{\partial}{\partial t}\left\{\iint_S \mathbf{B} \cdot d\mathbf{S}\right\}. \tag{1.10}$$

Referring to Figure 1.4, C is the closed contour in the direction of the differential segment $d\mathbf{L}$, the vector direction of $d\mathbf{S}$ is given by the right hand rule (fingers of the right hand in the direction of $d\mathbf{L}$ means the thumb is in the direction of $d\mathbf{S}$), and the open surface S is any surface subtended by the contour C. The final empirical relationship known by Maxwell, Ampère's Law (based on an extension of Oersted's experiments), was complete for the electrostatic case but contained a paradox in the time-varying case. In the static case (Figure 1.5a) the total current enclosed by the contour C has been expressed in terms of the current density vector, $\mathbf{J}$ (A/m):

$$\oint_C \mathbf{H} \cdot d\mathbf{L} = I_{\text{enclosed}} = \iint_S \mathbf{J} \cdot d\mathbf{S}. \tag{1.11}$$

As shown in Figure 1.5b, this relation must hold for any open surface S defined by C. Then the two surfaces, S_1 which has a non-zero conduction current density in the wire (i.e. charge flux density) and S_2 which has a zero conduction current density in the capacitor, must result in the same left hand side for equation (1.11); but they cannot. Because of this the four relations do not form a complete definition of the four vector fields in the general time-varying case.

Maxwell postulated the existence of a time-varying flux density in the capacitor which had the same effect as conduction current density, $\mathbf{J}$, and showed that such a flux density must have the form of the time derivative of $\mathbf{D}$. The complete self-consistent set of Maxwell's

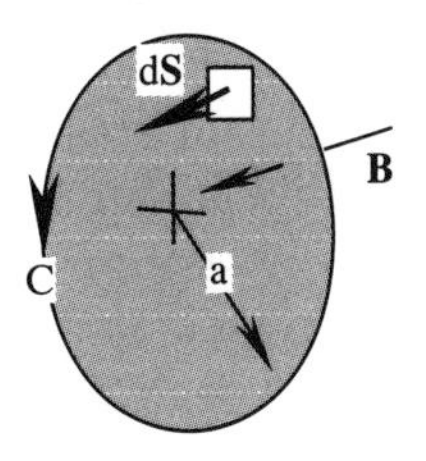

Fig. 1.4 *Faraday's Law integration contour.*

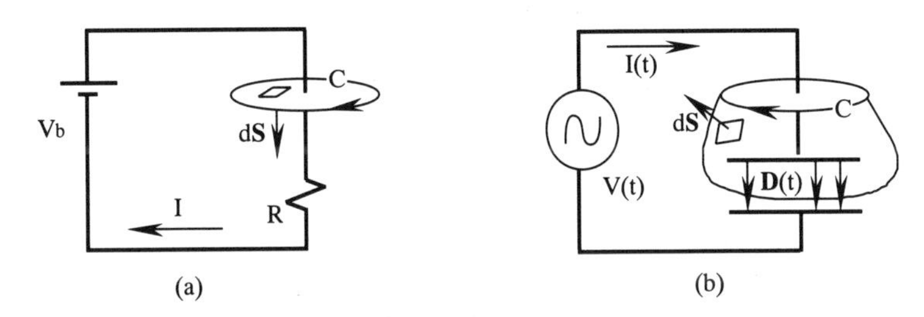

Fig. 1.5 *Faraday's Law applied to example circuits. (a) DC current example. (b) AC current example.*

equations in integral form are thus:

$$\begin{aligned}
&\oint_C \mathbf{E}\cdot d\mathbf{L} = -\frac{\partial}{\partial t}\left\{\iint_S \mathbf{B}\cdot d\mathbf{S}\right\}\\
&\oint_C \mathbf{H}\cdot d\mathbf{L} = \iint_S \mathbf{J}\cdot d\mathbf{S} + \frac{\partial}{\partial t}\left\{\iint_S \mathbf{D}\cdot d\mathbf{S}\right\}\\
&\oint_\Sigma \mathbf{D}\cdot d\mathbf{S} = \iiint_{\text{vol}} \rho_v \, dx\,dy\,dz\\
&\oint_\Sigma \mathbf{B}\cdot d\mathbf{S} = 0.
\end{aligned} \tag{1.12}$$

The equation set may also be expressed in point form by using the Divergence Theorem on Gauss' two laws and Stoke's Theorem on Faraday's and Ampère's Laws:

$$\begin{aligned}
&\nabla \times E = -\frac{\partial \mathbf{B}}{\partial t} \qquad \nabla \times \mathbf{H} = J + \frac{\partial \mathbf{D}}{\partial t}\\
&\nabla \cdot \mathbf{D} = \rho_v \qquad \nabla \cdot \mathbf{B} = 0
\end{aligned} \tag{1.13}$$

where ∇ represents the vector gradient operator.

In equations (1.12) and (1.13) the field quantities are general functions of space and time: $\mathbf{E}, \mathbf{D}, \mathbf{H}, \mathbf{B} = f(x, y, z, t)$. Usually, the generators in MW and RF systems are sinusoidal; so several simplifications can be obtained by using Fourier (frequency) domain analysis and complex sinusoidal representation: $\mathbf{E}, \mathbf{D}, \mathbf{H}, \mathbf{B} = f(x, y, z, \omega) = f(\mathbf{r}, \omega)$.

$$\mathbf{E}(\mathbf{r}, \omega) = \int_{-\infty}^{\infty} \mathbf{E}(\mathbf{r}, t) e^{-j\omega t}\, dt$$

so:

$$\mathbf{E}_0(\mathbf{r}) \cos(\omega t + \phi) \rightarrow \mathbf{E}_0(\mathbf{r}) e^{-(j\omega t + \phi)} \tag{1.14}$$

where $j = \sqrt{-1}$, $\mathbf{r}$ is the vector position, and ω is the angular frequency (radian/s). The most significant simplification is that the time domain differential equations become algebraic equations in the frequency domain. Taking the derivative in the time domain is equivalent to multiplication by $j\omega$ in the frequency domain — likewise, the integral over time is obtained by dividing by $j\omega$ in the frequency domain. Note that space derivatives, the divergence and curl, are unaffected by transformation into the frequency domain.

$$\begin{aligned} \nabla \times \mathbf{E} &= -j\omega\mathbf{B} \qquad & \nabla \times \mathbf{H} &= \mathbf{J} + j\omega\mathbf{D} \\ \nabla \cdot \mathbf{D} &= \rho_v \qquad & \nabla \cdot \mathbf{B} &= 0 \end{aligned} \tag{1.15}$$

An equivalent transformation can quickly be obtained for the integral form of Maxwell's equations. Whenever the frequency domain form of Maxwell's equations is used it must be understood that the results apply only when the generator signals are sinusoids. Of course, the equivalent time domain results are quickly obtainable by inverse Fourier transform in the case of non-sinusoidal signals.

There is one more important continuity relation, the law of conservation of charge, which states that in the general time-varying case the net charge flux for a control volume defined by a closed surface, Σ, must be stored within it (as on the plates of a capacitor, for example):

$$\begin{aligned} \oint_{\Sigma} \mathbf{J} \cdot d\mathbf{S} &= -\frac{\partial}{\partial t}\left\{\iiint_{vol} \rho_v\, dv\right\} \qquad & \nabla \cdot \mathbf{J} &= -\frac{\partial \rho_v}{\partial t} \\ \oint_{\Sigma} \mathbf{J} \cdot d\mathbf{S} &= -j\omega\left\{\iiint_{vol} \rho_v\, dv\right\} \qquad & \nabla \cdot \mathbf{J} &= -j\omega\rho_v. \end{aligned} \tag{1.16}$$

Note the substantial similarity between the time and frequency domain forms of the continuity equation.

1.4 CONSTITUTIVE RELATIONS

There are three constitutive relations which express each of the field strength vectors, $\mathbf{E}$ and $\mathbf{H}$, in terms of their three respective flux densities, $\mathbf{D}$, $\mathbf{B}$ and $\mathbf{J}$. The general form of the constitutive relation is that the flux density (·/m^2) = medium property (·/m) times the field strength (·/m):

$$\mathbf{D} = \varepsilon\mathbf{E} \qquad \mathbf{B} = \mu\mathbf{H} \qquad \mathbf{J} = \sigma\mathbf{E} \tag{1.17}$$

where ε is, again, the electric permittivity (F/m), μ is the magnetic permeability (H/m), and σ is the electrical conductivity (S/m). The last of these should be recognizable as the vector field form of Ohm's law. Each of the field vectors has three components, x, y and z in the Cartesian system. Consequently, the medium properties are, in general, tensors. As a specific example we will discuss the electrical conductivity:

$$\begin{bmatrix} J_x \\ J_y \\ J_z \end{bmatrix} = \begin{bmatrix} \sigma_{xx} & \sigma_{xy} & \sigma_{xz} \\ \sigma_{yx} & \sigma_{yy} & \sigma_{yz} \\ \sigma_{zx} & \sigma_{zy} & \sigma_{zz} \end{bmatrix} \begin{bmatrix} E_x \\ E_y \\ E_z \end{bmatrix}. \tag{1.18}$$

The system of equations represented in equation (1.18) describes the complete interaction between the electric field and current density. The physical significance of the conduction current density, **J**, is that it is the net *translational* motion of charge per unit area at a point. *Free charge* will be accelerated by an electric field in accordance with the Coulomb force law (equation (1.6))—that is, the force, and thus the acceleration vector, will be parallel to the electric field. In free space (i.e. a vacuum) the velocity of the charge may increase essentially without bound (up to relativistic limits); however, in a material medium the charge will accelerate until it collides with neighboring atoms or molecules. In semi-conducting material a finite average drift velocity of charge is established which depends on the mean free path between collisions and on the electric field. This topic will be treated in more detail in Chapter 7. At this point in the discussion it is sufficient to note that it is not possible for the y-direction electric field to accelerate charge in either the x- or z-directions. So, the off-diagonal terms in equation (1.18) must be zero. The system of equations then reduces to:

$$J_x = \sigma_x E_x \qquad J_y = \sigma_y E_y \qquad J_z = \sigma_z E_z. \tag{1.19}$$

If the material medium is linear the axial electrical conductivities will not depend on the electric field strength; σ_x, σ_y and σ_z are not functions of field strength in a linear medium. If the material is isotropic the electrical conductivity is not dependent on direction, so $\sigma_x = \sigma_y = \sigma_z = \sigma$. If the medium is homogeneous then the electrical conductivity has the same value everywhere. Electrolytic solutions are examples of media which may safely be assumed linear, homogeneous and isotropic with σ a simple scalar constant. This assumption also holds for most examples of bulk semiconductor materials (Si and Ge) but breaks down in a transistor where regional doping creates inhomogeneous conductivity. Whole muscle tissue *in situ* is linear (for reasonable electric field strengths), but certainly neither isotropic nor homogeneous, even after tissue death, owing to the highly layered structure and the presence of the low conductivity collagenous connective tissue supporting structure.

Turning now to the electric flux density, **D**, the physical significance of the permittivity, ε, in a material medium is that it is a measure of the work done in rotating or polarizing *bound charge* in atomic or molecular structures by an electric field. A similar argument regarding off-diagonal points in the tensor can be constructed for the electric permittivity, although in this case it is less clear that the off-diagonal terms must be zero, especially in a microscopic view of heterogeneous crystalline materials. Nevertheless, while there are certain materials for which the linear, homogeneous and isotropic assumptions may be questionable, a large number of materials may be assumed sufficiently simple that these assumptions will apply with acceptable accuracy. Subsequent derivations in Part 1 of this book will assume that ε is a simple scalar constant. We reserve a detailed discussion of polarization in more complex materials for Chapters 6 and 7.

The magnetic permeability relates the magnetic flux density, **B** (in essence the density of magnetically induced force in a material), to the magnetic field strength, **H**. Most practical materials (plastics, many metals, most body tissues, oils, wax, water and others) are essentially magnetically transparent; that is, they have magnetic permeabilities equal to free space,

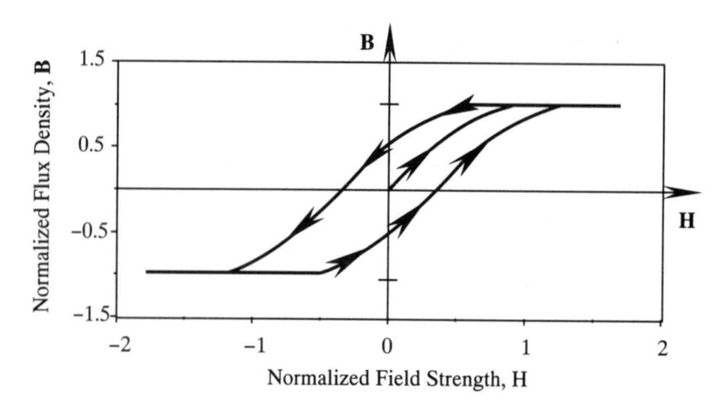

Fig. 1.6 *Normalized magnetic flux obtained, B, vs applied field strength, H, for a typical magnetic material. Saturation and hysteresis are shown.*

$\mu_0 = 4\pi \times 10^{-7}$ H/m. Notable exceptions include ferrites, ferrous metals, paramagnetic salts, oxygen (which is weakly paramagnetic) and cobalt. Magnetic permeabilities different from free space arise from net orientation of atomic and molecular magnetic moments (both electron spin moments and nuclear moments). In this book we will be primarily concerned with dielectrics and semiconductors, so most of the materials considered have the permeability of free space. Other materials will be treated as linear, homogeneous and isotropic even though real materials of this type exhibit saturation phenomena (see Figure 1.6). The material illustrated is saturated at high field strength because no increase in **B** can be realized once the maximum number of magnetic moments are oriented in parallel with **H**. In Figure 1.6 one can also see the hysteresis created by saturation effects, so materials of this type are only well behaved at very low magnetic field strengths.

1.5 BOUNDARY CONDITIONS

For the most interesting case of heterogeneous materials it is necessary to solve the differential (point) form of Maxwell's equations (equations (1.12)) in finite homogeneous regions subject to boundary conditions at the interfaces between the regions. Boundary conditions are also the forcing functions for integral equation solutions in homogeneous media. In heterogeneous media the working assumption is that the individual material regions are homogeneous and large compared to atomic or molecular dimensions. Of course, the point forms of the equations are not valid at discontinuities. We will apply the integral forms of Maxwell's equations at interfaces and discontinuities to derive the governing relations for the normal and tangential components of the magnetic and electric field strengths. Then, the point form equations (1.13), constitutive relations and boundary conditions comprise a complete and practically solvable system of equations for heterogeneous media.

1.5.1 Electric field boundary conditions

Imagine an interface between two media with respective electrical properties ε_1, σ_1 and ε_2, σ_2 as shown in Figure 1.7a. We establish a closed pillbox Gaussian surface of finite area, A, which is large compared to molecular dimensions and small compared to gradients in the

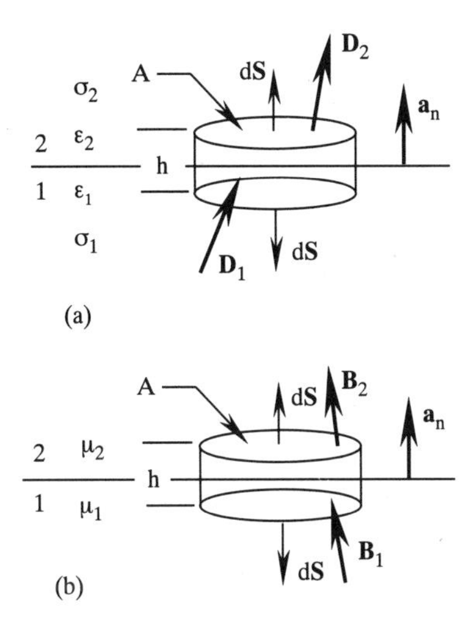

Fig. 1.7 *Geometry used for the determination of normal boundary conditions. (a) Gauss' electric law. (b) Gauss' magnetic law.*

D-field. Let the height of the pillbox, h, decrease until just the interface is enclosed within it; then the total charge enclosed is only the surface charge, ρ_s (C/m^2): $Q_{\text{enclosed}} = \rho_s A$. Remembering that the differential surface vector always points out of Σ gives the boundary condition for the normal component of the **D** and **E** fields:

$$\lim_{h\to 0}\left\{\oint_{\Sigma} \mathbf{D}\cdot d\mathbf{S}\right\} = D_{n2}A - D_{n1}A = \rho_s A$$

so that: (1.20)

$$\rho_s = D_{n2} - D_{n1} = \varepsilon_2 E_{n2} - \varepsilon_1 E_{n1}.$$

The n subscript refers to the component of the vector normal to the interface. Of course, a pure (ideal) dielectric has no free charge available (the charge in an ideal dielectric is bound charge only), so the only way a surface charge may exist at pure dielectric interfaces is if it is placed there by external means. When both media are pure dielectrics it is common to assume that $\rho_s = 0$. If either or both media are semiconductors a surface charge can exist. We will calculate its value presently. Finally, if medium 1 is a perfect conductor and medium 2 an ideal dielectric then $\mathbf{E}_1 = 0$ everywhere ($\mathbf{J}_1$ is unbounded because σ_1 is unbounded) and $D_{n2} = \rho_s$.

Returning to the semiconductor interface question, the normal electric field must also satisfy conservation of charge requirements:

$$\lim_{h\to 0}\left\{\oint_{\Sigma} \mathbf{J}\cdot d\mathbf{S}\right\} = J_{n2}A - J_{n1}A = -A\frac{\partial \rho_s}{\partial t}$$

so that: (1.21)

$$\frac{\partial \rho_s}{\partial t} = J_{n1} - J_{n2} = \sigma_1 E_{n1} - \sigma_2 E_{n2}.$$

The results of (1.20) and (1.21) may be quickly combined in the frequency domain to find the normal electric field boundary condition for semiconductor boundaries in both the time and frequency domains:

$$(\sigma_1 + \mathrm{j}\omega\varepsilon_1)E_{\mathrm{n}1} = (\sigma_2 + \mathrm{j}\omega\varepsilon_2)E_{\mathrm{n}2}$$

and:

$$\sigma_1 E_{\mathrm{n}1} + \varepsilon_1 \frac{\partial E_{\mathrm{n}1}}{\partial t} = \sigma_2 E_{\mathrm{n}2} + \varepsilon_2 \frac{\partial E_{\mathrm{n}2}}{\partial t} \tag{1.22}$$

The first-order Bernoulli differential equations implied by the corresponding time domain expressions mean that a surface charge placed at the interface of semiconducting media will be swept out by the dominant electric field (in the direction of **E**) according to a standard exponential solution:

$$\rho_{\mathrm{s}}(\mathrm{t}) = \rho_{\mathrm{s}}(0)\mathrm{e}^{-\mathrm{t}/\tau} \tag{1.23}$$

with time constant given by $\tau = \varepsilon/\sigma(\mathrm{s})$. So, the time average value of surface charge, the DC component in a sinusoidal electric field, is necessarily zero. Inspection of the frequency domain expression yields the condition at the interface for a static electric field $(\omega \approx 0)$: $\sigma_1 E_{\mathrm{n}1} = \sigma_2 E_{\mathrm{n}2}$. The frequency domain expression can be used to obtain the surface charge as a function of frequency:

$$\rho_{\mathrm{s}} = \left[\varepsilon_2 - \varepsilon_1 \frac{(\sigma_2 + \mathrm{j}\omega\varepsilon_2)}{(\sigma_1 + \mathrm{j}\omega\varepsilon_1)}\right] E_{\mathrm{n}2} = \left[\varepsilon_2 \frac{(\sigma_1 + \mathrm{j}\omega\varepsilon_1)}{(\sigma_2 + \mathrm{j}\omega\varepsilon_2)} - \varepsilon_1\right] E_{\mathrm{n}1}. \tag{1.24}$$

The above expression is only valid when both media are semiconducting. If one medium is an ideal conductor then equation (1.20) gives ρ_{s} and the surface current density, J_n, is then determined by σ.

We may employ Faraday's Law to determine the governing relation for the tangential component of the electric field. In Figure 1.8 the closed contour C in the clockwise direction determines the area S which has height h and width w. Again we let h become small so that the right hand side of Faraday's Law approaches zero (since S approaches zero) and for the tangential direction as shown:

$$\lim_{h\to 0}\left\{\oint_C \mathbf{E}\cdot \mathrm{d}\mathbf{L}\right\} = 0 = E_{t2} - E_{t1}. \tag{1.25}$$

That is, the tangential electric field is the same on both sides of the interface. This is the sensible result since one may imagine the interface as impedances in parallel in the tangential

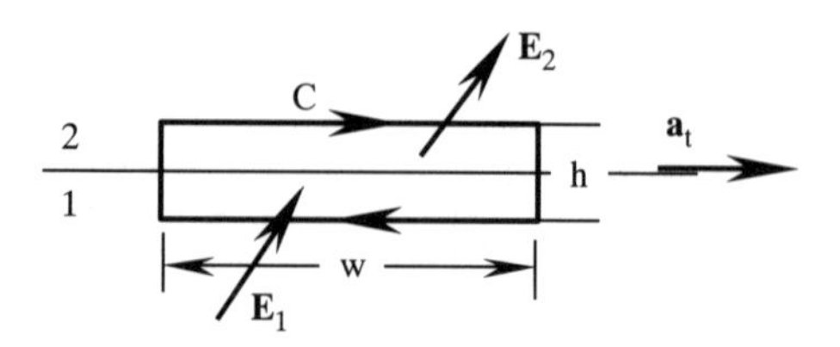

Fig. 1.8 *Geometry for determining the tangential electric field boundary condition.*

direction, so they must have the same voltage gradient. By the same token, the media are in series in the normal direction so they must have different voltage gradients.

1.5.2 Magnetic field boundary conditions

The magnetic field boundary conditions may be developed in an analogous way; first by using Gauss' magnetic law (in the geometry of Figure 1.7b) to quickly obtain the normal **B**-field condition:

$$\lim_{h \to 0} \left\{ \oint_{\Sigma} \mathbf{B} \cdot \mathrm{d}\mathbf{S} \right\} = 0 = B_{n2}A - B_{n1}A$$

so that: (1.26)

$$B_{n2} = B_{n1} \quad \text{or} \quad \mu_2 H_{n2} = \mu_1 H_{n1}.$$

So, for magnetic fields the normal component of the flux density is the same on both sides of the interface. Determination of the tangential **H**-field components requires a bit more care with regard to the geometry. Referring to Figure 1.9, we draw a closed contour, with which we intend to evaluate Ampere's law, similar to that in Figure 1.8; however, it is now evident that as the contour height shrinks to enclose just the interface we may enclose a finite surface current. This surface current density vector, **K** (A/m), may be thought of as a sheet charge, ρ_s, in motion for convenience. So, the tangential magnetic field strength components may be discontinuous by the value of the sheet current, if it exists:

$$\lim_{h \to 0} \left\{ \oint_{C} \mathbf{H} \cdot \mathrm{d}\mathbf{S} \right\} = \lim_{h \to 0} \left\{ \int_0^h \int_0^w [\mathbf{J} + \mathrm{j}\omega\mathbf{D}] \cdot \mathrm{d}\mathbf{S} \right\}$$

so that:

$$(\mathbf{H}_1 - \mathbf{H}_2) \cdot \mathbf{a}_t = \lim_{h \to 0} \left\{ \frac{1}{h} \int_0^h \int_0^w [\mathbf{J} + \mathrm{j}\omega\mathbf{D}] \cdot \mathrm{d}\mathbf{S} \right\} \quad (1.27)$$

and finally:

$$\mathbf{a}_n \times (\mathbf{H}_2 - \mathbf{H}_1) = \mathbf{K}.$$

Note that the only sheet current which contributes to the discontinuity is normal to the **H** components but still in the plane of the interface. For example, if the surface normal is in the z-direction ($\mathbf{a}_n = \mathbf{a}_z$) and the tangential **H**-fields were both in the x-direction, then any discontinuity in **H** must come from $\mathbf{K} = K_y\mathbf{a}_y$. In practice, a sheet current may only exist if one medium is an ideal conductor; consequently, including the displacement current

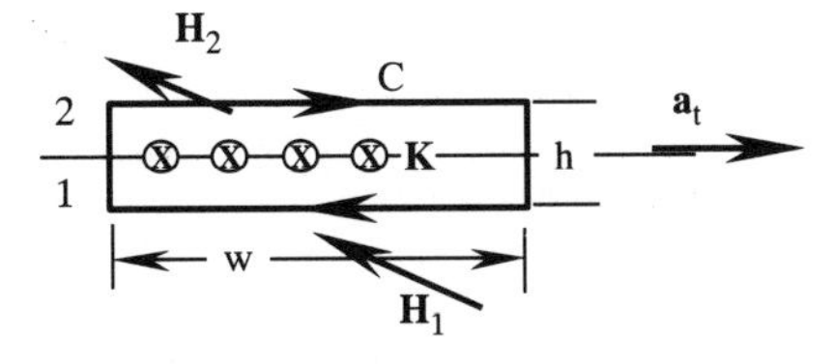

Fig. 1.9 *Geometry for determining the tangential magnetic field boundary condition.*

density in the calculation is not necessary — that is, **K** consists of conduction current alone. For convenience, real conducting walls can be approximated by sheet currents when the depth of penetration (the "skin depth") is very shallow compared to the dimensions of the problem. The skin depth, δ, is the depth in the direction of propagation in lossy media at which the field intensities are 36.8% (e^{-1}) of the values at the surface (in the next section of this chapter the skin depth is described in terms of the attenuation coefficient, α: $\delta = 1/\alpha$).

1.6 GENERAL SOLUTIONS OF MAXWELL'S EQUATIONS IN HOMOGENEOUS MEDIA

There is no universal solution for Maxwell's equations since each problem is driven by its boundary conditions. However, it may be shown that in linear homogeneous isotropic media, where ε and μ are real scalar constants, Maxwell's equations may be cast into two wave equations, one scalar and one vector in form, plus a few ancillary relations. We may describe the propagation of electromagnetic waves with several convenient parameters. The following discussion is limited to simple scalar media. We will treat more complex media in later chapters.

To generate the wave equations, and to facilitate other calculations, we define two potentials: (1) a scalar electric potential, V, for which the source is volume charge, ρ_V; and (2) a vector magnetic potential, **A**, which has current density, **J**, as its source. The potentials are *defined* quantities which make field calculation easier. In the general time-varying field case it will take some time for changes in charges, both $\rho_V(t)$ and $\mathbf{J}(t)$, to propagate to the point of observation — thus the wave nature of the solutions. That is, at a particular observation point, P as shown in Figure 1.10, one observes the effect of variations in charge and current originating at some time in the past — it takes r/u seconds for the disturbance to reach P where u is the speed of propagation. These potentials are referred to as retarded potentials because of the time delay and may be calculated from:

$$
\begin{aligned}
V(r,t) &= \iiint_{\text{vol}} \frac{\rho_V\,(t - r/u)\,\mathrm{d}v}{4\pi\varepsilon r} \\
\mathbf{A}(r,t) &= \iiint_{\text{vol}} \frac{\mu \mathbf{J}\,(t - r/u)\,\mathrm{d}v}{4\pi r}.
\end{aligned}
\tag{1.28}
$$

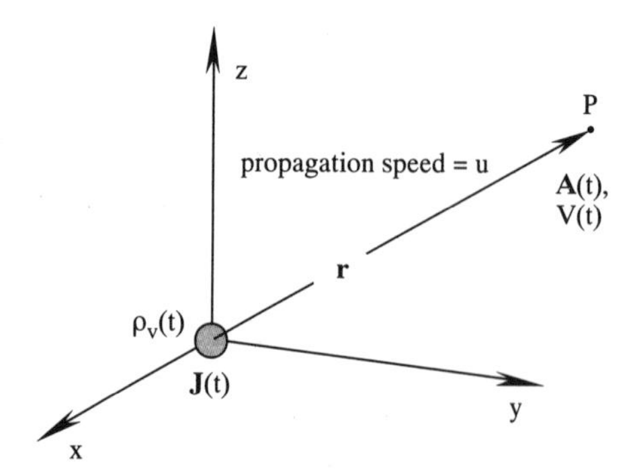

Fig. 1.10 *Time-dependent charge and current density located at the origin of coordinates with retarded potentials at point P.*

The scalar electric potential has a simple relationship to the electric field in the static case: $\mathbf{E} = -\nabla V$. In the time-varying case the relationship is slightly more complex: $\mathbf{E} = -\nabla V - \partial \mathbf{A}/\partial t$. The vector magnetic potential has been defined so that its curl is the magnetic flux density: $\mathbf{B} = \nabla \times \mathbf{A}$. A vector field is fully specified when *both* its divergence and curl are known, so we need a definition for $\nabla \cdot \mathbf{A}$. In deriving the general wave form of Maxwell's equations it turns out to be convenient to *define* the divergence of $\mathbf{A}$ using the Lorentz condition:

$$\nabla \cdot \mathbf{A} = -\mu\varepsilon \frac{\partial V}{\partial t}. \tag{1.29}$$

This definition is used to separate Maxwell's equations in terms of the potentials into two wave equations in the time domain:

$$\begin{aligned} \nabla^2 V - \mu\varepsilon \frac{\partial^2 V}{\partial t^2} &= -\frac{\rho_v}{\varepsilon} \\ \nabla^2 \mathbf{A} - \mu\varepsilon \frac{\partial^2 \mathbf{A}}{\partial t^2} &= -\mu \mathbf{J} \end{aligned} \tag{1.30}$$

and two analogous relations in the frequency domain:

$$\begin{aligned} \nabla^2 V + \omega^2 \mu\varepsilon V &= -\frac{\rho_v}{\varepsilon} \\ \nabla^2 \mathbf{A} + \omega^2 \mu\varepsilon \mathbf{A} &= -\mu \mathbf{J}. \end{aligned} \tag{1.31}$$

where ω is the angular frequency (r/s), $\omega = 2\pi f$, and f is the frequency in Hz. Note that the vector equation is really three separate equations, one for each of the A_x, A_y, and A_z vector components.

Solutions of these wave equations have several common parameters. The speed of propagation, u, is:

$$u = \frac{1}{\sqrt{\mu\varepsilon}} \qquad \text{so} \qquad c = \frac{1}{\sqrt{\mu_0\varepsilon_0}} = 3 \times 10^8 \text{ m/s} \tag{1.32}$$

where c is the speed of light in free space. For sinusoidal sources we may use complex notation with $\omega^2\mu\varepsilon$ as the square of the propagation phase constant, β (the parameter for constant phase fronts), and the wavelength, λ:

$$V(r, t) = V_0 \cos(\omega t - \beta r) \rightarrow V(r, \omega) = V_0 \mathrm{e}^{-\mathrm{j}\beta r} \mathrm{e}^{\mathrm{j}\omega t} \tag{1.33}$$

where:

$$\beta = \omega\sqrt{\mu\varepsilon} = \frac{2\pi}{\lambda} = \frac{\omega}{u} \qquad \text{so} \qquad \lambda = \frac{u}{f}.$$

A similar sequence of expressions can be written for the components of the vector potential, $\mathbf{A}$ (and, ultimately, the $\mathbf{E}$ and $\mathbf{H}$ fields). In fact, all four of the differential equations represented in equation (1.31) have the same general solution and differ only in the particular solution. In equation (1.33) note that the retarded potential implies propagation in the $+r$ direction. This convention will be important in the discussion of plane wave propagation. For propagation in simple lossy media, where all of the losses may be considered to be due to electrical conductivity alone, we may introduce a complex propagation constant, γ^2, in place of β^2. Also, for propagation in source-free media it may be shown that Maxwell's equations reduce to the vector Helmholtz equations:

$$\nabla^2\mathbf{E} - \gamma^2\mathbf{E} = 0 \qquad \nabla^2\mathbf{H} - \gamma^2\mathbf{H} = 0 \tag{1.34}$$

where:

$$\gamma^2 = \mathrm{j}\omega\mu(\sigma + \mathrm{j}\omega\varepsilon) = -\omega^2\mu\varepsilon + \mathrm{j}\omega\mu\sigma$$

and $\gamma = \alpha + \mathrm{j}\beta$, in general, with α the attenuation coefficient and β the phase coefficient. This simple model for lossy media can be modified to include dielectric losses by the introduction of a complex permittivity, $\varepsilon^* = \varepsilon' - \mathrm{j}\varepsilon''$. In that case ε' is inserted for ε above and the effective conductivity is $\sigma_{\mathrm{eff}} = \sigma + \omega\varepsilon''$. Whether or not dielectric losses are significant, we can obtain the real and imaginary parts of γ from:

$$\begin{aligned} \alpha &= \omega\sqrt{\mu\varepsilon}\left\{\frac{1}{2}\left[\sqrt{1+\left(\frac{\sigma}{\omega\varepsilon}\right)^2} - 1\right]\right\}^{1/2} \\ \beta &= \omega\sqrt{\mu\varepsilon}\left\{\frac{1}{2}\left[\sqrt{1+\left(\frac{\sigma}{\omega\varepsilon}\right)^2} + 1\right]\right\}^{1/2} \end{aligned} \tag{1.35}$$

and the wavelength and propagation speed are calculated by equation (1.33). The depth of penetration, the skin depth $\delta = 1/\alpha(\mathrm{m})$ in homogeneous lossy media. For the present we will describe propagation in simple media, where ε is a real scalar constant, and reserve our discussion of more complex media for Chapter 7.

1.7 UNIFORM PLANE WAVE EXAMPLE

One simple solution of Maxwell's equations is the case of a uniform plane wave propagating in a well behaved unbounded medium — we will choose the example of a wave polarized in the x-direction (the polarization is the direction of the electric field) with a y-directional **H**-field and propagation in the z-direction:

$$E_y = E_z = H_x = H_z = 0 = \frac{\partial}{\partial x} = \frac{\partial}{\partial y}$$

so:

$$\frac{\mathrm{d}^2E_x}{\mathrm{d}z^2} - \gamma^2E_x = 0 \quad \text{and} \quad \frac{\mathrm{d}^2H_y}{\mathrm{d}z^2} - \gamma^2H_y = 0. \tag{1.36}$$

Such a wave is termed a TEM (transverse electromagnetic) wave because both the electric and magnetic fields are normal to the direction of propagation. It is not possible to actually generate a uniform plane wave; however, the mathematics are simple enough that much can be learned about propagation from such a wave with minimal difficulty. Also, many real fields can be approximated by a plane wave with sufficient accuracy that it is a most useful formulation. The derivatives in equation (1.34) reduce to ordinary derivatives, as used in equation (1.36), and the solution (in the frequency domain) has the form:

$$E_x = Ae^{-\gamma z} + Be^{\gamma z}$$

where:

$$A = E^{+} \quad \text{and} \quad B = E^{-} \tag{1.37}$$

and we use E^+ to signify the magnitude of the wave propagating in the $+z$ direction and E^- the magnitude of the wave propagating in the $-z$ direction. In this, and subsequent, notation the $e^{j\omega t}$ term is understood, so that the full solution in the time domain would be written as:

$$\begin{aligned} E_x(z,t) &= \text{Re}\{E_x(z,\omega)\} \\ &= E_{x0}^{+}e^{-\alpha z}\cos(\omega t - \beta z) + E_{x0}^{-}e^{\alpha z}\cos(\omega t + \beta z). \end{aligned} \tag{1.38}$$

A similar expression can be written for the H_y component. The expression differs in the magnitude and also in that the sign before the $-z$ direction H_y term is negative for propagation in that direction; see equations (1.39) and (1.40) below. We may simplify the magnetic field expression by using the characteristic impedance of the medium, η:

$$\eta = \sqrt{\frac{j\omega\mu}{(\sigma + j\omega\varepsilon)}} = \sqrt{\frac{\mu}{\varepsilon}} \quad \text{for } \sigma = 0$$

which gives:

$$H_{y0}^{+} = \frac{E_{x0}^{+}}{\eta} \quad \text{and} \quad H_{y0}^{-} = \frac{-E_{x0}^{-}}{\eta}. \tag{1.39}$$

Note that **H** will only be in phase with **E** in a lossless medium where η is real.

Reflection will occur at a boundary between media, as in Figure 1.11, in order to satisfy the boundary conditions. In this case we assume a uniform plane boundary between lossless media and both **E** and **H** have only tangential components — i.e. normal incidence, or the wave is TEM (transverse electromagnetic) to the surface normal. We identify the incident wave by E_i ($+z$ direction), the reflected wave by E_r ($-z$ direction) and the transmitted wave by E_t. The electric field reflection coefficient is Γ and the transmission coefficient T. The total field in medium 1 is the sum of incident and reflected waves:

$$\begin{aligned} \mathbf{E}_1(z,\omega) &= E_i\left(e^{-j\beta_1 z} + \Gamma e^{j\beta_1 z}\right)\mathbf{a}_x \\ \mathbf{H}_1(z,\omega) &= \frac{E_i}{\eta_1}\left(e^{-j\beta_1 z} - \Gamma e^{j\beta_1 z}\right)\mathbf{a}_y \end{aligned} \tag{1.40a}$$

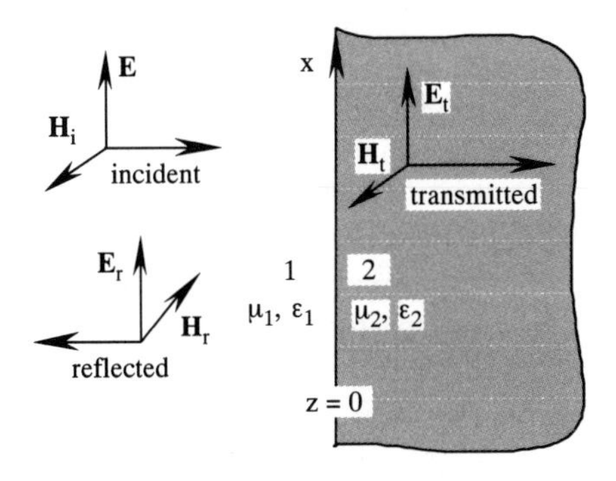

Fig. 1.11 *Normal incidence of a uniform plane wave.*

where:

$$\Gamma = \frac{E_\mathrm{r}}{E_\mathrm{i}} = \frac{\eta_2 - \eta_1}{\eta_2 + \eta_1}$$

and the field in medium 2 consists only of the transmitted wave:

$$\mathbf{E}_2(z, \omega) = E_\mathrm{i} T \mathrm{e}^{-\mathrm{j}\beta_2 z}\mathbf{a}_x$$
$$\mathbf{H}_2(z, \omega) = \frac{E_\mathrm{i}}{\eta_2} T \mathrm{e}^{-\mathrm{j}\beta_2 z}\mathbf{a}_y \tag{1.40b}$$

where:

$$T = \frac{E_t}{E_i} = \frac{2\eta_2}{\eta_2 + \eta_1}.$$

Note that $\Gamma + T$ is not 1—power (energy) is conserved at the interface, so: $(1 - \Gamma^2)/\eta_1 = T^2/\eta_2$.

We may also calculate reflection and transmission coefficients for oblique incidence of uniform plane waves, the Fresnel coefficients. However, there are two possible polarizations: (1) the electric field is parallel to the surface and perpendicular to the surface normal, a so-called TE (transverse electric) wave; and (2) the magnetic field is parallel to the surface and perpendicular to the surface normal, a so-called TM (transverse magnetic) wave. In fact, any plane wave can be decomposed into equivalent TE and TM components. The Fresnel reflection and transmission coefficients for the TE mode are:

$$\Gamma_{\mathrm{TE}} = \frac{\eta_2 \cos\theta_\mathrm{i} - \eta_1 \cos\theta_\mathrm{t}}{\eta_2 \cos\theta_\mathrm{i} + \eta_1 \cos\theta_\mathrm{t}} \qquad T_{\mathrm{TE}} = \frac{2\eta_2 \cos\theta_i}{\eta_2 \cos\theta_\mathrm{i} + \eta_1 \cos\theta_\mathrm{t}} \tag{1.41a}$$

and for the TM mode are:

$$\Gamma_{\mathrm{TM}} = \frac{\eta_1 \cos\theta_\mathrm{i} - \eta_2 \cos\theta_\mathrm{t}}{\eta_1 \cos\theta_\mathrm{i} + \eta_2 \cos\theta_\mathrm{t}} \qquad T_{\mathrm{TM}} = \frac{2\eta_2 \cos\theta_\mathrm{i}}{\eta_1 \cos\theta_\mathrm{i} + \eta_2 \cos\theta_\mathrm{t}}. \tag{1.41b}$$

In these expressions θ_i is the angle of incidence and θ_t the angle of transmission (the interface is assumed specular so the angle of reflection $\theta_\mathrm{r} = \theta_\mathrm{i}$). At normal incidence the expressions reduce to those of equations (1.40a) and (1.40b) with the exception of a sign in Γ—the difference is due to the assumed vector directions for **H** in the two different geometries of Figure 1.12. The incident angle and transmission angle are related by Snell's law of refraction (from the conservation of phase):

$$\beta_1 \sin\theta_\mathrm{i} = \beta_2 \sin\theta_\mathrm{t}. \tag{1.42}$$

There are two important incident angles which may be obtained from these relations. First, for an interface between low loss materials with permeability approximately μ_0 in the TM polarization only (**H** parallel to the surface) there is an angle of incidence at which there will be no reflection ($\Gamma = 0$), called the Brewster angle, θ_B:

$$\theta_B = \tan^{-1}\left\{\sqrt{\frac{\varepsilon_2}{\varepsilon_1}}\right\}. \tag{1.43}$$

Second, for similar materials and for the special case where $\varepsilon_1 > \varepsilon_2$, there is a critical incident angle, θ_C, which, if exceeded by θ_i, results in total internal reflection for both polarizations:

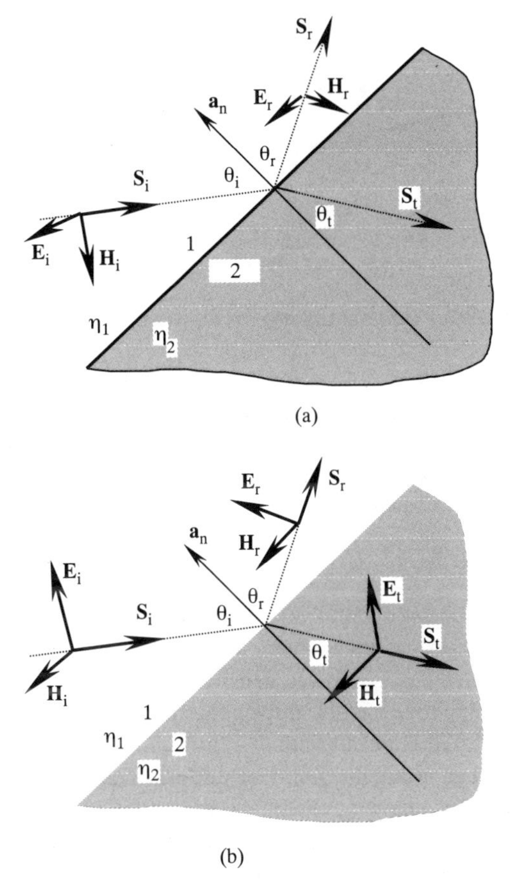

Fig. 1.12 *Oblique incidence of plane waves. (a) TE wave oblique incidence. (b) TM wave oblique incidence.*

$$\theta_C = \sin^{-1}\left\{\sqrt{\frac{\varepsilon_2}{\varepsilon_1}}\right\}. \tag{1.44}$$

Equations (1.41)–(1.44) are of considerable significance when using ray tracing models for determining heating distributions within objects, as we will see in Chapter 4.

1.8 POYNTING POWER THEOREM

In a static field, or at a particular instant in time in a dynamic field (in the special case where the materials are well behaved), the energy stored in an electric field, W_e (J), and in a magnetic field, W_m (J), are given by:

$$\begin{aligned} W_e &= \tfrac{1}{2}\iiint_{vol} \mathbf{E}\cdot\mathbf{D}\,dv = \tfrac{1}{2}\iiint_{vol} \varepsilon|\mathbf{E}|^2\,dv \\ W_m &= \tfrac{1}{2}\iiint_{vol} \mathbf{H}\cdot\mathbf{B}\,dv = \tfrac{1}{2}\iiint_{vol} \mu|\mathbf{H}|^2\,dv. \end{aligned} \tag{1.45}$$

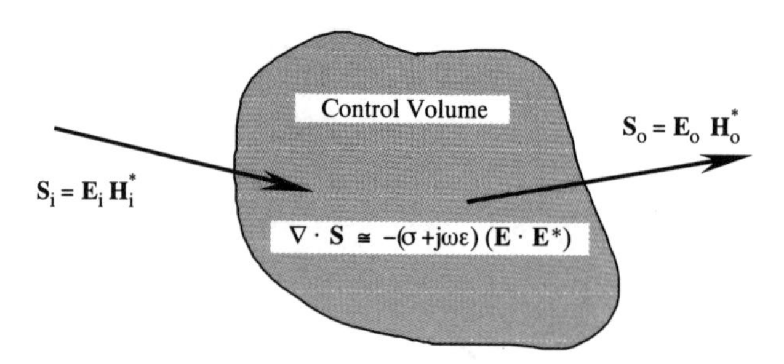

Fig. 1.13 *Control volume for Poynting power theorem.*

In linear network theory these relations give the familiar forms for the energy stored in a capacitor ($W_C = CV^2/2$) and an inductor ($W_L = LI^2/2$). The respective time derivatives of (1.45) yield the power in a material in the case of a lossless medium.

In order to describe the lossy medium case we define the Poynting vector, $\mathbf{P}(t)(W/m^2)$, in the time domain and $\mathbf{S}(\omega)$ in the frequency domain (called the "fluence rate" in optics):

$$\mathbf{P}(t) = \mathbf{E}(t) \times \mathbf{H}(t); \qquad \mathbf{S}(\omega) = \tfrac{1}{2}(\mathbf{E}(\omega) \times \mathbf{H}^*(\omega)) \tag{1.46}$$

where the * indicates a complex conjugate of the field. The Poynting vector describes the local power density in a propagating wave, and in this definition the complex Poynting vector represents the time-average power density. Its vector direction gives the direction of propagation and its magnitude is the power flux density. Figure 1.13 illustrates a wave propagating through a closed control volume—the control volume here is considered to be a small but finite portion of a larger volume of uniform material so no interface is implied, though in the general case one may exist. As the wave propagates through the control volume its fields interact with the material so that the total power flux across the boundary surface is less than zero if energy is absorbed from the wave:

$$-\oint_{\Sigma} \tfrac{1}{2}(\mathbf{E} \times \mathbf{H}^*)\,\mathrm{d}S = \iiint_{\mathrm{vol}} \tfrac{1}{2}(\mathbf{E} \cdot \mathbf{J}^*)\,\mathrm{d}v + \frac{\mathrm{j}\omega}{2}\iiint_{\mathrm{vol}} \left\{\mathbf{B} \cdot \mathbf{H}^* - \mathbf{E} \cdot \mathbf{D}^*\right\}\mathrm{d}v$$

$$-\nabla \cdot \mathbf{S} = \tfrac{1}{2}\{(\mathbf{E} \cdot \mathbf{J}^*) + \mathrm{j}\omega[\mathbf{B} \cdot \mathbf{H}^* - \mathbf{E} \cdot \mathbf{D}^*]\}. \tag{1.47}$$

The factor $\frac{1}{2}$ appears if the field magnitudes are expressed in terms of peak volts or amperes, if RMS volts and amperes are used it does not appear. The heating term, Q_{gen} (W/m^3) is the real part of $(-\nabla \cdot \mathbf{S})$ which will be due entirely to conductivity if the permittivity and permeability are both real. The minus sign indicates that energy is absorbed from the wave ($\mathbf{S}$ is the power flux density in the wave). The imaginary terms on the right hand side indicate stored energy, which does not contribute to heating—this is the vector field equivalent of reactive power in linear circuit theory. When, for example, the permittivity is complex then dielectric heating contributes to the real part of $\nabla \cdot \mathbf{S}$: $\mathrm{Re}(-\nabla \cdot \mathbf{S}) = (\sigma + \omega\varepsilon'')(\mathbf{E} \cdot \mathbf{E})$. This will be discussed in more detail in later chapters.

1.9 THERMAL GOVERNING EQUATIONS: THE FIRST LAW OF THERMODYNAMICS

In the general case of electromagnetic heating we sometimes consider an open control volume; for example, when a fluid flows through a waveguide, or when industrial materials move through a multimode cavity oven. An open control volume has mass flux across its boundaries. The total energy flux per unit time for the volume surface gives the storage rate within the volume: stored energy = influx − efflux. The control volume of Figure 1.14 is assumed large compared to molecular dimensions but small enough that the temperature and other thermodynamic state variables are uniform within it. Thermal energy is stored in the enthalpy enclosed by the volume:

$$h = \mathcal{U} + P\mathcal{V} = \rho c T + \frac{P}{\rho} \tag{1.48}$$

where h is the specific enthalpy (J/kg), $\mathcal{U}$ is the internal thermal energy (J/kg), P the pressure (Pa), $\mathcal{V}$ the specific volume (m^3/kg), c the specific heat (J/kg K), ρ the density (kg/m^3), and T the temperature (K). The total stored energy is $H = mh$, where m is the mass of the control volume. Energy may cross the boundary through mass flux, $\partial m/\partial t$, or by heat transfer, q:

$$\frac{\partial H}{\partial t} = \frac{\partial m_{\mathrm{i}}}{\partial t} h_{\mathrm{i}} - \frac{\partial m_{\mathrm{o}}}{\partial t} h_{\mathrm{o}} + Q_{\mathrm{cond}} + Q_{\mathrm{gen}} + Q_{\mathrm{ph}} \tag{1.49}$$

where the i and o subscripts refer to inlet and outlet flows, Q_{cond} refers to Fourier heat conduction across the volume boundaries, Q_{gen} is work done on the control volume by electromagnetic heating, and Q_{ph} is the enthalpy of phase change or material transformation (as in a chemical reaction).

In the usual situation, we are concerned with heating in an incompressible liquid or solid and often do not anticipate phase change. In that instance equation (1.49) can be expressed as a simple point form differential equation for infinitesimal control volumes:

$$\rho c \frac{\partial T}{\partial t} = \nabla \cdot (k \nabla T) + q_{\mathrm{gen}} + q_{\mathrm{losses}} \tag{1.50}$$

where k is the thermal conductivity (W/m K) and q_{losses} may be used to represent surface convection and radiation. The general form is given in (1.50) and the material has not been assumed isotropic — if the material is isotropic then k is a simple scalar constant and can be moved outside of the dot product. Again, q_{gen} is the electromagnetic power deposition. The surface loss terms can be included for control volumes on the surface in what appears

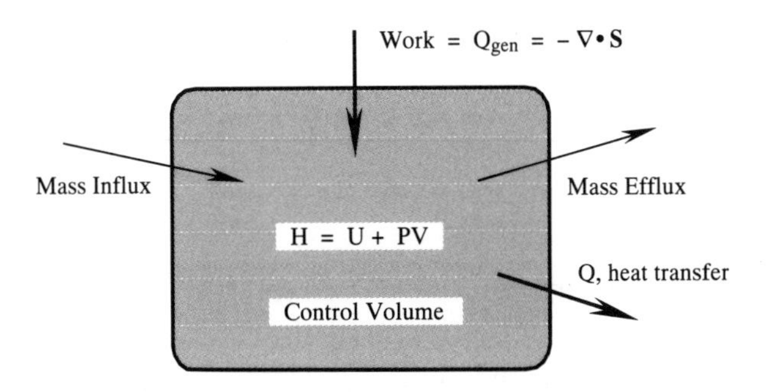

Fig. 1.14 *Control volume for energy balance.*

to be a simple form:

$$q_{\text{conv}} = \frac{h_c}{\Delta z}(T - T_\infty) \qquad q_{\text{rad}} = \frac{\varepsilon_{TH}}{\Delta z}\sigma_B(T^4 - T_\infty^4) \tag{1.51}$$

where we have assumed that the surface is the $\Delta x \Delta y$ face of the infinitesimal Δv control volume, h_c is the convection coefficient (W/m^2 K), T_∞ is the environment temperature (K), ε_{TH} is the emissivity of the surface, and σ_{B} is the Stefan–Boltzmann constant (5.67×10^{-8} W/m^2 K). This form is useful in many cases but suppresses several effects: (1) h_{c} varies over several orders of magnitude depending on the velocity and physical characteristics of the surrounding medium (for free convection on a flat horizontal surface in air it ranges from about 5 to 25); (2) h_{c} may be a sensitive function of surface temperature; (3) the simplest possible assumptions about the shape factors and view factors and the enclosure, which strongly affect thermal radiation, have been made (in many applications, notably at the high temperatures experienced in ceramic sintering applications, the surface radiation term is the most sensitive in the expression and must be treated with great care).

1.10 TIME-VARYING, STATIC AND QUASI-STATIC ANALYSIS

To this point the complete electromagnetic field equations have been used. For strictly static problems, $\partial/\partial t = 0$ and Ampere's and Faraday's Laws are completely decoupled. That is, we may calculate electric fields and magnetic fields separately; and the typical two cases are electrostatic and magnetostatic. In electrostatics we assume that $\mathbf{H} = \mathbf{B} = \mathbf{J} = 0$ which gives:

$$\begin{aligned}
&\oint_C \mathbf{E} \cdot d\mathbf{L} = 0 \qquad \text{or} \qquad \nabla \times \mathbf{E} = 0 \\
&\oint_\Sigma \mathbf{D} \cdot d\mathbf{S} = \iiint_{\text{vol}} \rho_{\text{v}}\, dv \qquad \text{or} \qquad \nabla \cdot \mathbf{D} = \rho_{\text{v}} \\
&V = \iiint_{\text{vol}} \frac{\rho_{\text{v}} dv}{4\pi\varepsilon r} \qquad \text{and} \qquad \mathbf{E} = -\nabla V.
\end{aligned} \tag{1.52}$$

The astute reader will identify the integral form of Faraday's Law under these assumptions as Kirchoff's Voltage Law from linear circuit analysis. The electric field is conservative; also, the point form of Gauss' electric law (1.30) reduces to either the Poisson ($\nabla^2 V = -\rho_{\text{v}}/\varepsilon$) or Laplace equations ($\nabla^2 V = 0$ in a source-free region), for which many standard solutions are already known.

For the case of magnetostatics, we set $\mathbf{E} = \mathbf{D} = \rho_{\text{v}} = 0$ and Maxwell's equations reduce to:

$$\begin{aligned}
&\oint_C \mathbf{H} \cdot d\mathbf{L} = \iint_S \mathbf{J} \cdot d\mathbf{S} \qquad \text{or} \qquad \nabla \times \mathbf{H} = \mathbf{J} \\
&\oint_\Sigma \mathbf{B} \cdot d\mathbf{S} = 0 \qquad \text{or} \qquad \nabla \cdot \mathbf{B} = 0.
\end{aligned} \tag{1.53}$$

In magnetostatics we assume that all current density is confined to perfect conductors and that no *net* charge exists—that is, positive charges moving in $+x$ direction contribute to $\mathbf{J}$

and are equal in number to negative charges moving in the $-x$ direction, so **J** is finite when $\rho_v = 0$.

The quasi-static, or electro-magnetostatic, case is of prime interest in this book, however. This case is equivalent to assuming that though time variations are large in magnitude they are very slow: equivalently, that the fields at all locations within the domain of the problem remain in phase. The electric and magnetic fields need not be in phase with *each other*, but each individual field at a particular instant in time has one phase at all points within the problem space. The electric and magnetic field relations are coupled through **J** alone (or when convenient we define **J** as the total current density: $\mathbf{J} = (\sigma + j\omega\varepsilon)\mathbf{E}$, so that **E** is still approximately conservative. Thus both (1.52) and (1.53) are assumed to apply simultaneously. This analysis is considerably simplified over the wave propagation approach and is preferable when it applies. We may use quasi-static analysis whenever a problem satisfies the governing criterion that the fields are in phase at all points in the geometry. Therefore, whenever the maximum dimension of a problem is short compared to a wavelength, quasi-static analysis may be used. Typically, if the maximum dimension is less than about $\lambda/20$ the error accrued by quasi-static analysis will be acceptable (where λ is the wavelength in the medium). This applies to a large number of RF analysis problems where the wavelengths range from about 1 to 60 m. It also applies in the calculation of MW fields around small inclusions up to diameters of the order of 1 mm.

1.11 SUMMARY

We have presented the structure and limiting assumptions for the macroscopic analysis of general time-varying electromagnetic fields. The field vector boundary conditions were derived and will prove of significant importance in the analysis and design of electromagnetic fields in real materials. Example solutions in simple plane wave geometries have been discussed along with simplified quasi-static analysis. Finally, the governing thermal equations have been introduced. The thermal relations are coupled to the electromagnetic field equations by the power generation term and by the effect of temperature rise on medium properties.

REFERENCES

The following texts will provide additional insight into the topics of this chapter.

Cheng, D. K. (1989) *Field and Wave Electromagnetics (2nd Ed.)* Addison Wesley, Reading, MA.

Hayt, W. H. (1989) *Engineering Electromagnetics (5th Ed.)* McGraw-Hill, New York.

Iskander, M. (1992) *Electromagnetic Fields and Waves* Prentice-Hall, Englewood Cliffs, NJ.

Kong, J. A. (1986) *Electromagnetic Wave Theory* Wiley-Interscience, New York.

Paris, D. T. and Hurd, F. K. (1969) *Basic Electromagnetic Theory* McGraw-Hill, New York.

Pozar, D. M. (1990) *Microwave Engineering* Addison-Wesley, Reading, MA.

Sadiku, M. N.O. (1989) *Elements of Electromagnetics* Saunders, New York.

Stillings, D. (1973) Artifact: Benjamin Franklin's Celebrated Kite Experiment *Medical Instrumentation*, **7**, 234.

2 Radio Frequency and Microwave Transmission

2.1 INTRODUCTION

This chapter discusses several examples of currently used radio frequency (RF) and microwave (MW) transmission lines. In practice an electromagnetic heating system consists of a generator, transmission line and applicator. The generator is the power source for the system and establishes the frequency of operation. The transmission line couples the generator to the applicator. The applicator is designed to create favorable EM field distributions around the material to be irradiated. In home microwave ovens and many industrial RF system designs the transmission lines are so short that they may be ignored in the analysis. In those cases the generator is located within about 1 wavelength of the applicator. It is not always convenient nor advisable to connect the applicator directly to the generator. Transmission lines and transmission elements are often used to provide impedance matching in order to obtain maximum power transfer between generator and load. Also, even if the transmission line is shorter than λ, we may use transmission line analysis as a convenience as long as the line is long compared to the extent of the fringing fields at each end. This chapter discusses transmission lines at both radio and microwave frequencies, waveguide propagation, and a few representative transmission line impedance matching techniques.

2.2 TRANSMISSION-LINE THEORY

The uniform plane wave discussed in Chapter 1 was an example of a TEM (transverse electromagnetic) wave because **E** and **H** were both mutually perpendicular to the direction of propagation, **S**. Propagation in transmission lines (T-L) at both RF and MW frequencies is also TEM (providing that the T-L interior dimensions are small compared to λ) and most of the relationships and concepts previously presented transfer directly to T-L problems with little or no modification. We will present the pertinent transmission line results and also review the use of a Smith chart in T-L design problems.

2.2.1 Characteristics of coaxial transmission lines

The classic transmission line geometry is that of a coaxial cable, Figure 2.1, where the inner conductor has radius a and the outer conductor (usually grounded or earthed) has inner and outer radii of b and c respectively. We ground the outer conductor so that the cable will make an effective shield for electric fields: signals on the center conductor are

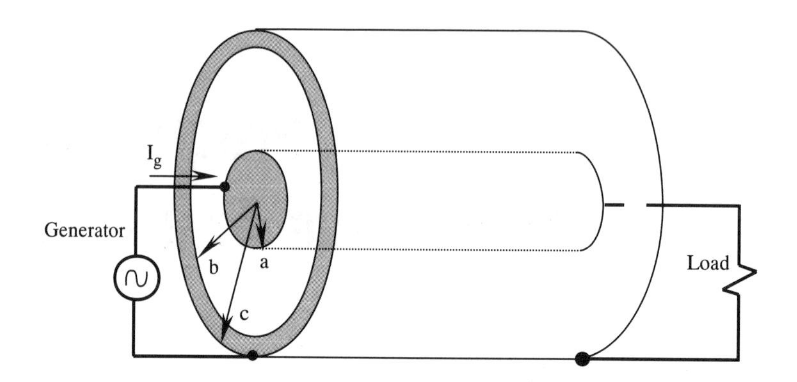

Fig. 2.1 *Coaxial transmission line with resistive load.*

shielded from the environment and, reciprocally, the environment is shielded from the center conductor's electric fields. The coaxial line is also an effective magnetic field shield since, when connected as shown in Figure 2.1, the magnetic field created by the current on the center conductor is canceled by that from the current on the outer conductor for $r > c$; to see this establish Amperian contours at two locations, $a < r_1 < b$ (for which $I_{\text{encl}} = I_{\text{g}}$) and $r_2 > c$ (for which $I_{\text{encl}} = 0$). The center conductor need not be solid since the skin effect will confine currents to near the surface of conductors; consequently, conductors need only be several skin depths thick in order to have the lowest impedance practically achievable.

By using either Gauss' electric law (this problem has sufficient symmetry) or the definition of the electric field in terms of charge density it may be shown that a single infinite cylindrical conductor of finite radius a has an electric field equal to that of a uniform infinitesimal line charge source, ρ_{L} (C/m), for $r \geqslant a$:

$$\mathbf{E} = \frac{\rho_{\text{L}}}{2\pi\varepsilon r}\mathbf{a}_r. \tag{2.1}$$

Because of the symmetry we may also use Ampere's law to quickly show that the magnetic field is also dependent on the reciprocal of radius for $r \geqslant a$:

$$\mathbf{H} = \frac{I}{2\pi r}\mathbf{a}_\phi \tag{2.2}$$

where I is the total current in the cylinder (assumed axi-symmetrically distributed, though not necessarily uniform in the cross-section). Both the $\mathbf{E}$ and $\mathbf{H}$ fields are zero for $r > c$ in Figure 2.1.

Using these two results we may calculate lumped parameter impedances for a short section of transmission line Δl long. We use Δl as the axial length segment rather than Δz as would ordinarily be done in cylindrical coordinates in order to avoid confusion between the axial coordinate and impedance z. The Δl segment must satisfy the criterion for quasi-static analysis. The capacitance of the Δl segment, $\mathcal{C}(\text{F}) = C\Delta l$, can be obtained from the charge and the potential difference for the segment. We may make the calculation by establishing a closed Gaussian surface within the cross-section at some radius, $b < r < a$

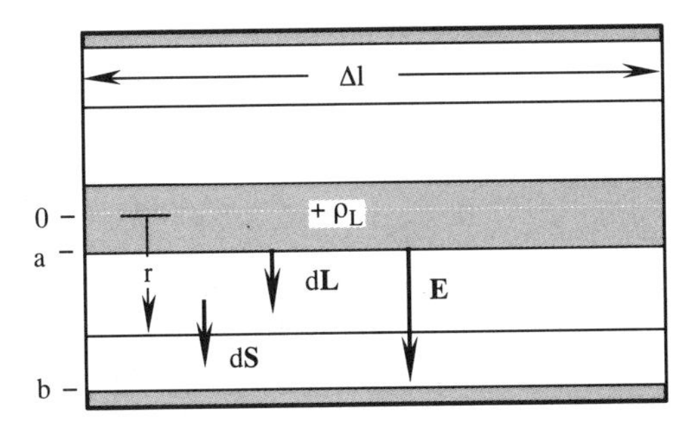

Fig. 2.2 *Transmission line segment.*

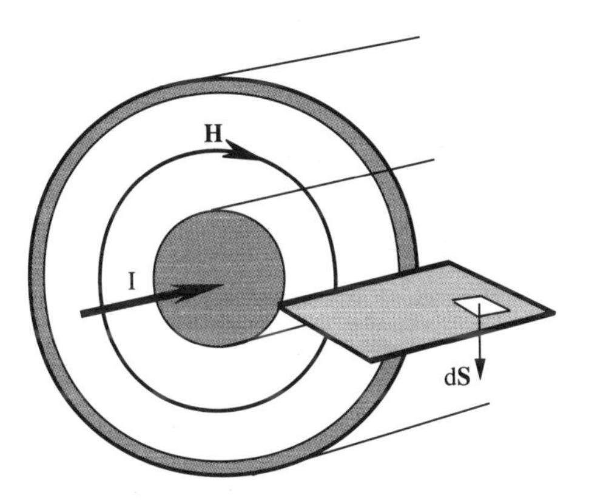

Fig. 2.3 *Geometry for estimation of T-L segment inductance.*

(as in Figure 2.2); then:

$$\mathcal{C} = \frac{Q}{V} = \frac{\oint_S \mathbf{D} \cdot \mathrm{d}\mathbf{S}}{\int_b^a \mathbf{E} \cdot \mathrm{d}\mathbf{L}} = \frac{\int_0^{\Delta l} \int_0^{2\pi} \varepsilon \dfrac{\rho_\mathrm{L}}{2\pi\varepsilon r}\, \mathrm{d}\phi \mathrm{d}z}{-\int_\mathrm{b}^\mathrm{a} \dfrac{\rho_\mathrm{L}}{2\pi\varepsilon r}\, \mathrm{d}r} = \frac{2\pi\varepsilon}{\ln\left\{\dfrac{b}{a}\right\}} \Delta l. \tag{2.3}$$

The potential has been evaluated from the reference (at $r = b$) to the center conductor and $\mathbf{dS}$ is parallel to $\mathbf{dL}$ in the figure. In transmission line analysis we are concerned with the capacitance per unit length, C (F/m) a characteristic parameter of the T-L design, so the above result is divided by Δl.

In a similar fashion we may estimate the inductance of the line segment, $\mathcal{L} = L\Delta l$ (H). The appropriate geometry is shown in Figure 2.3. The inductance is obtained from:

$$\mathcal{L} = \frac{N\Phi_\mathrm{M}}{I} = \frac{1}{I} \int_0^{\Delta l} \int_a^b \frac{\mu I}{2\pi r}\, \mathrm{d}r \mathrm{d}z = \frac{\mu}{2\pi} \ln\left\{\frac{b}{a}\right\} \Delta l \tag{2.4}$$

where the total flux is:

$$\Phi_M = \int\int_S \mathbf{B} \cdot d\mathbf{S}. \tag{2.5}$$

In this equation, N = the number of turns ($N = 1$ since we imagine that the T-L is connected to a load so that the outer conductor current is equal and oppositely directed to the inner conductor current, as in Figure 2.1); Φ_M is the magnetic flux "linked" (surrounded by) the turn, so $d\mathbf{S} = dr dz \mathbf{a}_\phi$ is parallel to $\mathbf{H} = H_\phi \mathbf{a}_\phi$; and μ is the permeability of the dielectric material. At low frequencies this relationship contains a small error (of magnitude $\mu/8\pi$) since the total flux linked is under-estimated because the current is distributed uniformly throughout the inner conductor. At the high frequencies and high powers with which we are concerned, however, equation (2.4) is more accurate since the skin effect will confine the current to the periphery of the center conductor, and the center conductors are usually hollow.

The transmission line may be lossy, so two other lumped equivalent impedances are required to fully characterize it. First, the dielectric between the conductors may have some semiconducting properties, resulting in a "leakage conductance", $\mathcal{G} = G\Delta l$ (S):

$$\mathcal{G} = \frac{I_\varepsilon}{V} = \frac{\int_0^{\Delta l}\int_0^{2\pi} \sigma \frac{\rho_L}{2\pi\varepsilon r} r \, d\phi dz}{-\int_b^a \frac{\rho_L}{2\pi\varepsilon r} dr} = \frac{2\pi\sigma}{\ln\left\{\frac{b}{a}\right\}} \Delta l \tag{2.6}$$

where I_ε signifies leakage current in the dielectric; so here σ is the electrical conductivity of the dielectric material. Second, the conductors may have a finite electrical conductivity, σ_c, which results in a series impedance, $\mathcal{R} = R\Delta l (\Omega)$. The form of R (Ω/m) depends on the frequency. At very low frequencies the current will be uniformly distributed in the conducting cross-sections so the segment resistance is simply:

$$\mathcal{R} = \frac{\Delta l}{\sigma_c}\left[\frac{1}{\pi a^2} + \frac{1}{\pi(c^2 - b^2)}\right] \tag{2.7}$$

where the simple series combination has been used. At higher frequencies the skin effect will limit the depth of penetration of the current, so the segment resistance per unit length, R, will be closer to:

$$R = \frac{1}{2\pi\delta\sigma_c}\left(\frac{1}{a} + \frac{1}{b}\right) \tag{2.8}$$

where $\delta = 1/\alpha$ is the skin depth (m) in the conductor and σ_c is the conductivity of the conductors — here we have assumed that the skin depth is small compared to inner conductor radius a and to outer conductor thickness $(b - c)$.

2.2.2 The telegrapher's equations and their solutions

Using these characteristic parameters, and enforcing the quasi-static assumption, we may formulate a lumped circuit model for the segment of transmission line, as shown in Figure 2.4. The voltages and currents are expressed in the frequency domain. Because the quasi-static assumption holds, we may use the conservative form of Faraday's Law

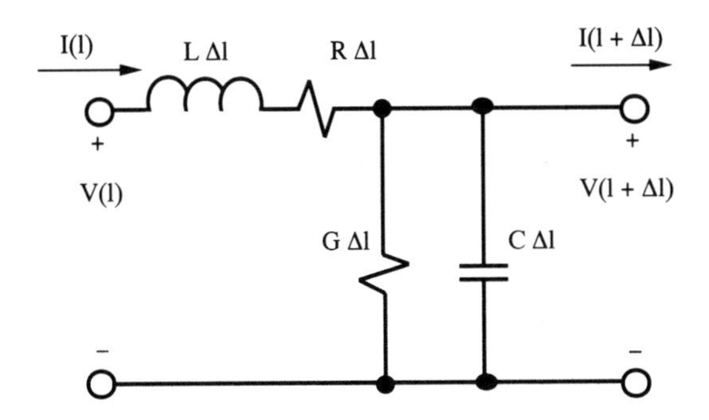

Fig. 2.4 *Lumped circuit model for a T-L segment.*

(Kirchoff's Voltage Law) to obtain:

$$V(l) - I(x)[R\Delta l + \mathrm{j}\omega L\Delta l] - V(l + \Delta l) = 0$$

or:

$$\frac{V(l + \Delta l) - V(l)}{\Delta l} = -I(l)[R + \mathrm{j}\omega L]. \tag{2.9}$$

We may also write Kirchoff's Current Law for the T-L segment:

$$I(l + \Delta l) = I(l) - V(l)[G\Delta l + \mathrm{j}\omega C\Delta l]$$

or:

$$\frac{I(l + \Delta l) - I(l)}{\Delta l} = -V(l)[G + \mathrm{j}\omega C]. \tag{2.10}$$

where it has been assumed that $V(l+\Delta l) \to V(l)$ to estimate the current in the dielectric. In the limit as $\Delta l \to 0$, equations (2.9) and (2.10) approach the respective partial derivatives. In order to decouple the equations we may take the derivative of each with respect to Δl and substitute the expressions in (2.9) and (2.10) as appropriate:

$$\begin{aligned} \frac{\partial^2 V}{\partial l^2} &= -\frac{\partial I}{\partial l}[R + \mathrm{j}\omega L] = V(l)[(R + \mathrm{j}\omega L)(G + \mathrm{j}\omega C)] \\ \frac{\partial^2 I}{\partial l^2} &= -\frac{\partial V}{\partial l}[G + \mathrm{j}\omega C] = I(l)[(R + \mathrm{j}\omega L)(G + \mathrm{j}\omega C)]. \end{aligned} \tag{2.11}$$

If a complex propagation coefficient γ^2 is defined, as was done for solutions of Maxwell's equations, we find that equations (2.11) are wave equations as well:

$$\gamma^2 = [(R + \mathrm{j}\omega L)(G + \mathrm{j}\omega C)]$$

gives:

$$\frac{\partial^2 V}{\partial l^2} - \gamma^2 V = 0 \qquad \text{and} \qquad \frac{\partial^2 I}{\partial l^2} - \gamma^2 I = 0. \tag{2.12}$$

Just as in the wave propagation case, $\gamma = \alpha + \mathrm{j}\beta$ where α is the attenuation coefficient and β the phase coefficient. The wavelength and speed of propagation on the transmission line may be calculated from β using equations (1.33). Comparing this result to the propagation constant in equation (1.34) suggests that C corresponds to ε, G to σ and L to μ in the unbounded medium, a reasonable pairing. Note that there is no wave equivalent for the series resistance, R, in the T-L conductors, which is also reasonable. The above development results in the so-called telegrapher's equations of propagation along a transmission line, equations (2.12).

As in the wave propagation solutions of Chapter 1, equations (2.12) have the following well known solutions:

$$V(l) = V_0^+ e^{-\gamma l} + V_0^- e^{+\gamma l}$$
$$I(l) = \frac{V_0^+}{Z_0} e^{-\gamma l} - \frac{V_0^-}{Z_0} e^{+\gamma l} \tag{2.13}$$

where Z_0 is the complex characteristic impedance of the transmission line (analogous to η in a medium):

in general: $$Z_0 = \sqrt{\frac{R + j\omega L}{G + j\omega C}}$$

for a lossless T-L: $$Z_0 = \sqrt{\frac{L}{C}}.$$

As in the previous discussion of the wave propagation solutions, the $e^{-\gamma l}$ term describes propagation in the $+l$ direction and the $e^{+\gamma l}$ term describes propagation in the $-l$ direction; and the sign inversion in the solution for current reflects the fact that for the $-l$ wave the center conductor current is of opposite direction to that shown in Figure 2.4. Note that Z_0 is only real for a lossless transmission line ($R = G = 0$). Though practical transmission lines are only approximately lossless, we often assume a low loss T-L to simplify calculations; i.e. that $R \approx 0$, $G \approx 0$. Also for a low loss transmission line, $\alpha \approx 0$ and $\beta \approx \omega\sqrt{LC}$.

The voltage and current at any point on the line may be determined once the amplitudes of the $+l$ and $-l$ waves are known. These are determined by the generator, its output impedance, Z_g, the transmission line impedance, Z_0, and the load impedance, Z_L as in the circuit in Figure 2.5. In this problem we assume that the line length L is substantial compared to wavelength $\lambda = 2\pi/\beta$. Consequently, points along the line have a phase angle different from either the generator or the load and the signal propagates with speed $u = \omega/\beta$. At every point along the line the instantaneous relationship between $V(l)^{\pm}$ and $I(l)^{\pm}$ is given by the characteristic impedance Z_0; however, it must be remembered that there are two waves propagating, so the apparent impedance is usually quite an interesting function of position. Table 2.1 summarizes the propagation characteristics of commercial coaxial transmission lines.

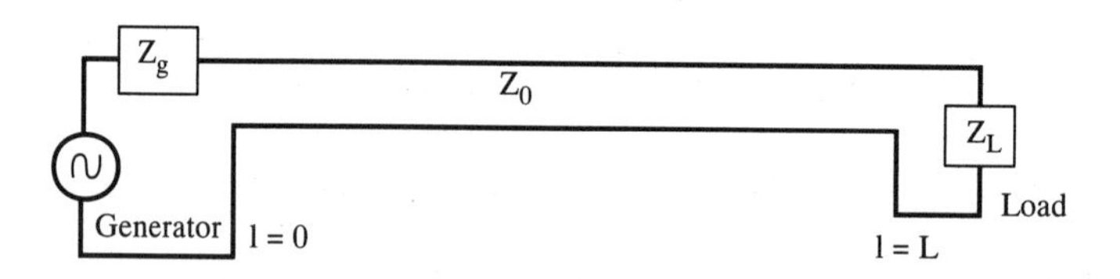

Fig. 2.5 *Loaded transmission line.*

Table 2.1 *Characteristics of commercial coaxial transmission lines.*

Designation	Z_0 (Ω)	C (pF/m)	a (cm)	b (cm)	V_{max} (V RMS)	P_{max} (kW)
Signal Coaxial Cables:						
RG-174/U	50	101		0.076	1500	
RG-59/U	75	67		0.185	2300	
RG-58/U	50	85.5		0.145	1900	
RG-8/U	50	82		0.362	600	
RS-225 Standard Air Lines (US; EIA Standard):						
$\frac{3}{8}$ inch	50		0.159	0.362		
$\frac{7}{8}$ inch	50		0.433	0.934		
$1\frac{5}{8}$ inch	50		0.831	1.94		
$3\frac{1}{8}$ inch	50		1.67	3.84	15 kV	80
$6\frac{1}{8}$ inch	50	67	3.30	7.70	30 kV	300

2.2.3 *The effect of load impedance*

First we should determine the relative amplitudes of the $+l$ and $-l$ waves. For purposes of discussion we consider the generator to have a real output impedance and to be matched to a low loss transmission line, $Z_g = Z_0 = R_0 = \sqrt{L/C}(\Omega)$. This means that reflections at the generator end can be neglected ($\Gamma_g = 0$). Assume that at $l = L$ the load impedance is a short circuit, $Z_L = 0$. Then it must follow that $V(L)^- = -V(L)^+$ so that $V(L) = 0$. The V^- wave is the reflected wave and its amplitude is determined by the complex reflection coefficient at the load, Γ_L:

$$\text{at the load:} \quad \Gamma_L = \frac{V_0^-}{V_0^+} = \frac{Z_L - Z_0}{Z_L + Z_0}$$

$$\text{at the generator:} \quad \Gamma_g = \frac{Z_g - Z_0}{Z_g + Z_0} \tag{2.14}$$

For the present case of a low loss line $\gamma \approx \mathrm{j}\beta$. We wish to write the solution so that the line voltage will be zero at $l = L$, so we introduce the distance from the end of the line, $d = l - L$; and the solutions for line current and voltage are:

$$V(d) = V_0^+ \mathrm{e}^{-\mathrm{j}\beta d} + V_0^- \mathrm{e}^{\mathrm{j}\beta d} = V_0^+[\mathrm{e}^{-\mathrm{j}\beta d} + \Gamma_L \mathrm{e}^{\mathrm{j}\beta d}]$$

$$I(d) = \frac{V_0^+}{Z_0}[\mathrm{e}^{-\mathrm{j}\beta d} - \Gamma_L \mathrm{e}^{\mathrm{j}\beta d}] \tag{2.15}$$

where, again, in phasor notation the $\mathrm{e}^{\mathrm{j}\omega t}$ term is understood. When the end of the transmission line is shorted $\Gamma_L = -1$ and a standing wave (a wave fixed on the transmission line) results:

$$V(d, \omega) = V_0^+[\mathrm{e}^{-\mathrm{j}\beta d} - \mathrm{e}^{\mathrm{j}\beta d}] = -j2V_0^+ \sin(\beta d)\mathrm{e}^{\mathrm{j}\omega t}$$

or:

$$V(d, t) = 2V_0^+ \sin(\beta d)\cos(\omega t). \tag{2.16}$$

The standing wave is identified by the sin (βd) term, a function of position only and independent of time. A shorted line represents a perfectly reflecting load (a mirror). The voltage has maxima whenever $\beta d = -(2n+1)\pi/4$, where $n = 0, 1, \ldots$, at which points the amplitude of the wave is twice that of the $+d$ wave voltage (constructive interference); and minima (zero amplitude) at $\beta d = -\text{n}\pi/2$. The solution for the line current, $I(d)$ has a similar form, but has maxima at $\beta d = -n\pi/2$ (in the short, for example) and minima (also zero amplitude) at $\beta d = -(2n+1)\pi/4$ (at maxima in the voltage):

$$I(d,\omega) = \frac{V_0^+}{Z_0}[\mathrm{e}^{-\mathrm{j}\beta d} + \mathrm{e}^{\mathrm{j}\beta d}] = 2\frac{V_0^+}{Z_0}\cos(\beta d)\mathrm{e}^{\mathrm{j}\omega t}$$

or:

$$I(d,t) = 2\frac{V_0^+}{Z_0}\cos(\beta d)\cos(\omega t) \tag{2.17}$$

The T-L and its load form the load impedance "seen" by the generator, Z_{eq} (Figure 2.6). The equivalent impedance depends on the point at which it is calculated; that is, on the length of the T-L. In the shorted line case Z_{eq} is imaginary at every point along the T-L. The loaded line equivalent impedance is periodic with a period of $\lambda/2$ on the d (or l) axis. Therefore, we may describe the "electrical length" of the line as the length modulo $\lambda/2$, i.e. the length in terms of an integral number of half-wavelengths. So a loaded T-L 8.75λ long has exactly the same equivalent impedance as a line 0.25λ or 2.75λ long.

For an open circuit load, $Z_{\text{L}} \to \infty$ and $\Gamma_L \to +1$, so the standing wave pattern is 90° out of phase with that of the shorted line since the end of the line is a voltage maximum rather than a current maximum:

$$V(d,\omega) = V_0^+[\mathrm{e}^{-\mathrm{j}\beta d} + \mathrm{e}^{\mathrm{j}\beta d}] = 2V_0^+\cos(\beta d)\mathrm{e}^{\mathrm{j}\omega t}$$

or:

$$V(\text{d},t) = 2\text{V}_0^+\cos(\beta d)\cos(\omega t) \tag{2.18}$$

and:

$$I(d,\omega) = \frac{V_0^+}{Z_0}[\mathrm{e}^{-\mathrm{j}\beta d} - \mathrm{e}^{\mathrm{j}\beta d}] = -\mathrm{j}2\frac{V_0^+}{Z_0}\sin(\beta d)\mathrm{e}^{\mathrm{j}\omega t}$$

or:

$$I(d,t) = 2\frac{V_0^+}{Z_0}\sin(\beta d)\cos(\omega t). \tag{2.19}$$

In practice it is impossible to obtain a true open circuit since some power will be radiated off the end of a cut transmission line, as in Figure 2.7a. However, a very accurate simulation

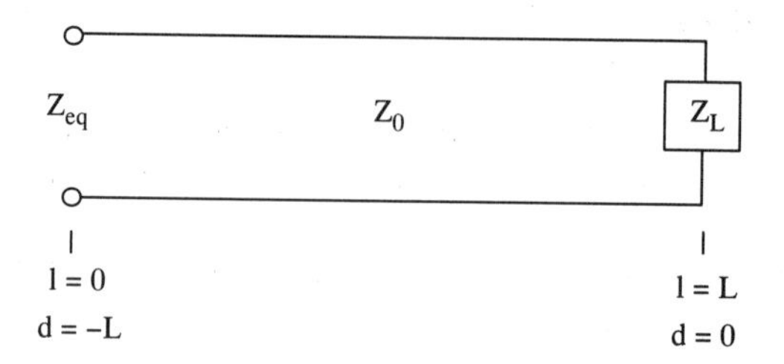

Fig. 2.6 *Equivalent impedance of transmission line of length L, with load.*

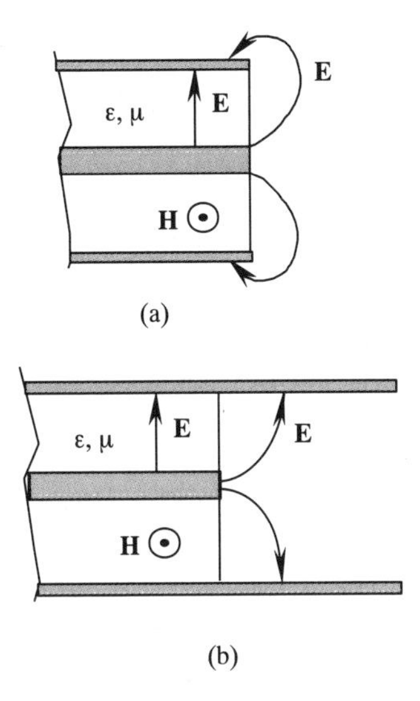

Fig. 2.7 *Termination of coaxial lines. (a) Radiating open-ended line. (b) Shielded "open" termination, non-radiating.*

of an open termination can be obtained by extending the outer conductor some distance beyond the inner conductor so that the fringing fields will terminate on the outer conductor rather than radiating significant energy into free space, as in Figure 2.7b.

Other load impedances at $d = 0$ ($l = L$) will result in a complex value for Γ_L such that the reflection coefficient magnitude is always $\leqslant 1$:

$$\Gamma = |\Gamma| e^{j\phi}. \tag{2.20}$$

The complex reflection coefficient means that the location of maxima and minima in voltage and current solutions (equations (2.15)) will not occur at the line terminus, but will be shifted by the electrical distance $\Delta\beta d = \phi/2$.

Finally, we must introduce a characterizing parameter, the Standing Wave Ratio (SWR)—or Voltage Standing Wave Ratio, VSWR—which is an indication of the degree of match or mismatch in the load impedance:

$$\text{VSWR} = \frac{V(d)_{\text{max}}}{V(d)_{\text{min}}} = \frac{1 + |\Gamma|}{1 - |\Gamma|}. \tag{2.21}$$

The range of SWR is from 1 to ∞; 1 for a perfectly matched load ($Z_L = Z_0$) and ∞ for a short or open termination. Operating lines and T-L devices (amplifiers, matching circuits, filters) are often characterized by their SWR.

2.2.4 Impedance at a selected point on the transmission line

We may write the equivalent impedance of the loaded length of transmission line in long form by taking the ratio of solutions in equation (2.15), expressing the reflection coefficient in terms of the impedances, and using Euler's complex exponential formula (we have assumed a low loss line):

$$Z(d)_{\text{eq}} = Z_0 \left[\frac{Z_\text{L} \cos(\beta d) - \text{j} Z_0 \sin(\beta d)}{Z_0 \cos(\beta d) - \text{j} Z_\text{L} \sin(\beta d)} \right]. \tag{2.22}$$

Remember that $d \leqslant 0$ in this discussion. The long form result can be simplified by dividing numerator and denominator by $\text{V}^+\text{e}^{-\text{j}\beta d}$:

$$Z(d)_{\text{eq}} = \frac{V(d)}{I(d)} = Z_0 \frac{[1 + \Gamma \text{e}^{\text{j}2\beta d}]}{[1 - \Gamma \text{e}^{\text{j}2\beta d}]}. \tag{2.23}$$

At this point it will be helpful to introduce two slight modifications. First, transmission line load problems are usually expressed in terms of a positive distance from the load and we have defined d such that it will always be negative (which was convenient for expressing voltage and current solutions). Now we introduce the (positive) distance from the load, $s = -d$, which results in only a slight modification to equations (2.22) and (2.23). Second, since the transmission line is fixed with impedance Z_0 and the variable of interest is the load, Z_L, it is convenient to use normalized impedances, $z(s) = Z(s)/Z_0$. In continuing the discussion upper case labels will be used to indicate impedance, $Z(\Omega)$, and admittance, Y (S), and lower case to signify normalized variables based on the transmission line Z_0 and $Y_0(= 1/Z_0)$, respectively. The normalized line impedance has real part, r, and imaginary part, x:

$$z(s)_{\text{eq}} = \frac{z_\text{L} \cos(\beta s) + \text{j} \sin(\beta s)}{\cos(\beta s) + \text{j} z_\text{L} \sin(\beta s)} = r + \text{j}x \tag{2.24}$$

or, equivalently:

$$z(s)_{\text{eq}} = \frac{[1 + \Gamma \text{e}^{-\text{j}2\beta s}]}{[1 - \Gamma \text{e}^{-\text{j}2\beta s}]} = r + \text{j}x. \tag{2.25}$$

The admittance is simply the complex reciprocal, $y(s) = 1/z(s)$. Using normalized impedance and admittance reduces the complexity of the class of solutions remarkably. The normalized impedance is periodic over $\beta s = \pi$ (i.e. as s goes from 0 to $\lambda/2$), as before.

2.2.5 *The Smith chart and its use*

Inspection of equation (2.25) suggests that a local reflection coefficient, $\Gamma(s)$ (at a distance s from the load), may be defined as:

$$\Gamma(s) = |\Gamma_\text{L}| \text{e}^{\text{j}\phi} \text{e}^{-\text{j}2\beta s} = |\Gamma_\text{L}| \text{e}^{\text{j}\psi}. \tag{2.26}$$

A little algebraic analysis reveals that there is a simple reciprocal relationship between the normalized impedance, $z(s)$, and $\Gamma(s)$:

$$z = r + \text{j}x = \frac{1 + \Gamma}{1 - \Gamma} \tag{2.27}$$

and:

$$\Gamma = p + \text{j}q = \frac{z - 1}{z + 1}. \tag{2.28}$$

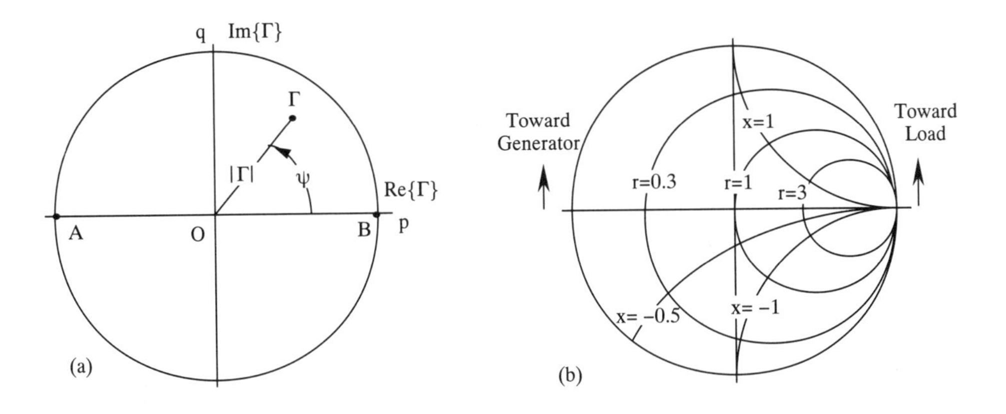

Fig. 2.8 *Smith chart geometry. (a) Polar plot of* Γ. *(b) Constant r and constant x loci.*

We can determine the impedance and the reflection coefficient at any point on the line as soon as we know one of them at one point: knowledge of Γ is sufficient for the determination of z, and vice-versa. So, both variables may be plotted on the same complex plane axes—such a plot is called the Smith chart, a method for graphically solving transmission line problems and for visualizing impedance effects. Figure 2.8a is an example polar plot of Γ, the locus of points for which the magnitude, $|\Gamma|$, is constant are on the perimeter of a circle which has its center at the origin. All the impedance values of passive circuits are represented by the points in the interior of the circle of radius $|\Gamma| = 1$. The point A corresponds to a short circuit $z(s) = 0$, where $\Gamma(s) = -1$, a voltage minimum; the point B corresponds to $z(s) = \infty$, where $\Gamma(s) = +1$, a current minimum. The origin corresponds to $z = 1(\Gamma = 0)$, for which $Z_L = Z_0$ and there is no reflection.

The transformation represented by equations (2.27) and (2.28) is a conformal transformation. The partitioning of the complex plane with curves r = constant and x = constant induces in the polar plot a partitioning of a family of orthogonal circles (Figure 2.8b), which comprise the Smith chart:

$$\begin{aligned} \left(p - \frac{r}{r+1}\right)^2 + q^2 &= \frac{1}{(r+1)^2} \\ (p-1)^2 + \left(q - \frac{1}{x}\right)^2 &= \frac{1}{x^2}. \end{aligned} \tag{2.29}$$

Note that when ψ increases, s decreases, so the position on the transmission line (equation (2.26)) has been moved "toward the load"; conversely, a decrease in ψ (increase in s) corresponds to movement along the T-L "toward the generator". Also, a movement of $s = \lambda/4$ is equivalent to a half-turn on the Smith chart, while moving $s = \lambda/2$ gives one complete revolution on the Smith chart, as expected from equation (2.26). The Smith diagram permits rapid graphical determination of the values of Γ, VSWR and the distances to the nearest voltage maximum (when $\psi = 0$) and voltage minimum (when $\psi = \pi$) from any load point (r, x). For example, if you plot the load impedance (r_L, x_L) on Smith axes then $\psi = \phi$ because $s = 0$, and the radius is the magnitude of Γ (see Figure 2.9). You must

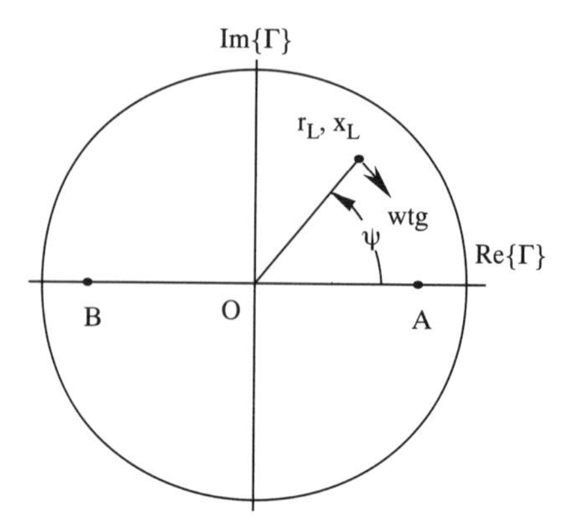

Fig. 2.9 *Example of an impedance plot. First voltage maximum occurs at point A wavelengths toward the generator (wtg) and voltage minimum at point B.*

move clockwise toward the generator (maximum of 0.5λ) to encounter the first voltage maximum at point A in the figure (if x is capacitive, i.e. $x < 0$, you will encounter a voltage minimum, point B, before encountering a voltage maximum). Note that, as expected, any movement of $\lambda/2$ wavelengths returns one to the same equivalent impedance.

The Smith chart was originally developed during the 1930s to facilitate transmission line and waveguide calculations. The recent advent of wide scale computer use and digital calculation prompted some technical "visionaries" to predict the imminent demise of the Smith chart and its companions from the analog world. While the slide rule has all but disappeared (except from museums and, well, our desks) the Smith chart is a well used and important device — for example, the fanciest of digital network analyzers use the Smith chart as a standard output. Its enduring value arises from the ease with which humans extract information from images — columns of numbers are nearly meaningless, but a clear Smith "picture" of what occurs on a T-L is invaluable to the engineer or scientist. So, rumors of the demise of the Smith chart are, at least, premature.

2.2.6 Stub tuning and disturbances on the line

Unmatched load impedances can be tuned to match the T-L impedance by using shorted lengths of transmission line — so-called "single-stub tuners". The approach is to note that a shorted section of T-L always has $|\Gamma| = 1$, which puts the locus of possible solutions on the perimeter of the Smith chart, where $r_{SC} = 0$ ($y_{SC} = \infty$). The point A in impedance coordinates in Figure 2.8a corresponds to point B in admittance coordinates; and for any arbitrary load $y = g + jb$ can always be obtained from $z = r + jx$ by rotating π radians on the Smith chart at constant $|\Gamma|$.

By carefully placing a shorted section (length s_{SC}) in parallel with a chosen length of the loaded line (s_L), as in Figure 2.10a, the combined line impedances may be made equal to Z_0 achieving a match to the T-L and no reflection toward the generator. Note that each section of T-L may have a high SWR (in fact the shorted section will have SWR $\rightarrow \infty$). This problem is easiest to illustrate in admittance coordinates. The match is achieved by selecting

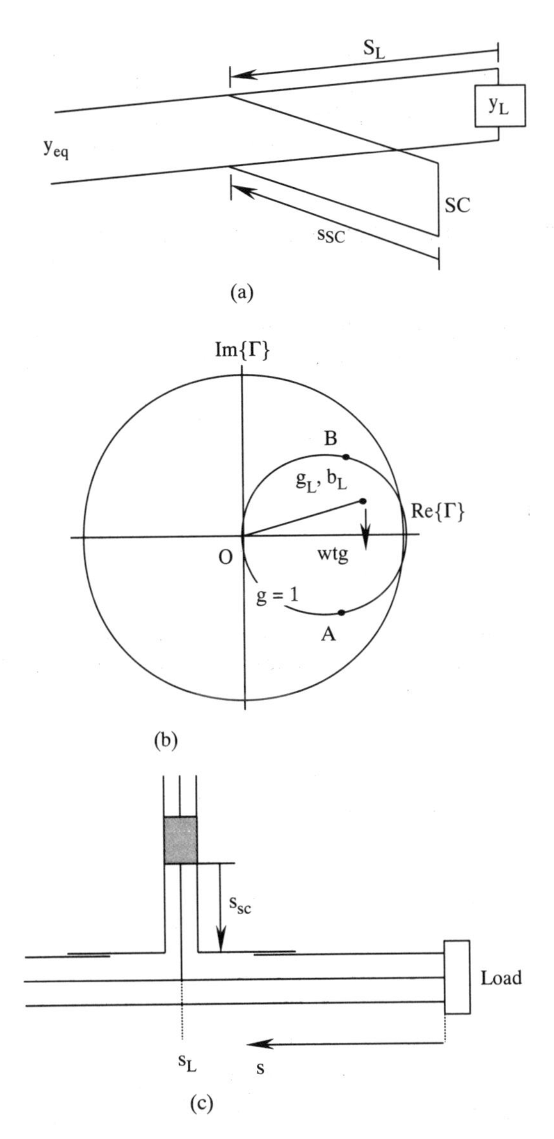

Fig. 2.10 *Example single-stub matching calculation. (a) Shorted length of transmission line used to match the load. (b) Location of stub using $g = 1$ circle at points A and B. (c) Practical single-stub tuner for a coaxial transmission line.*

s_L at a point where the T-L and load admittance, $y(s_L) = 1 + jb_L$. For every load there are two possible choices for s_L since there are two intersections of $y(s) = 1$ for each $|\Gamma|$, at points A and B in Figure 2.10b. Note that we are restricted to moving toward the generator (wtg). Then, select the stub length s_{SC} so that its admittance is $y_{SC}(s_{SC}) = -jb_L$ at s_L. Some advantage can be obtained by choosing the location which makes the stub length convenient, either for accuracy or physical size. Admittances in parallel add, so the loaded line and parallel stub will be matched to the T-L ($y_{eq} = 1 + j0$) at all locations to the left of s_L.

In fact, the effect of any disturbance on a transmission line may be nullified by stub tuners; however, the tuning is obtained at one frequency only and there are some regions in the Smith chart where uncertainties in stub length create unacceptable uncertainty in the match. A single-stub tuner must be very carefully placed, and frequency drift cannot be compensated. In a later section we will present multiple-stub tuners, which can be used at many frequencies and in nearly all regions of the Smith chart to achieve matched line conditions.

2.2.7 Other transmission line geometries

Twin lead lines

One inexpensive and popular form of transmission line is the twin lead configuration (Figure 2.11a). The inductance and capacitance per unit length are given by:

$$L = \frac{\mu}{\pi}\cosh^{-1}\left(\frac{d}{2a}\right) \tag{2.30}$$

and:

$$C = \frac{\pi\varepsilon}{\ln\left\{(d + \sqrt{d^2 - 4a^2})/2a\right\}}. \tag{2.31}$$

This type of line has a characteristic impedance which approaches that of free space ($\eta_0 = \sqrt{\mu_0/\varepsilon_0} = 120\pi\ \Omega = 377\ \Omega$)—typical values for Z_0 are around 300 Ω, and the line is

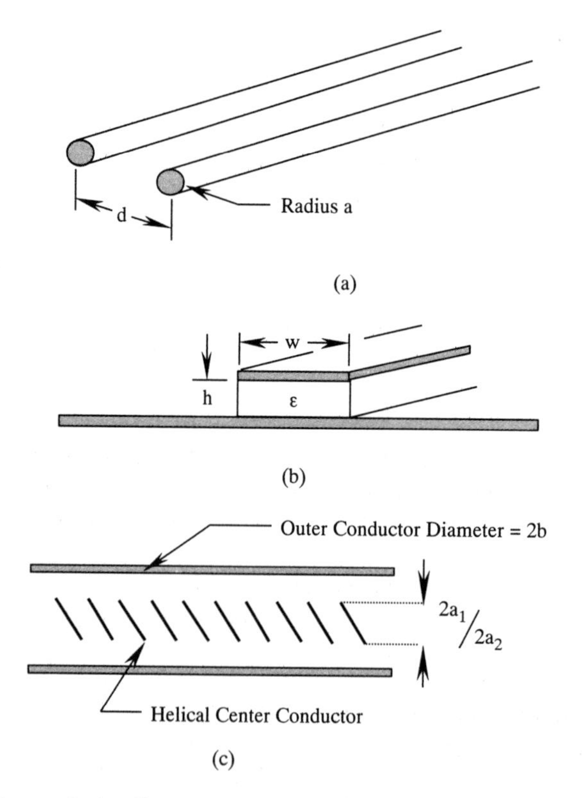

Fig. 2.11 *Other transmission line geometries. (a) Twin lead transmission line. (b) Microstrip transmission line. (c) Helical coil center conductor.*

fairly low loss. This geometry is often used to match V-shaped ("rabbit ear") television antennas (with antenna source impedance $Z_a = 300\ \Omega$) and is simple; however, it has the distinct disadvantage that the electromagnetic fields are exposed (outside of the conductors) so that disturbances created by clips holding the T-L may not be negligible, and rain coating the dielectric may also result in significant loss.

Microstrip lines

Recent advances in microcircuitry include the evolution of new small transmission lines termed microstrip lines. A typical microstrip line consists of a metallic electrode placed upon a pedestal of high permittivity, ε, which is itself mounted on metallic base, as shown in Figure 2.11b. The characteristic impedance is a function of the dimensions w and h, and of the dielectric constant of the ribbon. A particular characteristic impedance can be obtained by variations, for example, in the width of the ribbon. A very large number of useful low power microwave circuits can be fabricated using this technology. If $0.05 < w/h < 1$, then the characteristic impedance is:

$$Z_0 = \frac{120\pi}{\sqrt{\varepsilon_{eq}}} \log\left\{\frac{8h}{w} + \frac{w}{4h}\right\} \tag{2.32}$$

and for $1 < w/h < 20$:

$$Z_0 = \frac{120\pi}{\sqrt{\varepsilon_{eq}}} \left[\frac{w}{h} + 1.393 + 0.667 \log\left\{\frac{w}{h}\right\} + 1.49\right]^{-1} \tag{2.33}$$

where ε_{eq} is the equivalent permittivity as a function of ε_r for the dielectric, of width w and thickness h.

Helical center conductor lines

It is often necessary to use lines with high characteristic impedance values. A high impedance line can be obtained by replacing the central conductor of a coaxial cable with a helically-wrapped wire, as shown in Figure 2.11c, where b is the radius of the external conductor, a_1 and a_2 are inner and outer radii of the internal conductor, and n is the number of turns per meter. For a very long and tightly wound center coil the magnetic field outside will be negligible compared to the magnetic field inside, so the line inductance will be dominated by the central coil inductance. The line capacitance is approximately the same as that for a coaxial cable so:

$$L \cong \mu\pi a_1^2 n^2 \tag{2.34}$$

$$C = \frac{2\pi\varepsilon}{\ln\{b/a_2\}}. \tag{2.35}$$

2.2.8 Impedance matching strategies

There are important applications for radio frequency and microwave systems which require compensation of T-L impedance differences. First, consider a line without loss for which an obstacle creates a reflection of waves. Of course, reflection of waves means reflection of energy; which is undesirable (but in some sense unavoidable) in high power heating applications. The problem is to reduce the transmission line loss of energy due to obstacles or imperfections by using impedance matching methods. For example, for a coaxial line connector which has dimensions such that its characteristic impedance is different from Z_0, the equivalent circuit of the discontinuity is a capacitance, C_0. The capacitance is a function of the dimensions (see Figure 2.12) and therefore dependent upon the differences in characteristic impedance on both sides of the discontinuity.

Perturbations can be compensated by the introduction of other obstacles such that a complementary impedance is created to nullify the effect of the initial obstacle, as was seen in the stub tuner example above. For a given small obstacle on the line with normalized impedance z_B, the compensating obstacle (z_{BC}) may be chosen to have the same impedance z_B, and placed at a distance $s = \lambda/4$ (180° out of phase) in front of the initial obstacle to obtain $z_{BC} = -z_B$ so that for the two obstacles $z_{TOT} = z_B - z_B = 0$, as shown in Figure 2.13.

Another problem in transmission applications is to transfer energy without losses from a given generator to a given load. In the special case where the load and the generator are equal pure resistances, $Z_L = R_L = R_g$, the connection between the two can be made with a low loss line which has a characteristic impedance equal to both values:

$$Z_0 = R_L = R_g \tag{2.36}$$

so that $\Gamma_L = \Gamma_g = 0$. The connection can be made with a line of any length without introducing loss. Mismatch at the generator and/or at the load will introduce reflection which, if $|\Gamma_L| = |\Gamma_g|$, may be eliminated by clever choice of the length of the line. This is often not feasible (even when the condition is satisfied), especially in RF systems with expensive large diameter coaxial lines. When R_g is different from R_L (and both are sufficiently well behaved, i.e. real), matching can sometimes be obtained by a system of three transmission lines, as illustrated in Figure 2.14. The first section must have $Z_{01} = R_g$, and the third section

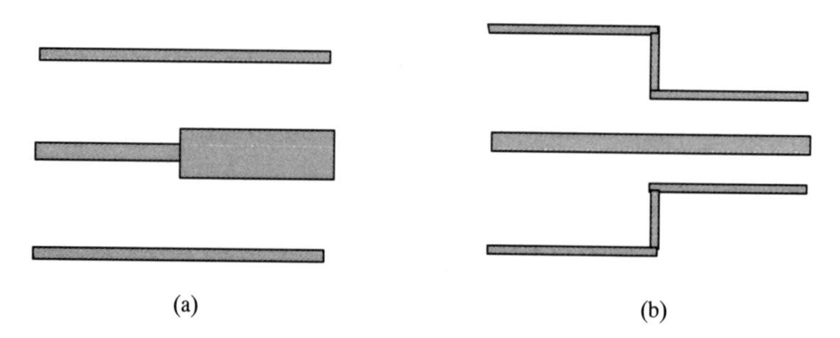

Fig. 2.12 *Discontinuities in coaxial transmission lines. (a) Center conductor step discontinuity. (b) Outer conductor step discontinuity.*

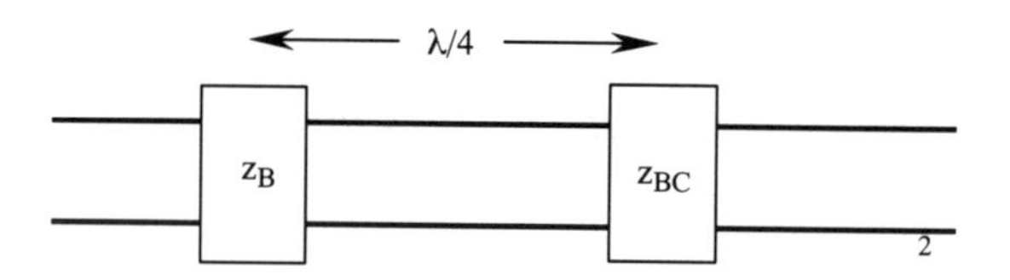

Fig. 2.13 *Compensating obstacle, twin lead line example.*

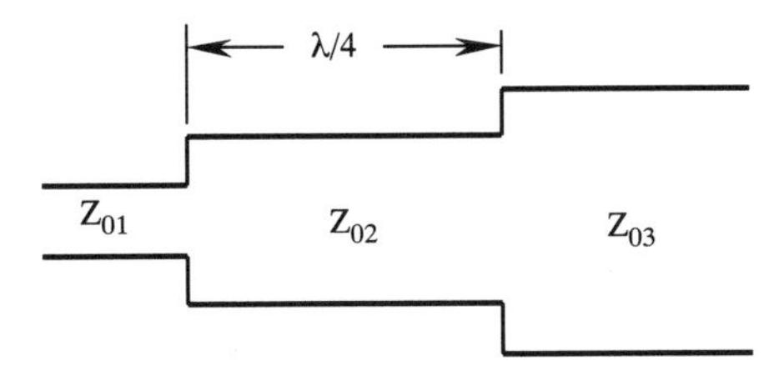

Fig. 2.14 *Quarter wave matching transmission line section.*

must have $Z_{03} = Z_L$ (if this can be arranged). If Z_L is complex and a proper Z_{03} cannot be found, then a match may still be made if s_L is selected so that the equivalent impedance, $z_3(s) = r_L(s) + j0$. Then, providing that it can be arranged to have $Z_{02}{}^2 = Z_{01}Z_{3eq}(s_L)$ the interconnection may be made with a section of length $\Delta s_2 = \lambda_2/4$. The length of section number 1 is arbitrary; however, the length of section number 3 may be critical if Z_L is complex and z_L winds up in an inconvenient location on the Smith chart.

If this is achievable, the matching section-2 transmission line (Z_{02}) is termed a quarter wave transformer — its action on the equivalent impedance of the section-3 T-L (Z_{03}) and load can be clearly seen in equations (2.22) or (2.24) where $\beta s = \pi/2$ so $\cos(\beta s) = 0$ and $\sin(\beta s) = 1$. This kind of impedance matching solution is the transmission line equivalent of an optical anti-reflective coating on camera lenses or of a non-reflecting radome covering for radar and radio antennas. Matching between the generator and the load occurs, of course, only for the domain of frequencies specified by λ_2, i.e. the impedance match is only successful over a narrow range of frequencies (a monochromatic anti-reflective coating). To increase the frequency domain, it is possible to use a sequence of intermediate sections, where each section satisfies:

$$Z_i^2 = (Z_{i-1})(Z_{i+1}) \tag{2.37}$$

and the length of each section is $\Delta s_i \approx \lambda_i/4$ — a multi-coated lens uses this approach to eliminate reflections over the whole range of visible optical wavelengths.

2.2.9 Discrete transmission lines

We will conclude the discussion of transmission lines with a consideration of discrete lines. The propagation equations can be rewritten in terms of a succession of many localized identical quadrupoles, as in Figure 2.15. This is termed an artificial line; however, in practice a useful line of this type may be synthesized from a sequence of integrated circuit transmission lines in a microstrip format. As with other lines, synthesized artificial lines are characterized by their propagation coefficients and by their characteristic impedance. Both of these generally complex numbers are obviously functions of the electrical parameters of the quadrupoles.

The transmission line voltage and the current have these general formulae:

$$V_n = AV_{n+1} + BI_{n+1} \qquad I_n = A'I_{n+1} + CV_{n+1} \tag{2.38}$$

or:

$$\begin{bmatrix} V_n \\ I_n \end{bmatrix} = \begin{bmatrix} A & B \\ C & A' \end{bmatrix} \begin{bmatrix} V_{n+1} \\ I_{n+1} \end{bmatrix}. \tag{2.39}$$

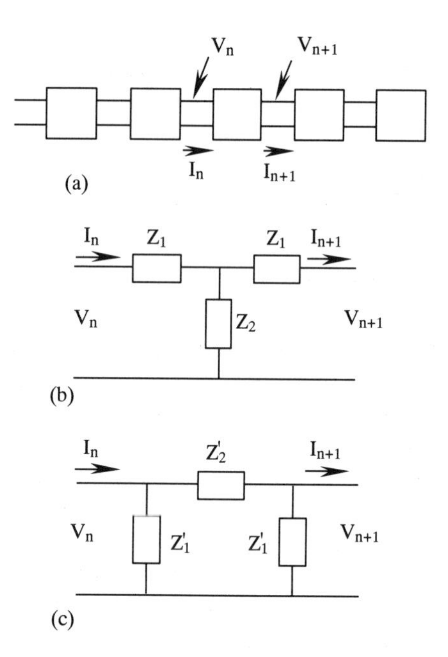

Fig. 2.15 *Quadrupole transmission line model. (a) Sequence of quadrupoles. (b) T-equivalent circuit model. (c) π-equivalent circuit model.*

The reciprocity principle (from linear network analysis) yields the relation:

$$AA' - BC = 1. \tag{2.40}$$

If each quadruple is symmetric $A = A'$; and if we assume a standard (discrete) complex exponential dependence for voltage and current:

$$V_n = \mathrm{V}_0 \mathrm{e}^{-\gamma n \Delta s} \quad \text{and} \quad I_n = I_0 \mathrm{e}^{-\gamma n \Delta s}. \tag{2.41}$$

One then obtains the T-L solution for the general equations of voltage and current along the line in discrete form:

$$\begin{aligned} V_n &= V_1 \mathrm{e}^{-\gamma n \Delta s} + V_2 \mathrm{e}^{+\gamma n \Delta s} \\ I_n &= \frac{V_1}{Z_\mathrm{c}} \mathrm{e}^{-\gamma n \Delta s} - \frac{V_2}{Z_\mathrm{c}} \mathrm{e}^{+\gamma n \Delta s}. \end{aligned} \tag{2.42}$$

These formulae are identical to those of the continuous line with a length $n\Delta s$. The characteristic impedance, Z_c, of the chain of quadrupoles is $Z_\mathrm{c} = B/C$ and the propagation constant γ is determined by $A = \cosh(\gamma)$. Each quadrupole introduces a delay τ, where $\omega\tau = \beta$ (remember that $\gamma = \alpha + \mathrm{j}\beta$), as for a continuous line. The n quadrupoles have a total delay of $n\tau$; the delay, τ depends directly upon ω, and in an indirect way upon β. Each quadrupole has an equivalent three-impedance T-section circuit model, as shown in Figure 2.15b.

If the quadrupoles are symmetric, the characteristic impedances are given by:

$$Z_\mathrm{c} = \sqrt{Z_1^2 + 2Z_1 Z_2} \quad \text{where} \quad \cosh(\gamma) = 1 + \frac{Z_1}{Z_2} \tag{2.43}$$

or:

$$Z'_c = \frac{Z'_1 Z'_2}{\sqrt{Z'^2_1 + 2Z'_1 Z'_2}} \qquad \text{where} \quad \cosh(\gamma) = 1 + \frac{Z'_2}{Z'_1} \tag{2.44}$$

for a π-section circuit representation, as shown in Figure 2.15c.

2.3 PROPAGATION IN A WAVEGUIDE

At elevated frequencies waveguides are more useful than transmission lines because of conductor losses. We remember that the governing assumption of the T-L geometry was that the characteristic dimensions (such as a and b) were small compared to λ even though the length of the line was long. Increasing the line dimensions reduces losses but introduces wavelength-dependent phase considerations in the cross-sectional fields, as well as in the direction of propagation. The equations describing waveguide function are similar in many respects to the T-L results and, in fact, the Smith chart is equally useful in waveguide analysis. This section summarizes waveguide construction and design considerations. Chapter 3 describes waveguide devices useful for high power applications.

2.3.1 Rationale for the use of waveguides

One advantage of transmission line propagation is that it is always TEM and thus relatively easy to analyze. The disadvantage is that at high power the current density on the center conductor (of a coaxial T-L, for example) is quite high and losses approach intolerable levels; especially as the frequency increases and the T-L dimensions become small in order to satisfy the quasi-static assumption which we made. The solution to the loss problem is to use a guided wave rather than a transmission line. For our purposes, a waveguide will be any long hollow conductor of constant cross-sectional dimension such that its pertinent dimensions are at least $\lambda/4$ at the frequency of operation. Thus, the size will be considerably larger than the comparable transmission line and transmission losses will decrease dramatically. Note that the number of possible solutions is very large since now a candidate solution need only satisfy the zero tangential E-field boundary conditions on the surfaces of the waveguide (assuming perfect conductors for the walls):

$$\begin{aligned} &\text{at} \qquad x = 0 \text{ or } a: \qquad E_y = E_z = 0 \\ &\text{at} \qquad y = 0 \text{ or } b: \qquad E_x = E_z = 0. \end{aligned} \tag{2.45}$$

Both surface charge and surface current may exist (if we assume that the guide walls are very good conductors). Also note that (purposefully) the quasi-static assumption is not valid for a waveguide cross-section. A very large number of possible cross-sections could be used to make a waveguide, but circular and rectangular cross-sections are the most practical. Since it is by far the most commonly used form of waveguide, we will consider the rectangular tube with metallic walls, illustrated in Figure 2.16, in some detail.

2.3.2 Rectangular waveguide TE and TM propagation

It is easy to verify that for propagation in the z-direction, the only interesting case, all applicable solutions for the rectangular cross-section can be obtained from:

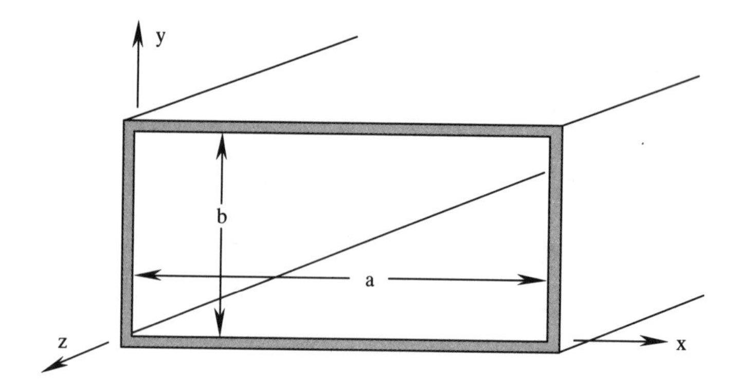

Fig. 2.16 *Rectangular waveguide geometry.*

$$E_x = -\frac{\gamma}{h^2}\frac{\partial E_z}{\partial x} - \frac{j\omega\mu}{h^2}\frac{\partial H_z}{\partial y} \qquad E_y = -\frac{\gamma}{h^2}\frac{\partial E_z}{\partial y} + \frac{j\omega\mu}{h^2}\frac{\partial H_z}{\partial x}$$
$$H_x = -\frac{\gamma}{h^2}\frac{\partial H_z}{\partial x} + \frac{j\omega\varepsilon}{h^2}\frac{\partial E_z}{\partial y} \qquad H_y = -\frac{\gamma}{h^2}\frac{\partial H_z}{\partial y} - \frac{j\omega\varepsilon}{h^2}\frac{\partial E_z}{\partial x} \tag{2.46}$$

where $h^2 = \gamma^2 + \omega^2\mu\varepsilon = A^2 + B^2$. There are many possible fields which satisfy (2.46); but there are two interesting classes of solutions which are mathematically simpler to manage—namely, TE propagation (the electric field is normal to the direction of propagation so $E_z = 0$) and TM propagation (where $H_z = 0$). Every possible solution can be decomposed into the linear combination of a TE and TM expression, so no loss in generality results from looking at each case separately. Three important results can be shown for either TE or TM modes in a rectangular guide (as in Figure 2.16): (1) $A = m\pi/a$, (2) $B = n\pi/b$, where m and n are integers, and (3) propagation only occurs when γ is primarily imaginary (remember that $\omega^2\mu\varepsilon = -[j\omega\sqrt{\mu\varepsilon}]^2$):

$$\omega^2\mu\varepsilon \geqslant \left(\frac{m\pi}{a}\right)^2 + \left(\frac{n\pi}{b}\right)^2 \tag{2.47}$$

So, there is a cutoff frequency, ω_c (in radians) corresponding to f_c (in Hz), obtained when the equality is satisfied in (2.47), below which the guide will not propagate. The resulting cutoff wavelength is an important parameter:

$$\lambda_c = \frac{c}{f_c} = \frac{1}{\sqrt{\mu\varepsilon} f_c} = \frac{2}{\sqrt{(m/a)^2 + (n/b)^2}} \tag{2.48}$$

so that:

$$\beta = \frac{2\pi}{\lambda_g} = \omega\sqrt{\mu\varepsilon}\sqrt{1 - (f_c/f)^2} \tag{2.49}$$

where c is the velocity of a wave in an unbounded medium composed of the same material as the contents of the waveguide (μ, ε), λ_c is the cutoff wavelength (associated with f_c), and λ_g is the guide wavelength ($\lambda_g > \lambda_c$, which will be explained) at the operating frequency, f (where $f > f_c$).

TE rectangular waveguide solutions

Any propagating TE electromagnetic field in the rectangular guide which solves Maxwell's equations may be found from the following two sets of relations:

$$\begin{aligned} E_x &= \frac{j\omega\mu}{h^2}\left(\frac{n\pi}{b}\right) C \cos\left(\frac{m\pi x}{a}\right) \sin\left(\frac{n\pi y}{b}\right) e^{-j\beta z} \\ E_y &= -\frac{j\omega\mu}{h^2}\left(\frac{m\pi}{a}\right) C \sin\left(\frac{m\pi x}{a}\right) \cos\left(\frac{n\pi y}{b}\right) e^{-j\beta z} \\ E_z &= 0 \end{aligned} \tag{2.50}$$

and:

$$\begin{aligned} H_x &= \frac{j\beta}{h^2}\left(\frac{m\pi}{a}\right) C \sin\left(\frac{m\pi x}{a}\right) \cos\left(\frac{n\pi y}{b}\right) e^{-j\beta z} \\ H_y &= \frac{j\beta}{h^2}\left(\frac{n\pi}{b}\right) C \cos\left(\frac{m\pi x}{a}\right) \sin\left(\frac{n\pi y}{b}\right) e^{-j\beta z} \\ H_z &= C \cos\left(\frac{m\pi x}{a}\right) \cos\left(\frac{n\pi y}{b}\right) e^{-j\beta z} \end{aligned} \tag{2.51}$$

where C is an arbitrary amplitude constant. Integer values for m and n will satisfy the point forms of Maxwell's relations, and those solutions are referred to as TE_{mn} modes, where m and n specify the mode of propagation — the mode tells the number of maxima and minima of each field in the waveguide.

The TE_{10} mode

By far the most commonly used waveguide mode is the TE_{10} mode where $m = 1$ and $n = 0$. This mode is the most fundamental mode which will propagate in a rectangular waveguide. For a particular rectangular cross-section in which $a = 2b$ the TE_{10} mode is the mode which has the lowest cutoff frequency, f_C. Waveguides can be sized to prevent other modes from propagating when the TE_{10} mode is the design goal by choosing $a = 2b$ and using a guide sized so that the frequency of interest is just above the cut off frequency for the TE_{10} mode. Commercial waveguides have $a = 2b$ and are available in diverse sizes with specified operating frequency ranges for each, assuming that TE_{10} is the design mode. Table 2.2 lists defined standard rectangular waveguides and their propagation characteristics at 2.45 GHz. For the TE_{10} mode the respective fields are:

$$E_x = E_z = H_y = 0 \qquad \text{from (2.50) and (2.51)}$$

Table 2.2 *Characteristics of standard rectangular waveguides.*

Designation	a (cm)	f_C (GHz)	λ_g (cm) (at 2.45 GHz)	P_{rating} (kW)
WR-284	7.21	2.077	23.2	2.2
WR-340	8.64	1.735	17.3	4.2
WR-430	10.9	1.372	14.7	6.5

$$
\begin{aligned}
E_y &= \frac{-\mathrm{j}\omega\mu a}{\pi} C \sin\left(\frac{\pi x}{a}\right) \mathrm{e}^{-\mathrm{j}\beta z} \\
H_x &= \frac{\mathrm{j}\beta a}{\pi} C \sin\left(\frac{\pi x}{a}\right) \mathrm{e}^{-\mathrm{j}\beta z} \\
H_z &= C \cos\left(\frac{\pi x}{a}\right) \mathrm{e}^{-\mathrm{j}\beta z}.
\end{aligned} \tag{2.52}
$$

These equations describe a wave propagating in the z-direction with electric field distribution depicted in Figure 2.17a and magnetic field distribution as in Figure 2.17b. The electric field originates and terminates on a surface charge distribution located, for this mode, at $y = 0$ and $y = b$. The H-fields are tangential to the waveguide walls and, if we assume a waveguide constructed of very high electrical conductivity materials, satisfy the surface "sheet" current boundary condition (equation (1.27)) there. There is one maximum in both the E-field and H_x-field across $0 < x < a$ located at $x = a/2$ but $\lambda_g/4$ apart in the z-direction (i.e. $E_{\max}$ occurs at $H_{x_{\min}}$); and two maxima in H_z at $x = 0$ and $x = a$. Neither field is dependent on y inside of the waveguide for this mode.

TE_{10} mode waveguide wall current distribution

Waveguide walls can be configured to create effective radiators by cutting slots of the appropriate geometry. Conversely, a slot may be cut which does not radiate and thus can be used either to sample the contents of the waveguide or to introduce a product to be irradiated. An understanding of the wall current distribution is essential in order to make use of these properties of a waveguide, so it is appropriate to look at it at this time.

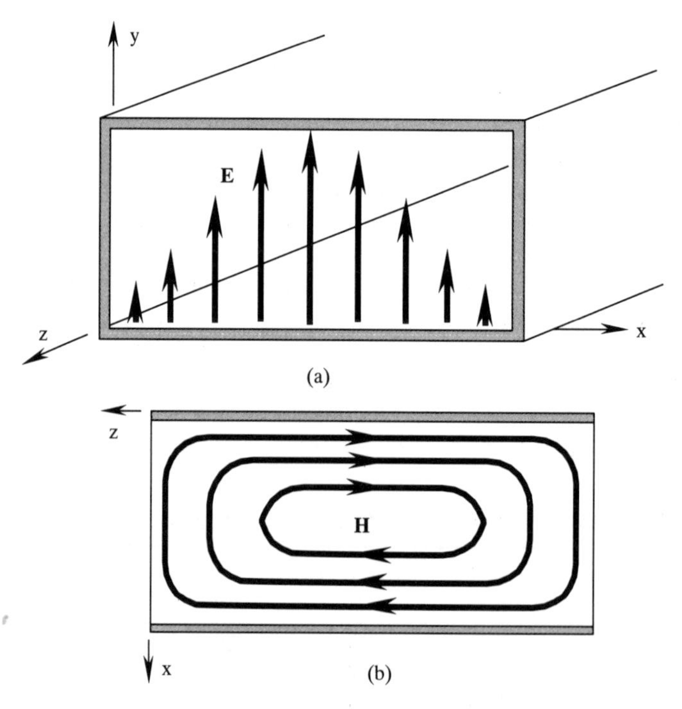

Fig. 2.17 *TE_{10} propagation fields. (a) Electric field lines. (b) Magnetic field lines.*

The propagation of a wave in a hollow tube is associated with the flow of "sheet" current, **K** (or its equivalent), in the metallic walls which exists to match the tangential magnetic field boundary conditions. In real waveguides, of course, the electrical conductivity of the walls, though very high, is by no means infinite. In practice, the wall currents will exist as a finite current density, **J**, within a very short distance of the surface essentially limited by the skin depth, δ, as was introduced in Chapter 1. The skin depth is in turn described by the finite conductivity of the wall, σ, assuming a "good" conductor:

$$\delta = \frac{1}{\alpha} \cong \sqrt{\frac{2}{\omega\mu\sigma}} \tag{2.53}$$

where a very good conductor has a large loss tangent:

$$\text{loss tangent} = \left(\frac{\sigma}{\omega\varepsilon}\right) \gg 1. \tag{2.54}$$

In fact, for typical waveguide wall materials σ is so large that ε is not measurable and displacement currents may be safely ignored over the frequencies of interest. If we designate the n-direction as normal to the wall surface (as we did in Chapter 1) the equivalent sheet current density is tangential to the wall (the t-direction) and is the depth integral of **J** in the wall:

$$\mathbf{K}_{\text{eq}} = k_t\mathbf{a}_t = \left[\int_0^\infty J(n)\,\mathrm{d}n\right]\mathbf{a}_t = \left[\int_0^\infty J(0)\mathrm{e}^{-\alpha n}\,\mathrm{d}n\right]\mathbf{a}_t = \delta J(0)\mathbf{a}_t. \tag{2.55}$$

Here we recall that the direction of the wall current is parallel to the surface but perpendicular to the tangential H-field just outside the wall (Figure 2.18a). The product $\delta J(0) = H_t$ in magnitude, though the direction is as described. The result is that wall currents will exist in the patterns shown in Figure 2.18b in the broad wall and in the narrow wall.

If the current is flowing without resistance, that is the walls are infinitely conductive,—$\alpha = \infty$ and $\delta = 0$; the wall current is then a sheet current and the wave propagates without attenuation. Under that condition the waveguide comprises a lossless transmission line. Attenuation in the propagating wave is generated by losses in the waveguide walls, alone, due to their finite conductivity. Often the combined effects of current density confined to a thin layer near the surface are lumped into a so-called "surface resistance", R_s (usually expressed in the somewhat unsatisfying units of "ohms per square"—square "whats" is what I always want to know, but, as you can see from the definition in equation (2.56), it is a futile metaphysical question since the length units cancel):

$$R_s = \frac{1}{\sigma\delta}. \tag{2.56}$$

The total power lost per unit length of waveguide in the TE_{10} mode, P_L (W/m), can be estimated from:

$$P_L = R_s C^2\left(b + \frac{a}{2} + \frac{a}{2}\frac{\beta^2 a^2}{\pi^2}\right) \tag{2.57}$$

and the resulting wall attenuation coefficient may be obtained from:

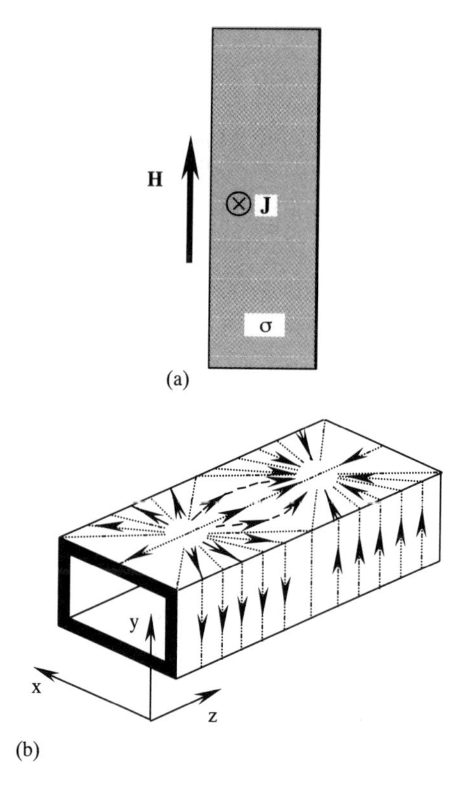

Fig. 2.18 *Waveguide wall currents. (a) Wall boundary condition, current density vector, J, is directed into the paper when H is as shown. (b) Waveguide currents in TE_{10} mode.*

$$\alpha = \frac{P_{\mathrm{L}}}{2P_{\mathrm{T}}} = \frac{R_{\mathrm{s}}(1 + 2bf_{\mathrm{c}}^2/af^2)}{b\sqrt{\mu/\varepsilon}\sqrt{1-(f_{\mathrm{c}}/f)^2}}. \tag{2.58}$$

The attenuation is, as should be expected, a frequency-dependent parameter (since δ is frequency dependent). The ultimate power handling capacity — the maximum allowable value of P_{T} — of a waveguide is dependent on two considerations: (1) the ability of the waveguide walls to dissipate heat from surface resistance losses, and (2) the breakdown electric field strength of the waveguide dielectric medium (as was used to obtain the maximum power values in Table 2.2).

Referring to Figure 2.18b, in the center of the broad wall of the waveguide, the current flow is parallel to the Z-axis. A slot can be cut in the guide along this center-line without cutting wall current lines, as long as it is not too wide. Because no wall current lines are cut such a slot is a very poor radiator, as wall slots go. Similarly, current flow is parallel to the y-axis on the side wall of the waveguide. Moderate slots can also be cut in the guide parallel to the y-axis and have no effect upon the propagation, i.e. negligible radiation. Such waveguide wall slots are often used to introduce samples into the guide for treatment. Slots purposefully oriented to cut waveguide wall current lines often make very effective radiators and are frequently used in so-called "leaky waveguide" applicators, which will be discussed in Chapter 4.

TM rectangular waveguide solutions

The TM mode solution set has many features in common with the TE solutions:

$$
\begin{aligned}
E_x &= \frac{-\mathrm{j}\beta}{h^2}\left(\frac{m\pi}{a}\right) C\cos\left(\frac{m\pi x}{a}\right)\sin\left(\frac{n\pi y}{b}\right)\mathrm{e}^{-\mathrm{j}\beta z} \\
E_y &= \frac{-\mathrm{j}\beta}{h^2}\left(\frac{n\pi}{b}\right) C\sin\left(\frac{m\pi x}{a}\right)\cos\left(\frac{n\pi y}{b}\right)\mathrm{e}^{-\mathrm{j}\beta z} \\
E_z &= C\sin\left(\frac{m\pi x}{a}\right)\sin\left(\frac{n\pi y}{b}\right)\mathrm{e}^{-\mathrm{j}\beta z}
\end{aligned}
\tag{2.59}
$$

and:

$$
\begin{aligned}
H_x &= \frac{\mathrm{j}\omega\varepsilon}{h^2}\left(\frac{n\pi}{b}\right) C\sin\left(\frac{m\pi x}{a}\right)\cos\left(\frac{n\pi y}{b}\right)\mathrm{e}^{-\mathrm{j}\beta z} \\
H_y &= \frac{-\mathrm{j}\omega\varepsilon}{h^2}\left(\frac{m\pi}{a}\right) C\cos\left(\frac{m\pi x}{a}\right)\sin\left(\frac{n\pi y}{b}\right)\mathrm{e}^{-\mathrm{j}\beta z} \\
H_z &= 0.
\end{aligned}
\tag{2.60}
$$

Because of the canonical form of these relations there are no non-trivial solutions when either m or n is 0. So, the most fundamental TM mode in a rectangular waveguide is TM_{11}.

Rectangular waveguide power calculations

The power in the wave propagating within the waveguide is given by the complex Poynting power flux density vector, **S**, as was introduced in equation (1.47). The time-average power in the *propagating* wave is given by the Z-component alone:

$$P_z = \tfrac{1}{2}\mathrm{Re}\{E_x H_y^* - E_y H_x^*\} = \frac{1}{2Z_0}(|H_x|^2 + |H_y|^2) \tag{2.61}$$

where: $Z_{0\,TE} = \dfrac{\omega\mu}{\beta}$ for TE modes

and: $Z_{0\,TM} = \dfrac{\beta}{\omega\varepsilon}$ for TM modes.

The Z_0 are the respective characteristic wave impedances and their product, $Z_{0TE}Z_{0TM} = \mu/\varepsilon = \eta^2$, the square of the characteristic impedance of the waveguide medium in unbounded propagation geometry. Integrating the cross-sectional field distribution over the rectangular waveguide aperture to get the total propagating power, P_T, for each of the modes is lengthy; however, the results are fairly simple: firstly

$$\text{for TM modes: } P_T = \frac{ab}{8}\sqrt{\frac{\varepsilon}{\mu}}\left(\frac{\lambda_c}{\lambda}\right)^2 C^2\sqrt{1-\left(\frac{\lambda}{\lambda_c}\right)^2} \tag{2.62}$$

and, for all TE modes except the two most fundamental, TE_{10} and TE_{01}:

$$P_T = \frac{ab}{8}\sqrt{\frac{\mu}{\varepsilon}}\left(\frac{\lambda_c}{\lambda}\right)^2 C^2\sqrt{1-\left(\frac{\lambda}{\lambda_c}\right)^2} \tag{2.63}$$

and finally, for the fundamental TE_{10} and TE_{01} modes:

$$P_T = \frac{ab}{4}\sqrt{\frac{\mu}{\varepsilon}}\left(\frac{\lambda_c}{\lambda}\right)^2 C^2\sqrt{1-\left(\frac{\lambda}{\lambda_c}\right)^2}. \tag{2.64}$$

Since it is not too difficult to measure forward and reflected wave power in a waveguide, one can also finally determine the proportionality constant, C, from the appropriate form of (2.62), (2.63) or (2.64), and thus the electric and magnetic field expressions. This sequence of relations is of prime importance when calculating the fields around samples in a waveguide, as is often done, from measurements of waveguide power.

Some additional insight into the interrelationships between guide wavelength and propagation velocity can be acquired from an inspection of the TE_{10} mode results. As may be seen from equation (2.49) the wavelength in the guide, λ_g, is longer than would be the case, λ, if the wave were propagating in an unbounded medium. Various waveguides have differing λ_g values depending on their cutoff frequencies: λ_g is longer in smaller waveguides at the same frequency. At first sight this would seem to suggest that the velocity of propagation was higher than in an unbounded medium of the same type; and thus that an evacuated waveguide (free space) might have velocities in excess of the speed of light. But, in fact, we do not violate relativity limits: the maximum velocity possible in the z-direction (the net propagation of energy or "information", as it were) is limited by the speed of light, but the phase velocity can exceed that value. To see why, we can compute the complex Poynting vector, **S**, for the TE_{10} mode from equations (2.52):

$$\mathbf{S} = \tfrac{1}{2}\mathbf{E}\times\mathbf{H}^* = \tfrac{1}{2}[E_y H_z^*]\mathbf{a}_x - \tfrac{1}{2}[E_y H_x^*]\mathbf{a}_z \tag{2.65}$$

which results in:

$$\mathbf{S} = -C^2\left(\frac{j\omega\mu a}{2\pi}\right)\sin\left(\frac{\pi x}{a}\right)\cos\left(\frac{\pi x}{a}\right)\mathbf{a}_x - C^2\left(\frac{\omega\mu\beta}{2}\right)\left(\frac{a}{\pi}\right)^2\sin^2\left(\frac{\pi x}{a}\right)\mathbf{a}_z \tag{2.66}$$

for which the $e^{2j\omega t}$ term is understood, as always. The Poynting vector, which ultimately describes the entire propagation, has both a time-dependent X- and Z-component which are 90° out of phase with each other (S_x is imaginary and S_z is real) and of unequal magnitude. The result is that if we follow a particular segment of a phase front it bounces from side to side as the wave propagates in a net z-direction, somewhat as shown in Figure 2.19a. The guide wavelength is actually the apparent wavelength due to the angle of propagation; as it were, a projection of λ. So, while the "phase velocity", u_p, is higher than $u = 1/\sqrt{\mu\varepsilon}$, the rate at which information is transmitted down the waveguide is less than u:

$$u_p = \frac{1}{\sqrt{\mu\varepsilon}}\frac{1}{\sqrt{1-(f_c/f)^2}} \tag{2.67}$$

and $f > f_c$ to achieve propagation.

The effect is akin to observing ocean waves intersecting the beach at an angle of incidence, θ_i (as in Chapter 1) in excess of, say, 30° as in Figure 2.19b. The wave peak travels along the beach with high phase velocity u_p but information on the wave moves at the propagation velocity, u. This is a rather simple empirical demonstration of the phenomenon, but one which can be made in the most pleasant of surroundings and is, therefore, commended to the reader on that basis alone. The beach front takes the role of the waveguide wall in the example. We will appreciate a postcard if the reader should

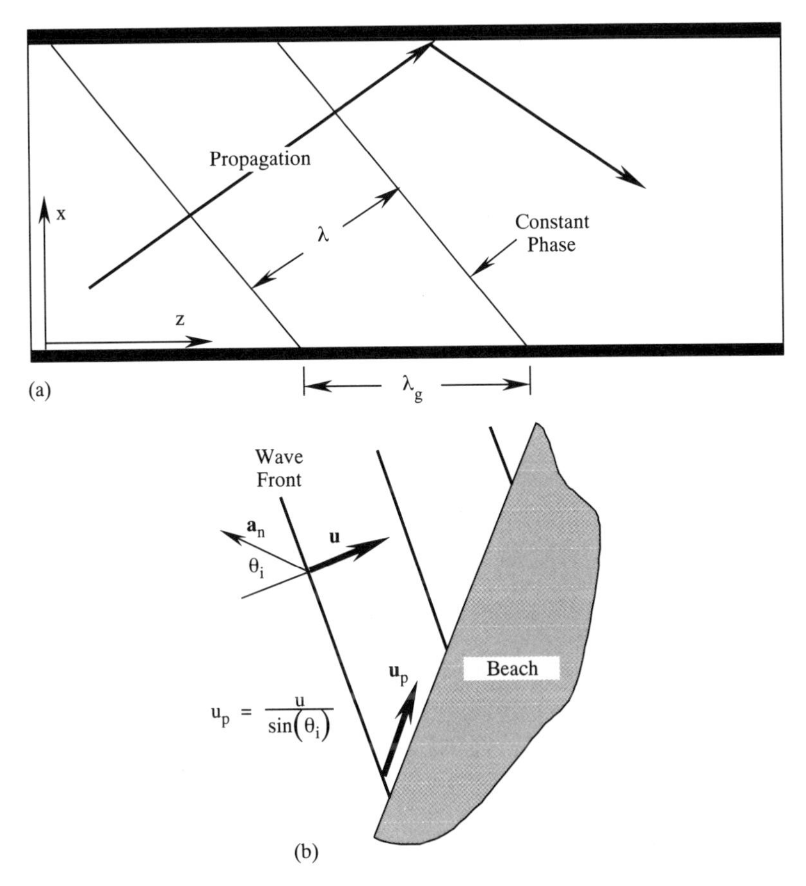

Fig. 2.19 *Waveguide propagation and phase velocity. (a) Constant phase fronts are not in the z-direction so that the apparent wavelength is longer than the free space value. (b) Apparent velocity along the beach is quite high owing to the effect of projecting the wave onto the shore.*

perform such an experiment. Note that if the angle of incidence approaches 0° the equivalent phase velocity increases without bound, but from the standpoint of the waveguide wall this would be the trivial case since $S_z = 0$.

2.3.3 Rectangular waveguide structures

All waveguides behave completely analogously to transmission lines. We can use Smith charts and the same describing parameters to analyze waveguide problems. So, for a loaded waveguide, the standing wave ratio and the reflection coefficients of the propagating wave are calculated in precisely the same way:

$$\Gamma = \frac{E_{\text{refl}}}{E_{\text{inc}}}$$

$$\text{so that} \qquad \text{SWR} = \frac{1+|\Gamma|}{1-|\Gamma|} \qquad \text{and} \qquad z = \frac{1+\Gamma}{1-\Gamma}. \tag{2.68}$$

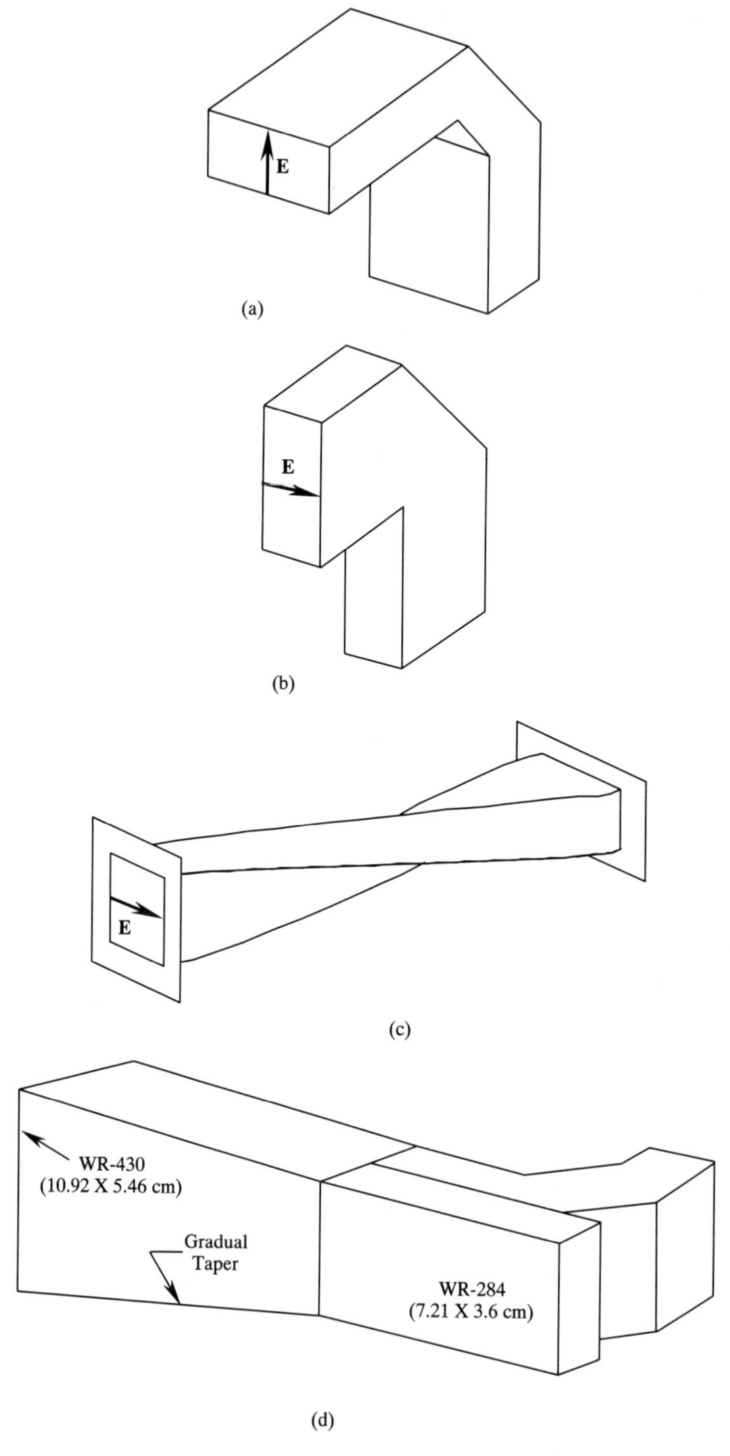

Fig. 2.20 *Simple waveguide structures. (a) E-plane bend. (b) H-plane bend. (c) Waveguide twist. (d) Bifurcation from WR-430 into two WR-284 waveguides.*

A metallic plate which closes the end of a waveguide is a close representation of a short circuit ($z = 0$). An open waveguide radiates the field into space and has a large, though by no means infinite, impedance. If the end of the guide is flanged, the impedance is much higher and $\Gamma \to +1$.

Corners, twists and bifurcations (fork junctions) can be made in a rectangular waveguide. Assuming TE_{10} propagation, examples of an E-plane bend, an H-plane bend, a twist, and a bifurcation are shown in Figure 2.20. Of the two bends shown, the E-plane bend creates the least distortion in the wall currents and is usually preferred when feasible. The electrical characteristics for these microwave circuit elements are represented by two-port, usually S-parameter, matrices; we will discuss S-parameter representation in Chapter 3. These example circuit elements are generally low loss and have small reflection coefficients. Other useful microwave waveguide circuit elements will be discussed in some detail in Chapter 3.

2.3.4 Cylindrical waveguide propagation

Rectangular waveguides are the most universally used transmission devices for microwaves. The largest variety of devices are available in rectangular form designed to be used in TE_{10} propagation with $a = 2b$. This shape is strong, lightweight and reasonably economical to fabricate. However, the 2:1 rectangular guide is not the only practically useful geometry. There are many situations in which a cylindrical geometry may be preferred, and a cylindrical waveguide may be indicated. While few commercial guides of this geometry are available, creditable waveguide structures can be fabricated from, for example, soldered copper pipe; so a short discussion of this geometry is a good investment at this point.

TE modes

For a cylindrical waveguide in the TE mode, the mathematical solution introduces Bessel functions of the internal radius, r, in response to the different forms of the spatial derivatives in the calculation of the curl (in Maxwell's equations). The solutions of the wave equations (2.46) for TE^O modes (the O superscript indicates a cylindrical waveguide solution) take the form:

$$E_r = \frac{\omega\mu}{\beta} H_\varphi \qquad E_\varphi = \frac{-\omega\mu}{\beta} H_r \qquad \text{and} \qquad E_z = 0 \tag{2.69}$$

and for the internal magnetic fields:

$$H_r = \frac{-\mathrm{j}\beta}{h} C \frac{\partial\{J_n(hr)\}}{\partial(hr)} \cos(n\varphi)\mathrm{e}^{-\mathrm{j}\beta z} \tag{2.70a}$$

$$H_\varphi = \frac{\mathrm{j}n\beta}{h^2 r} C J_n(hr) \sin(n\varphi)\mathrm{e}^{-\mathrm{j}\beta z} \tag{2.70b}$$

$$H_z = C J_n(hr) \cos(n\varphi)\mathrm{e}^{-\mathrm{j}\beta z} \tag{2.70c}$$

where $h^2 = \omega^2\mu\varepsilon - \beta^2 < \omega^2\mu\varepsilon$ to ensure propagation (as before), and J_n is a Bessel function of the first kind of order n (and real argument). To ensure that $E_\varphi = 0$ at $r = a$ (the waveguide radius) we must have:

$$J'_n(ha) = 0 \tag{2.71}$$

where J' is the derivative of the Bessel function (as in equations (2.70)), and (2.71) gives the value for h for the different modes. As a practical matter, we normally size the waveguide so that only the first few roots of J'_n are used (to avoid large numbers of possible modes). The roots of the derivative of the Bessel function occur at known locations and are listed in Table 2.3. Here again we designate the TE modes as TE_{nm}, where n is the order of the Bessel function and m is the particular root used to find h. So, TE_{01}, TE_{11}, etc. refer to the different modes.

It is interesting to note here that the fundamental mode is $\mathrm{TE}^{\mathrm{O}}_{11}$ since it has the lowest cutoff frequency:

$$f_{\mathrm{C}} = \frac{h_{nm}}{2\pi\sqrt{\mu\varepsilon}} = \frac{u}{2\pi}\frac{(ha)'_{1,1}}{a} = \frac{0.293}{a\sqrt{\mu\varepsilon}}. \tag{2.72}$$

For notational convenience we have used the simplification that $(ha)'_{1,1}$ is the 1,1 first root ($m = 1$) of the derivative of the Bessel function, $J'_n(ha)$, where $n = 1$. The other propagation parameters may be found from the cutoff frequency, as in equations (2.48), (2.49), and (2.67). A sketch of the resulting electric field lines is contained in Figure 2.21.

An axisymmetric TE field distribution can be obtained only from the $\mathrm{TE}^{\mathrm{O}}_{01}$ mode. In that case the waveguide wall current will have only a circumferential component with no component in the z-direction. This mode can also propagate in a hollow tube for which the external wall is a tightly wound solenoid. The disadvantage of the $\mathrm{TE}^{\mathrm{O}}_{01}$ mode is that it is not fundamental, and obstacles inside the guide will upset the stability of the mode and induce higher order modes. An interesting property of the $\mathrm{TE}^{\mathrm{O}}_{01}$ mode is that its losses decrease with increasing frequency. The losses are also small when the guide radius, a, is large. A large guide will support many propagation modes; however, mode filters can be used to eliminate unwanted modes, if necessary.

Table 2.3 *Roots of the derivative of J_n; designated $(ha)'_{nm}$.*

Root Number (m)	$n = 0$	$n = 1$	$n = 2$
1	3.832	1.841	3.054
2	7.016	5.331	6.706

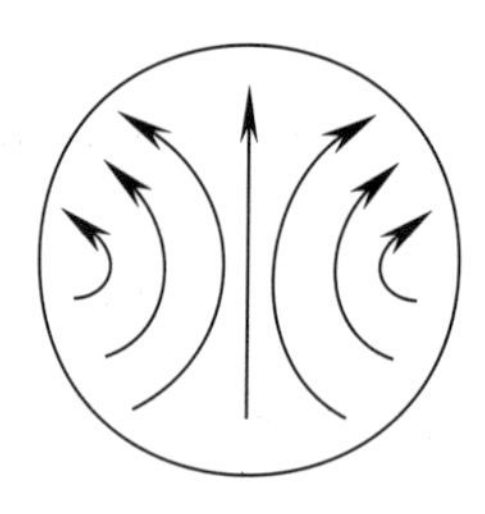

Fig. 2.21 *Fundamental $\mathrm{TE}^{\mathrm{O}}_{11}$ cylindrical mode electric field.*

TM modes in cylindrical waveguides

The solutions for TM modes in circular waveguides are similar to the TE modes in that they also depend on the derivative of the Bessel function. The electric fields are:

$$E_r = \frac{\beta}{\omega\varepsilon} H_\varphi \qquad E_\varphi = \frac{-\beta}{\omega\varepsilon} H_r$$

and (2.73)

$$E_z = C J_n(hr)\cos(n\varphi)\mathrm{e}^{-\mathrm{j}\beta z}$$

and the magnetic fields are:

$$H_r = \frac{-\mathrm{j}\omega\varepsilon n}{h^2 r} C J_n(hr)\sin(n\varphi)\mathrm{e}^{-\mathrm{j}\beta z} \tag{2.74a}$$

$$H_\varphi = \frac{-\mathrm{j}\omega\varepsilon}{h} C \frac{\partial J_n(hr)}{\partial(hr)}\cos(n\varphi)\mathrm{e}^{-\mathrm{j}\beta z} \tag{2.74b}$$

$$H_z = 0 \tag{2.74c}$$

where, as above, J_n is the Bessel function of the first kind and order n (with a real argument), and $h^2 = \omega^2\mu\varepsilon - \beta^2 < \omega^2\mu\varepsilon$ to ensure propagation. Since very large numbers of modes are usually avoided (the frequency of operation will not be many times the cutoff frequency), h will be small. The value of h is determined from the necessary boundary condition that:

$$J_n(ha) = 0 \tag{2.75}$$

for $E_z = 0$ at the waveguide wall, $r = a$. The first few zeroes of J_n are listed in Table 2.4 for various values of n.

With cylindrical TM modes, $\mathrm{TM}^{\mathrm{O}}_{nm}$ as with TE modes, the first subscript refers to the order of the Bessel function (which is equal to the number of full periods in the field components as φ varies, as can be seen in equations (2.73) and (2.74)) and the second to the particular root under consideration.

Cylindrical waveguides are not often used in commercial MW systems as a transmission line. However, cylindrical volumes make excellent resonant cavities. A cavity consists of a cylindrical waveguide of finite length (only a few λ_g at most) terminated at both ends with short circuits, which creates a standing wave pattern in the z-direction. In fact, commercial frequency meters for waveguides often consist of a cylindrical cavity with an adjustable length which operates in the TE_{111} mode—the first two subscripts indicate the fundamental mode in the cavity while the third subscript indicates how many maxima occur in the

Table 2.4 *Roots of J_n; designated $(ha)_{nm}$.*

Root Number (m)	$n = 0$	$n = 1$	$n = 2$
1	2.405	3.832	5.136
2	5.520	7.016	8.417

cavity length (i.e. the z-direction). The resonant cavity frequency meter will be discussed in Chapter 5 in more detail, while resonant cavity applicators will be treated in Chapter 4.

2.3.5 Other waveguide geometries

Ridged waveguides

There are many homogeneous waveguides with other cross-sections. In Figure 2.22 single-ridged and double-ridged guides are shown. The advantage of these waveguide sections is that their fundamental mode domain is larger than that of either rectangular or cylindrical guides.

The same reasoning can also be applied to open guides, such as in Figure 2.23. For this case, the TE, TM, TEM modes are determined by properties and shape of the dielectric slab.

Fig. 2.22 *Ridged waveguides. (a) Single-ringed waveguide cross-section. (b) Double-ridged waveguide cross-section.*

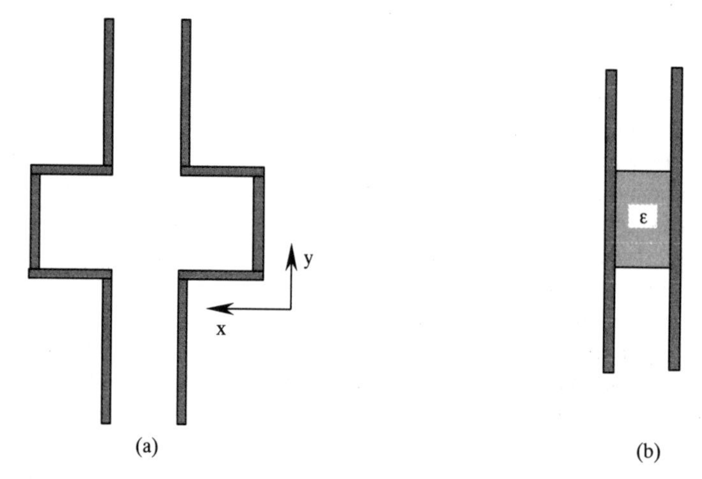

Fig. 2.23 *Open waveguide cross-sections. (a) Open wall. (b) Dielectric slab.*

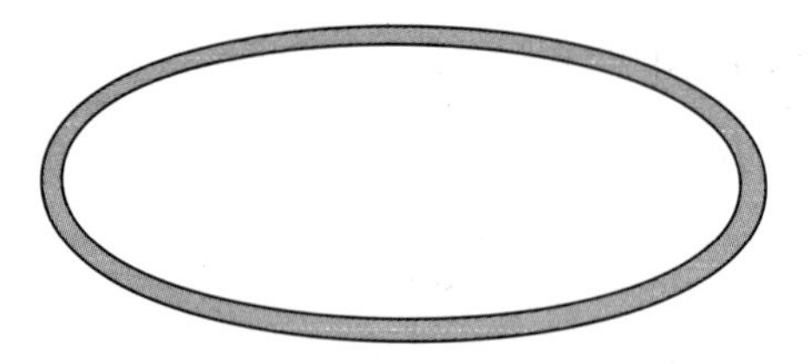

Fig. 2.24 *Elliptical waveguide.*

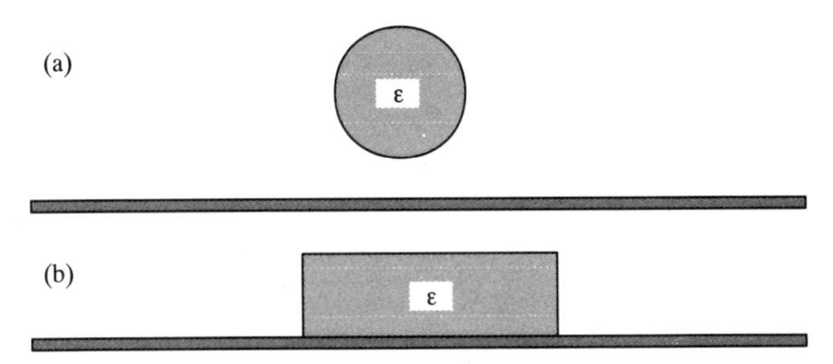

Fig. 2.25 *Other waveguide designs. (a) Round dielectric rod above ground plane. (b) Dielectric slab on a ground plane.*

Elliptical waveguides

Elliptical cross-section waveguides (Figure 2.24) can also be used. They are more mechanically stress-resistant, but can also be twisted easily.

Inhomogeneous waveguides

We conclude with a very brief look at waveguides which are completely enclosed by a metallic boundary in which there are two different dielectric materials. The mathematical problem is complicated because there are two regions of solution for Maxwell's equations: region 1, which has μ_1 and ε_1 and region 2, which has μ_2 and ε_2. From the mathematical viewpoint, the system is no longer homogeneous and the boundary conditions must be applied at the dielectric interface. The general solution is no longer a simple linear combination of individual solutions for which E_z or H_z is 0 over the whole section. The explicit labeling of modes will introduce quasi-TE, quasi-TM, LSTM, and LSTE modes.

In some simple cases, it is possible to obtain a distribution in medium 1 that is not largely different from the distribution of field that we would have if the entire section consisted of dielectric 1. The distribution is completed by some distribution of the field in medium 2, which is the sum of other components. Calculations for this type of guide can be found in the literature, where the propagation coefficients have been deduced. In some instances, for high power usage, we may use some of these calculations in order to take advantage of the tendency of a particular geometry to concentrate the field in a particular dielectric.

There are many other special types of waveguides. For example, a round rod dielectric over a metallic plane can support a propagating field inside the dielectric and around the surface. A dielectric bar placed upon a metallic plane can also support such a special distribution of field. These are illustrated in Figure 2.25.

2.4 SUMMARY

This chapter has described the functional aspects of transmission lines for RF and MW propagation. At lower frequencies TEM propagation is achieved in the transmission line by restricting the dimensions to values small compared to λ; thus a quasi-static analysis of the cross-sectional fields is valid. At high power and higher frequency conductor losses limit the usefulness of TEM propagation. Consequently, waveguides are preferred so that wall losses will be acceptably small since the cross-section is of the order of λ in size. The number of possible propagation modes is very large as a result, but can usually be broken down into dominant TE and TM components. Both rectangular and cylindrical geometries have been used, but most commercially available waveguides are rectangular in cross-section with $a = 2b$.

Additionally, we have illustrated some simple transmission line and waveguide devices constructed of T-L components. In Chapter 3 we will describe other RF and MW circuit elements which are in common use in high power systems.

REFERENCES

The following texts will provide additional insight into the topics introduced in this chapter.

Cheng, D. K. (1989) *Field and Wave Electromagnetics (2nd Ed.)* Addison Wesley, Reading, MA.

Harrington, R. F. (1961) *Time-Harmonic Electromagnetic Fields* McGraw-Hill, New York.

Harvey, A. F. (1963) *Microwave Engineering* Academic Press, New York.

Hayt, W. H. (1989) *Engineering Electromagnetics (5th Ed.)* McGraw-Hill, New York.

Iskander, M. (1992) *Electromagnetic Fields and Waves* Prentice-Hall, Englewood Cliffs, NJ.

Jackson, J. D. (1975) *Classical Electrodynamics (2nd Ed.)* John Wiley, New York.

Kong, J. A. (1986) *Electromagnetic Wave Theory* Wiley-Interscience, New York.

Liboff, R. L. and Dalman, G. C. (1985) *Transmission Lines, Waveguides and Smith Charts* Macmillan, New York.

Magnusson, P. C., Alexander, G. C. and Tripathy, V. K. (1992) *Transmission Lines and Wave Propagation (3rd Ed.)* CRC Press, Boca Raton, FL.

Paris, D. T. and Hurd, F. K. (1969) *Basic Electromagnetic Theory* McGraw-Hill, New York.

Pozar, D. M. (1990) *Microwave Engineering* Addison-Wesley, Reading, MA.

Sadiku, M. N. O. (1989) *Elements of Electromagnetics* Saunders, New York.

3 Microwave and Radio Frequency Circuit Design

This chapter addresses the power generation and delivery portions of an electromagnetic (EM) heating system. Discrete devices typically found in RF and MW heating apparatus are discussed in terms of operating principles and application in a system. As such, we will treat the applicator and its load as a complex lumped circuit element located at the end of the transmission line. In keeping with our original intent to focus on the heated materials themselves, it is not our goal to treat topics in this chapter in detail. Rather, we include a short description of the most useful devices and introduce relevant descriptive parametric representations. The interested reader is gently encouraged to consult the works listed at the end of this chapter for more detailed analyses.

We begin with parametric description of circuit elements and devices, continue with a review of typical RF generators and impedance matching circuits, and conclude with their microwave equivalents. We will see that, from a system design point of view, one of the most significant advantages that microwave systems have over RF systems is that any reflected power resulting from load–waveguide mismatches can be prevented from interacting with and harming the MW generator by a protection device known as a circulator. There is no RF equivalent for a circulator, so impedance matching is a more critical consideration in RF systems, and we have considered it in some detail.

3.1 INTRODUCTION TO CIRCUIT ELEMENTS AT HIGH FREQUENCY

At lower frequencies, typically less than about 100 MHz, it is not too difficult to construct circuits which are conveniently described as either primarily resistive, capacitive or inductive. Even so, in RF devices the small stray capacitances and inductances which we can normally ignore become extremely important to overall circuit performance, and must be included. As the frequency increases a wire is no longer simply a resistor (or even just an inductance with series resistance), and two small conductors near each other actually form a complex circuit. As a general rule, a circuit which has been designed to function as a self-inductance in the low frequency domain appears more capacitive at high frequency — primarily due to inter-winding capacitances. Conversely, a capacitance at low frequency transforms to an inductance at higher frequency, due in part to the current density in the wires and conducting planes. Thus, in the design of an RF oscillator the ultimate frequency of operation (and its stability) depends on vacuum tube stray capacitances (grid-to-plate, grid-to-cathode, screen-to plate, etc.), and on circuit element shape factors (winding radius and density in a coil, plate thickness in a capacitor) as much as the nominal L or C values.

In the microwave regime (above about 500 MHz) we will see that the shape, dimensions and location of a circuit element are more useful descriptors than lumped circuit parameters such as resistance, capacitance and inductance. This may be appreciated from our discussion of Smith charts in Chapter 2. A capacitance looks inductive at a position where its reactance has changed sign — the sign of the reactance indicates whether the circuit is inductive or capacitive, but this is not the sole criterion. A simple inductance or capacitance must also have a reactance which increases or decreases linearly with the frequency. This second characteristic is not often realized over a wide frequency range in many circuit devices. Consequently, at microwave frequencies we use wave descriptions and retain "inductive" and "capacitive" as useful impedance descriptors, realizing that an accurate description of device performance must be couched in other terms.

3.1.1 Parametric representation of circuit elements

Both RF and MW circuit elements are often more conveniently represented using parametric descriptions. The most common parametric representations are in terms of impedance or admittance parameters and so-called scattering, or S-parameters (primarily used for MW circuits). We briefly review parameter system notation in this section before discussing circuit element representation. The two general classes of devices are one-port and two-port devices. The one-port device is equivalent to a two-terminal circuit element, and the two-port device is equivalent to a four-terminal circuit element (which includes transfer parameters).

One-port device models

As we saw in Chapter 2, the characteristic admittance of a transmission line, $Y_0 = G_0 + \mathrm{j}B_0(S)$, is the reciprocal of its characteristic impedance, $Z_0 = R_0 + \mathrm{j}X_0(\Omega) : Y_0 = 1/Z_0$, where G is a conductance, B a susceptance, R a resistance and X a reactance.

For a lossless transmission line both Y_0 and Z_0 are real and make convenient normalizing reference values, as we used in Smith chart representation. The loaded transmission line segment is an example of a one-port device which we characterize by a local voltage and current, V and I, which consist of the forward wave (+) and reflected wave (−) components:

$$V = V^+ + V^- \qquad \text{and} \qquad I = I^+ - I^-$$

so that

$$Z = \frac{V}{I} = \frac{V^+ + V^-}{V^+ - V^-} Z_0. \tag{3.1}$$

The one-port equivalent of a loaded transmission line segment is the impedance, $Z_{eq} = Z$, which is dependent on the length of transmission line and the load at the termination, as we have seen. The application of this notation is inherently obvious in RF circuits, and indeed we used it without comment in the loaded RF line segment examples in Chapter 2. However, it may not be immediately clear what is meant in the case of waveguides. For the waveguide the characteristic impedance is the wave impedance (see equation (2.61)). At a point on a waveguide we may establish a transverse reference plane, t, as shown in Figure 3.1. There is

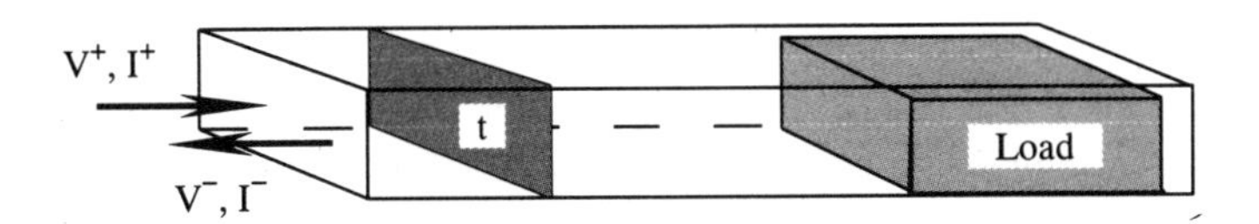

Fig. 3.1 *Loaded waveguide segment and transverse plane, t.*

net forward and reflected power flux across the plane, and in accordance with the Poynting power theorem (Chapter 1), the loaded waveguide volume has both dissipated power, the real part P_l, and stored energy in the electric (e) and magnetic (m) fields, represented by the net reactive power $W_m - W_e$:

$$\frac{1}{2}\oint_S (\mathbf{E} \times \mathbf{H}^*) \cdot d\mathbf{S} = P_l + 2j\omega(W_m - W_e). \tag{3.2}$$

By suitable manipulation (see Collin 1991) it may be shown that the relationship between measured power for the volume (the real and reactive parts) and an equivalent representation in terms of R_{eq} and X_{eq} is:

$$Z_{eq} = \frac{P_l + 2j\omega(W_m - W_e)}{\frac{1}{2}II^*} = R_{eq} + jX_{eq} \tag{3.3a}$$

where I is an equivalent current. Since P_l, W_m, and W_e are all linearly related to the current product, $II^* = |I|^2$, then R_{eq} and X_{eq} are dependent on geometry alone, as must be the case. Analogously, for the admittance representation of the one-port device we may obtain:

$$Y_{eq} = \frac{P_l - 2j\omega(W_m - W_e)}{\frac{1}{2}VV^*} = G_{eq} + jB_{eq}. \tag{3.3b}$$

The equivalent susceptance, B_{eq}, is capacitive (> 0) if $W_e > W_m$, which is the sensible result. We use the admittance representation in developing this observation since the waveguide "capacitance" may be regarded as being in parallel with its "inductance", and parallel admittances add.

Many interesting microwave structures may be regarded as approximately lossless. In that case $P_l \to 0$ and if, in addition, $W_m = W_e$ the loaded waveguide approaches two possible resonance conditions: (1) $I \neq 0$ due to a zero in the equivalent reactance (this is called series resonance), or (2) $V \neq 0$ due to a zero in the equivalent susceptance (parallel resonance). We saw this in the maxima and minima encountered in the Smith chart analysis of the shorted T-L segment. The same observations apply in waveguides, so a shorted waveguide segment would have analogous standing wave patterns to a shorted T-L.

There is an additional interesting property of the equivalent impedance functions: namely, that the real part of Z_{eq}, R_{eq}, is an even function of ω and the imaginary part, X_{eq}, is an odd function of ω. This is necessary to satisfy restrictions on the group velocity of propagating waves—for example, a real time-dependent driving function must yield a real response in the waveguide.

The S-parameters (dimensionless) are voltage ratios corresponding to reflection and transmission coefficients. An S-parameter representation for a one-port network utilizes a single parameter, $S = V^-/V^+$, as does the impedance parameter, Z_{eq}, above. The physical significance of S is that it is the voltage reflection coefficient, $\Gamma = V^-/V^+$ in the one-port case. The directional waveguide power is readily available from directional coupler measurements, as described later in this chapter, and in Chapters 2 and 5, so S is easily measurable. The power reflection coefficient is $|\Gamma|^2$, and so $|S| = |\Gamma|$ may be readily estimated from a simple measurement, if that is all that is desired. The full complex representation of S (magnitude and phase angle) can also be determined either from slotted-line measurements or a network analyzer (see Chapter 5).

N-port device models, with special attention to $N = 2$

We may generalize the one-port descriptions to multiple-port devices with little mathematical difficulty. In Figure 3.2 an N-port device has bi-directional power flux at each port reference plane so that:

$$\begin{bmatrix} V_1 \\ V_2 \\ \dots \\ V_N \end{bmatrix} = \begin{bmatrix} Z_{11} & Z_{12} & \dots & Z_{1N} \\ Z_{21} & Z_{22} & \dots & Z_{2N} \\ & \dots & & \\ Z_{N1} & Z_{N2} & \dots & Z_{NN} \end{bmatrix} \begin{bmatrix} I_1 \\ I_2 \\ \dots \\ I_N \end{bmatrix} \tag{3.4}$$

with an analogous relationship for admittance parameters:

$$\begin{bmatrix} I_1 \\ I_2 \\ \dots \\ I_N \end{bmatrix} = \begin{bmatrix} Y_{11} & Y_{12} & \dots & Y_{1N} \\ Y_{21} & Y_{22} & \dots & Y_{2N} \\ & \dots & & \\ Y_{N1} & Y_{N2} & \dots & Y_{NN} \end{bmatrix} \begin{bmatrix} V_1 \\ V_2 \\ \dots \\ V_N \end{bmatrix}. \tag{3.5}$$

There are N-1 cross-coupled power inputs to each port and one self-coupled power term, so the net power for the N-port device is:

$$\begin{aligned} \tfrac{1}{2}[I^*]^{\mathrm{t}}[V] &= \tfrac{1}{2}[I^*]^{\mathrm{t}}[Z][I] \\ &= \tfrac{1}{2}\sum_{n=1}^{N}\sum_{m=1}^{N} I_n^* Z_{nm} I_m = P_{\mathrm{l}} + 2\mathrm{j}\omega(W_{\mathrm{m}} - W_{\mathrm{e}}) \end{aligned} \tag{3.6}$$

where the superscript t indicates a transposed matrix. For a lossless device *all* of the Z_{mn} must be purely imaginary.

The N-port relations may be quickly reduced to the two-port device case. The number of required parameters may be further reduced by noting that the special relationship between the transfer impedances, Z_{12} and Z_{21}, means that any two-port network can be represented

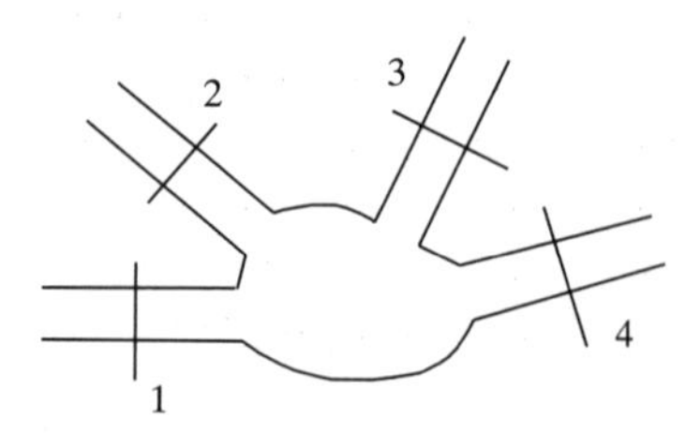

Fig. 3.2 *N-port electromagnetic device.*

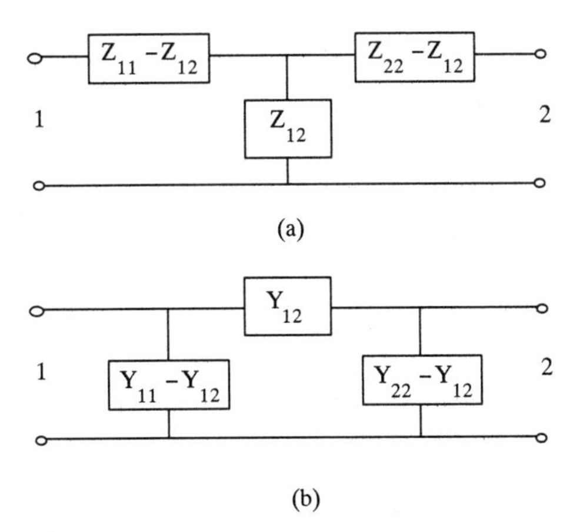

Fig. 3.3 *Two-port impedance models. (a) T-network model. (b) Π-network model.*

by an equivalent T-network (see Figure 3.3a) for which Z_{11}, Z_{22} and Z_{12} suffice. Analogously, an equivalent dual circuit, a Π-network, may be formulated in terms of admittances (Figure 3.3b) which utilizes Y_{11}, Y_{22} and Y_{12}. Several interesting equivalent two-port networks have been derived and we summarize them for the reader's convenience in Figures 3.4a and 3.4b.

Symmetric parameter matrices represent symmetric microwave devices which can be described as quadrupole circuits. We have already seen T-section and Π-section models for this class of devices. One important theorem in impedance matching is that any quadrupole placed before a normalized load impedance, $z_L = Z_L/Z_0$, will transform this impedance such that it appears to be z_L'. The normalized transformed impedance z_L' is a homographic function of the original impedance z_L:

$$z_L' = \chi(z_L) = (az_L + b)/(cz_L + 1) \tag{3.7}$$

The complex coefficients (a, b, c) introduced in the homographic function then specify the quadrupole. Inversely, the coefficients a, b, and c can be determined from the parameters

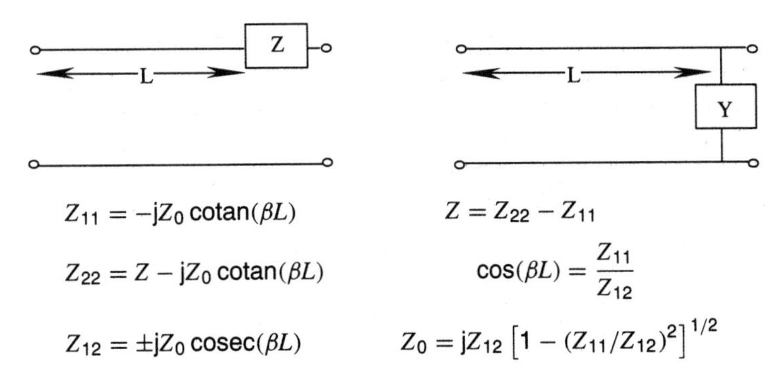

$Z_{11} = -jZ_0 \cot(\beta L)$ $\qquad$ $Z = Z_{22} - Z_{11}$

$Z_{22} = Z - jZ_0 \cot(\beta L)$ $\qquad$ $\cos(\beta L) = \dfrac{Z_{11}}{Z_{12}}$

$Z_{12} = \pm jZ_0 \operatorname{cosec}(\beta L)$ $\qquad$ $Z_0 = jZ_{12}\left[1 - (Z_{11}/Z_{12})^2\right]^{1/2}$

Fig. 3.4a *Representative T-network (left) and Π-network (right) two-port impedance models, for line of length L with series impedance and its dual, a parallel admittance.*

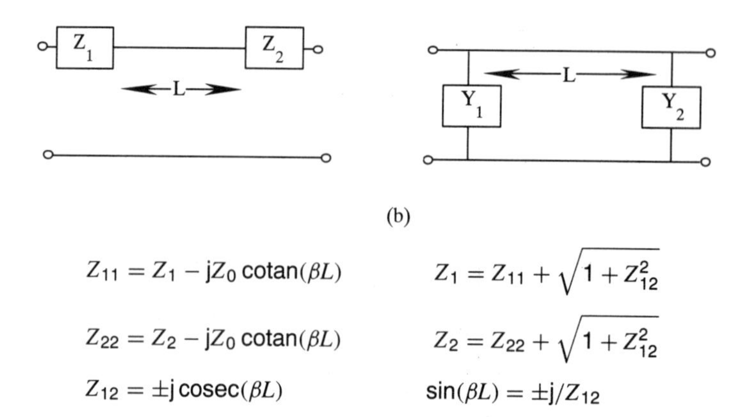

$$Z_{11} = Z_1 - jZ_0 \operatorname{cotan}(\beta L) \qquad Z_1 = Z_{11} + \sqrt{1 + Z_{12}^2}$$

$$Z_{22} = Z_2 - jZ_0 \operatorname{cotan}(\beta L) \qquad Z_2 = Z_{22} + \sqrt{1 + Z_{12}^2}$$

$$Z_{12} = \pm j \operatorname{cosec}(\beta L) \qquad \sin(\beta L) = \pm j/Z_{12}$$

Fig. 3.4b *Representative T-network (left) and Π-network (right) two-port impedance models, for line of length L between two series impedances and its dual.*

of the quadrupole. Three measurements are performed in order to determine a, b and c which involve the use of special values of z_L:

(i) $z_L = 0$, a short circuit, yields $z'_L = b$.
(ii) $z_L = \infty$, an open circuit, yields $z'_L a/c$.
(iii) $z_L = 1$, a matched load, yields $z'_L = (a + b)/(c + 1)$.

This calibration strategy is used in commercial network analyzers (Chapter 5) and is a very powerful method, indeed, since it permits one to transform the measurement plane to any point along a transmission line.

The S-parameter representation of a two-port device is composed similarly to the impedance and admittance representations:

$$\begin{bmatrix} V_1^- \\ V_2^- \end{bmatrix} = \begin{bmatrix} S_{11} & S_{12} \\ S_{21} & S_{22} \end{bmatrix} \begin{bmatrix} V_1^+ \\ V_2^+ \end{bmatrix}. \tag{3.8}$$

The off-diagonal terms represent voltage transmission coefficients and the diagonal terms reflection coefficients. Here it is convenient to choose the definitions of the voltages so that the power transmitted in the forward direction through a waveguide plane is given by: $\{|V_i^+|^2\}/2$—this is equivalent to assuming that the wave impedance is unity for both waveguide ports. There is no loss of generality as long as all of the waveguides connected to the device have the same wave impedance (we know of no exception to this among commercial microwave multi-port devices). Doing this gives a symmetrical scattering matrix with $S_{12} = S_{21}$, a considerable simplification. Also, the respective equivalent currents are actually irrelevant since the new definitions of V^+ and V^- are linear combinations of the V and I used above in order to satisfy this normalization condition:

$$V^+ = \frac{V + I}{2} \quad \text{and} \quad V^- = \frac{V - I}{2}. \tag{3.9}$$

The other variables may be easily recovered since $V = V^+ + V^-$ (as we must have) and $I = I^+ - I^- = V^+ - V^-$ because of the simplifying effect of the normalization.

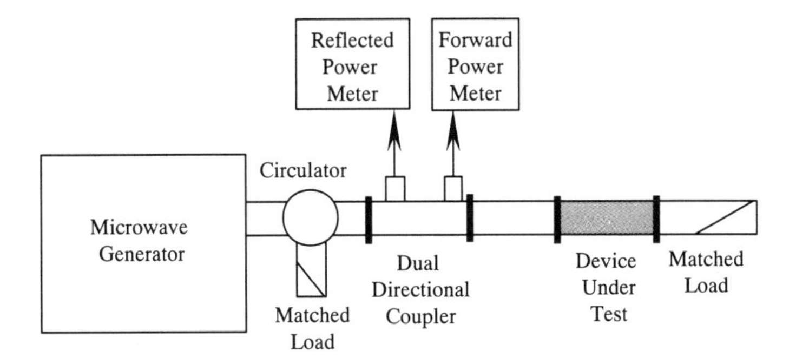

Fig. 3.5 *Experimental determination of S_{11} for a two-port device.*

The S-parameters are complex and their values also depend on the location of the respective measurement planes (assuming a fixed single frequency), just as are impedance and admittance parameters. Movement of the measurement plane along a waveguide (for no change in waveguide geometry) can be included in the computation by appending the respective electrical phase shifts. For an electrical displacement $\theta_i = \beta d_i$, where d_i is the distance the measurement plane has moved away from the device, we may use:

$$\begin{bmatrix} V_1^- \\ V_2^- \end{bmatrix} = \begin{bmatrix} e^{-j\theta_1} & 0 \\ 0 & e^{-j\theta_2} \end{bmatrix} \begin{bmatrix} S_{11} & S_{12} \\ S_{21} & S_{22} \end{bmatrix} \begin{bmatrix} e^{-j\theta_1} & 0 \\ 0 & e^{-j\theta_2} \end{bmatrix} \begin{bmatrix} V_1^+ \\ V_2^+ \end{bmatrix}. \tag{3.10}$$

The two-port S-parameters may be calculated for several simple cases, and we present some examples in this chapter. In general, they are determined experimentally. The first experiment determines $|S_{11}|$ by connecting the plane 1 waveguide port to a generator through a directional coupler and terminating the plane 2 waveguide port in a matched load (so that $V_{2^+} = 0$, as in Figure 3.5); then $V_{1^-} = S_{11}V_{1^+} + S_{12}V_{2^+} = S_{11}V_{1^+}$ as in the one-port device above. The magnitude of S_{12} may quickly be determined by measuring the power in match-terminated waveguide 2 during the same experiment. The second experiment determines $|S_{22}|$ by inverting the device in the same apparatus (i.e. by reversing the roles of plane 1 and plane 2). As a check, one should determine $|S_{21}|$ from this experiment and compare it to $|S_{12}|$ to verify reciprocity. Alternately, one may replace the generator and directional coupler with a network analyzer or slotted line system to make full complex determinations.

3.1.2 RF circuit models for simple elements

It may be instructive at this point to look at more detailed lumped-parameter circuit models for what we usually treat as simple circuit elements. We begin by empirically looking at wire-wound, carbon-composition and metal film resistors; move to typical capacitors, (electrolytic, silver-mica, ceramic and vacuum); and finish with a brief look at inductors. We conclude this section with a brief review of RF circuit design and construction principles.

Resistors

The resistance associated with a distributed quasi-static electric field can be calculated from the potential difference between boundary electrodes, V, and the resulting current in the semiconductor, I:

$$R = \frac{V}{I} = \frac{-\int_b^a \mathbf{E} \cdot \mathbf{dl}}{\oint_\Sigma \mathbf{J} \cdot \mathbf{dS}} \tag{3.11}$$

where the potential reference electrode is located at b and the closed surface, Σ, must completely enclose only one of the electrodes to find its total current. The commonly-used resistance formula, $R = L/(\sigma A)$ was derived from this relation based on an assumed uniform quasi-static electric field. At elevated frequencies for high conductivity materials (i.e. wires) skin effects may dominate the resistance; also, wires exhibit self-inductance. Consequently, wire-wound resistors might be expected to exhibit bizarre behavior at high frequency. We naturally expect that high values of resistance would be especially badly behaved, but it turns out that low resistances are also subject to significant reactance effects in the frequency range of interest.

Several illustrative examples will put this into perspective. First, we consider a 71 cm (28 inch) length of 0.25 mm diameter (20 gage, 0.010 inch) nichrome wire. This wire has a low frequency resistance of about 12.1 Ω (0.171 Ω/cm or 5.19 Ω/foot). We may measure the impedance in two configurations, a low inductance (very narrow) and higher inductance (very wide) loop, as shown in Figure 3.6a. The magnitudes of the measured impedance and the measured phase angles are compared for both configurations in Figure 3.6b. From the results it is apparent that the wire is effectively a simple resistor ($R = 12.1\ \Omega$) in series with an inductance:

$$Z_{\text{eq}} = |Z|\angle\theta = R + \text{j}\omega L. \tag{3.12}$$

For the lower inductance configuration L is about 65 nH ($\theta = 45^\circ$ at 29.5 MHz) and for the higher inductance configuration $L = 71$ nH ($\theta = 45^\circ$ at 27 MHz). A higher inductance is realized with the wide loop since more magnetic flux is enclosed (refer to the ensuing discussion of inductance). We would expect that a self-inductance for a straight wire of the same length, d, (assuming uniform current in the cross-section) would be $\mu d/8\pi(\text{H}) =$ 35.5 nH. The rest of the inductance must come from the enclosed flux in the geometry.

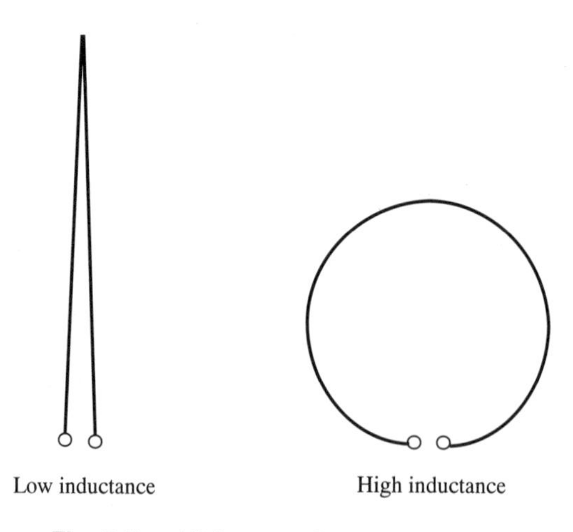

Fig. 3.6a *Nichrome wire geometries.*

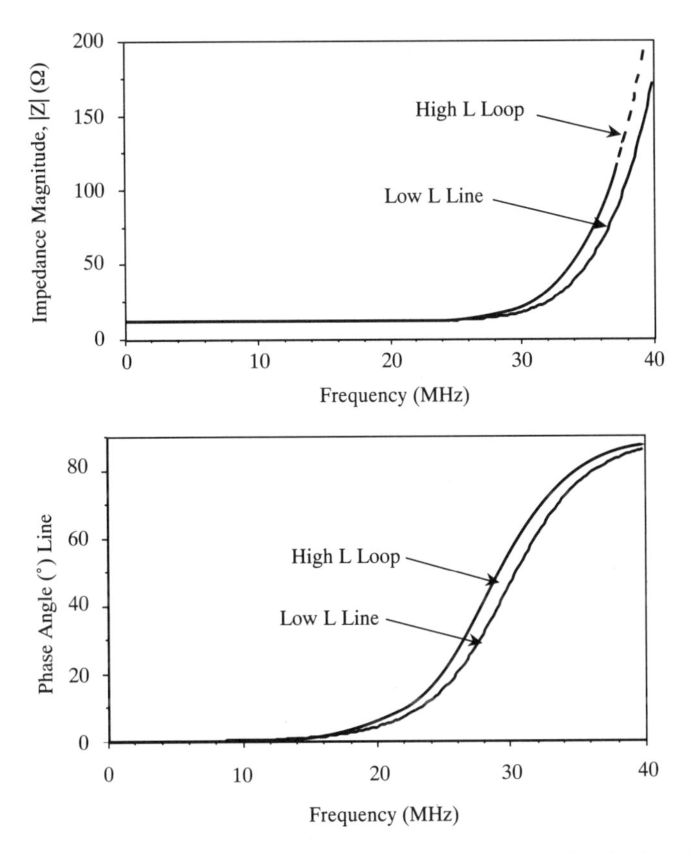

Fig. 3.6b *Comparison of impedance magnitudes and phase angles for low inductance line and high inductance loop configurations.*

A cylindrically wound wire resistor will also exhibit significant inductance effects. A standard cylindrical winding is illustrated in Figure 3.7a. An example resistor was made from 58.7 cm (23 inches) of the same nichrome wire: the resistor was 6.5 turns on a 2.54 cm diameter hardwood mandrel in the standard coil winding (Figure 3.7Aa). The inductance effects of a standard wire-wound resistor can be reduced by using a modified "low inductance" winding, as in Figure 3.7a. The same 20 gage nichrome wire was wound in 7 low-inductance "turns" on an identical hardwood mandrel. The measured impedance magnitude and phase for both configurations are compared in Figure 3.7b. The inductance of the coil winding is about 0.98 μH (45° phase angle at 1.85 MHz), while the low inductance winding is half that at 0.47 μH (45° phase angle at 3.54 MHz).

A commercial 10 Ω wire-wound resistor was also evaluated for comparison. The magnitude and phase angle plots are shown in Figures 3.8a and 3.8b, respectively. For this device the inductance is about 50 nH ($R = 9.7\ \Omega$ and $\theta = 45^\circ$ at 31 MHz). This is lower than the nichrome wire owing to the finer gage wire, its shorter length in the resistor and the smaller enclosed flux area.

Metal film resistors are often used in instrumentation for their low noise properties. They also have some favorable high frequency characteristics since the current density is lower than for a comparable wire-wound geometry and the effective resistance in the film

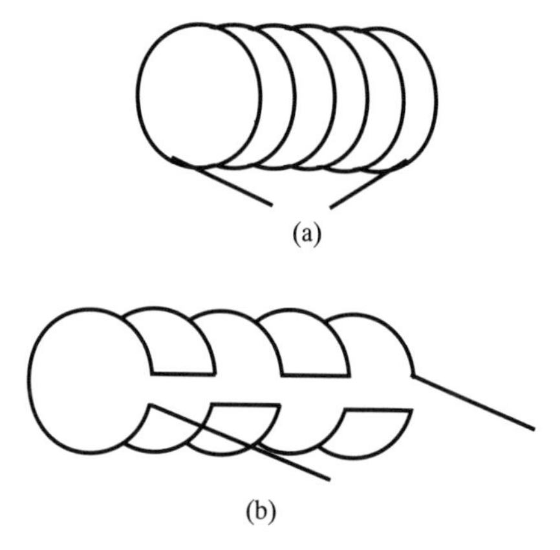

Fig. 3.7a *Wire-wound resistors having (a) standard cylindrical (solenoidal) winding, and (b) low inductance winding.*

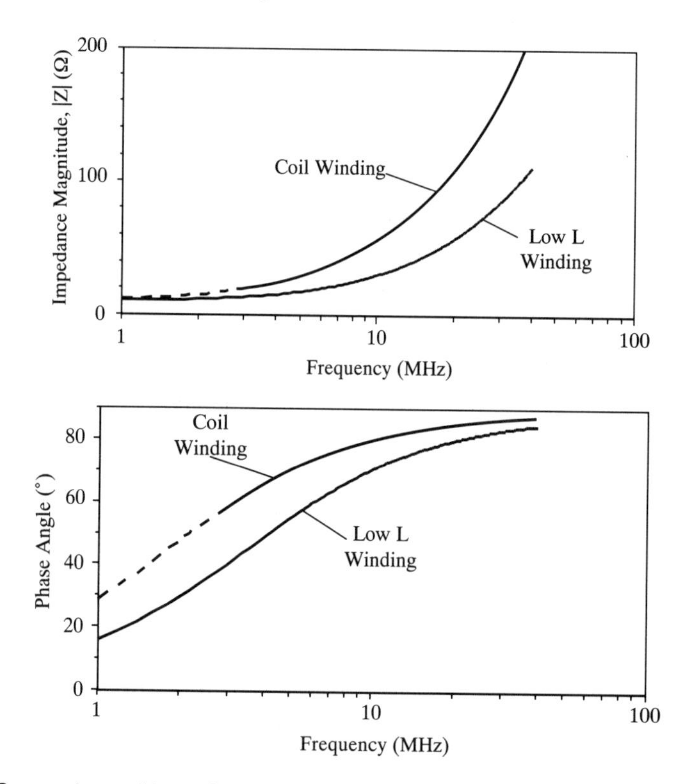

Fig. 3.7b *Comparison of impedance magnitudes and phase angles for standard cylindrical will winding and low inductance winding.*

approaches a sheet resistance, R_S (ohms per square, see Chapter 2). Thus, the effective electrical conductivity is low (giving full thickness penetration of the current in the film and reduced skin effect). A typical metal film geometry is shown in Figure 3.9a, and the impedance spectrum for a 10 Ω 0.25 W commercial metal film resistor in Figures 3.9b and

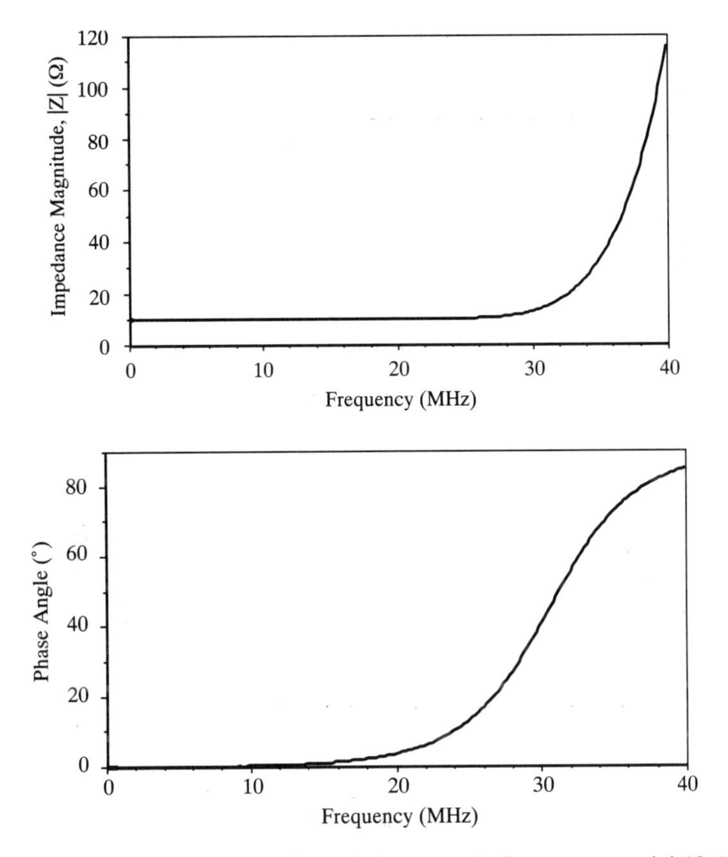

Fig. 3.8 *Plots of impedance magnitude and phase angle for a commercial 10 Ω wire-wound resistor.*

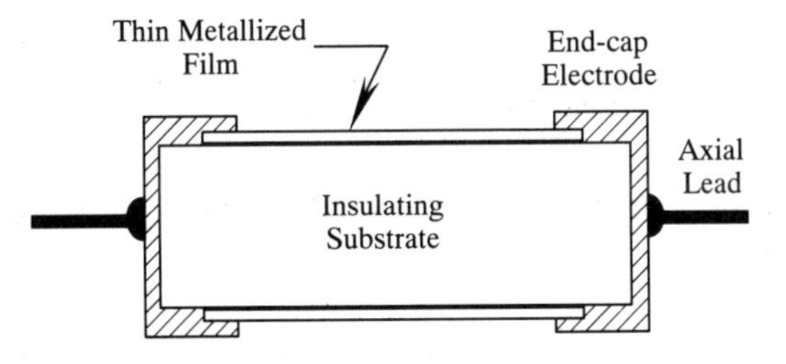

Fig. 3.9a *Typical geometry of metal film resistor.*

3.9c. We note that the inductance is about 28 nH ($R = 10.95\ \Omega$ and $\theta = 32.6^\circ$ at 40 MHz) for this device, much smaller than for the higher power wire-wound device.

A standard carbon-composition resistor consists of a semiconducting lump (of moderate electrical conductivity) with two cylindrical end electrodes, as sketched in Figure 3.10a. The material is semiconducting with non-negligible dielectric properties, so the electric field between the electrodes creates a parallel R-C network, $R_C \parallel C_C$ in the carbon-composition lump:

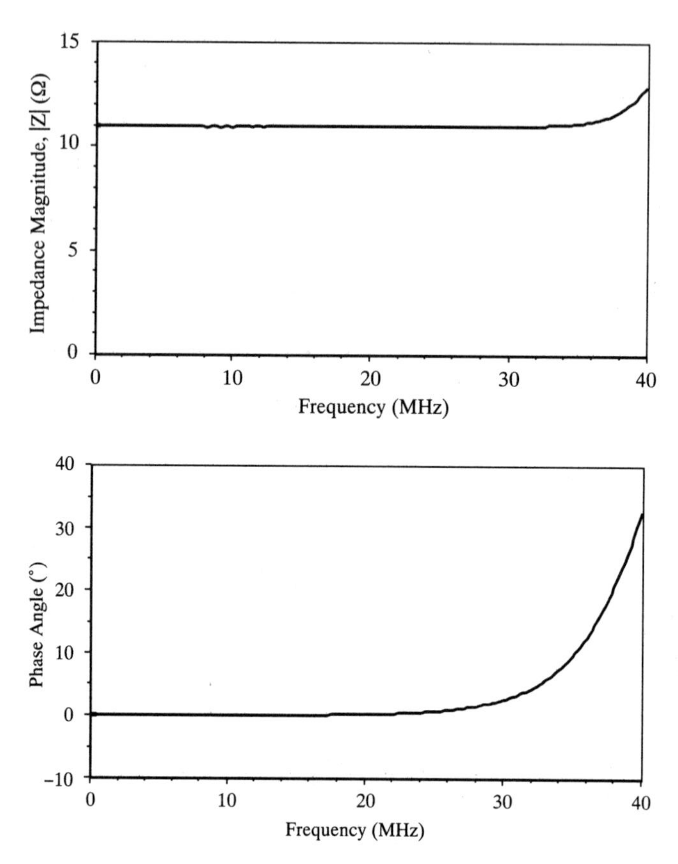

Fig. 3.9b *Plots of impedance magnitude and phase angle for a commercial 10 Ω 0.25 W metal film resistor.*

$$R_C \parallel C_C = \frac{R_C}{1 + j\omega R_C C_C} = \frac{R_C}{1 - \omega^2 R_C C_C} - j\omega \frac{R_C^2 C_C}{1 - \omega^2 R_C C_C}. \quad (3.13)$$

The lead wires have resistance, R_W, and significant self-inductance, L_W, in series with this parallel combination (Figure 3.10a):

$$R_W + j\omega L_W + R_C \parallel C_C = R_W + \frac{R_C}{1 - \omega^2 R_C C_C} - j\omega \left[L_W + \frac{R_C^2 C_C}{1 - \omega^2 R_C C_C} \right]. \quad (3.14)$$

The measured impedance spectrum of a commercial 10 Ω 2 W carbon-composition resistor is shown in Figure 3.10b. For this device $R_C + R_W = 11.4$ Ω and over this frequency range the capacitance of the carbon lump is negligible. The equivalent inductance of the lead wires and magnetic field in the carbon is 45.4 nH (since $\theta = +39.7°$ at 40 MHz), more than the metal film geometry but less than the wire-wound geometry. A 2 W resistor was chosen specifically for this example because it has a much lower capacitance than a lower power 0.25 W or 0.125 W carbon resistor.

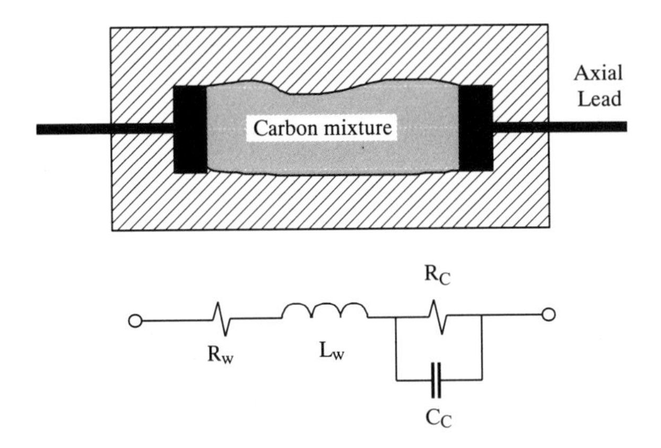

Fig. 3.10a *Structure (top) and approximate equivalent circuit (bottom) of a carbon-composition resistor.*

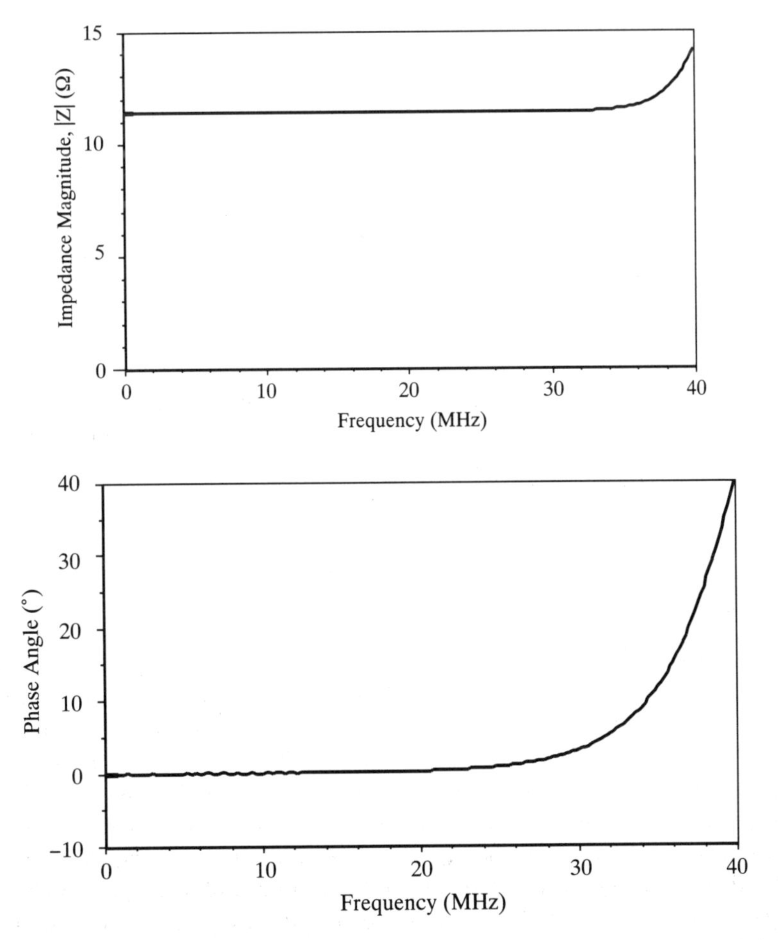

Fig. 3.10b *Plots of impedance magnitude and phase angle for a commercial 10 Ω 2 W carbon-composition resistor.*

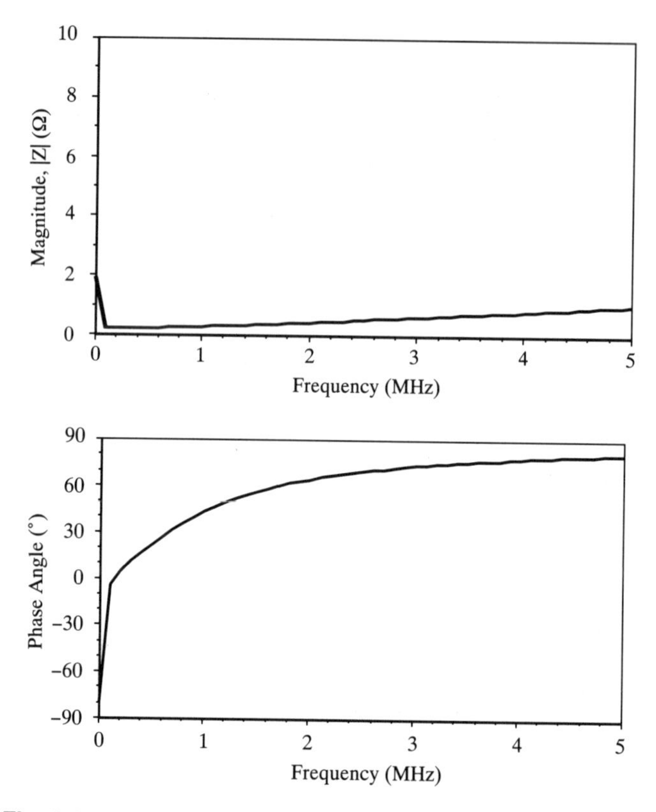

Fig. 3.11 *Impedance spectra of a 68 μF electrolytic capacitor.*

Capacitors

The capacitance of a distributed electric field can be calculated from the charge stored on one of the boundary electrodes, Q, and the potential between the electrodes, V:

$$C = \frac{Q}{V} = \frac{\oint_{\Sigma} \mathbf{D} \cdot d\mathbf{S}}{-\int_{b}^{a} \mathbf{E} \cdot d\mathbf{l}}. \tag{3.15}$$

As in the resistance calculations above, the reference electrode for potential is located at b. One of the most common capacitor technologies in instrumentation is the electrolytic capacitor. At high frequency these devices look inductive (see the impedance spectrum in Figure 3.11 for a 68 μF electrolytic capacitor). Even the improved solid-tantalum technology has comparable high frequency performance (see Figure 3.12 for a 68 μF tantalum capacitor). Better high frequency performance is realized from ceramic capacitors. We note in Figure 3.13, however, that the 50 nF ceramic capacitor shown is self-resonant at about 5 MHz for reasons which will be discussed. Because of the poor high frequency performance of ordinary power supply capacitors, the bypass capacitors in instruments and devices designed for video or RF use must consist of at least two parallel capacitances, one electrolytic or solid tantalum for low frequency bypass and one ceramic for high frequency

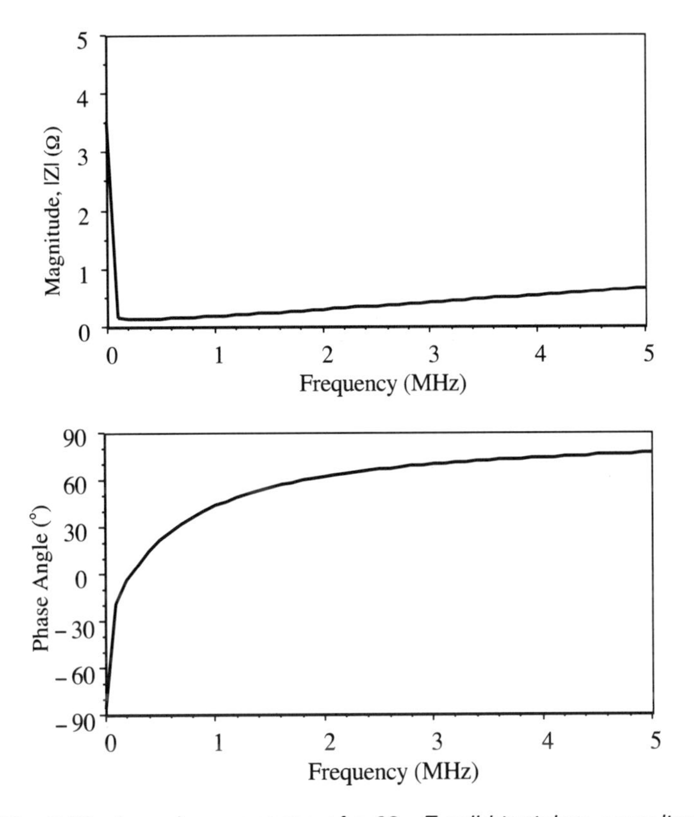

Fig. 3.12 *Impedance spectra of a 68 μF solid tantalum capacitor.*

bypass. Lower values of capacitance than 50 nF should be used, however, to stay below the self-resonant frequency.

In RF devices and equipment it is usually only necessary or desirable to have a small value of capacitance. Often we can make the required capacitor out of readily available materials. In Figure 3.14a we have designed a 20 pF capacitor out of double-sided copper-clad circuit board assuming that the relative electric permittivity of the fiberglass composite board material is in the neighborhood of 5.0. The capacitor is to be used in a voltage divider (see Chapter 5). Since the permittivity is significantly higher than the surrounding air, we expect the capacitance of the fringing field to be small, and so our estimated value of 20 pF corresponds reasonably well with the measured value, 23 pF. The impedance spectrum for this device is shown in Figure 3.14b. Here we see virtually no deviation from a pure capacitive impedance—the phase angle is −89°. In fact, the fiberglass dielectric has a finite but important loss factor—in use at 27 MHz above about 1 kV_{rms} the capacitor leads tend to de-solder due to a combination of resistive heating in the leads and dielectric heating.

As an alternative, one can use an air-dielectric capacitor provided the voltages are not sufficient to generate a breakdown electric field in the air gap. In the design of Figure 3.15a, the gap is adjustable in order to trim the capacitance to the desired value. The fringing fields add a small amount to the capacitance of the two finite discs. At a gap spacing of 3.81 mm (0.15 inch) we obtain a capacitance of about 6.5 pF. The impedance spectra in Figure 3.15b illustrate extremely good behavior of this capacitor at elevated frequencies compared to the

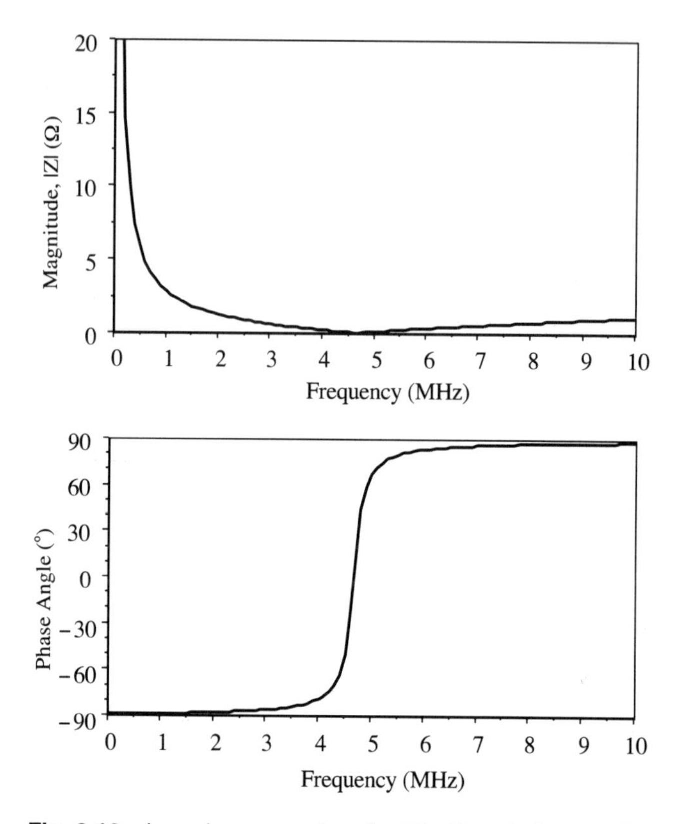

Fig. 3.13 *Impedance spectra of a 50 nF ceramic capacitor.*

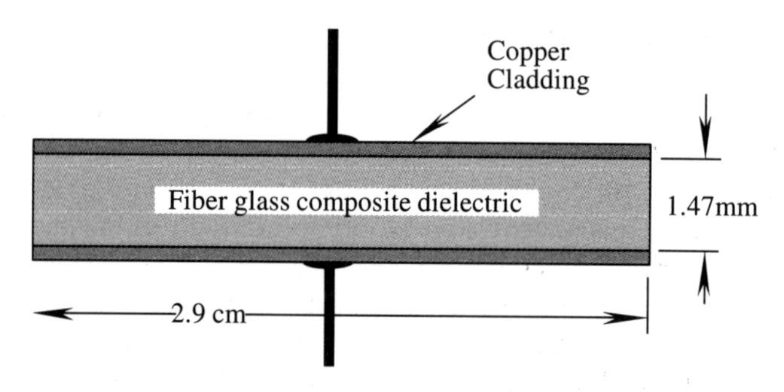

Fig. 3.14a *Approximately 20 pF capacitor made from double-sided circuit board.*

smaller-volume commercial devices. Expanding the volume occupied by the electric field between the plates greatly reduces the E-field magnitude and the self-inductance due to the time-varying E-field. We have used this capacitor to several kV_{rms} with no observed breakdown effects (it could be used at a higher voltage as well). As we will show in Chapter 4, the highest electric field strengths occur at the corners of the conducting disks; so, if it occurs, breakdown will initiate there.

The breakdown limitations inherent in air gap capacitors can be reduced by using vacuum capacitors. Two standard commercial RF designs are shown in Figure 3.16. These devices

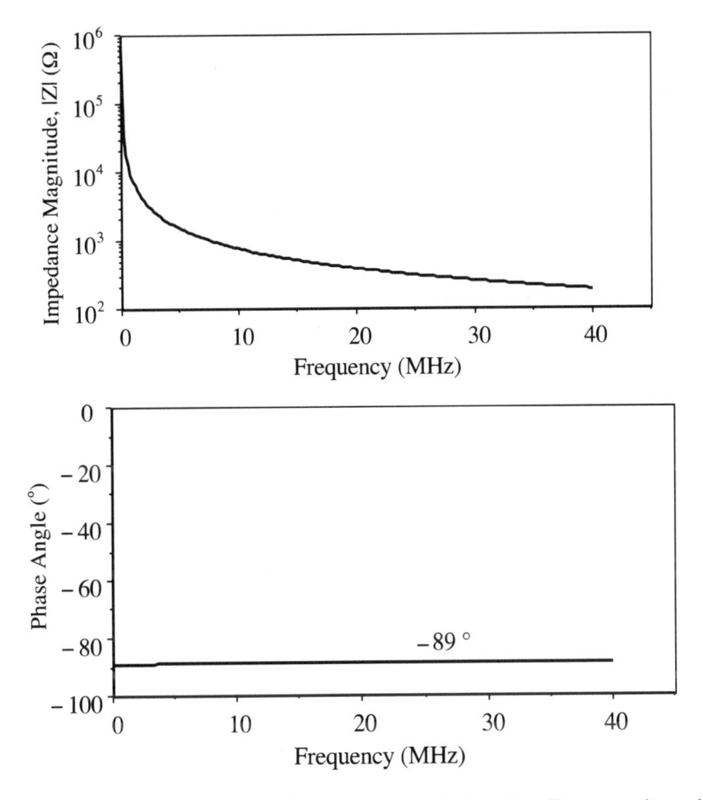

Fig. 3.14b *Impedance spectra of the approximately 20 pF capacitor illustrated in Figure 3.14a.*

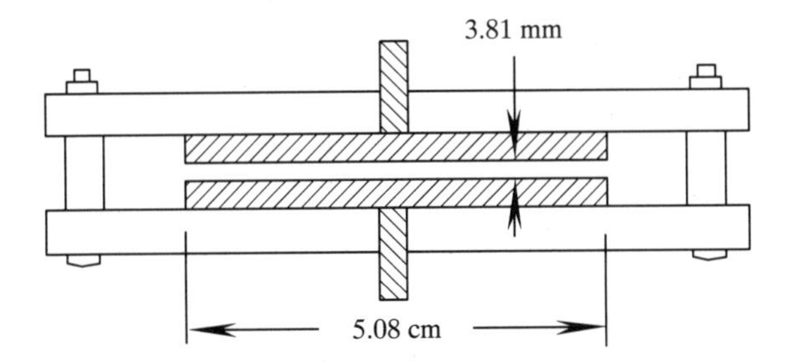

Fig. 3.15a *Adjustable gap capacitor with teflon supporting structure in cross-section.*

are obtainable in either fixed (3.16a) or variable capacitance (3.16b) forms—neither is inexpensive! The design of the electrodes in the variable capacitance configuration is of some interest (Figure 3.16c). Movement of the electrode changes the capacitance by modifying the interleaving of the "fingers". This geometry is used to give maximum variation in capacitance within a reasonable vacuum volume.

Vacuum capacitors do suffer from breakdown at high field strengths, however. Just as in a vacuum tube, electrons are continually emitted into the vacuum from the metal electrode surfaces owing to quantum effects in the metal. In the presence of a strong electric field (as we almost always have) they will be accelerated with little impedance from the cathode

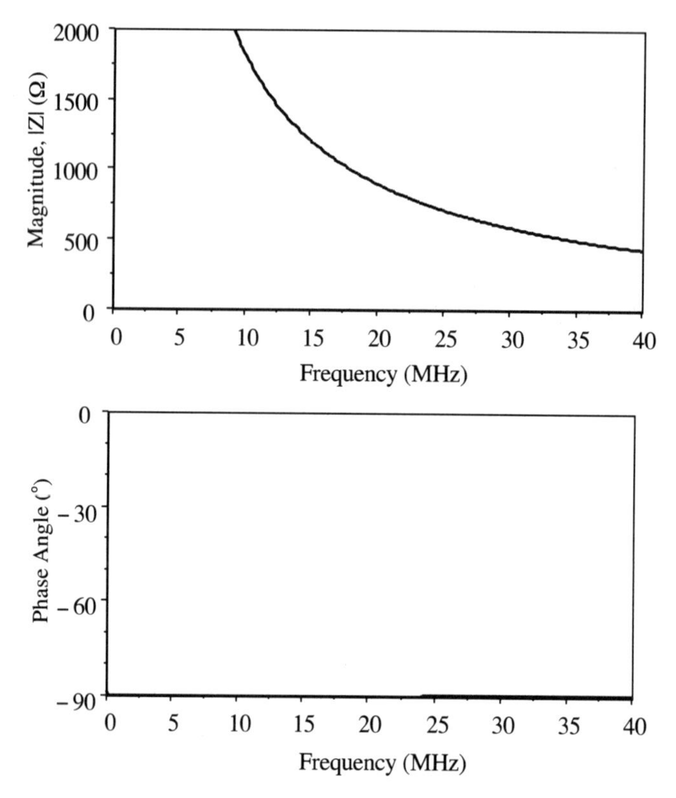

Fig. 3.15b *Impedance spectra of the 6.5 pF air gap capacitor illustrated in Figure 3.15a.*

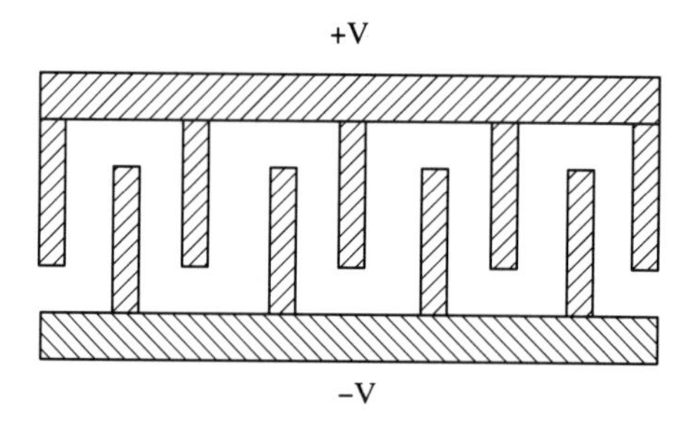

Fig. 3.16 *Interdigitated electrode typical of adjustable vacuum capacitors.*

(negative voltage) toward the anode (positive voltage). If they strike the anode with sufficient momentum (at very high field strengths) they will produce metal ions which will be accelerated toward the cathode. Metal ion generation may increase to avalanche multiplication current flow in the device and result in thermal damage to the electrode if the currents and internal fields are sufficiently high. If the vacuum capacitor has sufficient RF current that the electrode temperature rises this effect occurs at lower electric fields (the Boltzmann statistics governing the evolution of electrons are very sensitive to temperature). We learned this lesson in our laboratory in Austin on a very expensive (about \$3000 US) adjustable vacuum capacitor in a tuning circuit when the electrodes melted, resulting in a weld between them. So, some care is warranted in use, even if the rated maximum potentials are not exceeded.

Inductors

The self-inductance, L, of a distributed magnetic field can be calculated from the total magnetic flux "linked" by (i.e. surrounded by) a conductor, $N\Phi_M$, (where there are N turns in the inductor and each turn has total magnetic flux Φ_M) divided by the current, I, in the conductor which generates Φ_M:

$$L = \frac{N\Phi_M}{I} = \frac{N}{I} \iint_S \mathbf{B} \cdot \mathrm{d}\mathbf{S}. \tag{3.16}$$

Two self-inductances which share a common magnetic flux form a transformer. The transformer may conveniently be modeled by a T-network (Figure 3.17) which consists of the two self-inductances, L_1 and L_2 calculated from equation (3.16), with a mutual inductance, M_{12}, arising from the shared magnetic flux. The mutual inductance may in turn be calculated from:

$$\begin{aligned} M_{12} &= \frac{N_2\Phi_{12}}{I_1} = \frac{N_1\Phi_{21}}{I_2} = M_{21} \\ &= \frac{1}{I_1 I_2} \iiint_{\mathrm{vol}} (\mathbf{B}_1 \cdot \mathbf{H}_2)\, \mathrm{d}v \end{aligned} \tag{3.17}$$

where the volume of integration is only over the flux common to both coil 1 and coil 2. In both cases the inductance will be a function of geometry only, independent of I_1 and I_2.

We conclude this section with a few simple illustrative examples. First, it may be shown that the inductance per unit length of a long straight wire with a uniform current distribution is $\mu/8\pi$ (H/m). An idealized infinite solenoid coil (Figure 3.18a) has an easily calculable inductance per unit length. Since the coil is infinite in extent no magnetic flux escapes and the external magnetic field is zero. The internal field is approximately uniform with $H_z = NI/d$ where there are N turns in a length d—this may be obtained from the integral form of Ampère's law (Chapter 1) where the contour is rectangular with one leg inside the coil and one outside the coil ($\mathrm{d}\mathbf{l} = \mathrm{d}z\mathbf{a}_z$). The total magnetic flux linked by the solenoid is then $\Phi_M = \mu_0 H_z A = \mu_0 \pi r^2 NI/d$ where A is the coil cross-sectional area. The inductance of a length d of an ideal air-core infinite solenoid of radius r is thus:

$$L = \frac{\mu_0 N^2 \pi r^2}{d}. \tag{3.18}$$

Real finite coil inductors link only part of this flux due to spacing between the coils and end-of-coil effects. An empirical formula which has been used for finite solenoidal-wound single layer air-core coils is:

$$L = \frac{N^2 r}{[22.9d/r + 25.4]} (\mu\mathrm{H}) \tag{3.19}$$

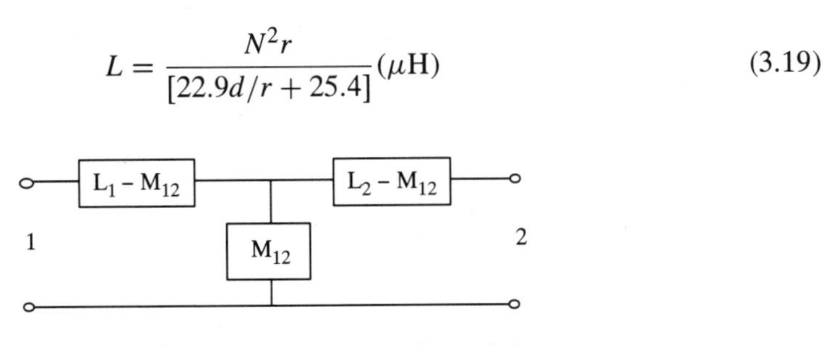

Fig. 3.17 *Transformer T-network model.*

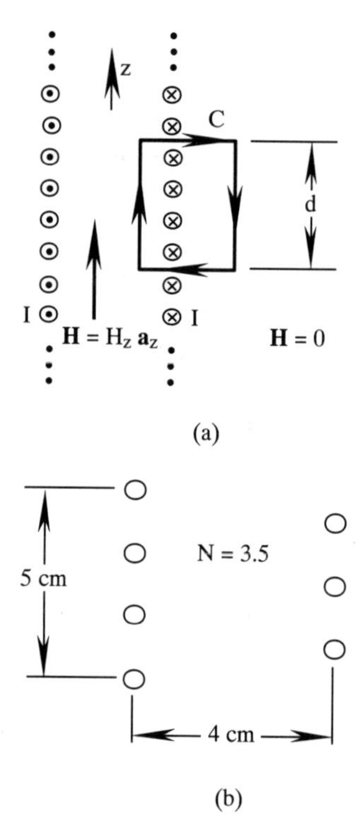

Fig. 3.18 *Inductance figures. (a) Cross-section of idealized infinite solenoid with current, I, Amperian contour, C, used to calculate L (H/m). (b) Dimensions of finite grid inductor from 8 kW RF generator.*

where r is, again, the coil mean radius (in cm for this equation) and d its length. An air-core coil 10 cm long with 10 turns of mean radius 1 cm would have an inductance of 395 nH by equation (3.18) and 393 nH by (3.19), very close to the same result (0.5% difference) because d/r is reasonably large. By comparison, the estimated inductance of a grid inductor from an 8 kW RF generator (dimensions in Figure 3.18b, $N = 3.5$ turns) would be 369 nH from equation (3.18) but 296 nH from equation (3.19) ($\approx$ 25% difference) owing to the smaller d/r ratio.

Complete models for RF circuit elements

Real circuit devices have all three types of impedance elements. A complete RF circuit model for a capacitor includes plate resistance and inductance (Figure 3.19a) which occur in series. The effective inductance of the capacitor at high frequency arises from the magnetic flux associated with the displacement currents in the dielectric material, as well as from the surface currents on and within the conductors themselves. Note that at RF frequencies the impedance of the inductive component opposes that of the capacitance giving a series resonant response with self-resonance frequency, f_{r}:

$$f_{\mathrm{r}} = \frac{1}{2\pi\sqrt{LC_0}} \tag{3.20a}$$

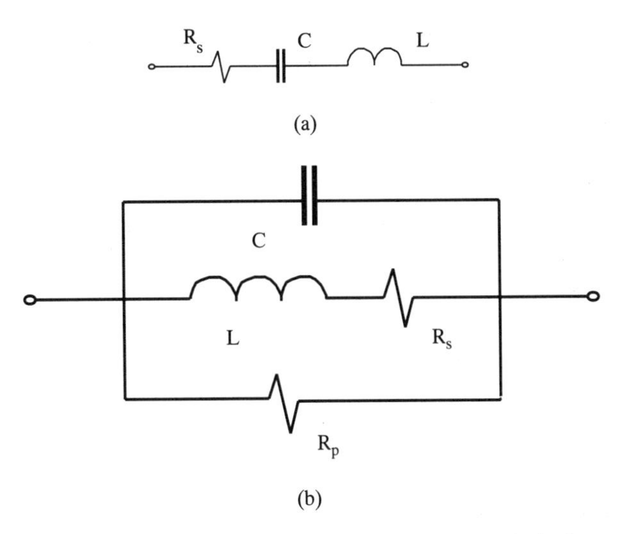

Fig. 3.19 *Complete RF circuit models for capacitors and inductors. (a) Capacitor (b) Inductor.*

where C_0 is the low frequency capacitance, and the minimum impedance, $|Z_{eq}|$ occurs at f_r. The effective capacitance, C_{eff}, is:

$$C_{eff} = \frac{C_0}{1 - (f/f_r)^2} \tag{3.20b}$$

which will be negative above f_r (inductive behavior); therefore, f_r is the limiting frequency of use. We saw just such behavior in the 50 nF ceramic capacitor of Figure 3.13 where $C_0 = 50$ nF and $f_r = 4.7$ MHz (consequently, we have $L \approx 23$ nH and $R \approx 0.04\ \Omega$ for that device).

Resonant circuits are conveniently characterized in terms of their quality factor, or Q-factor. The Q-factor is 2π times the ratio of the energy stored per cycle to the energy dissipated per cycle; equivalently, $Q = f_r/B$ where B is the 3 dB bandwidth in Hz. The series resonant circuit of Figure 3.19a has several equivalent expressions for the Q-factor which are each the ratio of reactance to resistance. We use the one expressed in terms of the effective capacitive reactance since we will always be using the capacitor at frequencies below the self-resonant frequency, f_r, in order to obtain a capacitive impedance:

$$Q = \frac{X_{eff}}{R_s} = \frac{1}{\omega_r R_s C_{eff}} = \frac{[1 - (f/f_r)^2]}{2\pi f_r R_s C_0}. \tag{3.20c}$$

Note that the quality factor is dependent on the frequency of operation (and on the temperature since C_0 is temperature dependent in general). The quality factor for the capacitor is the inverse of the loss tangent in a material medium or what is sometimes referred to as the "dissipation factor" for a device.

Practical inductors (Figure 3.19b) also exhibit self-resonance owing to parasitic capacitance between coil windings (C in the figure), series resistance in the coil (R_s) and a possible parallel resistance (R_p) due to losses in the medium surrounding the coil. The external losses are due to electric fields surrounding the coil which arise from the voltage gradient (∇V) along the coil and time-dependent vector potential ($\mathbf{A}$) around the coil, as we introduced in Chapter 1. The circuit exhibits behavior similar to a parallel resonant circuit in that the magnitude, $|Z_{eq}|$, has a maximum at the self-resonant frequency, f_r. When a coil is surrounded by air, as is the usual case in power RF circuits, R_p will be extremely large, and can safely be neglected. A significant exception to this is an inductive applicator in which the material to be heated is placed within a magnetic field and the magnetic field induces an electric field in the material. That case will be discussed in Chapter 4, so we will assume that $R_p = \infty$ in this discussion. Also, we must use the inductor below its self-resonant frequency, f_r, in order to obtain inductive reactance, just as with the capacitor above. If, in addition, the reactance of L is large compared to the inductor resistance, R_s, then:

$$Q_s = \frac{\omega L}{R_s} \gg 1 \qquad \text{so that} \qquad 1 + \frac{1}{Q_s} \approx 1 \tag{3.21a}$$

and:

$$L_{eff} = \frac{L}{1 - (f/f_r)^2}. \tag{3.21b}$$

Since R_p can be ignored the effective resistance will be:

$$R_{eff} = \frac{R_s}{1 - (f/f_r)^2}. \tag{3.21c}$$

so that the apparent Q-factor of the coil is lowered by the capacitance:

$$Q_{eff} = \frac{Q_s}{1 - (f/f_r)^2}. \tag{3.21d}$$

In designing and using RF circuit elements it is important to bear in mind that no reactive element is lossless and that the parasitic elements have strong influence at radio frequencies. Carefully designed and constructed vacuum capacitors are readily available from commercial sources. The RF circuit designer is almost always required to construct specific inductors. A very useful design method for single-layer solenoid coils may be found in Chapter 2 of the book by Abrie (1985) in which desired resonance frequencies, Q-factors and self-capacitances can be obtained in single layer solenoids.

Some circuit construction principles

In low frequency circuits the guiding principle of ground (or earth) connection is that one common connection point is preferrable to eliminate ground loop currents at 60 Hz. The opposite is true for RF circuit construction: one wants as many ground connections and as much distributed "ground" shielding as can be reasonably arranged. It should be borne in mind that "ground" in an RF sense is not a single potential, as it usually is at 60 Hz, but really a shield enclosure with potential (magnitude and phase) dependent on position. In RF circuits a large portion of the performance difficulties one encounters comes from parasitic elements, as we have just seen. The designer is almost never concerned about ground loops. Therefore, element shielding is the first consideration.

Also, it is important to keep all conductors to absolutely minimum lengths. Some time and consideration should be invested in arranging the components to achieve this. The designer can reduce the self-inductance of conductors by reducing the local current densities. That is, particularly at higher powers, one often uses large flat ribbons (singly or multiply) or hollow tubes rather than solid wires to combat the skin effect (Figure 3.20a). Adjacent parallel currents (x-direction, Figure 3.20b) exert an attracting force (y-direction) on each other:

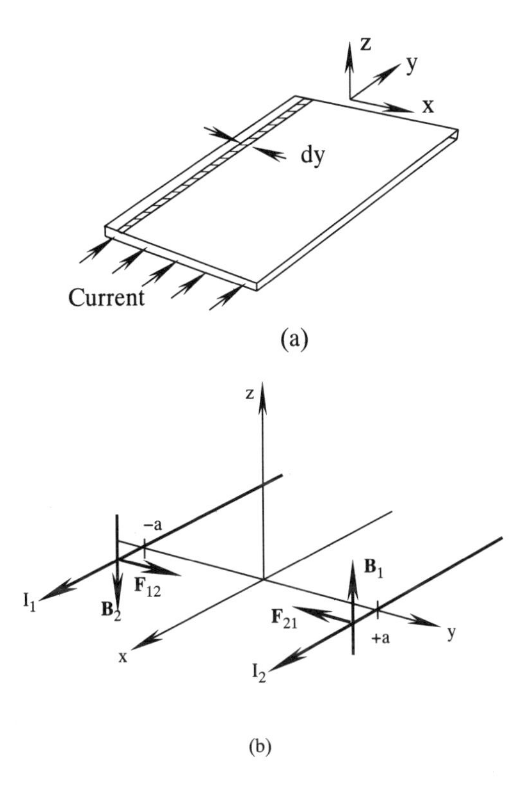

Fig. 3.20 *Current forces in ribbon conductors. (a) Ribbon conductors: total current is the sum of the differential strip currents. (b) Attraction forces generated by parallel currents at* $x = \pm a$.

$$\mathbf{dF}_{12} = \mathbf{J}_1 \times \mathbf{B}_2 \tag{3.22}$$

where $\mathbf{dF}_{12}$ is the incremental force per unit volume on current density $\mathbf{J}_1$ due to the magnetic flux density $\mathbf{B}_2$ generated by current element $\mathbf{J}_2$. The total magnetic force on $\mathbf{J}_1$ is the integral over all externally generated flux. Very strong repelling forces arise from the repulsion between charges of like sign. So, even if the thin metal ribbon has a uniform current density with depth (z-direction) there may be current density gradients in the y-direction. It is not uncommon to observe heat annealing at the edges of flat copper ribbon conductors used in RF circuits where the current density is higher. Hollow tube conductors avoid this phenomenon, but are mechanically less flexible than ribbons. When flexibility is not required soldered copper plumbing pipes and fixtures (elbows, flanges, etc.) may be used in RF circuits with good result.

3.1.3 Waveguide circuit element examples

By way of introduction to waveguide devices, we will very briefly consider a few representative waveguide circuit elements. We will create a perturbation of the local electromagnetic fields by inserting an arbitrarily shaped device or obstacle in a plane normal to the waveguide. This discontinuity will induce a reflected and a transmitted wave in both directions far upstream and/or downstream from the obstacle, necessitating a two-port parametric representation. In the near field of the obstacle, the incident and "scattered" (i.e. the obstacle-induced) fields combine to form so-called evanescent modes which are highly damped. The evanescent modes are therefore non-propagating and die out within several wavelengths of the obstacle. Consequently, we do not consider transmitted or reflected waves in the near field (in the presence of the evanescent modes), but only at some distance from the obstacle. The concept of reflection and transmission coefficients associated with an obstacle is thus only valid for waves measured in the so-called "far field" zone. Under this restriction, the concept of localized impedance for the representation of an obstacle in a transmission line is only valid if the T-L supports a single propagation mode. This is the significant advantage of sizing waveguides for the TE_{10} mode — it is the most fundamental mode, and is thus the most stable near obstacles. When multiple modes of propagation are possible an obstacle has a more complex equivalent circuit — the circuit model must include coupling factors to all possible transmitting modes. We will explore this in a little more detail in the discussion of multimode cavity applicators, and the effect which the load has on the modes realized, in Chapter 4. In the ensuing discussion, we will only consider effects far from the obstacle and assume that single modes are propagating in the waveguide.

E-plane step discontinuity in a TE_{10} mode rectangular waveguide

We begin with a discussion of the example step discontinuity in a TE_{10} mode waveguide in Figure 3.21. This will illustrate how the equivalent voltages and currents above relate to the propagation field quantities described in Chapter 2. Plane 1 is distance l_1 from the step (in the far field) and plane 2 is at l_2. In the far field:

$$\mathbf{E}_1 = \left[C_1^+ e^{j\beta l_1} + C_1^- e^{-j\beta l_1}\right] \sin\left(\frac{\pi x}{a}\right) \mathbf{a}_y \tag{3.23a}$$

$$\mathbf{H}_1 = -Y_0 \left[C_1^+ e^{j\beta l_1} - C_1^- e^{-j\beta l_1}\right] \sin\left(\frac{\pi x}{a}\right) \mathbf{a}_x \tag{3.23b}$$

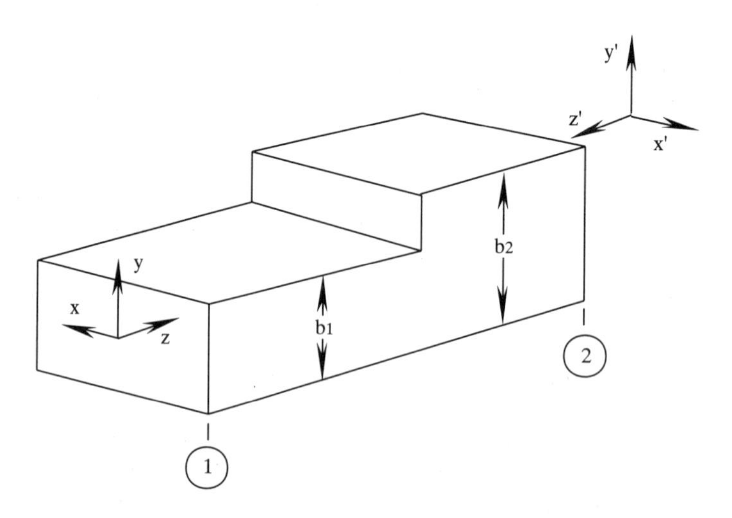

Fig. 3.21 *TE_{10} E-plane step discontinuity in waveguide.*

$$\mathbf{E}_2 = \left[C_2^+ \mathrm{e}^{\mathrm{j}\beta l_2} + C_2^- \mathrm{e}^{-\mathrm{j}\beta l_2}\right] \sin\left(\frac{\pi x'}{a}\right) \mathbf{a}_y \tag{3.23c}$$

$$\mathbf{H}_2 = -Y_0 \left[C_2^+ \mathrm{e}^{\mathrm{j}\beta l_2} - C_2^- \mathrm{e}^{-\mathrm{j}\beta l_2}\right] \sin\left(\frac{\pi x'}{a}\right) \mathbf{a}_x \tag{3.23d}$$

where we have used a normalized z-component magnetic field for the standard TE_{10} mode, $h_z = (\mathrm{j}\pi Y_0/\beta a)\cos(\pi x/a)$ (see equations (2.52)), in order to simplify the transverse field relations, and Y_0 is the waveguide characteristic admittance (see equations (3.43) and (3.44) below). The expressions for the equivalent voltages and currents are:

$$V_1^+ = K_1 C_1^+ \mathrm{e}^{\mathrm{j}\beta l_1} \qquad V_1^- = K_1 C_1^- \mathrm{e}^{-\mathrm{j}\beta l_1} \tag{3.24a}$$

$$I_1^+ = Y_0 K_1 C_1^+ \mathrm{e}^{\mathrm{j}\beta l_1} \qquad I_1^- = Y_0 K_1 C_1^- \mathrm{e}^{-\mathrm{j}\beta l_1} \tag{3.24b}$$

$$V_2^+ = K_2 C_2^+ \mathrm{e}^{\mathrm{j}\beta l_2} \qquad V_2^- = K_2 C_2^- \mathrm{e}^{-\mathrm{j}\beta l_2} \tag{3.24c}$$

$$I_2^+ = Y_0 K_2 C_2^+ \mathrm{e}^{\mathrm{j}\beta l_2} \qquad I_2^- = Y_0 K_2 C_2^- \mathrm{e}^{-\mathrm{j}\beta l_2} \tag{3.24d}$$

and we choose K_1 and K_2 to satisfy conservation of power requirements (the discontinuity is considered lossless here):

$$K_1 = \int_0^a \int_0^{b_1} \sin^2\left(\frac{\pi x}{a}\right) \mathrm{d}x\mathrm{d}y = \sqrt{\frac{ab_1}{2}} \tag{3.25a}$$

$$K_2 = \int_0^a \int_0^{b_2} \sin^2\left(\frac{\pi x}{a}\right) \mathrm{d}x\mathrm{d}y = \sqrt{\frac{ab_2}{2}}. \tag{3.25b}$$

From these expressions the respective scattering or impedance parameters may be determined analytically.

Thin metal obstacles

Thin metal obstacles can be used to form inductive and capacitive windows, or irises, in a waveguide. Figure 3.22 illustrates four examples of TE_{10} shunt inductive irises. The irises are referred to as inductive because the evanescent modes—which are excited in order to satisfy the boundary conditions (tangential E-field = 0 on the iris)—store energy primarily in the magnetic field. Of course, the physical significance of stored energy implies a purely imaginary impedance and admittance; so the ideal iris, like the step discontinuity, is lossless. The corresponding susceptance values for the inductive irises shown are first, for a center aperture of width d (Figure 3.22a):

$$B = Y_0 \frac{2\pi}{\beta a} \operatorname{cotan}^2\left[\frac{\pi d}{2a}\left(1 + \frac{a\kappa - 3\pi}{4\pi}\sin^2\left(\frac{\pi d}{a}\right)\right)\right] \tag{3.26a}$$

with

$$\beta = \sqrt{\omega^2 \mu_0 \varepsilon_0 - \left(\frac{\pi}{a}\right)^2} \qquad \kappa = \sqrt{\left(\frac{3\pi}{a}\right)^2 - \omega^2 \mu_0 \varepsilon_0}.$$

Second, for a side aperture, also of width d (Figure 3.22b):

$$B = Y_0 \frac{2\pi}{\beta a} \operatorname{cotan}^2\left[\frac{\pi d}{2a}\left(1 + \operatorname{cosec}^2\left(\frac{\pi d}{2a}\right)\right)\right] \tag{3.26b}$$

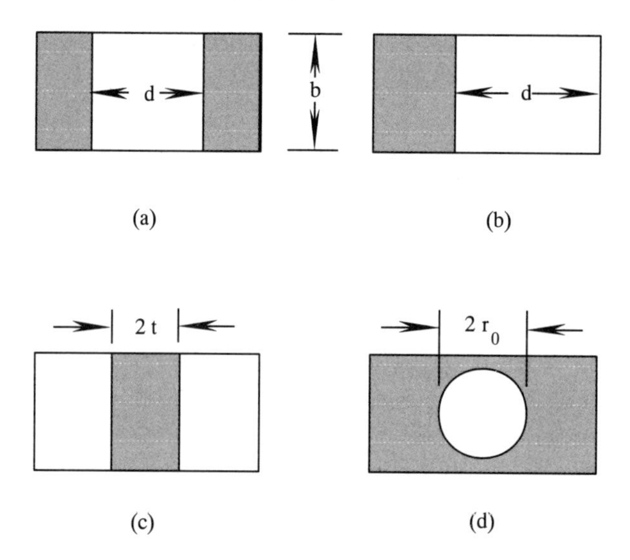

Fig. 3.22 *Inductive irises.*

where β has the same meaning. Third, for the inductive rod of radius t located along the centerline (at $x = a/2$) (Figure 3.22c):

$$B = \frac{Y_0 \dfrac{4\pi}{\beta a}}{\left[\ln\left(\dfrac{a}{\pi t}\right) - 1 + 2\left(\dfrac{a}{\pi t}\right)^2 \sum\limits_{n=3,5,\ldots}^{\infty} \left(\dfrac{\pi}{a\kappa_n} - \dfrac{1}{n}\right) \sin^2\left(\dfrac{n\pi t}{a}\right)\right]} \tag{3.26c}$$

where:

$$\kappa_n = \sqrt{\left(\frac{n\pi}{a}\right)^2 - \omega^2 \mu_0 \varepsilon_0}.$$

Finally, for the small circular aperture of radius r_0 at the center of the sheet (Figure 3.22d):

$$B = \frac{3abY_0}{8\beta r_0^3}. \tag{3.26d}$$

Orienting the iris along the x-direction gives a capacitive susceptance since the evanescent modes store energy in the electric field. Four examples are shown in Figure 3.23, but analytical expressions are only reasonable for the first two. The fourth configuration, a capacitive post, is often used as an inexpensive stub tuner, as we will see in Section 3.5 of this Chapter. First, the asymmetrical iris of width d (Figure 3.23a) has the susceptance:

$$B = Y_0 \frac{4\beta b}{\pi}\left[\ln\left(\operatorname{cosec}\left(\frac{\pi d}{2b}\right)\right) + \left(\frac{\pi}{b\kappa_1} - 1\right)\cos^4\left(\frac{\pi d}{2b}\right)\right] \tag{3.27a}$$

where:

$$\kappa_1 = \sqrt{\left(\frac{\pi}{b}\right)^2 - \beta^2}$$

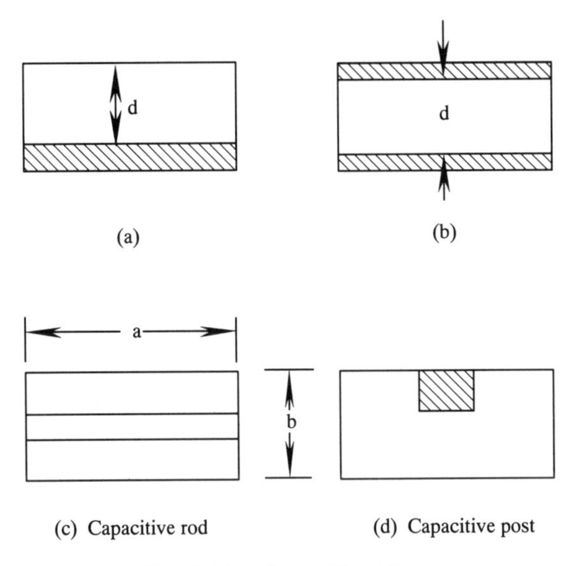

Fig. 3.23 *Capacitive irises.*

and β has the same form as in equation (3.26). The symmetrically placed iris (Figure 3.23b) has an equivalent susceptance of:

$$B = Y_0 \frac{2\beta b}{\pi} \left[\ln \left(\operatorname{cosec} \left(\frac{\pi d}{2b} \right) \right) + \left(\frac{2\pi}{b\kappa_2} - 1 \right) \cos^4 \left(\frac{\pi d}{2b} \right) \right] \quad (3.27b)$$

where:

$$\kappa_2 = \sqrt{\left(\frac{2\pi}{b} \right)^2 - \beta^2}$$

The capacitive rod of Figure 3.23c and post of Figure 3.23d do not have convenient expressions for their susceptance. The capacitive post may easily be made adjustable by using a finely threaded metallic screw. One may adjust the length of the screw protruding into the waveguide to obtain the desired effect on the circuit.

3.2 RADIO FREQUENCY GENERATORS

There are two major classes of RF generators: (1) a controlled frequency oscillator with power amplifier, and (2) a power oscillator in which the load is part of the resonant circuit. In regions where the operating RF frequency may vary (as in the US) the power oscillator circuit is widely used since it is a much lower cost approach. The power oscillator design is also able to achieve much higher overall efficiencies (load power compared to electric mains power) since the load is part of the tank circuit. In Europe where strict operating frequency limitations have the force of law it is usually necessary to use a controlled frequency oscillator and power amplifier to satisfy the requirements. Considerable work has been done (at Electricité de France and Thomsen-CSF) to develop stable RF tank oscillator tubes which employ feedback methods to stabilize the operating frequency (frequency drift is mostly attributable to thermal expansion in these tubes under load); however, practical appliances utilizing these devices are not yet readily available.

3.2.1 Vacuum tube fundamentals

Because of the high powers and high frequencies required of RF generators, the power amplifier or power oscillator stages are usually constructed with vacuum tubes rather than solid state devices. Recent engineering graduates may not have experience of vacuum tube function and design, so a very brief description is included here for completeness. A more complete description may be found in references at the end of this chapter (particularly in the ARRL Handbook). In terms of a working circuit model, the vacuum tube has voltage–current curves similar to depletion/enhancement mode MOSFET devices. There is no P-channel vacuum tube, since the charge carriers are exclusively electrons.

Referring to the "triode" three-element device shown in Figure 3.24a, a cathode (K) is heated to high temperature to increase the thermal energy of the loosely bound valence electrons sufficiently that they "boil off" at a high rate and are accelerated by the strong electric field established by the anode (A), or plate, as it is also called. The heater may

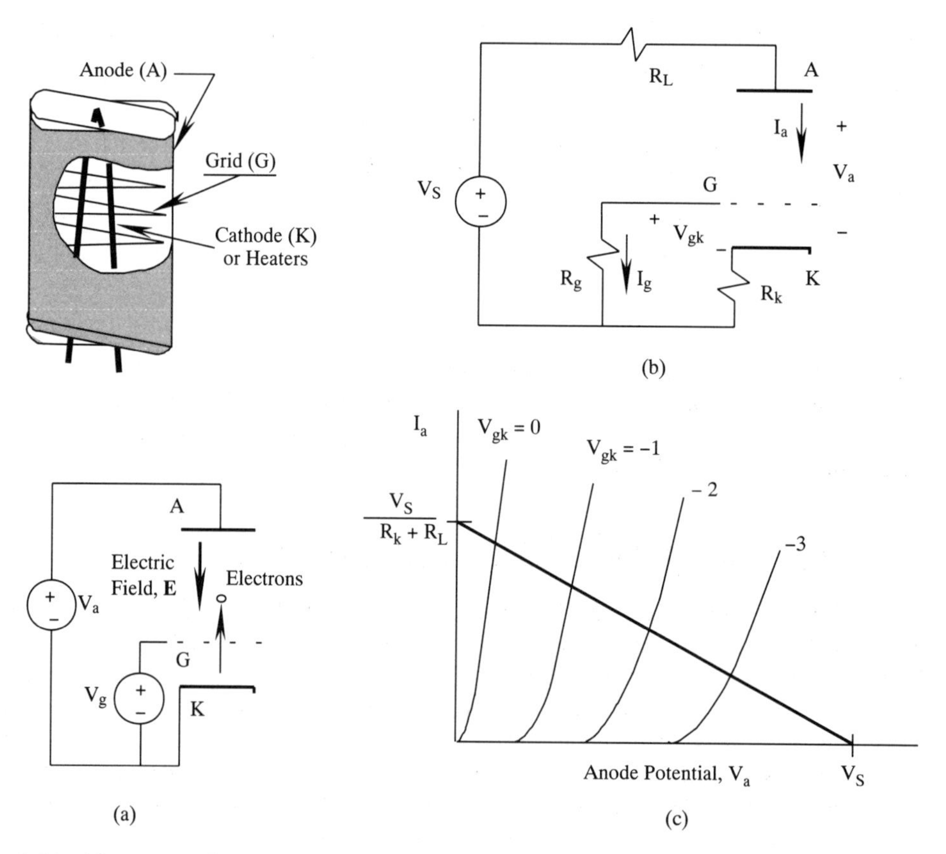

Fig. 3.24 *Vacuum tube fundamentals. (a) Vacuum triode structure and operation. (b) Practical bias circuit. (c) Typical triode characteristic curves and load line.*

be an element separate from the cathode (as in most small signal tubes) or one leg of the heater may be the cathode itself (many rectifiers and high frequency tubes have the second configuration). The very loosely wound grid wire (G) acts as a gate or valve: for negative grid voltages (with respect to the cathode) the electrons are repelled back to the cathode. For grid voltages less than the cutoff voltage all of the electrons are repelled back to the cathode and the tube is "cutoff". For example, a 5918A (a large RF vacuum tube) with anode voltage of $V_a = 15$ kV has a grid cutoff voltage of $V_g = -375$ V. In oscillator circuits the vacuum tube is biased below cutoff—as much as three times the cutoff voltage, or about -1200 V for the 5918A. As $V_{gk} = V_g$ approaches zero the tube current reaches the maximum cathode current. If the grid is biased positive with respect to the cathode it also becomes a current source for the anode current.

Small signal amplifiers are invariably operated with $V_{gk} < 0$ and anode current approximately constant (a class A amplifier, Figure 3.24b). Large signal amplifiers often employ two tubes on opposite ends of a transformer winding which conduct on opposite output signal half-cycles (class B or push-pull amplifier). In this configuration one tube is always in cutoff (not conducting) and harmonic distortion in the waveform occurs near the zero-crossings. The "conduction angle" for each tube is close to 180°. Some advantage can be obtained in tube efficiency by using smaller conduction angles—say 120° to 140°—and this is called class C amplifier operation. The grids must be driven positive to achieve this and, of course, considerable harmonic distortion results. Harmonic distortion is not of concern in RF power amplifier circuits since tuning networks are employed to limit the signal bandwidth. RF oscillators are often class C with smaller conduction angles than amplifiers, 10° to 60° might be typical depending on load conditions.

A practical bias scheme for a class A vacuum tube amplifier is shown in 3.24b in which the grid leakage current, I_g, is so small that $V_g = 0$ and the cathode bias voltage is approximately $V_{gk} = -I_a R_k$, as one encounters in the analogous MOSFET bias circuit. The quiescent point (I_a, V_a) may be determined graphically from the tube characteristic curves using a load line (Figure 3.24c) in which:

$$V_a = V_S - I_a (R_k + R_L) \tag{3.28a}$$

$$V_{gk} = -I_a R_k. \tag{3.28b}$$

The solution must lie on the load line and is usually determined iteratively by assuming an initial operating point which is updated by an improved estimate of I_a.

This triode circuit used as a common cathode ac amplifier is illustrated in Figure 3.25a where C_{in} and C_{out} provide dc blocking. The ac input signal, v_{in}, is applied to the grid across R_g, which determines the input impedance at low frequency. The output ac signal, v_{out}, is obtained from the anode, and R_L determines the output impedance. When there is no cathode capacitance, C_k, the slightly idealized small signal voltage gain for this amplifier configuration, $A_V = v_{out}/v_{in}$, is essentially determined by the ratio of the load (R_L) and bias (R_k) resistors:

$$A_V = -\frac{R_L}{R_k}. \tag{3.29a}$$

In this case the cathode resistor, R_k, comprises a feedback pathway which controls the gain, just as in comparable solid state circuits—the ac input to the tube is given by $v_{gk} = v_{in} - i_a R_k$, where i_a is the ac component of the total tube anode current. If one adds a sufficiently large bypass capacitance, C_k, in parallel with R_k (such that the impedance

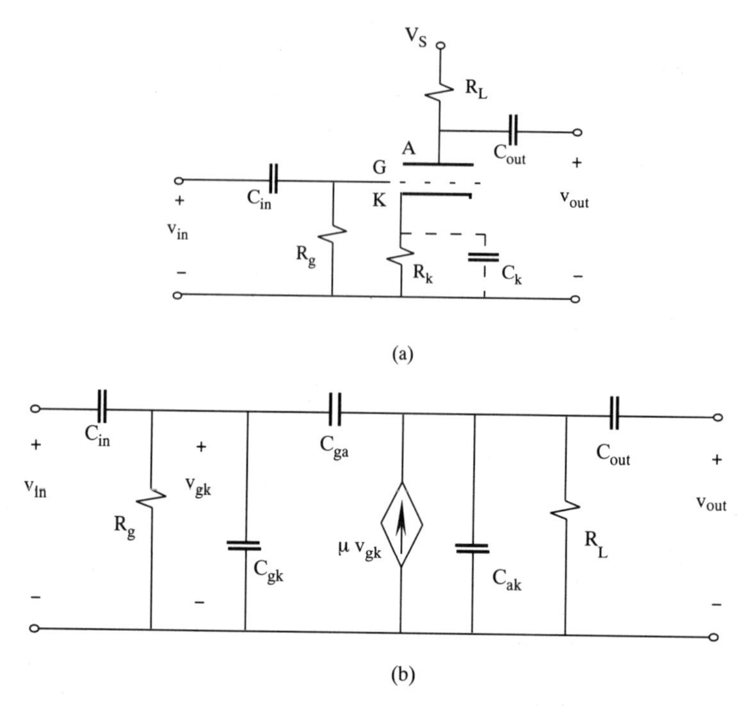

Fig. 3.25 *Vacuum triode common cathode amplifier stage. (a) Common cathode amplifier circuit. (b) High frequency circuit model assuming C_k is large.*

$Z_k \approx Z_{ck} \ll R_k$) the gain is increased ($v_{gk} = v_{in}$) at the expense of stability since the feedback path is eliminated. With C_k in place the small signal gain is:

$$A_V = -\mu R_L \tag{3.29b}$$

where μ is the transconductance, a vacuum tube characteristic. For most ordinary vacuum tubes the transconductance, μ, is in the neighborhood of 20 mS, but this value may vary widely.

Referring to the high frequency circuit model of 3.25b it may be seen that the phase shift in the ac plate voltage (equations (3.29)) gives a high effective input capacitance, C_{eq}, when the voltage gain ($A_V < 0$) is of large magnitude:

$$C_{eq} = C_{gk} + (1 - A_V)C_{ga} \tag{3.30}$$

At high frequencies the effective grid–anode capacitance, C_{ga}, may dominate all other impedances and reduce the gain. The effect of C_{ga} may be suppressed by a so-called screen grid located near the anode and biased with dc potential at or just slightly below the anode potential and bypassed with a large capacitor to the cathode. Such a tube is called a "tetrode" since it has four elements. The screen is a very porous shield which gives the primary acceleration to the cathode electrons. The high porosity means that the vast majority of electrons pass through without striking the screen and impinge on the anode. The ac ground (through the bypass capacitor) reduces the effect of C_{ga} by putting it in parallel with C_{gk} (and thus it is not multiplied by the gain). The effect of C_{ga} on the gain can also be neutralized at single frequencies (for example in IF amplifier stages) by placing a conjugate impedance (an appropriately sized inductor) in parallel, thus tuning the response for a single frequency. It turns out that grid–anode capacitance is an extremely important RF design consideration.

3.2.2 Example RF oscillator circuits

At RF operating frequencies we must include the tube element capacitances in the circuit model — in fact, an RF oscillator design often depends on these capacitances to initiate and sustain oscillation. In contrast to the simple amplifier case discussed above, the network connected to the grid (including C_{ga}) forms the feedback circuit. Oscillator designs take advantage of this relationship to influence the operating frequency and its stability. We will very briefly present a few of the common oscillator types in order to introduce the governing concepts. Of course, many additional types of oscillator circuits have been devised which we do not have the space to discuss.

Continuous wave oscillators are composed of four basic circuit elements: (1) a kinetic energy storage device, (2) a potential energy storage device, (3) a switching device to add energy to the kinetic or potential modes, and (4) a power source to provide energy to replace that dissipated during oscillation. Electrical oscillator circuits invariably use an inductor for kinetic energy storage (in the magnetic field) and a parallel capacitor for potential energy storage (in the electric field) which comprise the "tank" circuit. The loaded tank circuit Q-factor (quality factor) is that for a parallel R-L-C circuit (see equation (3.31) below), where R represents the losses in the load. Remember that the individual reactive elements are self-resonant and have their own Q-factors, as was previously discussed. The switching modes vary from design to design.

Power oscillator circuits

Power oscillators are directly coupled to the load so that the load is, in fact, part of the tank circuit. Consequently, the Q of the tank varies as the load changes, by drying for example, and the operating frequency, f_r, is also affected since C changes. These circuits are very efficient since transmission and coupling losses are minimized: however, the operating frequency varies considerably with changes in load and applicator configuration.

The spark gap oscillator The simplest (and historically the first) RF oscillator was a spark gap oscillator. The original form of radio telegraphy employed these oscillators. This is why the radio operator on board ship was called "Sparks" by shipmates. This circuit is of more than passing interest, however, since it succinctly summarizes oscillator circuit function and is rugged, dependable, and still in use in commercial electrosurgical generators. The basic circuit is shown in Figure 3.26a where RFC is an RF choke to keep the RF signal off of the mains, SG is the spark gap, L and C the circuit elements and R a simplified model of the oscillator load. On each of the power line half-cycles the spark gap is initially not conducting — point A on the spark gap V–I curve in Figure 3.26b. As the voltage rises the capacitor charges and the gap current remains negligible until the switch-over voltage, V_s, is reached. At this point (B) the electric field in the gap exceeds the breakdown field in air, the gap ionizes and becomes conductive discharging the capacitor through the inductor and load, R. The gap then enters a negative resistance region (C) where potential decreases as the current increases. A stable operating point (D) is reached with a small gap forward

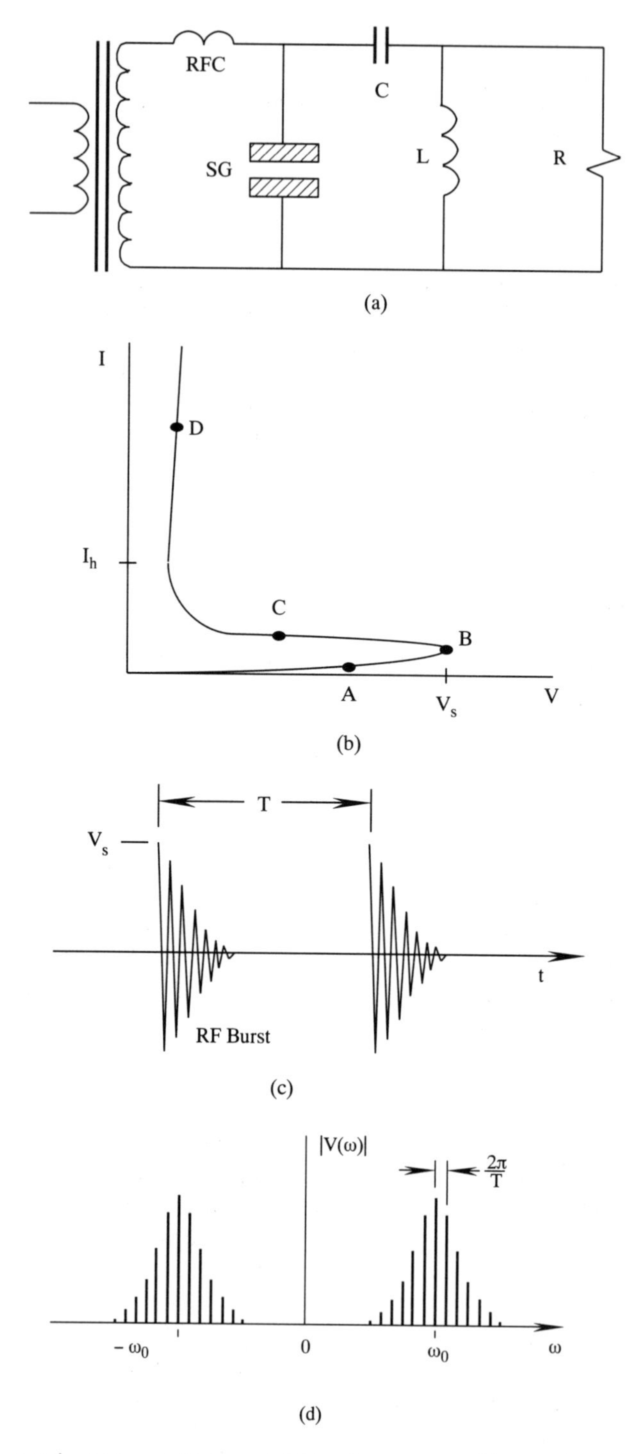

Fig. 3.26 *The spark gap oscillator. (a) Spark gap oscillator circuit. (b) Spark gap voltage–current diagram. (c) Output waveform, v(t). (d) Frequency spectrum of output, v(t).*

voltage and gap current, I_D, determined by the circuit connected to the gap. As long as the gap current exceeds the holding current, I_h, it remains switched on and energy exchange between the inductor and capacitor (oscillation) continues. When the current falls below I_h the gap switches off and oscillation ceases.

The switching occurs a minimum of 120 times per second on 60 Hz mains (100 on 50 Hz mains) and results in an interrupted exponentially-damped burst of RF as shown in Figure 3.26c. The duty cycle (on-time to total time ratio) of the waveform depends on the load, R (in this example). In this simple circuit when the gap is conducting the oscillator–load combination is a simple parallel R-L-C network (with each element far below its self-resonant frequency, f_r) and has a Q-factor of:

$$Q = \omega_r RC = \frac{R}{\omega_r L} = \frac{R}{2\pi f_r L} = R\sqrt{\frac{C}{L}}. \tag{3.31}$$

Obviously, if the load is a complex network rather than a simple resistor the frequency of operation (and Q-factor) depend critically upon it.

In addition to poor stability, this oscillator has an extremely wide frequency spectrum owing to two effects. First, the pulsed damped exponential waveform has a bell-shaped spectrum with width inversely dependent on the damping constant, α:

$$V(\omega) = \frac{2\pi V_s}{T}\frac{\alpha + j\omega}{(\alpha + j\omega)^2 + (\omega_0)^2} * \sum_{n=-\infty}^{\infty} \delta\left(\omega - \frac{2\pi n}{T}\right) \tag{3.32a}$$

where $V(\omega)$ is the Fourier transform of $v(t) = \Sigma V_s e^{-\alpha t}\cos(\omega_0 t)\delta(t-nT)$, the star symbol ($*$) denotes the convolution operation, $1/T$ is the RF burst repeat frequency (100 or 120 Hz) and ω_0 the angular oscillation frequency (see Figure 3.26d and note that $\omega_0 = \sqrt{\omega_r^2 - \alpha^2}$). The convolution operation results in spectral energy impulses at each repeat frequency interval (the frequency domain impulse train) with the overall spectrum envelope width determined by the damping (large α gives a wide spectrum, small α a very narrow spectrum). Second, the wide spectrum inherent in the impulsive nature of the spark gap itself ($v_{SG}(t) \approx \delta(t)$) spreads approximately white noise throughout the band:

$$V_{SG}(\omega) = 1. \tag{3.32b}$$

So, while an important type of RF generator, the spark gap device has too low a duty cycle and too wide a frequency spectrum to be practical for industrial heating systems.

The Hartley oscillator The modified Hartley oscillator in Figure 3.27a uses two triodes (V1 and V2) each oscillating on alternate half-cycles of the power line. The operation is thus as for a class B circuit with respect to the power line frequency, but is class C with respect to the RF frequency for reasons which will be discussed later. Here the grid feedback is tapped off the opposing side of the output inductor (which is 180° out of phase with the oscillating side) to give the positive feedback necessary to initiate and sustain oscillation. The resulting output waveform (Figure 3.27b) has an envelope determined by the mains frequency ($T = 1/2f$) since no filtering is used on the power supply. The oscillation frequency is determined primarily by L and C since the power transformer is isolated by the radio frequency chokes (RFC).

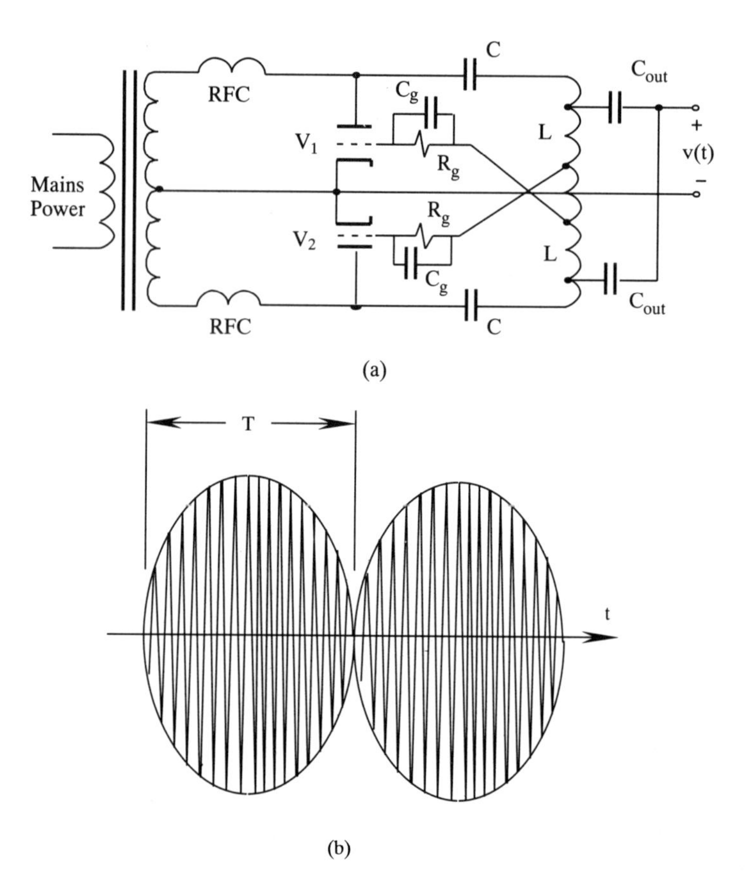

Fig. 3.27 *Modified Hartley oscillator. (a) Circuit. (b) Output waveform.*

The tuned-plate tuned-grid oscillator† A tuned-plate anode tuned-grid oscillator makes use of resonance in both the grid and plate circuits to obtain high efficiency. A typical two-triode design is illustrated in Figure 3.28a. The primary winding for the heater voltage, V_h, is not shown for clarity in the diagram. Here the tubes, V1 and V2, are connected in a class B configuration. In a typical high power RF tube the heater wires form the cathode, as shown here. The dc power supply (V_S) is connected to the center tap of a balanced tank inductor and differential capacitor, C (RF chokes are not shown). The tube anodes, connected to opposing ends of the inductor, conduct on alternate half-cycles of the oscillation waveform (at RF frequency rather than mains frequency), which is the essence of class B operation. The grids are connected in a similar fashion through an R-L circuit to ground to complete the feedback branch. Note that feedback is through the parasitic tube capacitance, C_{ga} in this case. The cathodes are grounded through the heater supply transformer—however, in practical applications RF bypass capacitors are usually connected from each heater wire to the ground plane. The output capacitors actually participate more in impedance matching than in oscillation, but are shown for completeness. This output stage is differential—neither

† Notes on the operation of the class C oscillator were kindly provided by John Zimmerly of PSC Inc. and made use of in assembling this discussion.

terminal is directly connected to the ground plane. A balanced (differential) output is not uncommon in commercial RF generators.

In Figure 3.28b a high power single-tube class C oscillator design is shown. This is by far the most common RF power oscillator design and is usually a single-ended (grounded) output configuration. The RF choke and dc blocking capacitors are shown, but the heater supply voltage (V_h) is not, for clarity. Here a second series L-C tuned circuit connected to the grid is used to complete the feedback pathway and optimize tube efficiency. The grid–anode capacitance is the major source of feedback signal to initiate oscillation. The adjustable grid capacitance C_g (usually a vacuum capacitor) is tuned to near-resonance with the anode $L - C$ oscillating tank to sustain oscillation.

Figure 3.28c illustrates the instantaneous grid and anode voltages and currents. The sequence of operation is as follows. The anode and grid are 180° out of phase. During the positive anode swing (above the dc power supply voltage, V_S) the grid is very negative and the tube is cutoff (the voltage swing is driven by stored energy in the inductor). During the negative swing in anode voltage, $v_a(t)$, the grid voltage is positive for a short time which results in a pulse of grid and anode current, i_g and i_a respectively. This pulse adds energy to the capacitor which makes up for energy dissipated in the load. A low resistance load will result in a longer current pulse (i.e. a large conduction angle because the amplitude of the grid voltage swing will be higher) and a high resistance load results in a shorter pulse, or small conduction angle. The conduction angle is always less than 180° and this is why the oscillator is in class C operation. The net result is that the tank capacitor voltage swing is about twice the dc supply voltage, V_S, as much as 25–30 kV in typical oscillators of this type. A more complete practical design is shown in Figure 3.28d where the grid and anode current meters have been added. Note that the anode supply current multiplied by the supply dc voltage, V_S, is a reasonably good estimate of power delivered to the load (actually to all of the loss mechanisms, load plus circuit) since it is a measure of the make-up current required to sustain oscillation.

The frequency of operation is not affected by the grid capacitor, only by the load, $f_0 \approx 1/\{2\pi\sqrt{LC}\}$. For example, in our 8 kW laboratory RF system of this type (in the US) when we use stray field rod electrodes (reasonably high impedance, lower C) we may lock in (resonate) at about 31 MHz, while comparably sized flat plate electrodes on the same load (relatively low impedance, higher C) resonate at around 24 MHz. The grid capacitance must be carefully adjusted to prevent less efficient oscillation at the second (or higher) harmonic frequency. Needless to say, careful attention to shielding is imperative here.

By way of interest, the physical dimensions of many of the elements in this generator are of critical importance to their function in the circuit. For example, the tank inductor is a 7/8 turn square-like coil approximately 40 cm on a side and made of 5.1 cm (2 inch) copper pipe. The RF choke coil is reasonably closely-wound (about 1.8 turns per cm) of 4.8 mm (3/16 inch) copper tubing on a 1.5 cm radius and 20 cm long. Equation (3.18) (idealized) predicts an inductance of 23.7 μH and equation (3.19) gives 22.4 μH. The grid inductor is likewise made of 4.8 mm (3/16 inch) copper tubing on a 2 cm radius and 5 cm long (3.5 turns), for which we obtain 369 nH by equation (3.18) and, more accurately, (296) nH by equation (3.19) (see also Figure 3.18b).

High power vacuum tubes are designed with careful attention to anode cooling. The average tube currents are quite high and the electrons striking the anode have very high energy (they have been accelerated to high velocity in the anode electric field) so high thermal fluxes result at the anode — much higher than at the cathode, which has to be heated

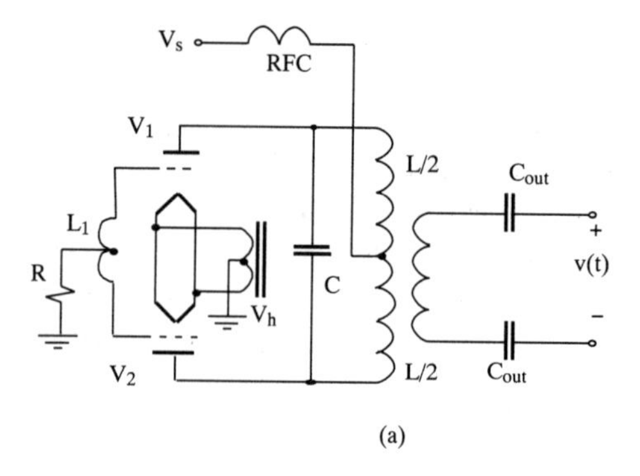

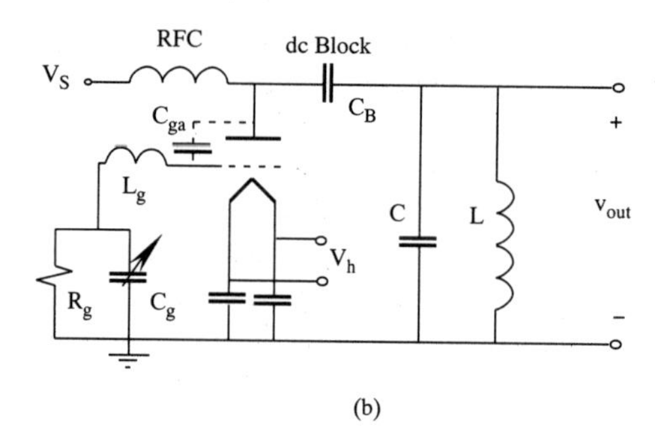

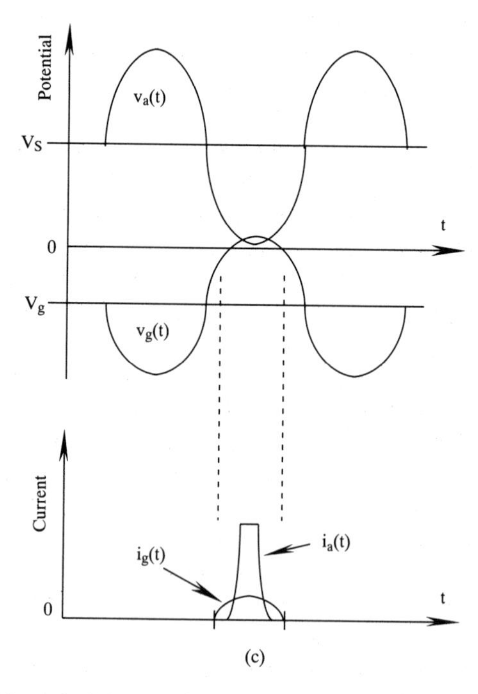

Fig. 3.28 *Examples of tuned-plate tuned-grid oscillators. (a) Two-triode design. (b) Simple single-triode design. (c) Instantaneous voltages and currents for a class C oscillator. (d) Practical oscillator circuit with anode and grid current measurement.*

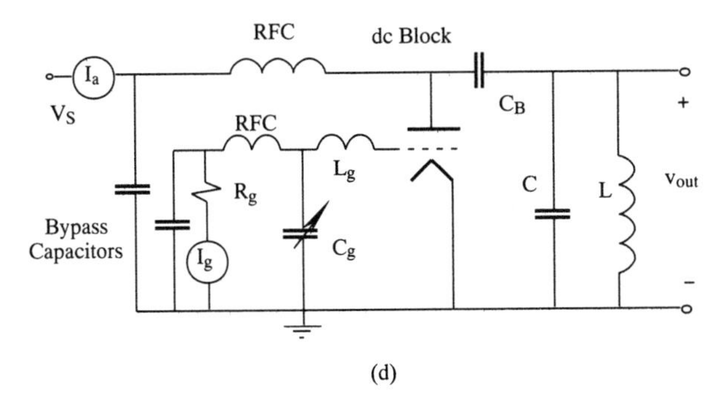

Fig. 3.28 *(continued)*

anyway. Moderate power RF tubes use large thick carbon anodes. There is no convection inside the vacuum envelope, so all of the cooling is provided by conduction heat transfer through the anode connection. High power tubes have metal and ceramic envelopes and carefully designed anode cooling arrangements. Air cooled vacuum tubes often employ a large ring of cooling fins and, as a result, look more like aircraft turbine engines than they do electronic devices. Water cooled anodes are more compact (because of the high specific heat and convection coefficients of water) and must be used for the highest power generators.

Controlled frequency oscillators

The frequency drift inherent in power oscillator circuits is often unacceptable. One approach to confining the frequency drift is to decouple the oscillating and load circuits. That is, a separate tank circuit is used which includes the load. This approach separates the anode tank, vacuum tube capacitance, C_{ga}, and grid network from the variable load, as shown in Figure 3.29a. Here the load capacitance, C_2, is complex because of losses in it. Power is coupled to the load through the mutual inductance between L_1 and L_2. In this configuration the frequency is not as sensitive to load variations. The load tank may be tuned by making L_2 adjustable (for example by using a "trombone" type loop, as in Figure 3.29b) or by adding an adjustable capacitance in parallel with the load. In many cases the load tank will be a fairly low-Q parallel $R - L - C$ network, so tuning in response to load changes may not even be necessary. Balanced (differential) output is common in inductively coupled oscillators.

When constant frequency is essential the oscillator may be used to drive a power amplifier to further isolate it from load variations. A simple single-triode low power oscillator circuit with adjustable frequency may be obtained with the same circuit as in Figure 3.28b if the load is not part of the tank (i.e. C is real) and an adjustable capacitor is placed in parallel with the tank capacitance, C, to provide frequency trim. The oscillator signal is usually inductively coupled to the power amplifier stage which is connected to the load. Because all of the circuit elements are fixed and the oscillator is well-isolated frequency drift is negligible. Also, we may quickly see that because the primary oscillator tank circuit is not connected to a lossy load its Q is much higher than that of a power oscillator (in equation (3.31) R is large so Q is large), virtually eliminating harmonic distortion. The reader should note at this point that the current between the tank inductor and capacitor, L_1 and C_1, is approximately Q times the vacuum tube anode current and may be very

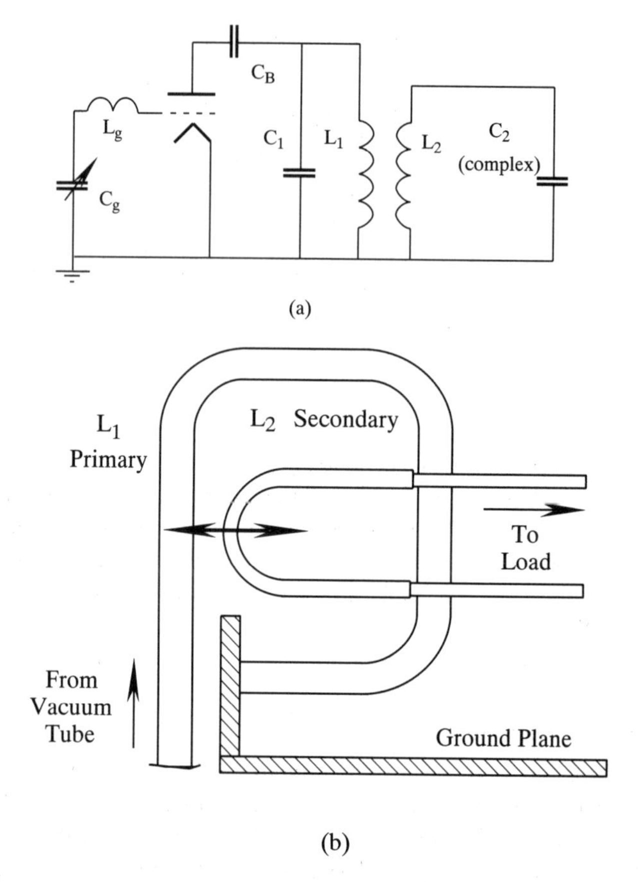

Fig. 3.29 *Auxiliary load tank circuit. (a) Load tank circuit separate from oscillator tank. (b) "Trombone" adjustable transformer.*

large indeed. The low power oscillator with power amplifier system is often used, and a typical power amplifier is described in the next section. The frequency of operation may be controlled through a motor-driven adjustment capacitor connected to a frequency-measuring feedback network, if necessary.

3.2.3 Example power amplifier circuit

Power amplifiers are also tuned to the frequency of operation to optimize their output power and minimize harmonic distortion. A typical design is shown in Figure 3.30 in simplified form. In this design the signal is inductively coupled into the grid inductance, L_g, which is part of a parallel tuned-input tank with C_g. The output power is adjusted by varying the coupling due to the mutual inductance between L_1 in the power stage and L_2 in the output matching stage (shown in Figure 3.31). In Figure 3.30 the large currents expected in the L_1–C_1 tank are emphasized by using heavy lines. The dc tube supply is connected to the center tap of L_1, and so C_1 is split into two parts. The C_1 halves must be carefully matched, so adjustable capacitors are in parallel with them (not shown in the drawing). The output stage is differential with adjustable load tuning in this circuit (Figure 3.31). The load impedance is intended to be primarily capacitive (but lossy).

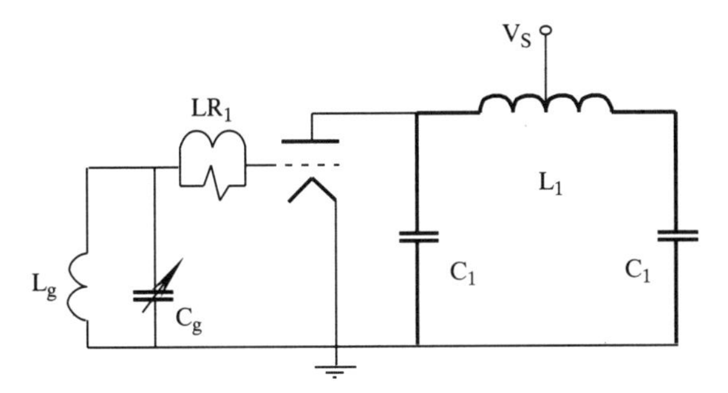

Fig. 3.30 *Example RF power amplifier circuit.*

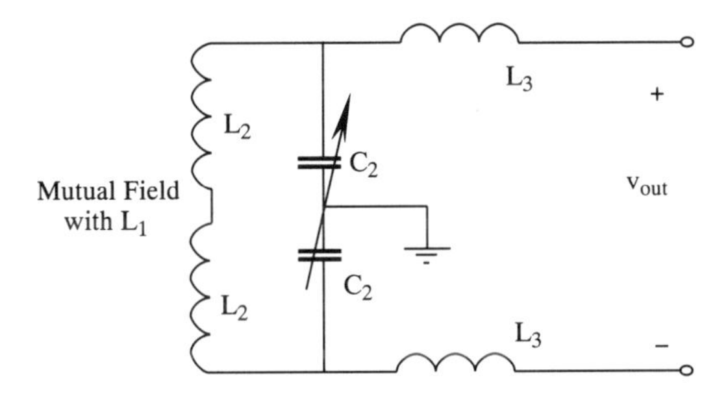

Fig. 3.31 *Output-matching network for power amplifier in Figure 3.30.*

3.3 RADIO FREQUENCY IMPEDANCE-MATCHING STRATEGIES

We have already looked very briefly at one example of a load-matching network, that of Figure 3.31. This network is an example of an approximately L-section network as shown in Figure 3.32 (the differential network may be considered as two single-ended networks). L-section tuning may be used to match loads, but more flexibility can be obtained from T-section and Π-section matching networks such as those in Figure 3.33. We begin with a short introduction to L-section matching and conclude by discussing Π-section and T-section networks for matching complex loads to real transmission lines.

By using a matching network we can make a complex load appear to match the transmission line and generator. One advantage is that by doing this we may reduce the VSWR on the generator–load connection line to near 1.0. If the T-L VSWR is low we can use simple and relatively inexpensive sensors to measure the forward and reflected power for the

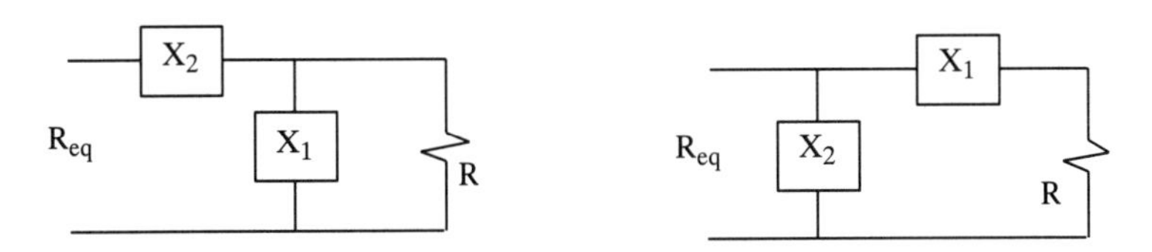

Fig. 3.32 *L-section impedance-matching networks. (a) Downward and (b) upward transformation of R.*

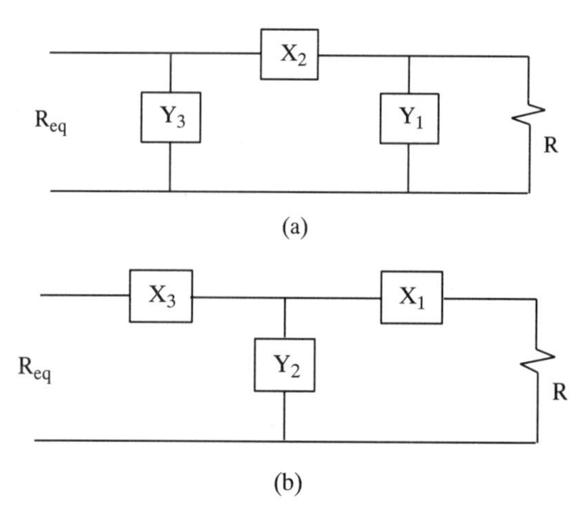

Fig. 3.33 *(a) Π-section and (b) T-section matching networks.*

RF system. Also, the generator will always be loaded with a known impedance, typically 50 Ω, so interchangeable components and modular design methods may be used. This design paradigm is termed 50 Ω technology and has been explored in depth by researchers at Electricité de France (see Marchand and Meunier 1990). The disadvantages are that a match can only be obtained at one frequency and the match point is usually very sensitive to load changes. Consequently, the frequency of oscillation must be carefully controlled to prevent drift and load changes must be compensated, preferably by feedback control. We begin a brief review of these matching networks by inspecting the L-section matching circuit and conclude with the T-section and Π-section approaches.

3.3.1 L-section matching networks

Referring to Figure 3.32, the reactances which form the matching network are numbered in sequence from the load (R in the initial case) toward the generator. Two forms of matching network are shown in the figure, one for transforming R to a lower value (Figure 3.32a) and one for transforming to a higher value (Figure 3.32b). The first element, X_1, transforms the resistance to the desired value, so Q_1 is the governing design parameter, and the second compensating reactance, X_2, is used to cancel the reactance introduced by X_1. There are two possibilities for the value of Q_1; first, if the transforming reactance is in parallel (Figure 3.32a):

$$Q_1 = \frac{R}{X_1} = \frac{R}{\omega L_1} = \omega R C_1 \tag{3.33a}$$

and second, if the transforming reactance is in series (Figure 3.32b):

$$Q_1 = \frac{X_1}{R} = \frac{\omega L_1}{R} = \frac{1}{\omega R C_1}. \tag{3.33b}$$

The resistance changes by a factor determined by Q_1 so that the equivalent resistance, R_{eq}, for a parallel transforming element is:

$$R_{\text{eq}} = \frac{R}{1 + Q_1^2} \tag{3.34a}$$

and for a series transforming element (after reactance cancellation by the parallel compensating element, X_2):

$$R_{\text{eq}} = R(1 + Q_1^2). \tag{3.34b}$$

The reactance also changes by a factor determined by Q_1 so that the equivalent reactance, X_{eq}, for a parallel transforming element is:

$$X_{\text{eq}} = \frac{X_1}{1 + 1/Q_1^2} \tag{3.35a}$$

and for a series transforming element:

$$X_{\text{eq}} = X_1(1 + 1/Q_1^2). \tag{3.35b}$$

Q_1 is thus the Q-factor of transformation. In the calculations we must have a sign for the transformation Q, that is, Q_1. The sensible rule is that Q is positive when a parallel X_1 is a capacitor or a series X_1 is an inductor, and negative for the opposite pairing.

Selection of the second reactance to obtain a purely resistive equivalent impedance, Z_{eq}, proceeds from two simple rules. First, when the first element is a parallel element:

$$X_2 = -\frac{X_1}{1 + 1/Q_1^2} = -\frac{R_{\text{eq}}}{Q_1}. \tag{3.36a}$$

Second, when the first element is a series element:

$$X_2 = -X_1\left(1 + \frac{1}{Q_1^2}\right) = -\frac{R_{\text{eq}}}{Q_1}. \tag{3.36b}$$

and the sign of Q_1 is given by the rule above. When the transformation Q is large the L-section network frequency response (near f_r) resembles that of a series or parallel resonant circuit, depending on which configuration is used. The Q-factor of the L-section network is approximately half of Q_1 when match is obtained.

An example solution will help visualize the design approach. Suppose at 27 MHz we have a typical RF load which is described by $Z = 1 - j10\ \Omega$ (in series $R = 1\ \Omega$ and $C = 59$ nF), as in the paper by Marchand and Meunier (1990). We choose as the example calculation the matching network with a series X_1. In order to transform the resistance up to 50 Ω we must have $|Q_1| = 7$, according to equation (3.34b). Thus the reactance, X_1 must be ± 7, by equation (3.33b). If the load is complex we may easily include the reactance by lumping it in with X_1. So, at this point we have two possibilities: (1) we may select $Q_1 = -7$ so that $X_1 = \omega L_1 - 1/\omega C = -7\ \Omega$ and $\omega L_1 = 3\ \Omega$, thus $L_1 = 18$ nH; or (2) we may select $Q_1 = +7$ so that $X_1 = \omega L_1 - 1/\omega C = +7\ \Omega$ and $\omega L_1 = 17\ \Omega$, thus $L_1 = 100$ nH. Now from equation (3.36b) X_2 has one of two values: either $X_2 = +7.14\ \Omega$ when $Q = -7$ (i.e. $L_2 = 42$ nH) or $X_2 = -7.14\ \Omega$ when $Q_1 = +7$ (i.e. $C_2 = 826$ pF). Both of these results appear in Figure 5 of the paper by Marchand and Meunier. The calculation for a parallel X_1 element follows the same logic with different arithmetic; and this calculation gives two other possible networks—both have a parallel L_1 first element and one has a series C_2 solution, following the same pattern as above.

While any of the four possible solutions is potentially viable, the variable nature of the typical RF load must be taken into account. Since variable inductors are not common (but could be built) and variable capacitors are commercially available, we should use the solution which puts the capacitor in a position to compensate for load variations. Thus with the L-section network we are not able to easily adjust for changes in the load resistance, R, since in both cases the transforming element must be an inductor. We may only compensate for changes in load reactance using either of the two solutions which require that X_2 be capacitive. We will see that a Π-section or T-section network may be used to compensate for both R and C variations in the load and either would be a much better choice.

3.3.2 Π-section matching networks

Generic Π-section and T-section networks were shown in Figure 3.3. We now assign reactances to series elements and admittances to parallel elements numbering from the load, R, as shown in Figure 3.33. We will begin by inspecting the Π-section network using the same philosophical approach as with the L-sections above. As in the previous section, a complex load may be easily included by lumping its reactive component into the first element.

General observations on three-element matching circuits

In either Π- or T-sections the first two elements are the transforming elements; one causes the resistance to increase and the other causes it to decrease. The third element is the compensating element which is used to cancel the load plus transformation reactance. Because the resistance is transformed twice there are two Q-factors of importance. In design the highest of these, Q_{max}, can be any value in excess of that required for the equivalent L-section network (7 in our previous example). Also, the Q-factor of the entire network will be approximately half of Q_{max} when Q_1 and Q_2 are different, so the network bandwidth is essentially determined by Q_{max}. Since Q_1 and Q_2 are independently selectable (as long as their ratio is preserved) we can tailor the bandwidth of the matching network—however, we should be mindful that the Q_1 and Q_2 values are ultimately limited by the Q-factors of the individual reactive devices used, so in an individual design there may be severe limitations on Q_{max}.

Π-section design example

For the Π-section network the first transformation of R is downward by a factor dependent on Q_1 to give the intermediate resistance R':

$$R' = \frac{R}{1 + Q_1^2} \tag{3.37}$$

and Q_1 has the same definition as before; that is, from equation (3.33a) since $Y_1(= 1/X_1)$ is in parallel. The second transformation raises R' by the analogous factor for a series element:

$$R_{eq} = R'(1 + Q_2^2) = R\frac{(1 + Q_2^2)}{(1 + Q_1^2)}. \tag{3.38}$$

Therefore, one may decrease R by setting $Q_1 > Q_2$ and increase R by setting $Q_2 > Q_1$. If there is no need to transform R (and we might wind up with $Q_1 \approx Q_2$) then the solution is much cheaper using simple reactance cancellation rather than Π-section matching.

The first step in a design is to select the two Q values. In the previous example the load $Z = 1 - j10\ \Omega = 10.05\ \underline{/-84.3^\circ}\ \Omega$ at 27 MHz, so $Y = 0.0995\ \underline{/+84.3^\circ}\ S = 0.0099 + j0.099$ S. We will lump the reactive part of Z in with the first element, Y_1 (parallel), so it will be helpful to recast Z into its equivalent parallel network: $Z = 101\ \Omega \parallel -j10.1\ \Omega$ ($R = 101\ \Omega$ and C = 584 pF) which corresponds to the load admittance $Y = 0.0099 + j0.99$. In the previous L-section circuit Q_1 was 7 since we needed to transform 1 Ω to 50 Ω, a resistance increase. However, in this design we must obtain a resistance decrease (from 101 Ω down to 50 Ω, a ratio of 2.02) so the equivalent L-section design would have a parallel first element. From equation (3.34a) $Q_{max} > 1.04$ is required (and $Q_1 > Q_2$) for this particular Π-section design, which is very easy to obtain. The design equations for a Π-section network which results in a resistance decrease (from Abrie 1985, Chapter 3) are:

$$Q_1 = Q_{max} \approx 2Q \tag{3.39a}$$

$$R_{eq} = R'(1 + Q_2^2) \tag{3.39b}$$

$$R' = \frac{R}{(1 + Q_1^2)} \tag{3.39c}$$

$$Y_1 = \frac{Q_1}{R} \tag{3.39d}$$

$$X_2 = R'(Q_1 + Q_2) \tag{3.39e}$$

$$Y_3 = \frac{Q_2}{R_{eq}}. \tag{3.39f}$$

We may arbitrarily choose Q-factors as long as $Q_{max} > 1.04$ and, for this case, $Q_1 > Q_2$. The ratio in equation (3.38) must be 2.02, so we arbitrarily set $|Q_1| = 20$ ($R' = 0.252\ \Omega$ from equation (3.39c)) and find that $|Q_2| = 14.05$ from equation (3.39b). From this $Y_1 = Q_1/R = \pm 0.198$ S and we remember that Y_1 includes +j0.099 S from the load. In order to obtain two adjustable capacitors in the design we need $Y_1 > 0$. So, for Y_1 (total) = +0.198 S (i.e. $X_1 = -5.05\ \Omega$ and $Q_1 = +20$ since it is a parallel capacitor) we will use an additional capacitance, C_1, of 584 pF (this is in addition to the capacitance in the load itself). Now X_2 comes from equation (3.39e) and can be either negative or positive. If we want X_2 to be an inductor (so that we end up with two adjustable capacitive elements C_1 and C_3), at this point it appears that we may use either sign, $Q_2 = \pm 14.05$, since $Q_1 = +20$. Consequently, X_2 should be either 1.5 Ω or 8.58 Ω, and L_2 is either 8.84 nH or 50.6 nH. However, we see that the design has less freedom than that because since Y_3 must be a capacitor, $Q_2 = +14.05$ ($L_2 = 50.6$ nH); so that $Y_3 = 0.281$ S, or $C_3 = 1.66$ nF (if we had used $Q_2 = -14.05$ then Y_3 would have been an inductor and we would have lost a convenient adjustment). The final design is shown in Figure 3.34b. Having two adjustable reactances means that we can compensate changes in both R and X in the load; and we have selected the various Q's to get two adjustable capacitors.

For completeness, the set of design equations required to obtain a resistance increase, $R_{eq} > R$, is (also from Abrie 1985, Chapter 3):

$$Q_2 = Q_{max} \approx 2Q \tag{3.40a}$$

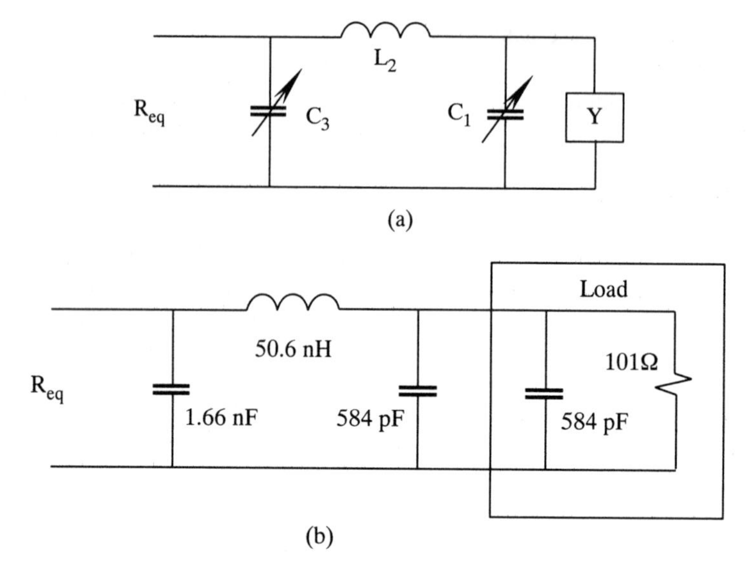

Fig. 3.34 *Π-section matching example (a) Practical Π-section network with two capacitors. (b) Solution to example load-matching network.*

$$R_{eq} = \frac{R'}{(1 + Q_2^2)} \tag{3.40b}$$

$$(1 + Q_1^2) = \frac{R}{R'} \tag{3.40c}$$

$$Y_1 = \frac{Q_1}{R} \tag{3.40d}$$

$$X_2 = R'(Q_1 + Q_2) \tag{3.40e}$$

$$Y_3 = \frac{Q_2}{R_{eq}} \tag{3.40f}$$

and the design approach follows the same method. It is interesting to note that the relations for Y_1, X_2 and Y_3 are the same for both cases.

3.3.3 T-section matching networks

A T-section network is the dual of a Π-section network. The typical practical T-section RF matching network utilizes two adjustable capacitors, as shown in Figure 3.35a. The design procedure and strategy are similar to those of the Π-section matching case. The relevant design relations for increasing the load resistance are:

$$Q_1 = Q_{max} \approx 2Q \tag{3.41a}$$

$$R_{eq} = \frac{R'}{(1 + Q_2^2)} \tag{3.41b}$$

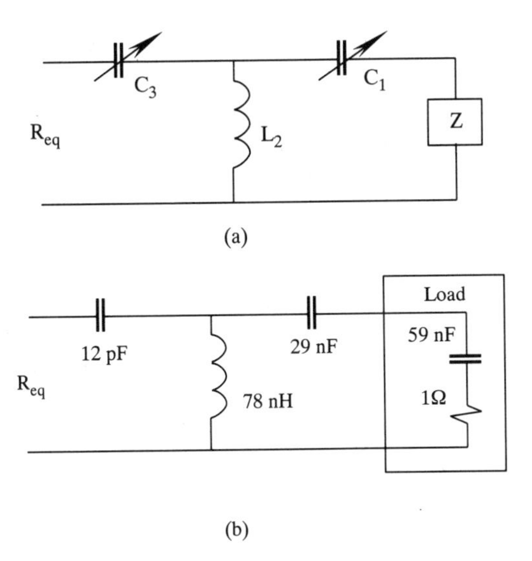

Fig. 3.35 *T-section load-matching example. (a) Practical T-section matching network with two capacitors. (b) Solution to example load matching network.*

$$R' = R(1 + Q_1^2) \tag{3.41c}$$

$$X_1 = Q_1 R \tag{3.41d}$$

$$Y_2 = \frac{(Q_1 + Q_2)}{R'} \tag{3.41e}$$

$$X_3 = Q_2 R' \tag{3.41f}$$

where all parameters are as previously defined. Note that the relations differ from those for the resistance-increasing Π-section design (equations (3.40)) since the T-section is the dual of a Π-section. Incidentally, the analogous relation for equation (3.41a) in Abrie (1985) has a small typographical error which has been corrected here — Q_1 must be Q_{max} in order to get a resistance increase.

The example RF load which we have been working with, $Z = 1 - j10\ \Omega$ at 27 MHz ($Z = 1\ \Omega$ in series with 59 nF), will now be matched using the T-section approach. Because the first element is a series element it is most convenient to retain the series equivalent form for the load. Therefore, the calculated X_1 reactance must include the $-j10\ \Omega$ from the load; and since X_1 will be a capacitor we will have to be careful in the choice of Q_1. We will use the increasing-resistance equations (3.41) since we wish to transform 1 Ω to 50 Ω. Following the previous approach, $|Q_{max}|$ (i.e. $|Q_1|$) must be greater than 7. We arbitrarily choose $|Q_1| = 20$ so that X_1 will be at least twice the load reactance (and we will have some latitude in the adjustable capacitor, C_1). Since the first element is a series element, equation (3.41d) gives X_1 (total) $= Q_1 R = \pm 20\ \Omega$. We want C_1 to be a capacitor so $Q_1 = -20$ and the reactance of C_1 (alone) should be $-j10\ \Omega$, or C_1 = 59 nF. Note that we could obtain a smaller C_1 value by selecting $|Q_1|$ lower. However, the limit is $Q_1 = -10$ since C_1 (alone) would be 0 to get the required $X_1 = -10\ \Omega$, the load reactance — for example, $Q_1 = -15$ means C_1 = 29 nF, etc. Also, for this case any Q_1 less than 10 (but greater than 7) would require an inductor as a part of X_1. So, we will continue the design with $Q_1 = -20$.

By equation (3.41c) the intermediate resistance, R', is 401 Ω. Thus to get $R_{eq} = 50\ \Omega$ we must have $|Q_2| = 2.65$. Inspecting equation (3.41f), we want X_3 to be a capacitor so we will use $Q_2 = -2.65$ and use an inductor to get $Y_2 = -56.5$ mS, or $L_2 = 104$ nH. This will make $X_3 = -1062\ \Omega$ and C_3 will be 5.55 pF, a very small capacitance indeed—in fact, not achievable in an adjustable design. Also, having $C_1 = 59$ nF is stretching the adjustable capacitor range considerably, especially at high voltage. So, our original choice of $Q_1 = -20$ ought to be modified downward in magnitude to, say, −15.

If $Q_1 = -15$ then $C_1 = 29$ nF (still high), but now $R' = 226\ \Omega$ and $Q_2 = -2.12$ (not very much change). From this $Y_2 = -75.8$ mS and $L_2 = 77.8$ nH (not much change) and $X_3 = -479\ \Omega$ thus $C_3 = 12.3$ pF (a significant change since it is now achievable).

For completeness, the T-section design equations for decreasing the load resistance are (see Abrie 1985, Chapter 3):

$$Q_2 = Q_{max} \approx 2Q \tag{3.42a}$$

$$R_{eq} = \frac{R'}{(1 + Q_2^2)} \tag{3.42b}$$

$$R' = R(1 + Q_1^2) \tag{3.42c}$$

$$X_1 = Q_1 R \tag{3.42d}$$

$$Y_2 = \frac{(Q_1 + Q_2)}{R'} \tag{3.42e}$$

$$X_3 = Q_2 R' \tag{3.42f}$$

where all parameters are as previously defined. There is, we believe, a small typographical error in the relation analogous to equation (3.42f) in Abrie's book—X_3 must depend on R', not R_{eq}.

In summarizing the impedance matching design strategy we note that for our "typical" RF load of $Z = 1 - j\ 10\ \Omega$ at 27 MHz we can achieve a good, but relatively inflexible, match with several realistic L-section designs. The Π-section and T-section networks can both compensate for changes in load R and C values, but in this instance the Π-section matches with more realistic values of capacitors. It may be seen that the Π-section and T-section topologies are better suited to different regions of the Smith chart. The Π-section covers capacitive reactance regions with more realistic values for the adjustable capacitors, while the T-section covers the inductive load region more effectively. This may also be seen in the results of Marchand and Meunier (1990) (specifically, in Figure 6 of their paper).

3.4 MICROWAVE GENERATORS

This section represents a very brief overview of the functional principles of magnetron tubes from a practical point of view. We choose not discuss them in detail because several very clear and effective treatments may be found—most especially in Metaxas and Meredith (1983) and in Thuery (1983)—and they have treated this topic far more skillfully than we are able. Also, we will not present other microwave sources such as traveling wave tubes (TWTs), Gunn diodes, Klystrons or MASERs (the microwave equivalent of the laser which, in fact, was invented first!). These devices have low efficiencies or power outputs (too low to be of practical use in industrial systems), poor reliability, high replacement cost, high operating cost, or some combination of these. The magnetron is, to date, the single device which makes industrial microwave applications possible owing to its simplicity, compactness, reliability, low cost and high efficiency. Common commercial high power magnetrons are supplied with a co-axial output stub designed to be used in a TE_{10} wave launcher (discussed in the next section), so we begin with a short description of these devices. The magnetron itself makes use of resonant cavities in generating the waves, which we do not discuss until Chapter 4; this discussion is contained in Subsection 4.3.4, if it is needed.

3.4.1 Magnetron tubes

A pictorial view of the functional aspects of a typical magnetron vacuum tube is given in Figure 3.36. As in other vacuum tubes, the anode is at high potential with respect to the cathode (the electron current source) so cathode electrons are accelerated toward the anode by the electric field. As in other vacuum tube devices, the cathode is heated to high temperatures to boil off the electrons. In almost all magnetrons the anode is near ground potential and the cathode is at a large negative potential. The difference between the magnetron and other vacuum tubes is that the electron stream travels in a spiral; created by an external dc magnetic field, *B*, in the figure, which is upwardly directed for the electron pathway shown. The electron cloud passes resonant cavities (formed by the vanes in the picture) many times on its way to the anode. These resonant cavities act as Helmholtz resonators and set up oscillations at a fixed frequency. The action is the electrical equivalent of blowing across the mouth of a bottle to create an audible sound. Every schoolboy knows that the pitch (i.e. the frequency) of the sound is determined by the dimensions of the cavity — small cavities give higher frequency resonance, large cavities lower frequencies. The actual interior shape of the cavity (Coke bottles, Orangina bottles, Pepsi bottles, etc.) has less influence on the standing wave frequency than its depth.

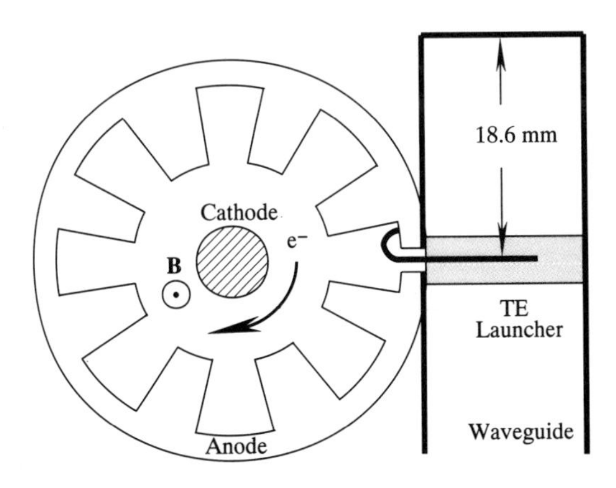

Fig. 3.36 *Functional diagram of a magnetron tube.*

In a magnetron the cavity oscillations are usually collected by a circular loop antenna and fed to the output stub (right hand side of the anode in Figure 3.36). So, the magnetron operating frequency is determined entirely by the dimensions of the resonant cavities. At 2.45 GHz the tubes are necessarily much smaller than their 894 MHz or 915 MHz counterparts. The power output is determined by its operating point: the dc cathode current and external magnetic field strength. One may control the output power by controlling tube current or dc magnet field strength. We have found that by controlling dc magnet current using a high gain feedback circuit we can achieve very stable output powers between about 50 W and 6 kW in our 6 kW magnetron.

The ultimate limit on tube output power is in turn determined by anode temperature limitations. There are several practical limits to the power output of a typical magnetron: (1) at 2.45 GHz air cooled anodes can be used up to about 1.5 kW, (2) the next commonly available step at 2.45 GHz is a 6 kW magnetron which is water cooled to achieve higher power densities, (3) the smallest commercially available magnetron for 915 MHz is 50 kW. Higher power generators are obtained by using multiple magnetrons. As a practical matter, at 2.45 GHz individual power sources are limited to about 25 kW each, while at 915 MHz much higher powers can be obtained in a single unit.

3.4.2 TE_{10} wave launchers

In the TE_{10} mode the electric field has a single maximum located midway across the broad wall of a 2:1 rectangular waveguide at $x = a/2$:

$$E_y = E_0 \sin\left(\frac{\pi x}{a}\right) \mathrm{e}^{-\mathrm{j}\beta z} \tag{3.43a}$$

for which the other wave field components are:

$$H_x = H_0 \sin\left(\frac{\pi x}{a}\right) \mathrm{e}^{-\mathrm{j}\beta z} = E_0 Y_w \sin\left(\frac{\pi x}{a}\right) \mathrm{e}^{-\mathrm{j}\beta z} \tag{3.43b}$$

$$H_z = H_0 \frac{\pi}{\mathrm{j}\beta a} \cos\left(\frac{\pi x}{a}\right) \mathrm{e}^{-\mathrm{j}\beta z} = E_0 Y_w \frac{\pi}{\mathrm{j}\beta a} \cos\left(\frac{\pi x}{a}\right) \mathrm{e}^{-\mathrm{j}\beta z} \tag{3.43c}$$

where we have used the electric field amplitude, E_0, as the reference quantity (in V/m rms), β is the propagation constant, and Y_w is the wave admittance for this mode in the waveguide ($Z_w = 1/Y_w$ is the wave impedance). The relationship between E-field and measured waveguide power depends on the wave admittance and waveguide dimensions, a and b, for the TE_{10} mode:

$$P_{10} = abY_\mathrm{w}[E_0]^2 \tag{3.44a}$$

where the wave impedance for any TE mode is:

$$Z_w = \frac{\eta}{\sqrt{1 - (f_\mathrm{c}/f)^2}}. \tag{3.44b}$$

Remember that the cutoff frequency can be determined from equation (2.47). For a WR284 waveguide, where $a = 7.21$ cm $= 2.84$ inches, it is about 2.08 GHz. Consequently, we expect a wave impedance of about $Z_\mathrm{w} = 377/0.528 = 713\ \Omega$ in WR284 (filled with free space) at 2.45 GHz—the wave impedance for TE modes (in a lossless dielectric) is real and larger than the characteristic impedance of the waveguide dielectric. The wave admittance for WR284 at 2.45 GHz is 1.4 mS.

Single stub TE_{10} launcher

The standard (simplest) TE_{10} launcher consists of a single probe or cylindrical stub in the center of the broad wall (at $x = a/2$) fed from a coaxial source, either a cable or a magnetron probe, as shown in Figure 3.37a. The designer sets the distance, l, from the end shorting wall, the probe length, d, and the probe diameter (although this is not critical) to optimize power transfer from the cable to the waveguide. The junction will have reactive properties from the evanescent modes excited by the probe. The position of the shorting wall is selected to cancel the junction reactance and obtain a good match between the probe and coaxial cable—empirically, 18.6 mm is typical. The probe in the waveguide at $z = +1$ with a shorting wall at $z = 0$ is equivalent to having an image probe located at $z = -1$ in an infinite waveguide, as in Figure 3.37b. The current in the launching probe is:

$$I(y) = I_0 \sin[k_0(d - y)] \tag{3.45}$$

where k_0 describes the standing wave in the coaxial line. Note that the current in the image probe must be 180° out of phase with $I(y)$ so that the tangential field is zero at $z = 0$ (the plane of the short).

It may then be shown (see Collin 1966) that the electric field magnitude and total radiated power are given by:

$$E_0 = \frac{I_0 Z_w}{abk_0}(e^{-2j\beta l} - 1)(1 - \cos k_0 d) \tag{3.46a}$$

$$P = \frac{I_0^2 Z_w}{4abk_0^2}|e^{-2j\beta l} - 1|^2(1 - \cos k_0 d)^2. \tag{3.46b}$$

Define the input impedance to the probe at its base (where $I = I_0 \sin k_0 d$) as Z_{in}, then:

$$Z_{in} = R_0 + jX = \frac{P + 2j\omega(W_m - W_e)}{\frac{1}{2}II^*} \tag{3.47}$$

where P is the radiated power (as in equation (3.46b)), W_m and W_e signify the magnetic and electric field evanescent mode stored energy, and I is the probe base current. We may use equations (3.46) to find R_0:

$$R_0 = \frac{2P}{I_0^2 \sin^2(k_0 d)} = \frac{Z_w}{2abk_0^2}|1 - e^{-2j\beta 1}|^2 \tan^2\left(\frac{k_0 d}{2}\right). \tag{3.48}$$

This is the radiation resistance of the probe. By changing l and d we may adjust R_0 to match the characteristic impedance of the coaxial line to achieve optimum power coupling. The reactance, X, must be cancelled by selection of l alone. The relations for X in terms of ω, a, b, l, and d are very intricate and may be found in Collin (1991). In practical wave launchers at 2.45 GHz in a WR284 waveguide l is usually 18.6 mm, reasonably near $\lambda_g/8$ or so. Note that the TE wave impedance of about 713 Ω must be transformed downwards to about 50 Ω to match a typical coaxial line.

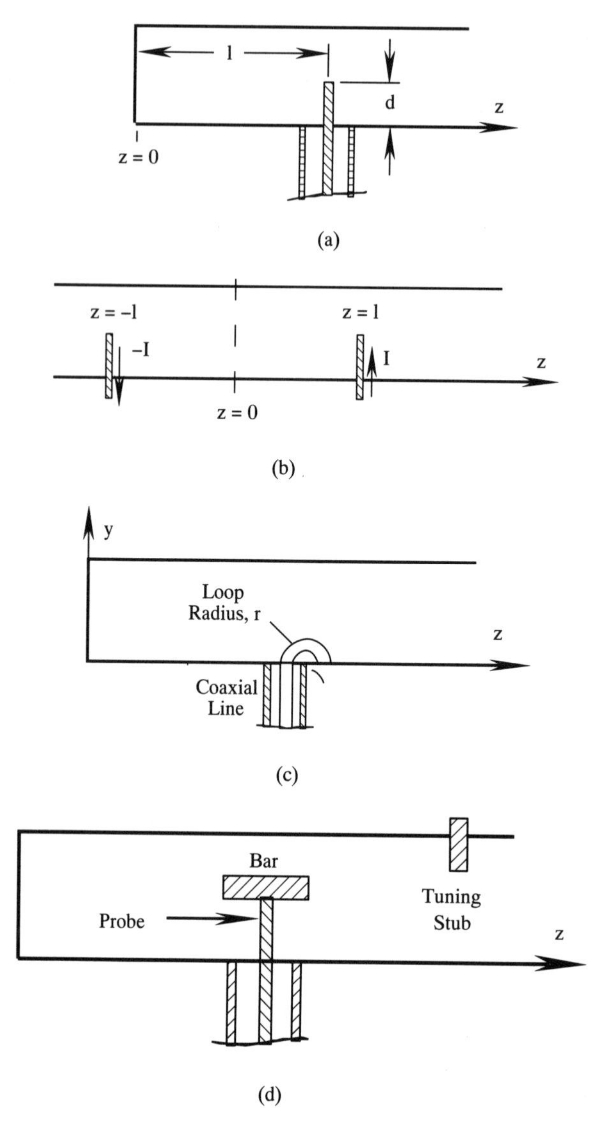

Fig. 3.37 *Typical TE_{10} wave launchers. (a) Simple probe TE_{10} wave launcher. (b) Infinite waveguide equivalent with image probe at $z = -l$. (c) Current loop TE_{10} launcher. (d) Probe plus horizontal bar design.*

Launching with a current loop

We may also launch TE_{10} waves with a current loop instead of a straight probe, as illustrated in Figure 3.37c. If the area of the loop ($S_0 = \pi r^2/2$, r = loop radius) is small enough that the magnetic field for our mode is approximately constant over it (the usual case for a launcher) then the loop acts as a magnetic dipole. The dipole moment, $\mathbf{M}$, has magnitude IS_0 and points in the direction given by the right hand rule. In our case TE_{10} has $\mathbf{H}_{10} = \mathbf{H}_x\mathbf{a}_x$ at the loop site since $\mathbf{H}_z = 0$ at $x = a/2$. The vector dot product, $\mathbf{H}_{10}\cdot\mathbf{M}$, should be maximized, so the loop is oriented in the Y–Z plane (as shown in the figure) to launch TE_{10} propagation. The power coupled into the waveguide is given by $\omega\mathbf{H}_{10}\cdot\mathbf{M}$.

Alternate probe with bar launcher

An alternative launching scheme which utilizes a probe attached to a horizontal rectangular bar (Figure 3.37d) is commercially available (from Gerling Laboratories, and perhaps others). Here the remaining reactance is canceled by a stub tuner near the output plane.

3.5 MICROWAVE IMPEDANCE-MATCHING STRATEGIES

Impedance matching in microwave circuits is very similar in philosophy to RF impedance matching with one notable exception: we can accomplish complex impedance transformations with a single simple device, the multiple stub tuner. Consequently, gross mismatches, even at high power, can be compensated with a single series of adjustments on a standard off-the-shelf unit rather than requiring specially designed networks which have a load-dependent topology. We begin with a brief phenomenological description of matched loads, continue with stub tuners, and conclude the chapter with multiport circuit elements.

3.5.1 Matched loads for waveguides

A matched load is an essential element for assembling microwave circuits which propagate traveling waves in the absence of standing waves (i.e. without reflections from the end of the waveguide). The required devices are reasonably simple to make for TE_{10} mode propagation if one pays attention to a few rules. Since we are operating at elevated power levels the matched loads must be able to absorb all of the waveguide power. This requirement means that the practical matched loads contain flowing water. Water is usually plentiful and a very effective microwave absorber, so total absorption is not too difficult to achieve. As a bonus, the flowing water gives the user a very inexpensive calorimetric measure of waveguide power, as described in Chapter 5.

Wedge loads

Perhaps the most fundamental matched load for high power use is the horizontal lossy wedge, as shown in Figure 3.38a. The configuration is essentially a lossy tapered waveguide. In the TE_{10} mode the electric field is normal to the inclined dielectric plane and is very small in the water to satisfy the normal E-field boundary condition — the magnitude ratio is about 9:1 for de-ionized water. Baffles are used to distribute the water flow evenly and to support the dielectric plane. The flow direction should be against gravity to ensure complete filling of the absorber. The reflection coefficient is minimal when the taper angle, θ, is small. An overall length of one or two λ_g (about 25 to 50 cm in WR284, shorter in WR340 and WR430) is sufficient to obtain a VSWR less than 1.01.

The second type of wedge, a simple slab of resistive material, shown in Figure 3.38b is only useful for very low power applications. There is just not a large enough volume to accommodate flowing water for heat removal.

The wedge may just as well be oriented so that the electric field is tangential to the dielectric interface, as illustrated in Figure 3.33c. The vertical wedge configuration is easier to construct and contains a larger volume of water than the horizontal wedge of Figure 3.38a. Again, we would prefer a shallow wedge angle in order to minimize the VSWR. Some scattering will probably be generated by the leading corner and a tuning stub may be needed (as discussed in a subsequent section).

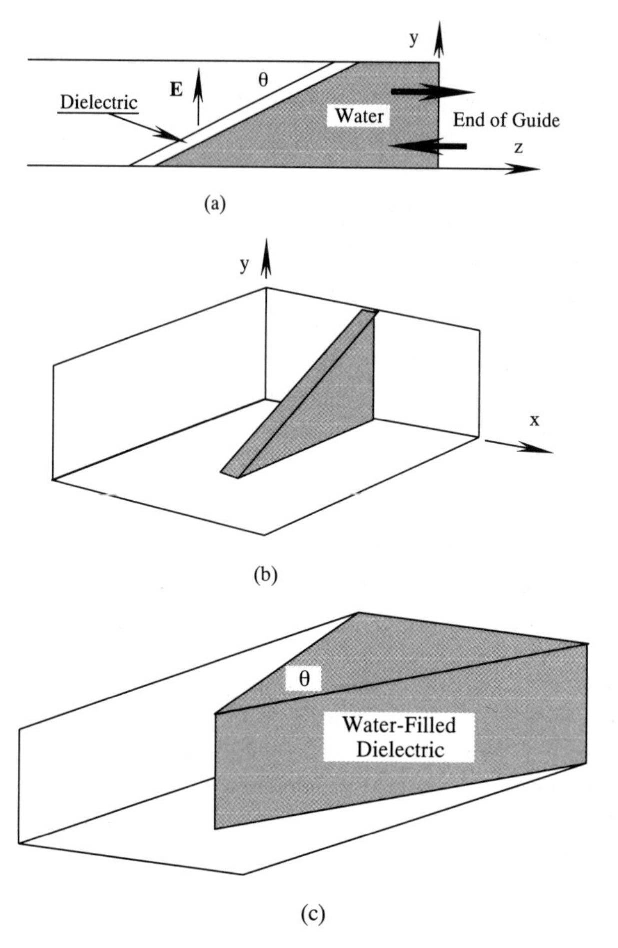

Fig. 3.38 *Wedge-shaped matched load designs. (a) Horizontal wedge matched load. (b) Resistive slab wedge absorber. (c) Vertical wedge matched load.*

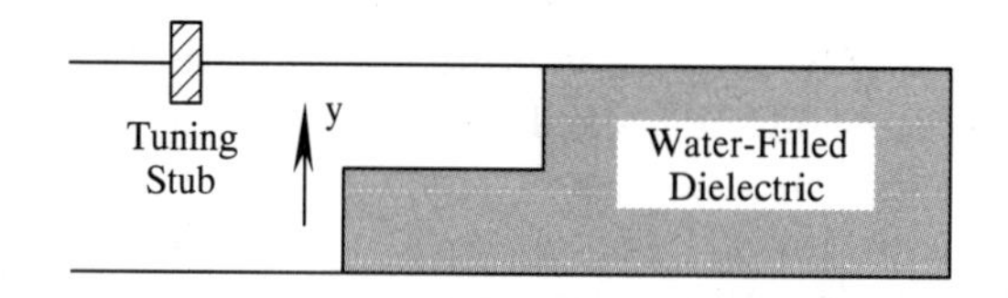

Fig. 3.39 *Horizontal-step matched waveguide load.*

Step loads

The absorber can also be constructed of a lossy dielectric step, as in Figure 3.39. Here the reactance of the sharp corner is, once again, compensated by a tuning stub near the inlet plane. This kind of matched load may be made much shorter than a wedge load and is thus very attractive when space is limited.

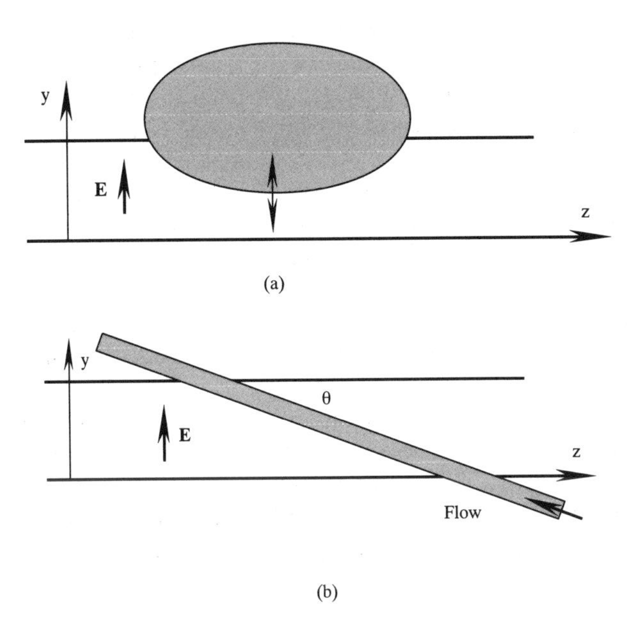

Fig. 3.40 *Attenuator designs (a) Thin resistive slab as a variable attenuator inserted through the broad wall at $x = a/2$. (b) High power attenuator obtained by flowing lossy dielectric fluids through the tube penetrating the broad wall at $x = a/2$.*

Attenuators

Attenuators can be constructed by inserting a thin resistive object in the middle of the guide. The most efficient shape is a rounded resistive card in order to reduce reflection, as in Figure 3.40a. The resistive slab may be inserted through a non-radiating slot in the center of the broad wall (at $x = a/2$) to varying depths to achieve variable attenuation at low power. High power attenuators can be constructed by inserting a dielectric tube through the center of the broad wall (again at $x = a/2$), inclined along the principal axis at a shallow angle of about 10°, through which water can flow, as in Figure 3.40b. The actual angle shown in the sketch is 20°, to give some sense of perspective. This angle will reduce the VSWR but necessitates a long attenuator section. Variable attenuation may be obtained by changing the permittivity of the fluid. A heat exchanger (to air or to water depending on power absorption) can be used to close the primary fluid loop. As in the matched loads above, fluid flow should be directed against gravity to ensure complete filling of the tube.

3.5.2 Adjustable short circuits

In many applications, particularly in low loss materials, it is necessary to establish a standing wave pattern in order to obtain adequate heating. By terminating the waveguide in a short circuit (i.e. a conducting plane) we realize $\Gamma = -1$ and the reflected field gives an E-field minimum and H-field maximum at the location of the short, $z = L$, and at every multiple of $\lambda_g/2$ in front of the short, as we saw in Chapter 2. There is also an E-field maximum at $z = L - (n + 0.5)\lambda_g/2$ for each value of n, as has been mentioned. We wish to place the

E-field maximum at the location of the sample to be heated, and an adjustable short will do this. The electric field is $2E_0$ at the location of the maximum which results in four times the wave power (four times the heating rate) at this point for the same magnetron power.

This type of applicator will be treated in more detail in Chapter 4. Our purpose here is to discuss the design of the short circuit. We should note at this point that the presence of the sample will generate evanescent modes resulting in an apparent phase shift in the wave fields which we will not be able to anticipate, in general. So, an adjustable sliding short is essential to obtain enhanced heating at the sample since we will not be able to place the short *a priori* except for extremely small samples of the lowest electric permittivity materials. In fact, we may use this property to measure the electric permittivity by measuring the shift in location of the E-field maximum caused by the sample. This technique will be discussed in Chapter 5.

The obvious approach is to insert a sliding metallic block into the waveguide, as shown in Figure 3.41a. This design exhibits very erratic behavior because of the high currents in the waveguide walls at the short. That is, we will have very large y-direction surface currents in order to match H-field boundary conditions (remember that the short is an H-field maximum) which will be severely disrupted by variations in wall contact as the short is moved.

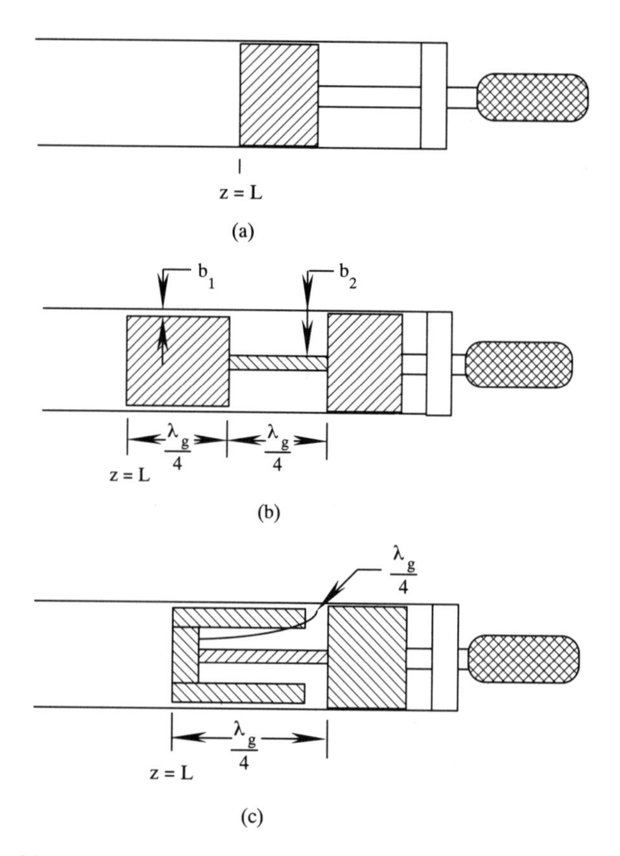

Fig. 3.41 *Adjustable short circuit designs. (a) Adjustable sliding block short circuit. (b) Multiple quarter-wave section design. (c) Quarter-wave transformer section design.*

There are two essentially equivalent choke-type geometries which overcome this limitation. The first is shown in Figure 3.41b. Here the plunger is formed in three sections, a lead section which is $\lambda_g/4$ long with small non-zero gaps $\Delta y = b_1$ and $\Delta x = a_1$ (not shown in the cross section) so that it does not touch the walls. The next section is also $\lambda_g/4$ long, and has a gap b_2 which must be as large as possible without sacrificing dimensional stability for the lead section. The a_1 and a_2 gaps are of no real consequence. The final section is dimensioned to achieve a close sliding fit on the waveguide walls. We will still experience some high impedance points as the final section slides along the walls, but the overall apparent impedance will be much less affected. That is, the characteristic impedances of the $\lambda_g/4$ sections are improved by ratios of $2b_1/b$ and $2b_2/b$ relative to the input guide. Consequently, the apparent impedance of the short, Z_{eq}, will be improved over that of the final sliding section, Z_{s}, by the factor:

$$Z_{\text{eq}} = \left(\frac{b_1}{b_2}\right)^2 Z_{\text{s}}. \tag{3.49}$$

By careful attention to mechanical design we may achieve b_i ratios of 10 or more ($b_2 \gg b_1$) so the apparent impedance of the sliding short is much reduced.

The second approach is a choke plunger, as in Figure 3.41c. This design is frequently used owing to its short length. The lead section is a folded quarter-wave transformer. The inner section transforms the second (sliding) plunger impedance from a short in its plane (at $z = L + \lambda_g/4$) to an open circuit. The axial current is zero at the sliding junctions, and intermittent wall contact does not affect the apparent impedance of the section very much. The outer $\lambda_g/4$ section transforms the open into a short at $z = L$. This is a very effective design and is commonly used in commercial adjustable short circuit devices.

3.5.3 Stub tuners

We have already discussed an example of using a single-stub tuner for impedance matching on transmission lines (see Chapter 2, Section 2.2.6). The stub in the example was a shorted section of transmission line. We have also presented a partially inserted metallic rod through the center of the broad wall as a capacitive post (see Figure 3.23d) for which no convenient analytical expression can be written. The broad wall post is the waveguide approximate equivalent of a single-stub line. By making the post movable along a broad wall center slot we can obtain a desired stub impedance by length and position adjustment. The circuit is a local capacitance up to an insertion depth $l = \lambda_g/4$, after which it becomes an inductance. Of course, the narrow wall is only $\lambda_g/4$ in dimension for the TE_{10} mode, so the broad wall post will always be capacitive in this case. The broad wall post is only an approximation of a line stub because the range of susceptances obtainable with a capacitive post is much smaller than that of a true shorted section of transmission line. We will begin the discussion of multiple stubs with a double-stub tuner, look briefly at triple- and quadruple-stub tuners and conclude this section with the E–H plane tuner.

Double-stub tuner

Some of the limitations of a single-stub match can be overcome by using a double-stub tuner. As in the previous examples we place the first stub, with susceptance jB_1 (S), an electrical distance s toward the generator from the load which is at $z = L$, and $0 \leqslant s \leqslant \lambda_g/2$. The

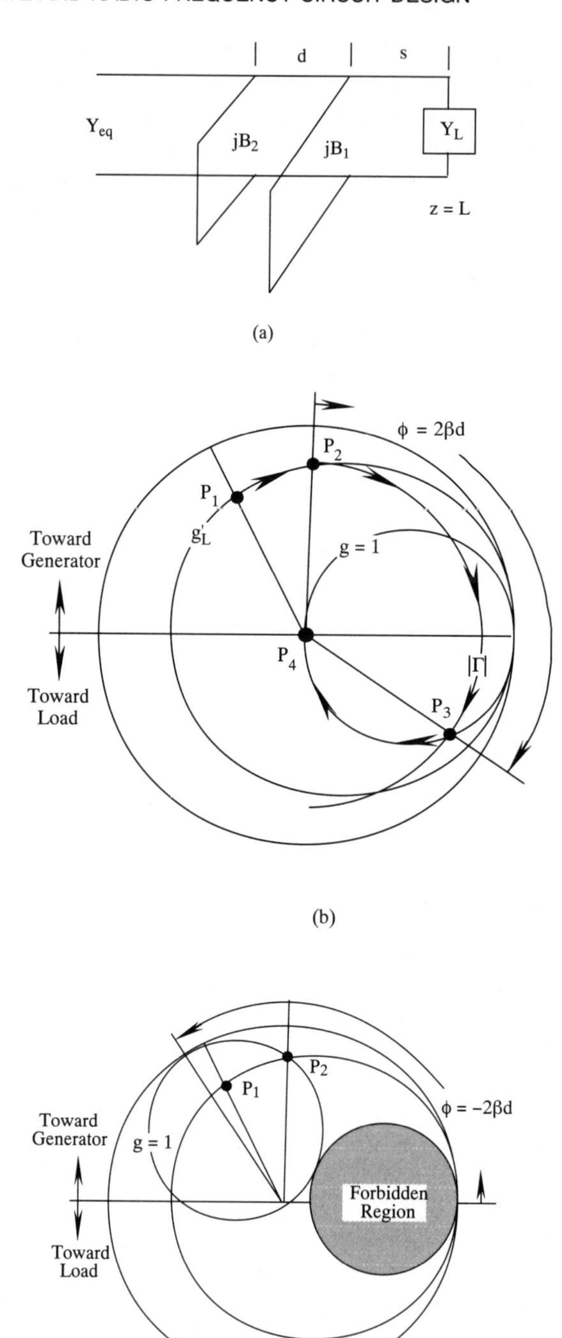

Fig. 3.42 *Double-stub tuner matching process. (a) Circuit elements. (b) Smith chart diagram. (c) Finding P_2 by rotating the $g = 1$ circle through $\phi = -2\beta d$ and the resulting forbidden match zone.*

second stub, susceptance jB_2 (S), is a fixed distance d toward the generator from the first stub, as in Figure 3.42a. We will work with the Smith chart in admittance coordinates since we are creating parallel admittances in this example. We may just as well work the problem by transforming the load Y_L to an equivalent admittance, Y'_L for which $s = 0$, with no loss in generality. Thus we may begin with the normalized (lower case) $y'_L = g'_L + jb_{L'}$ at point P_1 in Figure 3.42b. The first admittance moves the load susceptance toward the generator along a constant g'_L conductance circle to P_2. The effect of the separation, d, between stubs b_1 and b_2 is to move the admittance through an angle of $2\beta d$ toward the generator along a constant radius circle (i.e. a constant $|\Gamma|$ circle) to the point P_3. The admittance at the point $z = L - d$ must be selected so that P_3 lies on the $g = 1$ circle. The second stub (jb_2) will be used to move from P_3 along the $g = 1$ circle to the point P_4 where a match is achieved since $g_{eq} = 1$ and $b_{eq} = 0$.

But it is not yet clear how to find P_2. We may do this graphically by rotating the $g = 1$ circle through the angle $\phi = 2\beta d$ counter-clockwise (toward the load), as in Figure 3.42c. Now the point P_2 should be located at the intersection of the rotated $g = 1$ circle and the g'_L circle. That setting for P_2 will allow the jb_2 adjustment to follow the actual $g = 1$ circle down to P_4. There is always a forbidden match zone (as shown in the figure) which arises because d must be fixed in a practical double-stub tuning device—a typical value is $d \leqslant \lambda_g/8$ so that $\phi \leqslant \pi/4$ and the forbidden zone is not too large.

Triple-stub tuner

The forbidden zone may be avoided by using a triple-stub tuner, Figure 3.43a. Here the three tuning stubs are separated by a common distance, d, such that the two shift angles are each $\phi = 2\beta d$. The network may be thought of as a double-stub (formed by admittances jb_2 and jb_3) with load impedance determined by the first stub, jb_1. The action of the first stub is to move the equivalent load impedance, y'_L, so that the double-stub load (i.e. y^*) lies outside the forbidden zone for a double-stub tuner. We must obtain $y' = y'_L + jb_1$ (at $z = L$) such that transformed load for the 2-3 double-stub, y^*, has a conductance outside the forbidden circle—that is that $g^* < g_0 = \text{cosec}^2(\beta d)$, as in Figure 3.43b. The other two stubs then act as a double-stub tuner and shift the impedance as described in the previous section. The two stubs, jb_2 and jb_3, will then be able to match y^* so that $y_{eq} = 1.0$. We may see where y' ought to be by rotating the forbidden circle, $g_0 = \text{cosec}^2(\beta d)$, through $\phi = -2\beta d$ radians—g' must lie outside the rotated circle, as shown in Figure 3.43c.

The typical triple-stub tuner as realized in a waveguide has three capacitive posts with $d = \lambda_g/8$ or $3\lambda_g/8$ so that $2\beta d = \pi/4$ or $3\pi/4$, respectively, as shown in Figure 3.43d. The triple capacitive post tuner can match a wide range of impedances, but because the post is not truly a shorted transmission line section the waveguide triple-stub is less universal than its transmission line counterpart and there are some regions of the Smith chart which are not easily tunable.

Quadruple-stub tuner

We may compensate for the limited range of a waveguide triple-stub tuner by using a quadruple-stub tuner. A practical commercially available quadruple-stub tuner in WR284 waveguide ($\lambda_g = 23$ cm) has two different separation distances, 5.7 cm ($\approx \lambda_g/4$) and 8.5 cm ($\approx 3\lambda_g/8$), as in Figure 3.44. The network effectively forms two double-stub tuners each with $\lambda_g/4$ separation so that $\phi = 2\beta d = \pi$. The quadruple-stub tuner is able to match

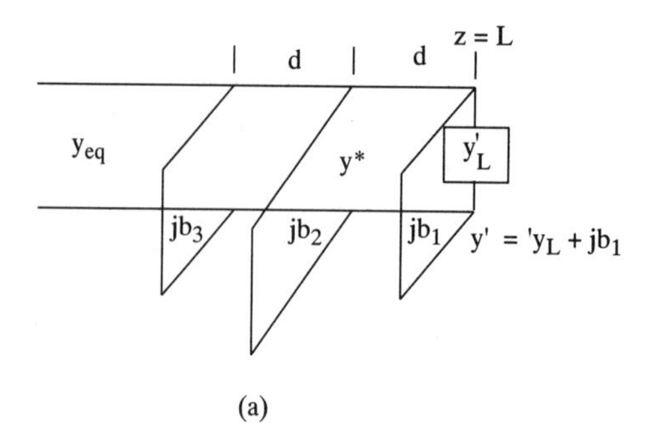

(a)

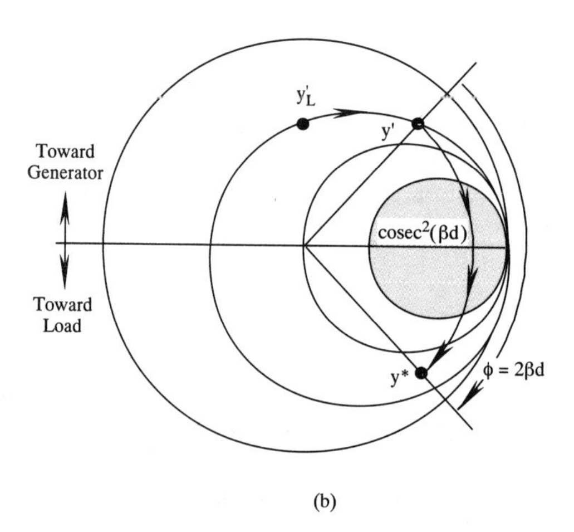

(b)

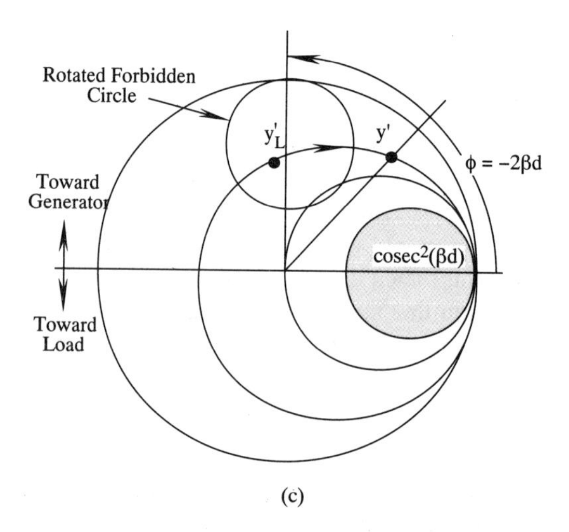

(c)

Fig. 3.43 *Triple-stub tuner matching example. (a) Circuit with equivalent load. (b) Transformation of y' into y^*. (c) Determining the location of y' by rotating forbidden circle $\phi = -2\beta d$. (d) Practical waveguide triple-stub tuner.*

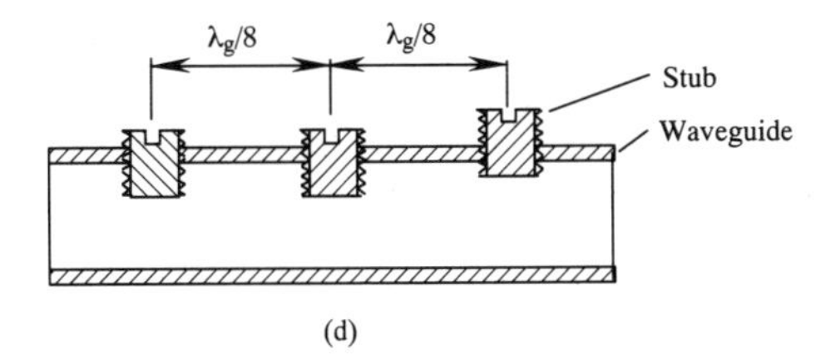

Fig. 3.43 *(continued)*

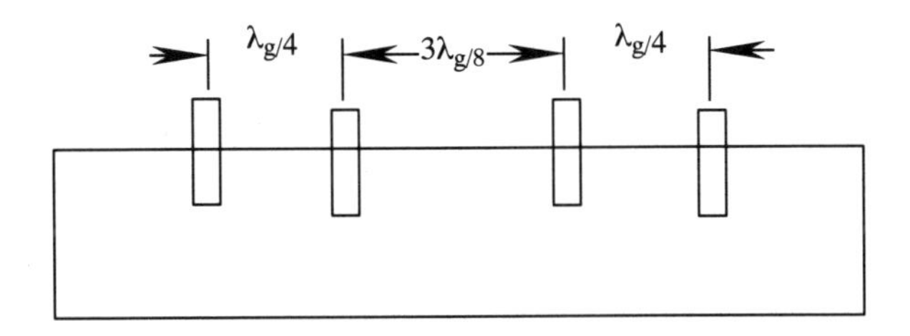

Fig. 3.44 *Four-stub waveguide tuner.*

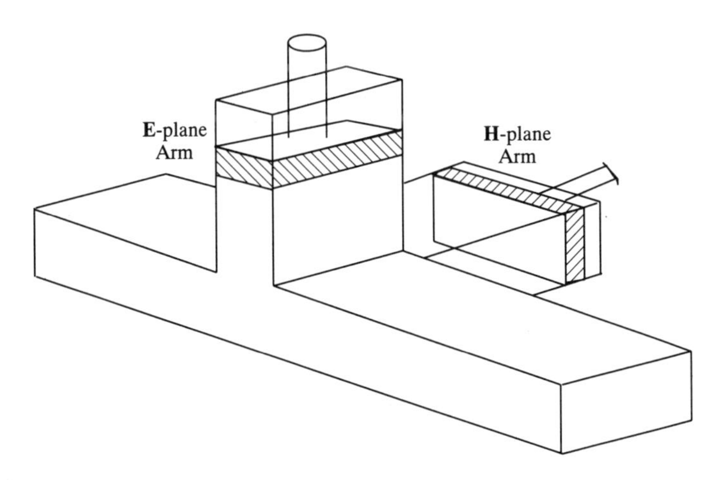

Fig. 3.45 *E–H plane tuner.*

throughout all regions of the Smith chart. However, the interactions of the stub adjustments make it almost necessary to visualize the effect of the various adjustments on a network analyzer to achieve a useful match.

E–H plane tuner

The E–H plane tuner, Figure 3.45, is a waveguide circuit with the true equivalent of shorted transmission line segments. Thus a wide range of impedances can be matched with this device. The behavior of this configuration is more complicated than the relatively simple transmission line model, however. This is because, as will no doubt be remembered, a transmission line by definition must have all of its transverse dimensions small compared to a wavelength (Chapter 2). A waveguide does not meet this condition. So, even though we can often apply transmission line concepts directly to waveguides without modification, this is a case where we must be more careful. This distinction is somewhat to be expected since the sliding shorts are not really far from the aperture and the evanescent modes are likely to be quite important. So, our description of waveguide wall currents, particularly in the shorting block, is likely to be inaccurate in this case. Nevertheless, the shorted waveguide sections are reactive only since no power is dissipated in the sliding short. Consequently, the device may be used (empirically) to match a variety of impedances in the loaded waveguide section.

3.5.4 Microwave circuits having more than two branches

Multiple-branch circuits are often used in microwave transmission systems. The most common device at high power is the circulator, which is used to protect a magnetron from a mismatched load. There are many other examples, particularly in microstrip transmission line devices for communication circuits. We will describe two particularly useful multi-branch devices: directional couplers and circulators.

Directional couplers

Directional couplers are used to measure forward and reflected power. Directional couplers are four-port devices in which power may flow in the directions shown in Figure 3.46a. The port 1 wave couples to ports 2 and 3 but not to 4. Likewise, 2 couples to 1 and 4, 3 couples to 1 and 4, and 4 to 2 and 3—ports 1 and 4 are not coupled and 2 and 3 are not coupled. There are two figures of merit, the coupling factor, C:

$$C_{13} = 10\log_{10}\left(\frac{P_1}{P_3}\right) \tag{3.50}$$

which determines the portion of power coupled from 1 to 3 (and there are three more analogous relations), and the directivity, D:

$$D_{14} = 10\log_{10}\left(\frac{P_1}{P_4}\right) \tag{3.51}$$

which indicates the degree of isolation of the "uncoupled" ports, and there are likewise three more of these relations. Note that the above two relations apply when the only incident power is that of port 1 (i.e. the other three ports are terminated in matched loads).

Power coupling is accomplished by precisely dimensioned and positioned holes between two waveguides with a common broad wall, as shown in Figure 3.46b. There are many possible hole combinations which may be used, but perhaps the most convenient is the standard two-hole design shown in the figure. Here the two apertures are separated by $\lambda_g/4$ so that the field at the second aperture is $\beta d = \pi/2$ out of phase with the first aperture. We might expect constructive and destructive interference between the coupled waves, and this is indeed what happens. Referring to Figure 3.46c, imagine a forward wave from port 1 at the first aperture of field strength 1 (V/m) which generates coupled waves of E-field amplitude E_f in the forward direction and E_b in the backward direction. If we assume that the amount of power coupled through the aperture is small (as is typically the case) then the port 1 wave has about the same amplitude under the second aperture but is shifted in angle

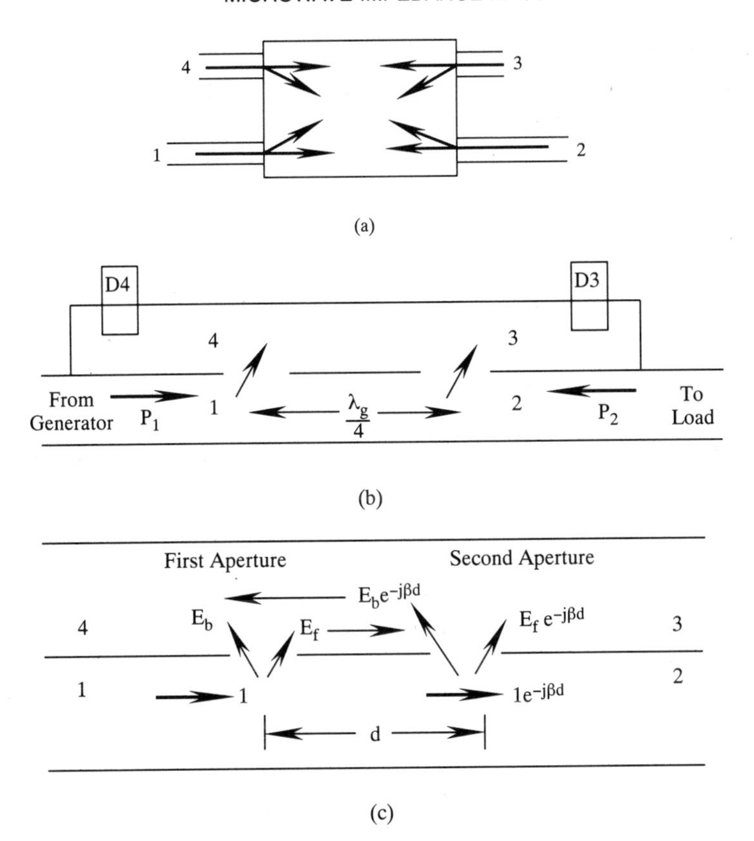

Fig. 3.46 *(a) Four-port model of a directional coupler. (b) Coupling apertures in a directional coupler. (c) Two-hole design phase relations.*

by $-\beta d$. Therefore, the two coupled fields at the second aperture are $E_f e^{-j\beta d}$ and $E_b e^{-j\beta d}$. The backward wave propagates toward the first aperture and has a phase angle of $-2\beta d$ with respect to the incident field when it arrives there. When $d = \lambda_g/4$, as in our case, the net backward wave at the first aperture is zero because the two waves are π radians out of phase there and cancel each other. Meanwhile, the two forward-coupled waves arrive at the second aperture in phase (no matter what d is) and add together. The coupling factor is easily seen to be:

$$C = -20\log_{10}(2|E_f|). \tag{3.52}$$

The directivity consists of two terms; the first is dependent on the aperture geometry alone (and is close to 0 dB) and the second is the "array factor" for the two holes (and contributes large negative numbers, as is desired):

$$\begin{aligned} D &= 20\log_{10}\left(\frac{2|E_f|}{|E_b||(1+e^{-2j\beta d})|}\right) \\ &= 20\log_{10}\left(\left|\frac{E_f}{E_b}\right|\right) + 20\log_{10}(|\sec(\beta d)|). \end{aligned} \tag{3.53}$$

Note that the coupling factor depends on aperture geometry alone and is relatively insensitive to frequency, but the directivity is extremely frequency-sensitive since it contains the secant (which approaches 0 for d = $\lambda_g/4$).

The S-parameter representation for a directional coupler has many terms:

$$\begin{bmatrix} V_1^- \\ V_2^- \\ V_3^- \\ V_4^- \end{bmatrix} = \begin{bmatrix} S_{11} & S_{12} & S_{13} & S_{14} \\ S_{21} & S_{22} & S_{23} & S_{24} \\ S_{31} & S_{32} & S_{33} & S_{34} \\ S_{41} & S_{42} & S_{43} & S_{44} \end{bmatrix} \begin{bmatrix} V_1^+ \\ V_2^+ \\ V_3^+ \\ V_4^+ \end{bmatrix}. \tag{3.54}$$

By reciprocity $S_{ij} = S_{ji}$ and several of the terms are zero: $S_{11} = S_{22} = S_{33} = S_{44} = S_{14} = S_{41} = S_{23} = S_{32} = 0$, assuming an ideal directional coupler. There are also symmetry relations which apply: $|S_{13}| = |S_{24}|$, $|S_{12}| = |S_{34}|$ in the ideal coupler. Finally, it may be shown that:

$$|S_{12}|^2 + |S_{13}|^2 = 1 \qquad \text{and} \qquad |S_{12}|^2 + |S_{24}|^2 = 1. \tag{3.55}$$

For coupling in which we choose the phase in port 1 so that $S_{12} = c_1$, a real constant, and the phase reference in port 3 so that S_{13} is a positive imaginary quantity, $S_{13} = -jc_2$, and we enforce conservation of power, then:

$$[S] = \begin{bmatrix} 0 & c_1 & jc_2 & 0 \\ c_1 & 0 & 0 & jc_2 \\ jc_2 & 0 & 0 & c_1 \\ 0 & jc_2 & c_1 & 0 \end{bmatrix}. \tag{3.56}$$

This result means that any fully adapted lossless four-port device must be a directional coupler.

Circulators

Circulators are three-port devices which allow power propagation to ports in sequence only. That is, for a typical three-port circulator, as in Figure 3.47a, power may propagate from port 1 to 2, from 2 to 3, and from 3 to 1. Other propagation is blocked. The resulting S-parameter matrix is:

$$[S] = \begin{bmatrix} 0 & 0 & S_{13} \\ S_{21} & 0 & 0 \\ 0 & S_{32} & 0 \end{bmatrix}. \tag{3.57}$$

The most common form of such a circulator uses a ferrite post in the center of a waveguide junction, Figure 3.47b. At high power the post must be water cooled. In a microwave circuit (Figure 3.47c) the circulator is placed at the generator output with port 1 connected to the generator, port 2 to the load and port 3 to a matched load. In microwave waveguide circuits at high power a circulator is an essential element to protect the magnetron from reflected power from the load and applicator. High reflected powers will drastically shorten the life of a magnetron tube. In fact, when working with high reflection coefficient loads (such as resonant cavities) it may be prudent to use two circulators in series since the directivity of a single high power device may not provide adequate protection for the magnetron.

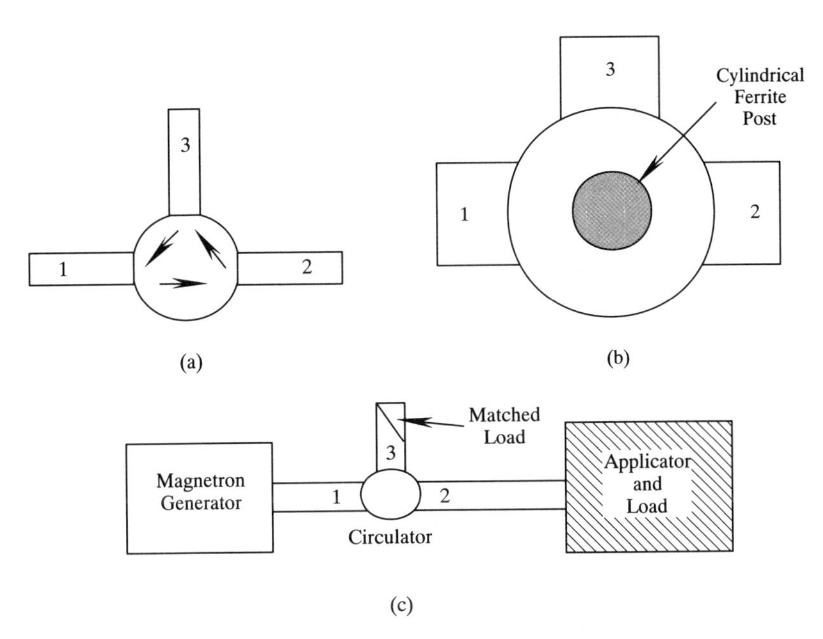

Fig. 3.47 *Three-port circulator and its use. (a) Circulator function. (b) Ferrite post circulator design. (c) Use of a circulator in a microwave circuit.*

3.6 SUMMARY

This chapter has presented device models and parametric characterization methods for use in both the RF and MW frequency ranges. Generating and impedance matching methods have also been discussed using a lumped complex impedance to represent the applicator and its load. The next chapter will focus on the applicator and load to complete the picture.

REFERENCES

Arrl 1986 Handbook for the Radio Amateur. The American Radio Relay League, 1986.

Abrie, P. L. D. (1985) *The Design of Impedance Matching Networks* Artech House, Dedham, MA.

Collin, R. E. (1966) *Foundations for Microwave Engineering* McGraw-Hill, New York.

Collin, R. E. (1991) *Field Theory of Guided Waves (2nd Ed)*. IEEE Press, Piscataway, NJ.

Dicke, R. H. and Purcell, E. M. (1948) *Principles of Microwave Circuits* McGraw-Hill, New York, pp 105–8 and (updated by Collin (1966) on pp 168–9).

Marchand, C. and Meunier, T. (1990) Recent developments in industrial radio-frequency technology *Journal of Microwave Power and Electromagnetic Energy*, pp 39–46.

Metaxas, A. C. and Meredith, R. J. (1983) *Industrial Microwave Heating* Peter Perigrinus, London.

Thuery, J. (1983) *Les Microondes et leurs Effets sur la Matiére* Technique et Documentation, Lavoisier, Pans.

High Power Applicators and Loads

4.1 INTRODUCTION

In industrial RF or MW applications, the applicator is an important part of the system design. The dimensions of the object or the volume of material to be treated with the electromagnetic field are significant factors in the design of the applicator. This usually means that large objects are better treated by RF, and small objects by MW heating. The total power which is to be applied will also affect the choice of frequency range. If low power is desired (1–20 kW), microwaves work well. For high power applications (> 200 kW), RF systems are usually required. If low to intermediate power is desired (1–100 kW), either frequency range may be used, depending on product geometry, thermal and electrical properties. The financial aspects (capital and operating costs) must also be considered in the choice of a method; however, they are remarkably similar when considered *en toto* on a per watt basis.

This chapter describes field analysis techniques for MW and RF industrial applications. The discussion will center on the applicator and its load (the workpiece, or the material to be heated) with three overall objectives: (1) to describe how the workpiece geometry interacts with the applicator design to obtain a particular field distribution within the material, (2) to examine applicator–workpiece combinations with respect to advantages and disadvantages, and (3) to provide the foundations of a one-port description of the workpiece and applicator suitable for use in conjunction with the methods of Chapter 3. All discussions in this chapter treat the workpiece as composed of linear homogeneous isotropic materials of known properties. Further, the discussion centers around the macroscopic distribution of electromagnetic fields. We will address microscopic theory and data analysis methods (including dielectric spectroscopy) in Part 2 of the book, Chapters 6 to 9.

The chapter begins with a discussion of RF applicators in which the quasi-static assumption can often be applied. Electric field heating is considered first, followed by heating of dielectric and semiconducting materials in a magnetic field. The chapter concludes with a consideration of microwave applicator–load combinations. Classical closed-form solutions are presented, where helpful, with extensions to more complex geometries using numerical analysis methods.

4.2 RADIO FREQUENCY ELECTRIC FIELD APPLICATORS

Because the wavelengths of RF systems are often much larger than the maximum dimensions of the objects to be heated, the quasi-static assumption usually applies. Consequently, the electric and magnetic fields may be thought of separately. There is still weak coupling between them in Ampere's law, but, at any instant in time, phase differences from point to point are negligible in many cases. The coupling between generator and load is either dominantly electric field or dominantly magnetic field. We describe two classes of electric field and two classes of magnetic field applicators with examples of how analytical methods may be applied to predict the field distribution in the load.

4.2.1 Quasi-static analytical methods—FDM, FEM and MoM

The phase is not a function of position under the quasi-static assumption, so retarded potentials are not required and we may use simplified forms of the volume integrals to calculate the potentials (see Chapter 1):

$$V(x, y, z) = \iiint_{\text{vol}} \frac{\rho_{\text{v}}(x', y', z')\,\mathrm{d}x'\mathrm{d}y'\mathrm{d}z'}{4\pi\varepsilon\sqrt{(x-x')^2+(y-y')^2+(z-z')^2}} \tag{4.1}$$

$$\mathbf{A}(x, y, z) = \iiint_{\text{vol}} \frac{\mu\mathbf{J}(x', y', z')\,\mathrm{d}x'\mathrm{d}y'\mathrm{d}z'}{4\pi\sqrt{(x-x')^2+(y-y')^2+(z-z')^2}}. \tag{4.2}$$

The potentials are defined quantities which make the calculation of electric and magnetic fields simpler in many cases. Specifically, we have defined the scalar electric potential, V, so that under the quasi-static assumption its gradient gives the electric field:

$$\mathbf{E} = -\nabla V - \frac{\partial\mathbf{A}}{\partial t} \cong -\nabla V \tag{4.3}$$

and the vector magnetic potential, $\mathbf{A}$, so that its curl is the magnetic flux density:

$$\mathbf{B} = \nabla \times \mathbf{A}. \tag{4.4}$$

At low frequency the frequency-domain form of Maxwell's equations, expressed in terms of the potentials (equations (1.31)), reduce to:

$$\nabla^2 V = -\frac{\rho_{\text{v}}}{\varepsilon} \tag{4.5a}$$

$$\nabla^2\mathbf{A} = -\mu\mathbf{J}. \tag{4.5b}$$

Equation (4.5a) is the point form of Gauss' electric law expressed in terms of the scalar potential, and is called the Poisson equation. Equation (4.5b) may be used to calculate the vector magnetic potential. We reserve its use for a later section since we are initially concerned with the electric field.

If the medium is source free, which is the usual case in a passive material heated by the fields, we may quickly reduce the Poisson equation to the Laplace equation:

$$\nabla^2 V = 0. \tag{4.6}$$

While a Poisson equation solution is driven by the charge distribution, which must be specified, the Laplace equation is driven entirely by boundary conditions (potentials and fluxes) and belongs to the class of differential equations known as boundary value problems. Both the Laplace and Poisson equations are "point form" differential equations and thus may not be used at an interface (a discontinuity). At interfaces and surfaces the boundary conditions apply because, as will no doubt be remembered, they were obtained from the integral form of Maxwell's equations in Chapter 1. There are two kinds of boundaries for the Laplace solution space: boundaries of fixed potential (Dirichlet boundaries), and boundaries with a specified flux or flux density (if the flux is zero the boundary is a Neumann boundary). Many classical solutions are available for the Laplace equation. We will review several of the analytical solutions most generally applicable to heating problems. In all of the analytical solutions we will treat the materials as linear homogeneous and isotropic, even though these assumptions may be questionable.

We will also discuss several numerical approaches. There are two generally useful numerical modeling philosophies for solving partial differential equations, the Finite Difference Method (FDM) and the Finite Element Method (FEM). Each method has its "true believers"; and, as might be expected, each has its advantages and disadvantages. FDM formulations are simpler to program and yield results in less time; however, they require uniform node-grids, larger amounts of memory (many more nodes) to obtain accurate results, and yield information about the potential only at the specified node point locations. Consequently, they are limited to simple well behaved geometries. In the FDM the calculation of gradient quantities — electric field and current density, the very fields we care most about — have a much lower accuracy than the potential, especially for a coarse node-grid spacing. FEM models represent the continuous potential field by basis functions (usually either linear or quadratic) with random node position. Two advantages arise: (1) the node potentials are the coefficients of the basis functions so a continuous representation of the potential results in much better accuracy in estimates of **E** and **J**, and (2) the model may easily be used to represent very complex geometries. There is a third commonly applied numerical technique in electric field calculations, the Method of Moments (MoM), which has some philosophy in common with the FEM. Each method is sufficiently complex that it is intractable to present in detail in a single chapter, such as we have available. We present, therefore, a very brief description of the principles of each of the methods, and recommend that the reader consult one of the many references on the individual subjects. The volume edited by Kung and Morgan (1990) is especially informative, as is the compendium edited by Itoh (1989).

The finite difference method

The fundamental principle of the finite difference method is that the appropriate partial or total differentials are approximated by their numerical equivalent. A one-dimensional function, $f(x)$, Figure 4.1a, may be estimated at the points $(x + \Delta x)$ and $(x - \Delta x)$ from a Taylor series expansion:

$$f(x+\Delta x) = f(x) + \Delta x \left[\frac{\mathrm{d}f}{\mathrm{d}x}\right]_x + \frac{\Delta x^2}{2!}\left[\frac{\mathrm{d}^2 f}{\mathrm{d}x^2}\right]_x + \frac{\Delta x^3}{3!}\left[\frac{\mathrm{d}^3 f}{\mathrm{d}x^3}\right]_x + \cdots \tag{4.7a}$$

$$f(x-\Delta x) = f(x) - \Delta x \left[\frac{\mathrm{d}f}{\mathrm{d}x}\right]_x + \frac{\Delta x^2}{2!}\left[\frac{\mathrm{d}^2 f}{\mathrm{d}x^2}\right]_x - \frac{\Delta x^3}{3!}\left[\frac{\mathrm{d}^3 f}{\mathrm{d}x^3}\right]_x + \cdots \tag{4.7b}$$

By adding or subtracting (4.7a) and (4.7b) and ignoring higher order terms we may obtain numerical approximations for the first and second derivatives at x:

$$\left[\frac{\mathrm{d}f}{\mathrm{d}x}\right]_x \cong \frac{f(x+\Delta x) - f(x-\Delta x)}{2\Delta x} \tag{4.8a}$$

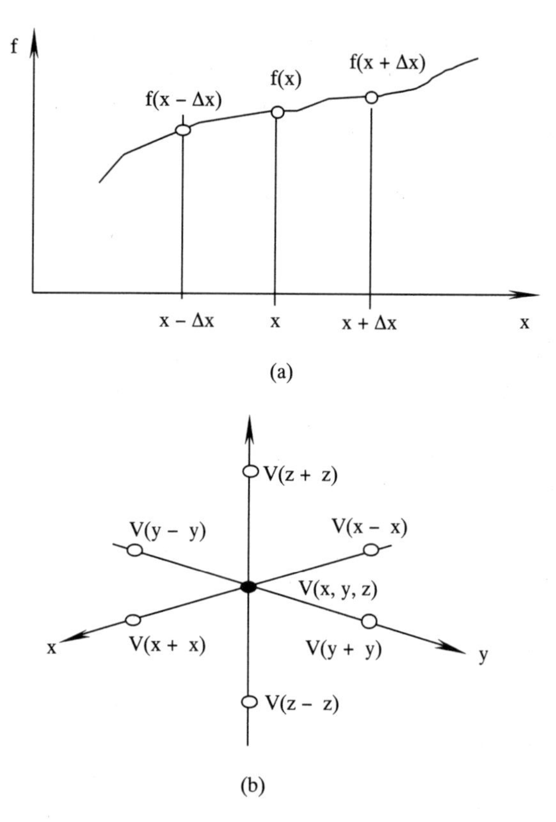

Fig. 4.1 *Finite difference function approximations. (a) Sampled one-dimensional function. (b) Three-dimensional finite difference grid.*

$$\left[\frac{d^2 f}{dx^2}\right]_x \cong \frac{f(x+\Delta x) + f(x-\Delta x) - 2f(x)}{\Delta x^2}. \tag{4.8b}$$

Equation (4.8a) is called the central difference formula for the first derivative. The potential at the central node may be estimated from its neighbors (in three dimensions) by applying (4.8b) to the Laplace equation (Figure 4.1b):

$$\begin{aligned} V(i, j, k) \simeq \tfrac{1}{6}\{&V(i+1) + V(i-1) + V(j+1) + V(j-1) \\ &+ V(k+1) + V(k-1)\}. \end{aligned} \tag{4.9}$$

Here we have used i as the index on the x-axis, j on the y-axis, and k on the z-axis, the notation on the right hand side has been shortened for clarity, and we have also assumed that $\Delta x = \Delta y = \Delta z$. The factor 1/6 would change to 1/4 for a two-dimensional Cartesian coordinate model (correspondingly, with four unknown potentials on the right hand side). The assumption of uniform node spacing is not necessary in the FDM, but certainly simplifies the formulation.

Alternately, one may recast the Laplace problem in Finite Control Volume (FCV) form using the integral equations, Gauss' electric law in a dielectric (4.10a) and/or the continuity equation in a semiconductor (4.10b) or (4.10c):

$$\oint_\Sigma \mathbf{D} \cdot \mathbf{dS} = 0 \Rightarrow \varepsilon \oint_\Sigma \mathbf{E} \cdot \mathbf{dS} = 0 \tag{4.10a}$$

$$\oint_\Sigma \mathbf{J} \cdot \mathbf{dS} = 0 \Rightarrow \sigma \oint_\Sigma \mathbf{E} \cdot \mathbf{dS} = 0 \tag{4.10b}$$

$$\oint_\Sigma \mathbf{J} \cdot \mathbf{dS} = -\mathrm{j}\omega \left\{ \oint_\Sigma \mathbf{D} \cdot \mathbf{dS} \right\} \Rightarrow (\sigma + \mathrm{j}\omega\varepsilon) \oint_\Sigma \mathbf{E} \cdot \mathbf{dS} = 0. \tag{4.10c}$$

If the control volume is homogeneous and the electric field can be obtained in each direction from the appropriate form of equation (4.8a), then equation (4.9) results from any of the three forms of equation (4.10). So, the FCV method is equivalent to the FDM approach.

At Dirichlet boundaries of the solution space one treats the node potential as a known constant. For a Neumann boundary on the z-axis at, say, $k = k_{\max}$ the unknown node potential for the $k_{\max} + 1$ node (outside the model space) is set equal to $V(k_{\max} - 1)$, which forces $dV/dz = 0$ at $k_{\max}$ (refer to equation (4.8a)). At an interface between media one must apply the integral equations; and the boundary conditions obtained in Chapter 1 do just that. The expression for a node potential on an interface (Figure 4.2) may be obtained in a similar fashion to the above:

$$V(i, j, k) \cong \frac{\varepsilon_1}{2(\varepsilon_1 + \varepsilon_2)} V(k+1) + \frac{\varepsilon_2}{2(\varepsilon_1 + \varepsilon_2)} V(k-1) + \tfrac{1}{6}[V(j+1) + v(j-1) + V(i+1) + V(i-1)] \tag{4.11}$$

where the interface is an x–y plane at $z = k$. The same relation can be obtained from equations (4.10) if the appropriate medium admittance—σ, or $(\sigma + j\omega\varepsilon)$—is used in place of ε in (4.11). Of course, using (4.10c) means that you must solve the Laplace equation in the frequency domain (that is, that the source is sinusoidal). Interestingly, we may solve the frequency domain Laplace equation by the FDM, but not the full time-varying wave

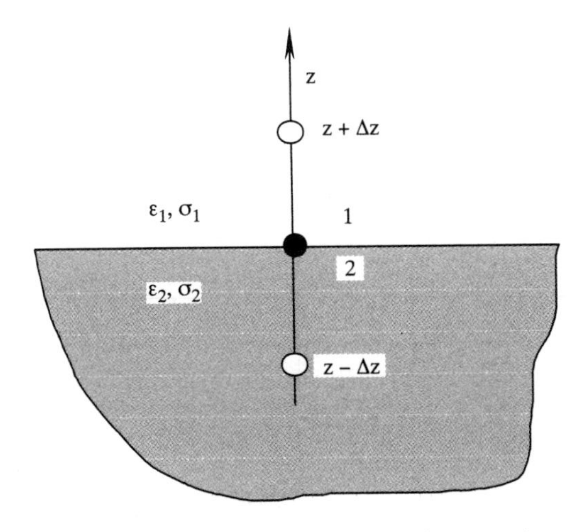

Fig. 4.2 *Interface between media in x–y plane.*

equation form (equation (1.31)) since the phase term, $\omega^2\mu\varepsilon V$, leads to a singular parameter matrix which may not be inverted.

Two approaches may be used in solving the resulting system of equations numerically: a linear system formulation, and successive approximation. The linear system formulation works well for a problem with relatively few nodes, while successive approximation is easier and faster for large systems of nodes. If one assigns node indices with some care, the linear system formulation typically results in a sparse and banded parameter matrix. An example arbitrary 2-D geometry is shown in Figure 4.3, and the resulting system of equations might look something like:

$$\begin{bmatrix} 4 & 1 & 0 & 0 & 0 & & 1 & & & 0 \\ 1 & 4 & 1 & 0 & 0 & & & 1 & & 0 \\ 0 & 1 & 4 & 1 & 0 & & & & & 0 \\ 0 & 0 & 1 & 4 & 1 & & & & & 0 \\ \cdots & & & & & & & & & \end{bmatrix} \begin{bmatrix} V_1 \\ V_2 \\ V_3 \\ \cdots \\ V_N \end{bmatrix} = \begin{bmatrix} 0 \\ 0 \\ 0 \\ -10 \\ \cdots \\ 0 \end{bmatrix}. \tag{4.12}$$

One may determine the unknown node potentials, V_i, by multiplying both sides by the inverse of the banded constant parameter matrix.

The successive approximation method works by refining the guess of each of the unknown node potentials. For example, for the initial potential of the unknown nodes one may use: (1) zero, (2) the weighted average of all of the boundary potentials, (3) some number between the maximum and minimum boundary potential, (4) a simple analytic solution (say, a linear function between boundaries), or (5) the result of a coarse grid approximation to the problem (which will converge very quickly). Of these, the fifth option yields accurate convergence in the fewest iterations, but requires easily scalable geometries. Once an initial guess is set an improved guess would be determinedat each point using equation (4.9)—to avoid bias this should be done non-recursively, that is the "new" node potential at (i, j, k) is determined using "old" values at *all* neighbor nodes. At interfaces between media equation (4.11) would be used. The resulting error in equation (4.6) is called the "residue", $R: \ \nabla^2 V = R$, and in general $R \neq 0$. One continues to iteratively sweep through the node matrix until the residue becomes small—until the residue "relaxes" acceptably close to zero.

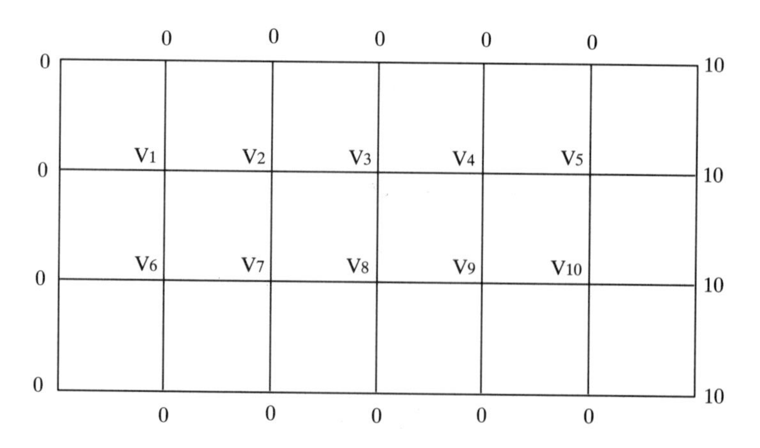

Fig. 4.3 *Example node assignments for linear system formulation.*

We can accelerate the convergence by "over-relaxing", the SOR or successive over-relaxation method. In this method the estimated change in potential for the node, ΔV, always has the correct sign, but could be larger without sacrificing stability. The new guess for the node potential is accelerated by an over-relaxation factor, w: $V_{\text{new}} = V_{\text{old}} + w\Delta V$, where $1 < w < 2$ for stability. Convergence may also be significantly accelerated by option (5): first iterating on a coarse grid to obtain the initial guess for the fine grid. The improved first guess drastically reduces the number of iterations required for the residue to be "swept out" (as much as a factor of 10 or more). Complex geometries — for example, several layers of differing material properties — may require a surprising number of iterations to converge adequately as the residue may have to sweep several times across the interfaces in order to settle adequately.

A note of caution is necessary for solutions in cylindrical and spherical geometries. If the point $r = 0$ is included in the solution space one must be careful not to locate any nodes there since this would result in a singularity (divide by zero) in the numerical approximations for the Laplacian:
cylindrical:

$$\nabla^2 V = \frac{\partial^2 V}{\partial r^2} + \frac{1}{r}\frac{\partial V}{\mathrm{d}r} + \frac{1}{r^2}\frac{\partial^2 V}{\partial \varphi^2} + \frac{\partial^2 V}{\partial z^2} \tag{4.13a}$$

spherical:

$$\nabla^2 V = \frac{1}{r^2}\frac{\partial}{\partial r}\left(r^2\frac{\partial V}{\partial r}\right) + \frac{1}{r^2 \sin\theta}\frac{\partial}{\mathrm{d}\theta}\left(\sin\theta\frac{\partial V}{\partial \theta}\right) + \frac{1}{r^2\sin^2\theta}\frac{\partial^2 V}{\partial \varphi^2}. \tag{4.13b}$$

An example node location scheme in cylindrical coordinates is shown in Figure 4.4. Here the radius $r = (n - 0.5)\Delta r$ at each location is never zero, where n is the integer radial coordinate index, as we used in the Cartesian expressions above. In subsequent discussions it will be clear from the context of a problem whether the cylindrical or spherical r is intended.

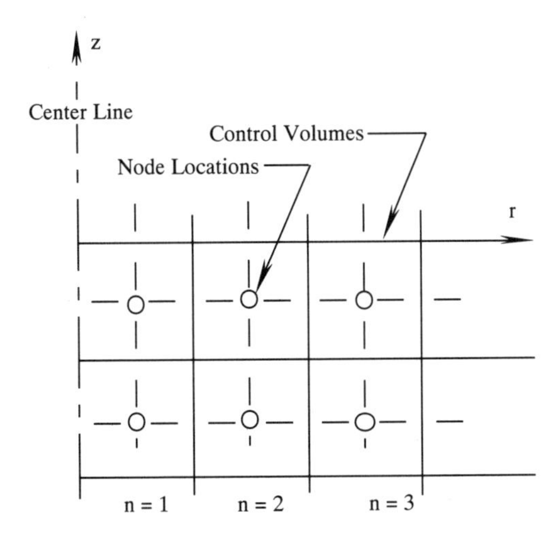

Fig. 4.4 *Node locations in cylindrical coordinates.*

The finite element method

The FEM approach differs from FDM/FCV in that a continuous representation for $V(x, y, z)$ is obtained from continuous basis functions, $\phi_i(x, y, z)$ structured so that their coefficients are the node potentials, V_i:

$$V(x, y, z,) \cong \sum_{i=1}^{N} V_i \phi_i(x, y, z) \tag{4.14}$$

This is obtained by forcing the ϕ_i to be 1 at node i and 0 at all other nodes. The power of this method comes from the separation of the spatial dependence of the potential from the unknowns. That is, the basis functions are known *a priori* and are specified continuous functions of space—the only unknowns are their coefficients. So, once the V_i are determined, the electric field (for example) has a continuous representation in space:

$$\mathbf{E}(x, y, z) \cong -\sum_{i=1}^{N} V_i \nabla \phi_i(x, y, z) \tag{4.15}$$

and we may find the Laplacian by taking the derivative of the basis functions as well:

$$\nabla^2 V(x, y, z) \cong \sum_{i=1}^{N} V_i \nabla^2 \phi_i(x, y, z) = 0. \tag{4.16}$$

Of course, the accuracy of both $\mathbf{E}$ and V representations is determined by the order of the basis functions.

With this method, the node spacing may be increased (decreasing the number of unknowns to determine) without sacrificing accuracy; but this must be balanced by consideration of the error in approximation of the potential function. For example, we may use linear, quadratic or cubic basis functions. Higher order polynomials will result in a large number of calculations during each iteration, but wider node spacings can be used, and the error in the result will not be as large as for a low order polynomial. Low order polynomials are calculationally simple, but we must (apparently) be careful that at least two derivatives are possible (in equation (4.16)), so it appears that quadratic basis functions are required. Yet, linear basis functions are very attractive, and it turns out that we can use them, as will be shown.

We must somehow determine the best possible choices for the node potentials. A very common optimization method is the Method of Weighted Residuals (MWR) in which we select the V_i to minimize the weighted residue, $R(x, y, z)w(x, y, z)$, over the entire solution space:

$$\iiint_{\text{vol}} R(x, y, z)w(x, y, z)\,\mathrm{d}x\mathrm{d}y\mathrm{d}z = 0 \tag{4.17a}$$

$$\sum_{i=1}^{N} V_i \iiint_{\text{vol}} \nabla^2 \phi(x, y, z)w(x, y, z)\,\mathrm{d}x\mathrm{d}y\mathrm{d}z = 0 \tag{4.17b}$$

where the w are weighting functions yet to be selected. Since the node potentials are not space dependent they have been moved outside of the integration. The weighting or "test" functions, $w(x, y, z)$, are chosen to sample the residue in a fashion which is advantageous.

It still appears that at least quadratic bases are required to be able to satisfy the relationship; but by clever application of integration by parts we may reduce the minimum order of the basis functions to linear bases.

To see how the test functions and basis functions are formulated, it is easiest to consider a one-dimensional example. Suppose a Laplace problem is 1-D Cartesian on the closed interval [0, 1] with $V(0) = 0$ and $V(1) = 5$. By simple integration we know that $V(x) = 5x$; however, we will use this as an example of the FEM. Suppose that we break the 1-D [0, 1] interval into 4 "elements", as shown in Figure 4.5. Five linear basis functions, ϕ_0 to ϕ_4 span the space and are used to represent $V(x)$ with coefficients V_0 to V_4. We choose the basis functions to be 1 at the location of their node voltage and 0 at other nodes so that equation (4.14) may be satisfied. Note that we already know that $V_0 = 0$ and $V_4 = 5$ by the boundary conditions. The system of equations as presently formulated involves the second derivative of the ϕ_i and no derivatives of w:

$$\sum_{i=0}^{4} V_i \int_0^1 \frac{\mathrm{d}^2\phi_i(x)}{\mathrm{d}x^2} w(x)\,\mathrm{d}x = 0. \tag{4.18}$$

We can obtain a first derivative equivalent using integration by parts:

$$\int_a^b f\,\mathrm{d}g = [fg]_a^b - \int_a^b g\,\mathrm{d}f. \tag{4.19}$$

If we set $f = w(x)$ and $\mathrm{d}g = d^2\phi/\mathrm{d}x^2$ then:

$$\int_0^1 \frac{\mathrm{d}^2\phi_i}{\mathrm{d}x^2} w\,\mathrm{d}x = \left[w\frac{\mathrm{d}\phi_i}{\mathrm{d}x}\right]_0^1 - \int_0^1 \frac{\mathrm{d}\phi_i}{\mathrm{d}x}\frac{\mathrm{d}w}{\mathrm{d}x}\,\mathrm{d}x. \tag{4.20}$$

We have some freedom of choice over the test functions, w, so we will restrict the set of test functions to those which are zero at $x = 0$ and $x = 1$. There is no loss in generality of the solution because we already know V_0 and V_4 and do not need test functions there. We assign the remaining three test functions at each appropriate node, w_j. Now the system of equations has only first derivatives and linear bases are feasible:

$$\sum_{j=1}^{3}\sum_{i=0}^{4} V_i \int_0^1 \left(\frac{\mathrm{d}\phi_i}{\mathrm{d}x}\right)\left(\frac{\mathrm{d}w_j}{\mathrm{d}x}\right)\mathrm{d}x = 0. \tag{4.21}$$

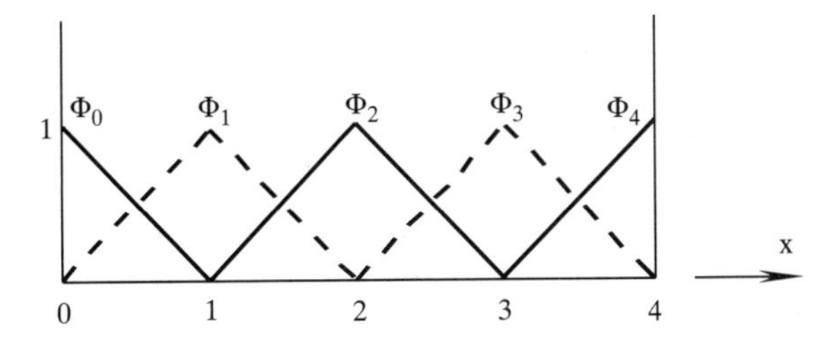

Fig. 4.5 *Finite element basis functions, one-dimensional example.*

The different manifestations of the FEM come from the choice of test functions. If the w_j are Dirac deltas, which are active at each node, the approach is called the "co-location" method (this formulation is equivalent to the FDM and is not recommended). If the w_j are constant over half of the element (say from $x = 0.125$ to $x = 0.375$) the approach is called the "sub-domain" method. In both of those cases we must use equation (4.18) since the w_j do not have useful derivatives. For our case the Galerkin method is preferable.

The Galerkin method is a very commonly used one in which the w_j are represented by the same ϕ_i basis functions used for representing V_i. The coefficients of the ϕ_j bases are 1 everywhere, consequently the system of equations is:

$$\sum_{i=0}^{4} V_i \sum_{j=1}^{3} \int_0^1 \left(\frac{\mathrm{d}\phi_i}{\mathrm{d}x}\right)\left(\frac{\mathrm{d}\phi_j}{\mathrm{d}x}\right) \mathrm{d}x = 0. \tag{4.22}$$

The derivatives of the basis functions are required in specific intervals which we list in Table 4.1. Note that all of the geometry-dependent calculations are performed before the unknown node potentials are obtained.

We may evaluate each of the integrals separately and express it as a parametric constant, a_{ji}, so the system of equations becomes:

$$\begin{bmatrix} a_{10} & a_{11} & a_{12} & a_{13} & a_{14} \\ a_{20} & a_{21} & a_{22} & a_{23} & a_{24} \\ a_{30} & a_{31} & a_{32} & a_{33} & a_{34} \end{bmatrix} \begin{bmatrix} 0 \\ V_1 \\ V_2 \\ V_3 \\ 5 \end{bmatrix} = \begin{bmatrix} 0 \\ 0 \\ 0 \end{bmatrix}. \tag{4.23}$$

Many of the a_{ji} are zero since the basis functions overlap in only a few of the elements. For example, it is inherently obvious that ϕ_1 only contributes in the first and second elements so $a_{13} = a_{14} = 0$—by extension, $a_{20} = a_{24} = a_{30} = a_{31} = 0$. The values of a_{j0} need not be calculated since $V_0 = 0$; however, for completeness we include them. The reader may quickly verify that:

$$a_{11} = \int_0^{0.25} (4)(4)\,\mathrm{d}x + \int_{0.25}^{0.5} (-4)(-4)\,\mathrm{d}x = 8 = a_{22} = a_{33} \tag{4.24a}$$

Table 4.1 *Derivatives of basis functions in the 1-D FEM example.*

Interval	Basis function	Derivative
0–0.25	ϕ_0	−4
	ϕ_1	4
0.25–0.5	ϕ_1	−4
	ϕ_2	4
0.5–0.75	ϕ_2	−4
	ϕ_3	4
0.75–1.0	ϕ_3	−4
	ϕ_4	4

$$a_{10} = \int_0^{0.25} (4)(-4)\,\mathrm{d}x = -4 = a_{12} = a_{21} = a_{23} = \cdots \tag{4.24b}$$

so that the system of equations gives a banded parameter matrix:

$$\begin{bmatrix} -4 & 8 & -4 & 0 & 0 \\ 0 & -4 & 8 & -4 & 0 \\ 0 & 0 & -4 & 8 & -4 \end{bmatrix} \begin{bmatrix} 0 \\ V_1 \\ V_2 \\ V_3 \\ 5 \end{bmatrix} = \begin{bmatrix} 0 \\ 0 \\ 0 \end{bmatrix} \tag{4.25}$$

and $V_1 = 1.25$, $V_2 = 2.5$, $V_3 = 3.75$, as expected. The 1-D example was homogeneous, but it should be noted that each element can, in fact, have different electrical properties (the property is included in the a_{ji} constant). We have already shown the equivalence between the Laplace equation and FCV integral equation formulations. The problem is simply re-formatted using an integral equation approach to obtain an arbitrary electrical geometry.

The approach in two or three dimensions follows the same philosophy except that all of the integration is two- or three-dimensional for each element. A simple 2-D example may help to put the technique into perspective. Sadiku (1989) describes a typically used canonical form for triangular elements which yields a well ordered parameter matrix. In Figure 4.6a we see an arbitrary triangular element. The ordering of the nodes (counter-clockwise) is used to preserve the canonical form of the linear equations. We assume that the element is small enough that the electric field is constant over the element. There is one linear basis

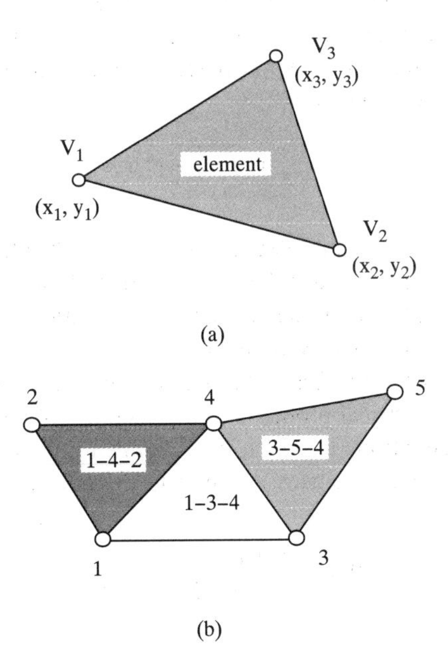

Fig. 4.6 *Two-dimensional finite element grids. (a) Clockwise single triangular element. (b) Arbitrary clockwise three-element model space.*

function for each node potential $V_i = a_e + b_e x_i + c_e y_i$, so for element e:

$$\begin{bmatrix} V_1 \\ V_2 \\ V_3 \end{bmatrix} = \begin{bmatrix} 1 & x_1 & y_1 \\ 1 & x_2 & y_2 \\ 1 & x_3 & y_3 \end{bmatrix} \begin{bmatrix} a_e \\ b_e \\ c_e \end{bmatrix} \tag{4.26a}$$

and the element geometry contained in a, b, and c is:

$$\begin{bmatrix} a_e \\ b_e \\ c_e \end{bmatrix} = \begin{bmatrix} 1 & x_1 & y_1 \\ 1 & x_2 & y_2 \\ 1 & x_3 & y_3 \end{bmatrix}^{-1} \begin{bmatrix} V_1 \\ V_2 \\ V_3 \end{bmatrix} \tag{4.26b}$$

with:

$$\begin{bmatrix} 1 & x_1 & y_1 \\ 1 & x_2 & y_2 \\ 1 & x_3 & y_3 \end{bmatrix}^{-1} = \frac{1}{2A} \begin{bmatrix} (x_2 y_3 - x_3 y_2) & (x_3 y_1 - x_1 y_3) & (x_1 y_2 - x_2 y_1) \\ y_2 - y_3 & y_3 - y_1 & y_1 - y_2 \\ x_3 - x_2 & x_1 - x_3 & x_2 - x_1 \end{bmatrix} \tag{4.26c}$$

where A is the element area: $A = [(x_2 - x_1)(y_3 - y_2) - (x_3 - x_1)(y_2 - y_1)]$. For notational simplicity we will denote the matrix in equation (4.26c) by $[G]^{-1}$ because it is the inverse of the geometry. Once the node potentials are known we may determine the potential at any point in the element from equation (4.14) where the basis functions ϕ_i are:

$$\begin{bmatrix} \phi_1(x, y) \\ \phi_2(x, y) \\ \phi_3(x, y) \end{bmatrix} = [1 \quad x \quad y][G]^{-1}. \tag{4.27}$$

Obviously, the system of equations may be solved by the Method of Weighted Residuals as above (using the Galerkin method since we have linear bases). To illustrate an alternative approach, we may also use a minimum energy criterion. Energy minimization works because it may be shown that all solutions of the Laplace equation minimize the energy stored in the region surrounded by the boundaries:

$$W_e = \tfrac{1}{2} \iint_S \varepsilon |\mathbf{E}|^2 \, dS \tag{4.28}$$

where W_e is the energy stored in the electric field (as in Chapters 1 and 3) and S is the domain of the solution. Using the static field assumption, $\mathbf{E} = -\nabla V = -\Sigma V_i \nabla[\phi_i(x, y)]$. When we apply this in (4.28) the signs cancel and the integration applies only to the gradient of the basis functions, as in the 1-D example above.

This formulation is similar to the Galerkin method, but here it is more obvious that each element may have its own electrical properties. Representing (4.28) with basis functions we obtain a linear system of equations in which the unknown potentials are "coupled" to each other in the double sum by coefficients a_{ij} similar to the ones above:

$$W_e = \sum_{j=1}^{N} \sum_{i=1}^{N} \frac{\varepsilon}{2} V_i V_j a_{ij} \tag{4.29a}$$

where:

$$a_{ij} = \iint_S (\nabla\phi_i(x, y)) \cdot (\nabla\phi_j(x, y)) \, dx dy \tag{4.29b}$$

and the full system of equations which results is:

$$W_e = \frac{\varepsilon}{2}[V_i]^t[a_{ij}][V_j] \quad (4.29c)$$

where the superscript t denotes the transpose. For a solution space which has N nodes (Figure 4.6b):

$$W_e = \frac{\varepsilon}{2}[V_1 \cdots V_N] \begin{bmatrix} a_{11} & a_{12} & \cdots & a_{1N} \\ a_{21} & a_{22} & \cdots & a_{2N} \\ \cdots & & & \\ a_{N1} & & & a_{NN} \end{bmatrix} \begin{bmatrix} V_1 \\ V_2 \\ \cdots \\ V_N \end{bmatrix}. \quad (4.29d)$$

We expect the a_{ij} to be a sparse matrix. This formulation has assumed a homogeneous medium. We may include the general inhomogeneous case by inserting the proper value for ε in each element and including the permittivity in the calculation of the a_{ij} coupling coefficients. For convenience we will henceforth assume that ε has been included in the a_{ij}.

Now that we have a formal representation for the energy we must select the node potentials to minimize it. We set the partial derivative of the energy with respect to each node potential equal to zero which gives N simultaneous equations:

$$0 = \frac{\partial W_e}{\partial V_1} = 2V_1 a_{11} + V_2 a_{12} + \cdots + V_N a_{1N} \quad (4.30a)$$

$$0 = \frac{\partial W_e}{\partial V_2} = V_1 a_{21} + 2V_2 a_{22} + \cdots + V_N a_{2N} \quad (4.30b)$$

$$\text{for node potential } k, \qquad 0 = V_k a_{kk} + \sum_{i=1}^{N} V_i a_{ik}. \quad (4.30c)$$

We may use matrix inversion for smaller systems of equation (4.30); however, in larger systems we can use an iterative scheme such as described above in the discussion of the FDM. That is, we have an initial guess for each node potential (we *may not* start with zero everywhere since it trivializes equation (4.30). Then a new guess for each potential can be obtained from:

$$[V_k]_{\text{new}} = \frac{-1}{a_{kk}} \sum_{i=1}^{N} V_i a_{ki}. \quad (4.31)$$

As in the FDM we stop iteration when the node potentials cease changing—when the residue relaxes.

The method of moments

The method of moments has some of its guiding philosophy in common with the finite element method: the spatial dependence of the potential is separated from the unknowns in the linear system. The major difference is that the method does not use the boundary conditions themselves as the forcing function for the problem (as in the FDM and FEM approaches). Rather, one calculates, for example, the surface charge which must exist on all of the conductors in order to satisfy the boundary conditions. Then, the resulting surface charge distributions are used in the general solution of the Poisson (equation (4.1) is the general solution) with $\rho_v \, dv' = \rho_s \, dS'$ to give a continuous volumetric representation for

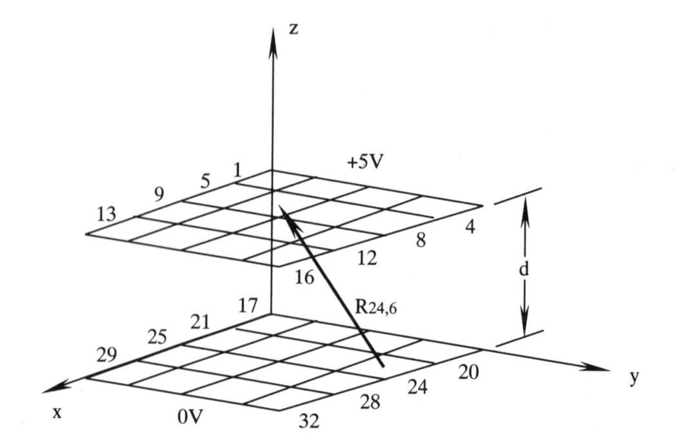

Fig. 4.7 *Method of moments applied to a parallel plate capacitor.*

$V(x, y, z)$. This method is especially powerful when a finite 3-D geometry must be modeled, or when the model space has all Neumann boundaries.

To illustrate the approach, we present the simple case of a finite parallel plate capacitor, Figure 4.7, with plates separated by d on the z-axis. Here the upper and lower plates are each subdivided into 16 finite areas of uniform size, $\mathrm{d}S = \Delta x \Delta y$. For convenience we number the areas with 1 next to the z-axis in the upper plate and 32 at the extremity of the lower plate. Each area is forced to have a uniform surface charge, ρ_{si}, and potential, V_i. Of course, the conductors have known potentials: $V_1 = V_2 = \cdots = V_{16} = +5$ V and $V_{17} = V_{18} = \cdots = V_{32} = 0$ V. The potential at each element is due to the superposition of the influences of each of the surface charge elements:

$$V_1 = \iint_1 \frac{\rho_{s1}\,\mathrm{d}x\mathrm{d}y}{4\pi\varepsilon R_{1,1}} + \iint_2 \frac{\rho_{s2}\,\mathrm{d}x\mathrm{d}y}{4\pi\varepsilon R_{2,1}} + \cdots + \iint_{32} \frac{\rho_{s32}\,\mathrm{d}x\mathrm{d}y}{4\pi\varepsilon R_{32,1}} \tag{4.33}$$

where $R_{2,1}$ is the scalar distance from element 2 to element 1, etc. A linear system of equations results:

$$\begin{bmatrix} a_{1,1} & a_{1,2} & a_{1,3} & \cdots & a_{1,32} \\ a_{2,1} & a_{2,2} & a_{2,3} & \cdots & a_{2,32} \\ \cdots & & & & \cdots \\ a_{32,1} & \cdots & & & a_{32,32} \end{bmatrix} \begin{bmatrix} \rho_{s1} \\ \rho_{s2} \\ \cdots \\ \rho_{s32} \end{bmatrix} = \begin{bmatrix} 5 \\ 5 \\ \cdots \\ 0 \\ 0 \end{bmatrix}. \tag{4.34}$$

The $a_{i,j}$ parameters are evaluated from the plate geometry. Note that none of the $a_{i,j}$ are 0, but by geometric symmetry $R_{2,1} = R_{1,2}$ and the parameter matrix must be symmetric. The parameters contain all of the geometry and are evaluated prior to solving, as in the FEM. In calculating the $a_{i,j}$ parameters for the off-diagonal terms one may treat the surface charge as an equivalent point charge, $q_j = \rho_{sj}\Delta x \Delta y$, located at the center of its element (the $\Delta x \Delta y$ is part of $a_{i,j}$ and the ρ_{sj} is outside of the integral):

$$a_{32,1} = \iint_{32} \frac{\mathrm{d}x\mathrm{d}y}{4\pi\varepsilon R_{32,1}} = \frac{\Delta x \Delta y}{4\pi\varepsilon\sqrt{9\Delta x^2 + 9\Delta y^2 + d^2}} = a_{1,32}. \tag{4.35}$$

However, special care must be given to the diagonal terms since the equivalent point charge would be located at the center of the element creating a singularity since $R_{i,i} = 0$. For diagonal elements we use a transformed integral

$$a_{i,i} = \frac{1}{4\pi\varepsilon\Delta x\Delta y}\int_0^{\Delta y}\left[\int_0^{\Delta x}\frac{dx}{\sqrt{x^2+y^2}}\right]dy$$
$$= \frac{1}{4\pi\varepsilon\Delta x\Delta y}\int_0^{\Delta y}\log\left[\frac{\Delta x+\sqrt{\Delta x^2+y^2}}{y}\right]dy \tag{4.36}$$

which is tedious to evaluate but in this Cartesian case (see Iskander 1992) if $\Delta x = \Delta y = \Delta$ it boils down to:

$$a_{i,i} = \frac{\Delta}{\pi\varepsilon}\ln[1+\sqrt{2}] \cong 0.8814\frac{\Delta}{\pi\varepsilon}. \tag{4.37}$$

Once the conductor surface charges are known the potential can be calculated at any location in the volume by application of equation (4.1). The integral of (4.1) need only be calculated over the surface charges. The reader should note at this point that for the finite plate capacitor example the surface charges will not be constant—far from it! In order to maintain a constant potential on the plates and satisfy the **E**-field boundary conditions, it turns out that edge elements (2, 3, 5, 12, ...) must have higher charge magnitudes than central elements (6, 7, 10, 11, ...); and that corner elements (1, 4, 13, 16, ...) have even larger charge magnitudes. The finite nature of the geometry is important in this problem.

Of course, in more complex geometries it is necessary to evaluate the parameter integrals numerically, being careful about singular points. This method is inherently 3-D but, as presented, applies only to homogeneous media. Heterogeneous media may be treated by replacing the medium with equivalent induced sources. That is, the inhomogeneity is replaced by an equivalent polarization charge distribution, which adds several equations to the linear system. This approach will be discussed later in this chapter.

The quasi-static methods described in this section may be used to study small segments of material in an RF field with acceptable accuracy as long as neither the depth nor area dimensions approach fractions of a wavelength (say $\lambda/20$, that is, less than 5% phase error). If a large area workpiece is to be analyzed, the full time-varying analysis (Section 4.4 of this chapter) must be used to handle the planar potentials and fields; but if the material is thin compared with λ the quasi-static calculation will be accurate along the depth axis. Consequently, quasi-static calculations may be safely applied, even in most very large RF dryer calculations, for example, but it should be noted that large plates may not represent a constant potential surface because of phase and amplitude variations.

4.2.2 Approximately uniform electric field applicators

This section describes the interaction between the workpiece and the external field when the heating is dominated by nearly uniform electric fields. We discuss several examples illustrating the effect of workpiece geometry. Classical analytical solutions of the Laplace equation are available for planar, cylindrical (in the transverse orientation), and spherical shapes. Other shapes may require numerical solutions.

Parallel plate applicator

Perhaps the simplest and most common RF applicator is a pair of large parallel plates, Figure 4.8a. There will be fringing fields at the edges of the plates which make the surface charge distribution non-uniform, as in the above capacitor example. Also, in practice for large dimension loads with large dimension plates (of the order of meters) the plate voltage will not be constant: in general it "sags" (decreases in magnitude) away from the feed points. This is because elemental capacitance, inductance and plate resistance make the large plate behave as a loaded transmission line (Chapter 2). There are many "tricks" which can be used to perk up the plate potential at the ends of the plates—one may, for example, distribute several low capacitance pH to μH inductors (3 or 4 turns, 10 to 20 cm in diameter and in height, as an example) around the plate edge. For purposes of discussion, however, we

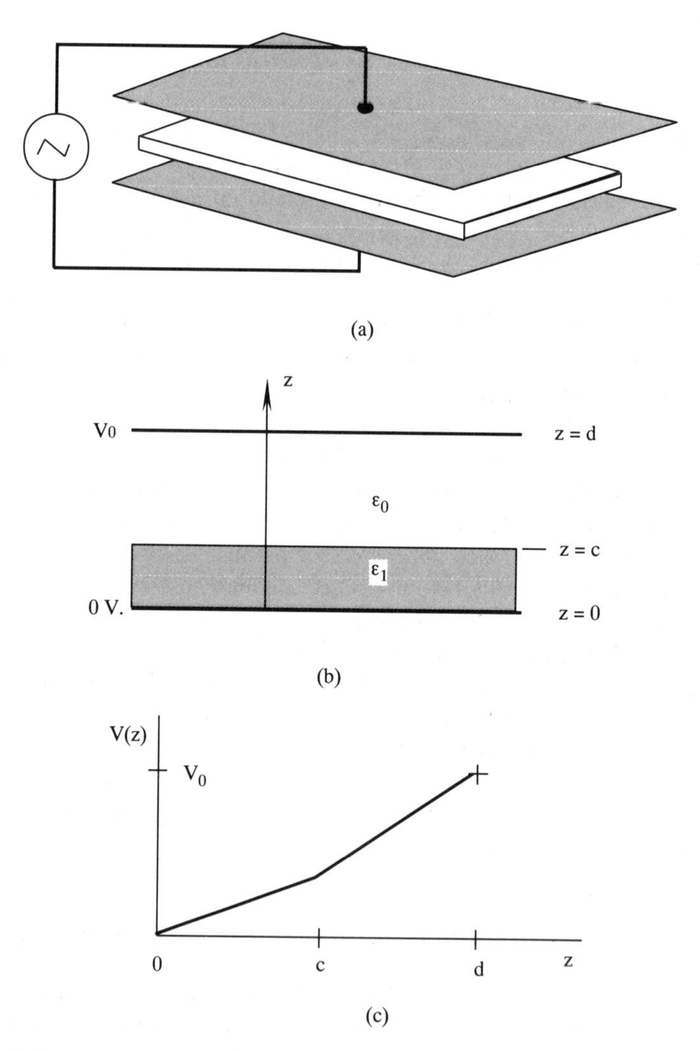

Fig. 4.8 *Parallel plate applicator. (a) Applicator with load in place; load is just inside of fringe field. (b) One-dimensional model of planar workpiece and applicator. (c) Laplace equation planar workpiece example solution.*

need to focus on the load itself. So, we will treat the parallel plate applicator as though it is ideal and has a uniform potential. Also, we will ignore the fringing fields and assume that the electric field is uniform between the plates. For most load configurations, where the workpiece is inside the plate perimeter, these assumptions are quite realistic.

Planar workpiece

Under the uniform field assumption a continuous homogeneous plane of thickness c between the plates separated by d on the z-axis, Figure 4.8b, results in a 1-D Laplace solution for $V(z)$:

$$\nabla^2 V = \frac{d^2V}{dz^2} = 0 \qquad \text{means that} \qquad V(z) = az + b. \tag{4.38}$$

The solution is only valid within a homogeneous region and there is a discontinuity at $z = c$. Consequently, there are two solution spaces, $V_1(z)$ for $0 \leqslant z \leqslant c$ and $V_2(z)$ for $c \leqslant z \leqslant d$, and four unknown coefficients, a_1, b_1 and a_2, b_2. The boundary conditions are thus:

$$V_1(0) = 0 \tag{4.39a}$$

$$V_2(d) = \mathrm{V_0} \tag{4.39b}$$

$$V_1(c) = \mathrm{V}_2(c) \tag{4.39c}$$

$$\varepsilon_1 \left[\frac{dV_1}{dz}\right]_{z=c} = \varepsilon_2 \left[\frac{dV_2}{dz}\right]_{z=c} \tag{4.39d}$$

where equations (4.39a) and (4.39b) are the Dirichlet boundaries, (4.39c) comes from the continuity constraint on potential and (4.39d) is the normal electric field boundary condition (assuming no free charge at the interface). We may easily include the case of a lossy dielectric material by using a complex permittivity, ε^* (F/m). In the usual case there is free space between the plates, so $\varepsilon_2 = \varepsilon_0$. We may apply (4.39a) to see that $b_1 = 0$, and (4.39d) to see that $a_2 = a_1\varepsilon_1/\varepsilon_0$. From (4.39c) $b_2 = a_1 c(\varepsilon_0 - \varepsilon_1)/\varepsilon_0$. Finally, from (4.39b): $a_1 = V_0/\{c + (d-c)\varepsilon_1/\varepsilon_0\}$. So, the potential and electric field in the workpiece are given by:

$$V_1(z) = \frac{(\varepsilon_0/\varepsilon_1)V_0 z}{(d-c) + (\varepsilon_0/\varepsilon_1)c} \tag{4.40a}$$

$$\mathbf{E}_1 = -\frac{(\varepsilon_0/\varepsilon_1)V_0}{(d-c) + (\varepsilon_0/\varepsilon_1)c}\mathbf{a}_z \tag{4.40b}$$

and the capacitance per unit area of material is:

$$C_{\text{total}} = \left[\frac{1}{C_1} + \frac{1}{C_2}\right]^{-1} = \frac{\varepsilon_0}{(d-c) + (\varepsilon_0/\varepsilon_1)\mathrm{c}}(\mathrm{F/m^2}). \tag{4.40c}$$

The potential is shown in Figure 4.8c assuming that $\varepsilon_1 = 2\varepsilon_0$ and $d = 2c$. This analysis applies equally well when the workpiece is suspended between the plates with two air gaps of thickness t_1 and t_2 where $(d-c) = t_1 + t_2$.

Ordinarily $\varepsilon_1 > \varepsilon_0$ and three practical design trends may be seen. First, much of the voltage drop is in the air, as would be expected. Second, for a constant V_0 as the air gap is thinned $(d-c)$ approaches 0 and the workpiece field increases to V_0/c; however, the impedance presented to the generator decreases: $Z_{\text{eq}} = V_0/\{\mathrm{j}\omega\varepsilon_1 E_1 \Delta x \Delta y\}$ where $\Delta x \Delta y$ is

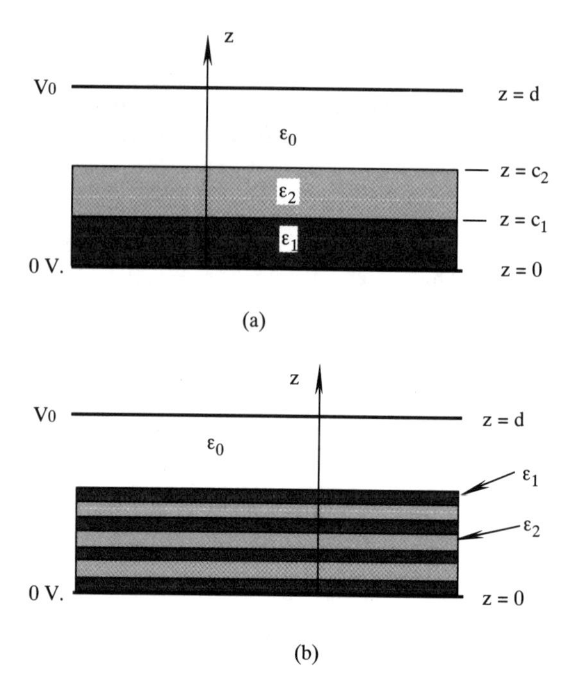

Fig. 4.9 *Multi-layer planar load in parallel plate applicator (a) Three-layer solution geometry. (b) Multiple-layer, two-property planar load.*

the plate area. Third, the equivalent admittance increases linearly with plate area. Admittances in parallel add, so we may conveniently express the admittance per unit area as $Y_{\text{eq}} = I/V = \{\text{j}\omega\varepsilon_1 E_1\}/V_0$ (S/m^2).

We may easily extend the result to a three-medium system (Figure 4.9a) where medium 1 (thickness c_1) and medium 2 (thickness $c_2 - c_1$) form a sandwich to be heated. As before, $\varepsilon_3 = \varepsilon_0$ and the potential must be solved for three regions and must match at both $z = c_1$ and $z = c_2$. Again $b_1 = 0$; also, $\varepsilon_1 a_1 = \varepsilon_2 a_2 = \varepsilon_0 a_3$, and the useful solutions are:

$$a_1 = \frac{V_0}{(\varepsilon_1/\varepsilon_0)(d - c_2) + (\varepsilon_1/\varepsilon_2)(c_2 - c_1) + c_1} \tag{4.41a}$$

$$\mathbf{E}_1 = \frac{-V_0 \mathbf{a}_z}{(\varepsilon_1/\varepsilon_0)(d - c_2) + (\varepsilon_1/\varepsilon_2)(c_2 - c_1) + c_1} \tag{4.41b}$$

$$\mathbf{E}_2 = -\frac{(\varepsilon_1/\varepsilon_2) V_0 \mathbf{a}_z}{(\varepsilon_1/\varepsilon_0)(d - c_2) + (\varepsilon_1/\varepsilon_2)(c_2 - c_1) + c_1} \tag{4.41c}$$

As in the previous case, this solution also applies to a multiple-layer two-medium sandwich like plywood (Figure 4.9b) where the total thickness of all of the medium 1 layers is c_1 and the total thickness of all of the medium 2 layers is $(c_2 - c_1)$. As in the previous case the admittance per unit area (S/m^2) is the current divided by the voltage:

$$Y_{\text{eq}} = \frac{\text{j}\omega\varepsilon_1 E_1}{V_0} = \frac{\text{j}\omega\varepsilon_2 E_2}{V_0} = \frac{\text{j}\omega}{\dfrac{1}{\varepsilon_0}(d - c_2) + \dfrac{1}{\varepsilon_2}(c_2 - c_1) + \dfrac{1}{\varepsilon_1}c_1}. \tag{4.42}$$

Spherical workpiece

Next, we inspect the case of a uniform spherical workpiece between the plates, for which an analytical solution is also possible. We will construct the solution in two stages by first considering a conducting sphere between the plates and then a dielectric sphere. This classical solution is accurate whenever the sphere radius, a, is small compared with λ, and may thus be used both in RF fields for ordinary dimensions of spheres and at microwave frequencies for droplets and small inclusions. The perfectly conducting sphere solution will prove useful in Chapter 8 and the dielectric sphere here and in Chapter 6. The dielectric sphere result can be generalized to lossy media by noting that the complex permittivity can be substituted (and the potentials are then phasors).

Conducting sphere in a uniform field Assume a uniform externally applied electric field between the plates $\mathbf{E}_0 = -E_0\mathbf{a}_z$, with a perfectly conducting sphere at the origin (Figure 4.10a). Because the electric field is $\mathbf{E} = -\nabla V$, we must have $V(z) = E_0 z$ far from the sphere. In spherical coordinates $z = r\cos(\theta)$; consequently, far from the sphere the potential is $V(\infty, \theta) = E_0 r\cos(\theta)$. It is convenient to choose $V = 0$ in the x–y plane. Therefore, $V(r, \pi/2) = 0$, and $V(a, \theta) = 0$ since the sphere is conducting. The problem is

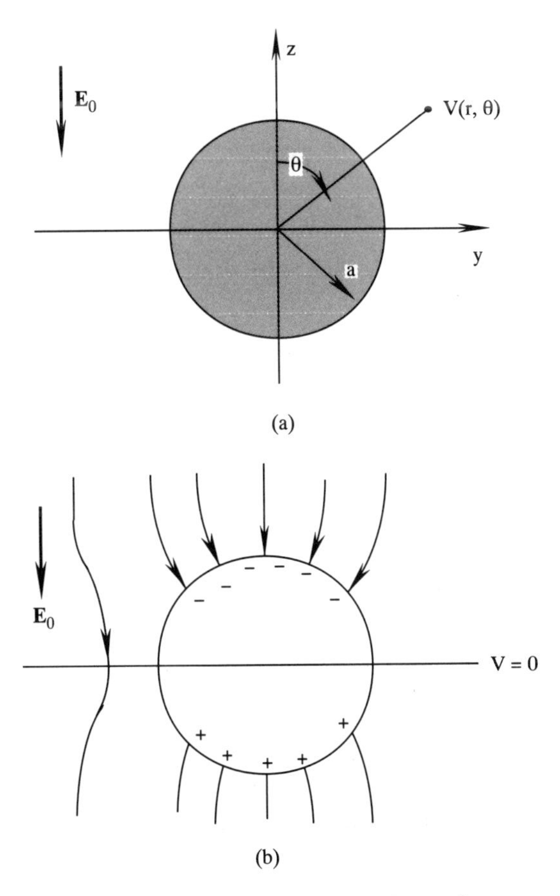

Fig. 4.10 *Laplace equation solution for the conducting sphere problem. (a) Geometry. (b) Field lines around the conducting sphere.*

axisymmetric, so $E_\varphi = 0 = \partial/\partial\varphi$ everywhere. The electric field inside the sphere must be 0, so just outside the sphere at $r = a^+$ we have: $E_\theta = 0$, $\varepsilon E_r = \rho_s$ and the external field terminates (or originates) on the surface.

The general solution of the axisymmetric spherical Laplace equation is:

$$V(r,\theta) = \sum_{k=0}^{\infty} \left(A_k r^k + \frac{B_k}{r^{(k+1)}} \right) [C_{1k} P_k(\theta) + C_{2k} Q_k(\theta)] \tag{4.43}$$

where the P_k and Q_k are Legendre polynomials of the first and second kind, respectively. Since $\theta = 0$ and $\theta = \pi$ are included in the solution space the C_{2k} must all be zero to avoid a singularity in the Q_k. The solution is then:

$$V(r,\theta) \sum_{k=0}^{\infty} \left(A_k r^k + \frac{B_k}{r^{(k+1)}} \right) P_k(\theta) \tag{4.44}$$

and the Legendre polynomials are as given in Table 4.2.

We keep only as many terms in the series as are required to satisfy the boundary conditions, and thus only $k = 0$ and $k = 1$ are kept:

$$V(r,\theta) = A_0 + \frac{B_0}{r} + \left(A_1 r + \frac{B_1}{r^2} \right) \cos(\theta). \tag{4.45}$$

At large r there is no constant term so $A_0 = 0$ and we must have $A_1 = E_0$ (the B_k terms disappear at large r).

At $r = a$ we must satisfy $V(a,\theta) = 0$ at every value of θ:

$$0 = \frac{B_0}{a} + \left(E_0 a + \frac{B_1}{a^2} \right) \cos(\theta) \tag{4.46}$$

so $B_0 = 0$ and $B_1 = -E_0 a^3$. The solution for $V(r,\theta)$ is:

$$V(r,\theta) = E_0 r \left[1 - \left(\frac{a}{r} \right)^3 \right] \cos(\theta). \tag{4.47}$$

Table 4.2 *Legendre polynomials of the first kind.*

$P_0(\cos\theta) = 1$
$P_1(\cos\theta) = \cos\theta$
$P_2(\cos\theta) = \dfrac{3\cos^2\theta - 1}{2}$
$P_3(\cos\theta) = \dfrac{5\cos^3\theta - \cos\theta}{2}$
...
$P_n(x) = \dfrac{1}{2^n n!} \dfrac{d^n (x^2-1)^n}{dx^n}$

The electric field at every location outside of the conductor may be calculated from the gradient of V:

$$\mathbf{E} = -E_0\left[1+2\left(\frac{a}{r}\right)^3\right]\cos(\theta)\mathbf{a}_r + E_0\left[1-\left(\frac{a}{r}\right)^3\right]\sin(\theta)\mathbf{a}_\theta. \quad (4.48)$$

Finally, the electric field just outside the sphere (at $r = a^+$) and surface charge are:

$$\mathbf{E}(r = a^+) = -3E_0\cos(\theta)\mathbf{a}_r \quad (4.49a)$$

$$\rho_s = -3\varepsilon E_0\cos(\theta). \quad (4.49b)$$

The presence of the sphere distorts the uniform field (Figure 4.10b). The field lines terminate on the negative surface charge induced on the upper hemisphere and originate on the positive surface charge of the lower hemisphere. Note that $\rho_s = 0$ on the equator and that (as must be the case) the net charge for the sphere is zero.

In this case the concept of equivalent admittance is not as clear as in the previous examples because the field is not uniform in the $x-y$ plane. For discussion we will study a cylindrical volume of diameter $10a$ and height $10a$. This box fully characterizes the sphere effects since contributions of the fields due to the sphere (the a^3/r^3 terms) are negligible at about 0.8% on the Cartesian axes at $\pm 5a$. If no metal sphere was present the total "admittance" (S) of the volume would be:

$$Y_{eq} = \frac{I}{V} = \frac{\pi r^2 j\omega\varepsilon_0 E_0}{10aE_0} = \frac{25\pi a^2 j\omega\varepsilon_0}{10a} = j\omega\varepsilon_0 2.5\pi a. \quad (4.50)$$

With the sphere in place we should calculate the current at a location where the electric field is normal to a convenient plane. The current is displacement current and the $x-y$ plane will do nicely. There is no current within the sphere ($\mathbf{E} = 0$) so the total $x-y$ plane ($\theta = \pi/2$) current is found by integrating outside of the sphere:

$$I = \iint \mathbf{J}\cdot d\mathbf{S} = \int_0^{2\pi}\int_a^{5a} j\omega\varepsilon_0 E_0\left[1-\left(\frac{a}{r}\right)^3\right] r\,dr d\varphi \quad (4.51a)$$

$$Y_{eq} = \frac{I}{V} = \frac{22.4\pi a^2 j\omega\varepsilon_0}{10a} = j\omega\varepsilon_0 2.24\pi a. \quad (4.51b)$$

The sphere results in an admittance decrease of about 10% which may at first appear paradoxical since one would normally expect an increased admittance due the conducting sphere load. While this might be explained by the observation that we have made some approximations in the calculation which may not be strictly valid, it should also be remembered that the energy required to establish the surface charge came from the external field, which decreases the energy stored in it. So the decreased admittance over the equivalent free space volume is not entirely unexpected.

Another way to visualize the effect of the sphere is to imagine a scattered field from the sphere charge distribution which interacts with $\mathbf{E}_0$ to give the total field of equation (4.48). The sphere with its induced charges looks like an electric dipole, and the scattered field calculation may be conducted using this approach. We will apply the scattering description to the case of a dielectric sphere as well. At high frequency the effect of the scattered field is strongly frequency dependent.

Dielectric sphere in a uniform field Now we consider the case of a dielectric sphere of permittivity ε_2 in a medium of permittivity ε_1 (usually $\varepsilon_1 = \varepsilon_0$) in the same uniform field, $\mathbf{E}_0$. We define the potential in regions 1 and 2 by V_1 and V_2 respectively. The potential is axisymmetric in each region (independent of φ) and the general solution is now given by:

$$V_1 = \sum_{k=0}^{+\infty} \left(A_k r^k + \frac{B_k}{r^{k+1}} \right) P_k(\cos\theta) \tag{4.52}$$

$$V_2 = \sum_{i=0}^{+\infty} \left(C_i r^i + \frac{D_i}{r^{i+1}} \right) P_i(\cos\theta) \tag{4.53}$$

where again $P_j(\cos\theta)$ are the Legendre polynomials of the first kind and degree j. The four coefficients for each term A_k, B_k, C_i, D_i must be calculated from the boundary conditions. As previously, far from the embedded sphere:

$$\lim_{r\to+\infty} V_1 = -E_0 z = E_0 r \cos\theta. \tag{4.54}$$

The potential is a continuous function of position, so:

$$\lim_{r\to a+} V_1 = \lim_{r\to a-} V_2. \tag{4.55}$$

The normal component of the vector **D**-field is continuous because no free surface charge is available from the dielectrics:

$$\varepsilon_1 \lim_{r\to a+} \frac{\partial V_1}{\partial r} = \varepsilon_2 \lim_{r\to a-} \frac{\partial V_2}{\partial r}. \tag{4.56}$$

The reference surface has both V_1 and $V_2 = 0$ on the $x-y$ plane, which implicitly satisfies the tangential electric field boundary condition. The implications are as follows. Equation (4.54) implies that all of the coefficients A_k are zero with the exception of A_1. If V_2 is zero at the center of the sphere, all of the D_i coefficients must be zero. Equations (4.55) and (4.56) imply that B_n and C_n are zero except when $n = 1$. It follows, then, that the coefficients are determined from:

$$\frac{B_1}{A_2} - E_0 a = C_1 a \tag{4.57}$$

$$\varepsilon_1 \left(\frac{2B_1}{a^3} + E_0 \right) = -\varepsilon_2 C_1 \tag{4.58}$$

where

$$B_1 = \frac{\varepsilon_2 - \varepsilon_1}{\varepsilon_2 + 2\varepsilon_1} a^3 E_0 \tag{4.59}$$

and

$$C_1 = \frac{3\varepsilon_1}{\varepsilon_2 + 2\varepsilon_1} E_0. \tag{4.60}$$

It also follows that outside of the sphere:

$$V_1 = \left(\frac{\varepsilon_1 - \varepsilon_2}{\varepsilon_2 + 2\varepsilon_1}\frac{a^3}{r^3} + 1\right) E_0 r \cos(\theta) \tag{4.61a}$$

$$\mathbf{E}_1 = -E_0\left[1 + 2\frac{\varepsilon_2 - \varepsilon_1}{\varepsilon_2 + 2\varepsilon_1}\left(\frac{a}{r}\right)^3\right]\cos(\theta)\mathbf{a}_r + E_0\left[1 - \frac{\varepsilon_2 - \varepsilon_1}{\varepsilon_2 + 2\varepsilon_1}\left(\frac{a}{r}\right)^3\right]\sin(\theta)\mathbf{a}_\theta. \tag{4.61b}$$

The potential and heating field inside the sphere are:

$$V_2 = \frac{3\varepsilon_1}{\varepsilon_2 + 2\varepsilon_1} E_0 r \cos(\theta) \tag{4.62a}$$

$$\mathbf{E}_2 = \frac{3\varepsilon_1}{2\varepsilon_1 + \varepsilon_2}\mathbf{E}_0 = \frac{3\varepsilon_0}{2\varepsilon_0 + \varepsilon_2}\mathbf{E}_0 \tag{4.62b}$$

when surrounded by free space. The internal field is uniform, and is proportional to the externally applied electric field. If we make a calculation of the admittance in the same $10a$ by $10a$ cylindrical volume as above, we see that now the dielectric sphere has contributed a displacement current of:

$$I_2 = \mathrm{j}\omega\varepsilon_2 \int_0^{2\pi}\int_0^a \frac{3\varepsilon_0 E_0}{2\varepsilon_0 + \varepsilon_2} r\,\mathrm{d}r\mathrm{d}\varphi = \pi a^2 \mathrm{j}\omega\varepsilon_2 E_0 \frac{3\varepsilon_0}{2\varepsilon_0 + \varepsilon_2} \tag{4.63a}$$

and in the space outside the sphere:

$$I_1 = \mathrm{j}\omega\varepsilon_0 \int_0^{2\pi}\int_a^{5a} E_{1\theta} r\,\mathrm{d}r\mathrm{d}\varphi = \pi a^2 \mathrm{j}\omega\varepsilon_0 E_0\left[24 - \frac{\varepsilon_2 - \varepsilon_0}{\varepsilon_2 + 2\varepsilon_0}\frac{8}{5}\right] \tag{4.63b}$$

so that the equivalent admittance is now:

$$Y_{\mathrm{eq}} = \frac{I_1 + I_2}{V} = \frac{I_1 + I_2}{10aE_0}$$

$$Y_{\mathrm{eq}} = \frac{\mathrm{j}\omega\pi a\varepsilon_0}{10}\left[\frac{3\varepsilon_2 - \frac{8}{5}(\varepsilon_2 - \varepsilon_0)}{\varepsilon_2 + 2\varepsilon_0} + 24\right] \tag{4.63c}$$

which is, of course, larger than that of the free space volume alone, as we expect. We will make use of the dielectric sphere results in future chapters.

Cylindrical workpiece

There is also a convenient analytical solution for conducting and dielectric cylinders oriented transversely to the uniform external field (Figure 4.11). In this case we orient an infinite cylinder of radius a along the z-axis and assume that far from the cylinder $\mathbf{E} = -E_0\mathbf{a}_x$. The solution for a perfectly conducting cylinder follows the same development as for the conducting sphere above; however, the solution for the Laplace equation in cylindrical coordinates where $V(r, \phi, z) = V(r, \phi)$ involves trigonometric functions rather than

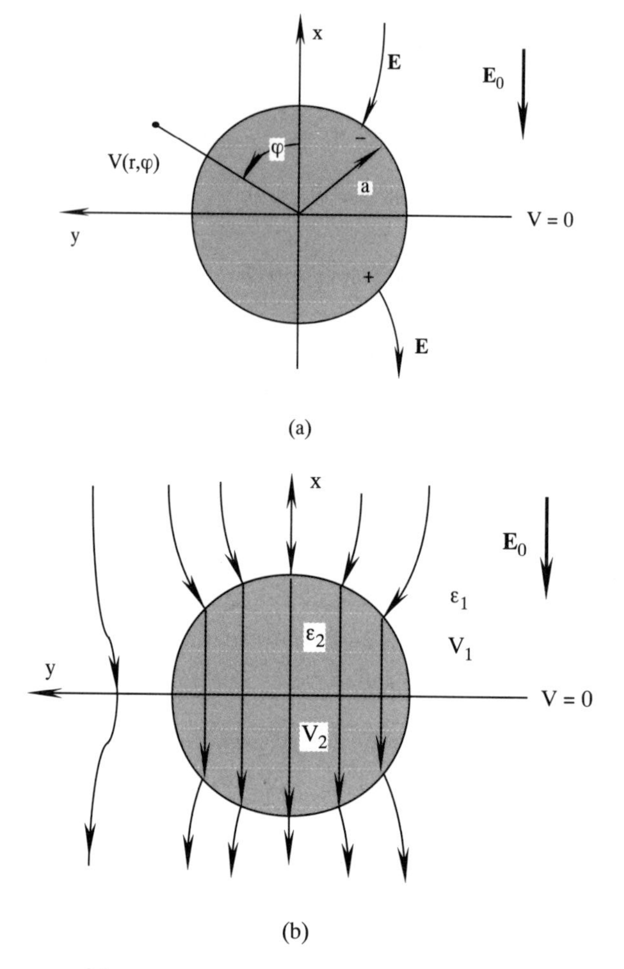

Fig. 4.11 *Transverse cylinder geometry. (a) Conducting cylinder transverse to the electric field. (b) Dielectric cylinder transverse to the uniform field.*

Legendre polynomials:

$$V(r,\varphi) = \sum_{n=0}^{\infty} \left[A_{1n} r^n + A_{2n} r^{-n}\right] \left[B_{1n} \cos(n\varphi) + B_{2n} \sin(n\varphi)\right] \tag{4.64}$$

We choose the y–z plane for the potential reference ($x = 0$). Far from the cylinder $V(x) = E_0 x$, so $V(r,\varphi) = E_0 r \cos(\varphi)$; consequently the B_{2n} terms must all be zero, the B_{1n} terms may then be lumped with the A_{jn} constants with no loss in generality, and $n = 0$ and $n = 1$ are sufficient to satisfy the boundary conditions.

For the first case, a conducting cylinder (Figure 4.11a), there is one potential function; that is, in the medium surrounding the cylinder. The boundary condition is that $V(a,\varphi) = 0$:

$$V(a,\varphi) = A_{10} + \left[A_{11} a + \frac{A_{21}}{a}\right] \cos(\varphi) = 0 \tag{4.65}$$

from which $A_{10} = 0$ and $A_{21} = -A_{11}a^2$. Applying the boundary condition at large r gives $A_{11} = E_0$ and:

$$V(r,\varphi) = E_0\left[r - \frac{a^2}{r}\right]\cos(\varphi) \tag{4.66a}$$

$$\mathbf{E} = -E_0\left[1 + \frac{a^2}{r^2}\right]\cos(\varphi)\mathbf{a}_r + E_0\left[r - \frac{a^2}{r}\right]\sin(\varphi)\mathbf{a}_\varphi \tag{4.66b}$$

and the surface charge $\rho_s = \varepsilon E_r(a) = -2\varepsilon E_0\cos(\varphi)$. We omit the admittance estimate in this case.

A dielectric cylinder with permittivity ε_2 suspended in a medium of permittivity ε_1 (Figure 4.11b) with uniform electric field at large $r(\mathbf{E}_0 = -E_0\mathbf{a}_x$, as above) has two solution regions, $V_1(r,\varphi)$ and $V_2(r,\varphi)$. As in the previous case, at large radius the potential is: $V_1(r,\varphi) = E_0 r\cos(\varphi)$. The interior field will be parallel to E_0 so $V_2(r,\varphi) = C_2 r\cos(\varphi)$. Note that $V_1(a,\varphi) = V_2(a,\varphi)$ and the solutions must satisfy the normal electric field boundary condition at $r = a$:

$$V_1(a,\varphi) = V_2(a,\varphi) \quad \text{means that} \quad C_2 a = E_0 a + \frac{A_{21}}{a} \tag{4.67a}$$

$$\varepsilon_2\left[\frac{\partial V_2}{\partial r}\right]_{r=a} = \varepsilon_1\left[\frac{\partial V_1}{\partial r}\right]_{r=a} \quad \text{means that} \quad \varepsilon_2 C_2 = \varepsilon_1\left[E_0 - \frac{A_{21}}{a^2}\right] \tag{4.67b}$$

from which the coefficients are:

$$C_2 = E_0\frac{2\varepsilon_1}{\varepsilon_1 + \varepsilon_2} \quad \text{and} \quad A_{21} = a^2E_0\frac{\varepsilon_1 - \varepsilon_2}{\varepsilon_1 + \varepsilon_2} \tag{4.68}$$

and the potential and electric field within the cylinder are:

$$V_2(r,\varphi) = E_0\frac{2\varepsilon_1}{\varepsilon_1 + \varepsilon_2}r\cos(\varphi) = E_0\frac{2\varepsilon_1}{\varepsilon_1 + \varepsilon_2}x \tag{4.69a}$$

$$\mathbf{E}_2(x,y) = -\nabla V_2(x,y) = -\frac{\partial V_2}{\partial x}\mathbf{a}_x = -E_0\frac{2\varepsilon_1}{\varepsilon_1 + \varepsilon_2}\mathbf{a}_x. \tag{4.69b}$$

As in the previous case for the sphere, the electric field inside the cylinder is uniform. We may realize such a field distribution if the plate electrodes are sufficiently far apart that the $1/r$ terms vanish. This usually results in a very high electrode impedance and low heating rate in the load. If the plates are moved closer to the electrode a uniform external field is not achieved (thus the internal field is also non-uniform) and the top and bottom edges (at $r = a, \varphi = \pm 90°$) heat preferentially. So, it is not particularly instructive to look at the uniform field admittance, as we did in the case of a sphere. The non-uniform heating may be partially alleviated by arranging the electrodes in a Vee configuration around the load (Figure 4.12). Optimum separation and Vee angle are dependent on load permittivity.

A cylindrical dielectric disk of radius a and thickness h placed normal to the field direction (Figure 4.13) introduces additional field distortions at the corners owing to the collision of tangential and normal field requirements. In this axisymmetric configuration the Laplace solutions for $V_1(r,z)$ and $V_2(r,z)$ involve Bessel Functions:

$$V(r,z) = (Ae^{\lambda z} + Be^{-\lambda z})(CJ_0(\lambda r) + DY_0(\lambda r)) \tag{4.70a}$$

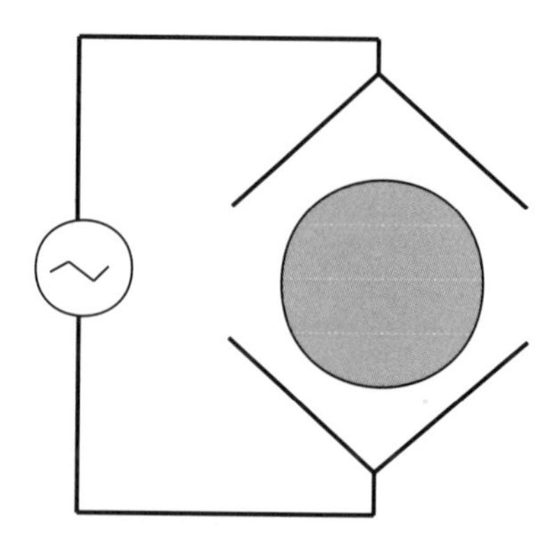

Fig. 4.12 *"Vee" electrode configuration.*

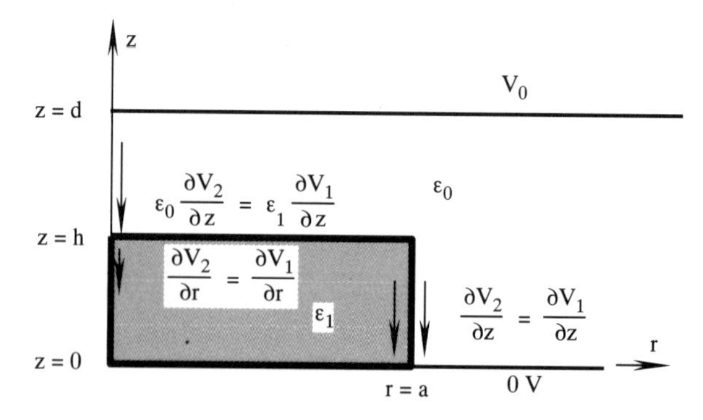

Fig. 4.13 *Boundary conditions for the cylindrical disk between parallel plates.*

or, for the disk:

$$V_1(r, z) = (A_1 e^{\lambda_1 z} + B_1 e^{-\lambda_1 z}) J_0(\lambda_1 r) \tag{4.70b}$$

and for the air gap:

$$V_2(r, z) = (A_2 e^{\lambda_2 z} + B_2 e^{-\lambda_2 z}) J_0(\lambda_2 r) \tag{4.70c}$$

since the D coefficients are zero because $r = 0$ is part of the solution space. The boundary conditions are complicated by their regional differences. First, $V = 0$ for $z = 0$; and $V = V_0$ for $z = d$. Second, by continuity of potential $V_1(r, h) = V_2(r, h)$ for $r \leqslant a$; and third $V_1(a, z) = V_2(a, z)$ for $z \leqslant h$. From the second relation $\lambda_1 = \lambda_2$; and from the third $A_1 e^{\lambda z} + B_1 e^{-\lambda z} = A_2 e^{\lambda z} + B_2 e^{-\lambda z}$. For $r \gg a$ we must have $V_1(r, z) = V(z) = V_0 z/d$. Fourth, at $z = h$ the normal and tangential electric field boundary conditions apply:

$$\varepsilon_1 \left(\frac{\partial V_1}{\partial z}\right) = \varepsilon_0 \left(\frac{\partial V_2}{\partial z}\right) \quad \text{and} \quad \left(\frac{\partial V_1}{\partial r}\right) = \left(\frac{\partial V_2}{\partial r}\right) \tag{4.71a}$$

$$\varepsilon_1 \lambda (A_1 e^{\lambda h} - B_1 e^{-\lambda h}) = \varepsilon_0 \lambda (A_1 e^{\lambda h} - B_1 e^{-\lambda h}) \tag{4.71b}$$

$$(A_1 e^{\lambda h} + B_1 e^{-\lambda h}) \frac{\mathrm{d}(J_0(\lambda r))}{\mathrm{d}r} = (A_2 e^{\lambda h} + B_2 e^{-\lambda h}) \frac{\mathrm{d}(J_0(\lambda r))}{\mathrm{d}r}. \tag{4.71c}$$

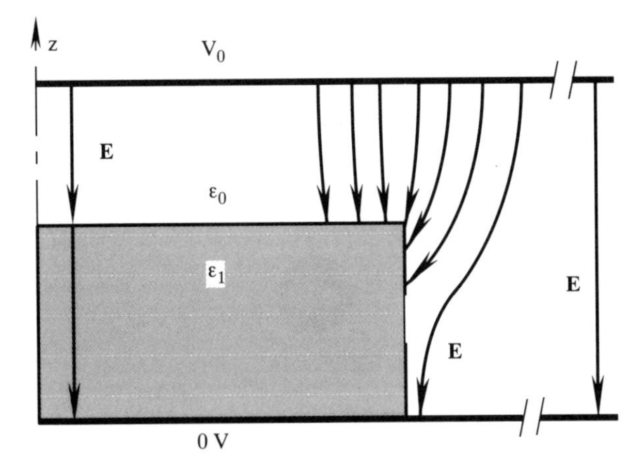

Fig. 4.14 *Sketch of field concentrations at corner for cylindrical disk between parallel plates.*

This suggests that a solution of the form of $\sin(kz)$ must exist in region 2, the air gap, and that λ is complex. A similar set of relations may be written at $r = a$. At $z = h$ for $r = 0$ the external field is normal to the surface, and at $r = a$ for $z = 0$ the external field is tangential to the interface. To solve this in closed form we would superpose all possible solutions using the Fourier–Bessel transform:

$$V(r, z) = \int_0^\infty A(\lambda) e^{-\lambda |z|} J_0(\lambda r)\, d\lambda. \tag{4.72}$$

Ample experimental evidence shows that the solution will result in electric field concentrations at the corner of the disk. This is because at the corner the solution must satisfy conflicting boundary conditions, and the effect of the second order discontinuity at the corner is to concentrate the fields (Figure 4.14). The field must be relatively undistorted near the centerline (an axis of symmetry), though the field strength inside the disk is less than in the air above it due to the normal field boundary condition. No distortion is seen at large radius, and the field is uniform there, as well.

4.2.3 Rod arrays

The parallel plate applicator of the previous section suffers from the limitation that the electric field is primarily normal to the upper surface of the workpiece. The required normal electric field boundary conditions mean that in high loss materials the air gap electric field must be very large to obtain heating — arc formation may be encountered as a result. Also, adequate heating may be difficult to obtain in thin objects. An alternative applicator configuration which addresses both of these disadvantages is the class of rod electrodes. There are two typically used configurations, the "stray field" configuration (Figure 4.15a) and the "staggered through-field" configuration (Figure 4.15b). Both of these arrangements result in an electric field more nearly parallel to the workpiece upper surface. This has three advantages: (1) the dominant electric field boundary condition is the tangential boundary condition so the internal **E**-field approaches the external **E**-field, (2) lower electrode voltages for the same heating rates mean less chance of arc formation, and (3) thin sheets are more effectively heated in this type of field. The disadvantage of this configuration is that when

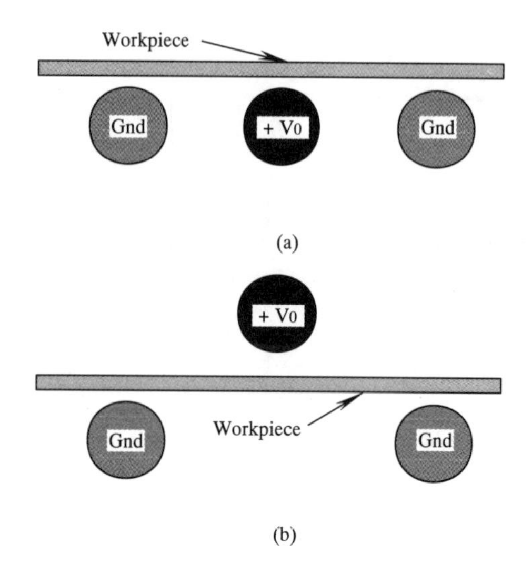

Fig. 4.15 *Typical RF rod electrode configurations. (a) Stray field configuration. (b) Staggered through-field configuration.*

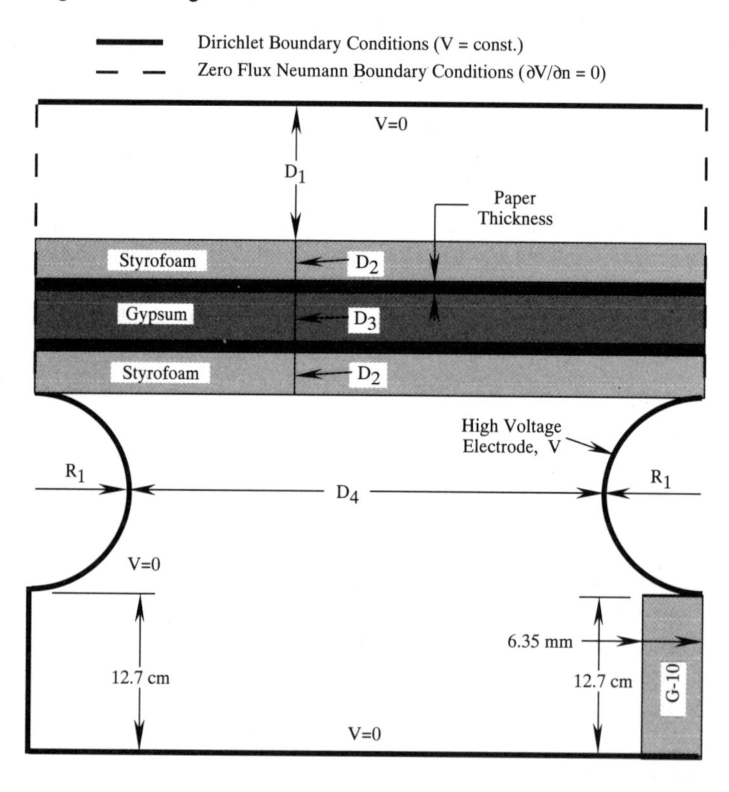

Fig. 4.16 *FEM model of round stray field rod electrodes. (a) Gypsum RF heating experiment geometry. (b) FEM grid used to model experimental geometry.*

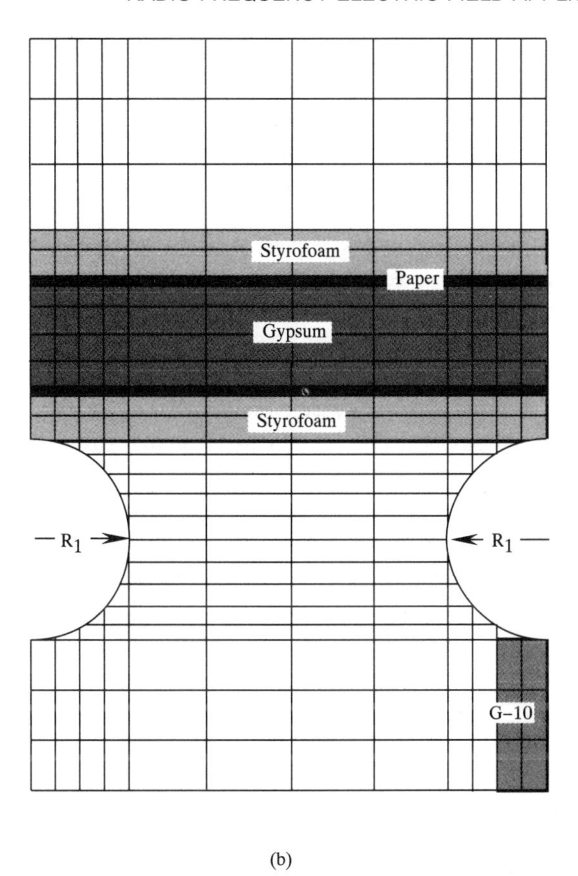

(b)

Fig. 4.16 *(continued)*

moving the material to be heated or dried with a belt over the electrodes the residue of previous loads accumulates on the belt or the electrodes due to gravitational effects. The result is that electrode impedance may change significantly during a run of a single product, confounding tuning and possibly resulting in field concentrations and arcs. Except for those effects, however, the rod electrode configuration can be an extremely effective heating configuration. Field analysis in this configuration must be conducted by numerical techniques because of the complex geometry.

We assume that the voltage distribution is symmetric and two-dimensional between electrode centers. Under these assumptions, the voltage distribution between the electrodes is governed by the two-dimensional Cartesian Laplace equation:

$$\nabla^2 V = \frac{\partial^2 V}{\partial x^2} + \frac{\partial^2 V}{\partial y^2} = 0. \tag{4.72}$$

Equation (4.72) is solved using the finite element method as described above. The electric field, $\mathbf{E}(x, y)$, is obtained from the gradient of the potential and the volume power density, $Q_{\text{gen}}(x, y)$, is calculated from the electric field, $\mathbf{E}(x, y)$, as we have seen before:

$$Q_{\text{gen}} = 2\pi f \varepsilon'' |\mathbf{E}|^2 \tag{4.73}$$

where f is the frequency (Hz), ε'' is the loss factor (F/m), and **E** is the electric field in the material (V/m). The temperature distribution (as described in Chapter 10) can then be determined from the energy balance of Chapter 1.

To eliminate the need to keep the global arrays in computer memory and to enable the programs to be run on personal computers, the frontal solution technique described by Irons (1970) and by Becker *et al.* (1981) was used. For typical grids of 3200 nodes and 780 quadratic elements, the execution time on a CDC Dual Cyber system was 860 CPU seconds, while on a Macintosh Plus the execution time was approximately 6 hours. The stray field electrode experiment which was modeled is shown in Figure 4.16a and the grid used for the FEM calculations is sketched in Figure 4.16b. G-10 indicates the support structure for the high voltage rod, the other rod is at ground potential. Styrofoam and paper layers were included as separate materials in the model. Figure 4.17 shows the round rod staggered through-field electrode experiment (4.17a) and FEM model grid (4.17b). The ground and high potential electrodes were each suspended from a metallic plate, which turned out

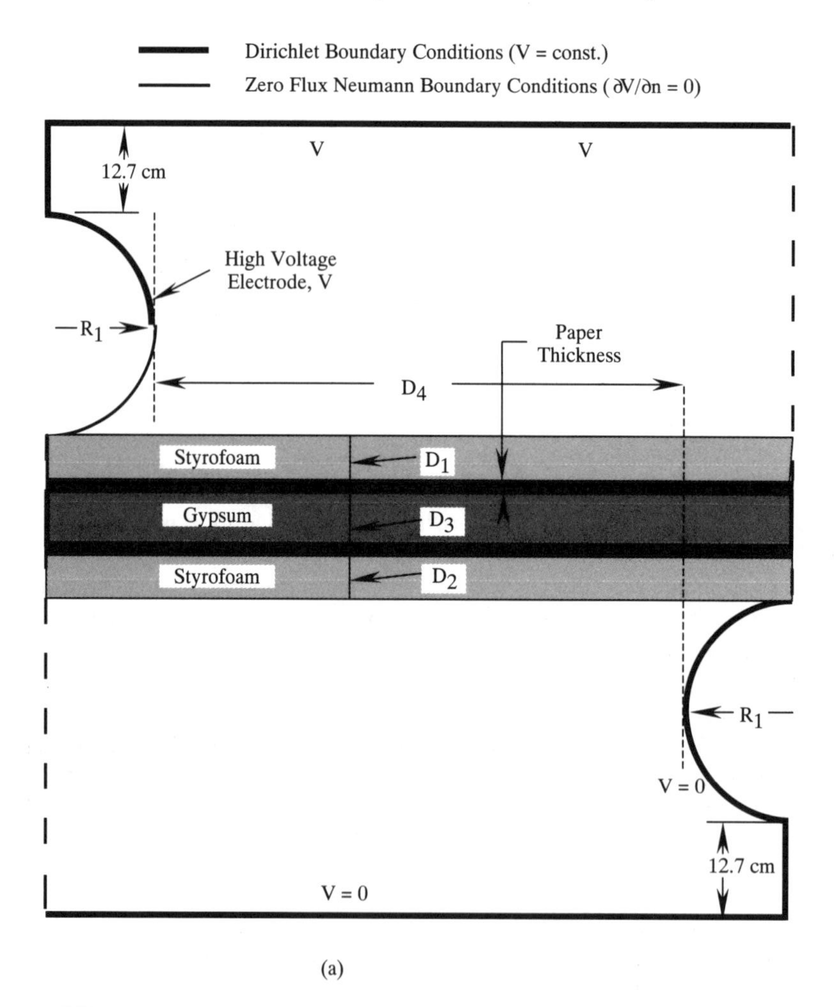

Fig. 4.17 *FEM model of round staggered through-field rod electrodes. (a) Gypsum RF heating experiment geometry. (b) FEM grid used to model experimental geometry.*

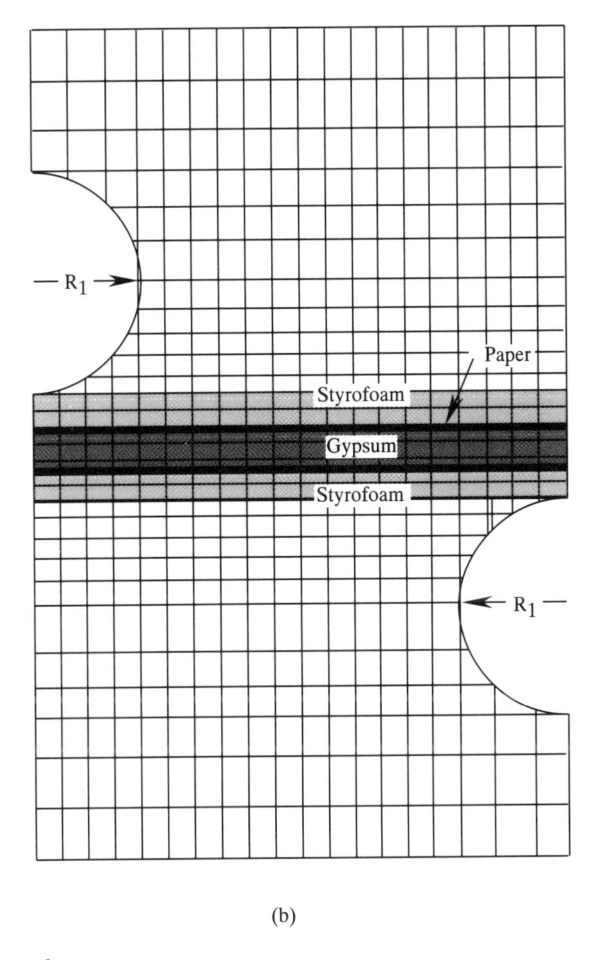

(b)

Fig. 4.17 *(continued)*

to have an important contribution to the field distribution. The electrode supports were included as Dirichlet boundaries The unit space shown in the figures is bounded by planes of symmetry, so the heating pattern is periodic with a spatial period equal to the electrode center separation distance, X_0 (m).

The calculated temperature distributions compare closely with thermographic images of RF heated gypsum. A discussion of these model results is contained in Chapter 10 under the section on gypsum drying. A good match was obtained when a permittivity of $\varepsilon_r = 5.0 - j0.25$ was used for the gypsum, which is within the range of published values.

4.3 RADIO FREQUENCY MAGNETIC FIELD APPLICATORS

Heating is obtained from magnetic fields in two possible ways. First, there can be direct absorption through the imaginary part of the magnetic permeability, μ'' (see the discussion of the Poynting Power Theorem in Chapter 1). In that instance knowledge of the properties and magnetic field strengths are sufficient to make the calculation. However, few materials of interest for industrial processes have significant magnetic dipole moments or permeabilities different from that of free space — this includes most metals and semiconductors, as well. Yet significant heating of metals, lossy dielectrics and semiconductors can be obtained in magnetic fields even in the absence of measurable externally-applied electric fields (remember that under the quasi-static assumption we may consider the electric and magnetic fields separately). Why is that so? The answer involves consideration of Faraday's Law of induction. The second way that magnetic fields may heat dielectrics is by inducing an electric field within the material (according to Faraday's Law), and the heating is actually due to the induced electric field. Remember that by Lenz' law the induced field (resulting in induced current in the material) acts to create a magnetic field which opposes the external magnetic field which induces it. We will consider induced electric fields in solenoidal and ring current applicators and some typical loads in this section. We will not treat direct magnetic field absorption by μ''.

4.3.1 Approximately uniform magnetic field applicator

By working inside of a long helical-wound coil (a solenoid, see Figure 4.18) one can expose the workpiece to an approximately constant magnetic field. Heating is accomplished through magnetic field coupling by a different mechanism than in the electric field case, so the patterns obtained follow different functional forms. For example, in contradistinction to

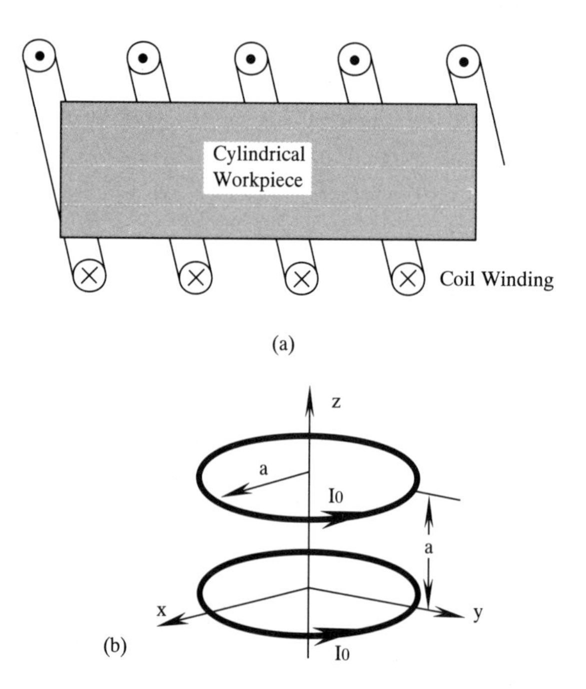

Fig. 4.18 *Two common coil geometries for induction heating. (a) Long helical-wound solenoid coil in cross-section: × is current into the page while · signifies current out of the page. (b) Helmholtz coils: spacing equal to radius gives a large relatively homogeneous field near the center.*

the electric field case, even if the magnetic field is uniform, the volume power density most decidedly, is not. We will now explore several simple examples of magnetic field heating in uniform magnetic fields. The final section will introduce a typical non-uniform field source, the pancake coil.

It is only necessary for either σ or ε'' to be non-zero to obtain heating in a magnetic field; while μ may be complex for some minerals and ceramics this is not the case for most materials which are heated in a magnetic field. Recall from Faraday's law of induction in Chapter 1, that an induced electric field is determined by the time rate of change of the enclosed magnetic flux.

$$\oint_C \mathbf{E}\cdot \mathrm{d}\mathbf{L} = -\frac{\partial}{\partial t}\left\{\iint_S \mathbf{B}\cdot \mathrm{d}\mathbf{S}\right\}. \tag{4.74}$$

The induced electric field will result in a current density, **J**, which (by Lenz's law) counteracts the magnetic field which induces it. Thus a workpiece may absorb energy from a magnetic field even if μ is real.

Solenoid coil field distribution

No real coil will yield a uniform magnetic field. We may calculate the actual magnetic field either by direct integration of the Biot–Savart law (equation (1.7)) or by first calculating the magnetic vector potential, **A** (equation (4.2)), and then taking its curl. In integrating the Biot–Savart law it is much simpler to make the calculations in Cartesian coordinates. This is because the Cartesian unit vectors always point in the same direction; which is not true for the cylindrical and spherical unit vectors. As an example calculation, we have determined the **H**-field component strengths for a solenoid of 8.5 turns of radius 3.8 cm and 12.7 cm in overall length from the Biot–Savart law by numerical integration. The result is plotted in Figure 4.19. We may easily see the dependence of H_z on z and on radius in the figure. Even so, the field is close to uniform if the workpiece is within the solenoid length, and we will make this assumption in the following discussion.

We may improve the uniformity of the **H**-field in a solenoid by increasing the number of turns per unit length; however, the disadvantage is that this increases both the self-inductance and the inter-winding capacitance. A high self-inductance and inter-winding capacitance decreases the self-resonance frequency of the coil, and is not usually desirable.

Cylindrical axial workpiece

The power absorbed from an RF magnetic field depends on the electrical conductivity of the material. When a magnetic field is loaded with a high conductivity load the induced currents drastically reduce the magnetic field strength in the material. A low conductivity load, on the other hand, has very small induced currents, and the magnetic field may be considered essentially unaffected by the presence of the load. For a cylinder placed in a co-axial uniform external magnetic field, $\mathbf{H}_0 = \mathbf{H}_0\mathrm{a}_z$, as shown in Figure 4.20, we initially assume that the effective conductivity, $\sigma_{\text{eff}} = \sigma + \omega\varepsilon''$, is low enough that the induced current density in the material does not affect the local magnetic field. The right hand rule indicates that H_z induces E_φ (only), and the problem is axisymmetric so that E_φ is constant. We may easily see that:

$$\int_0^{2\pi} E_\varphi r\,\mathrm{d}\varphi = 2\pi r E_\varphi = -\mathrm{j}\omega\int_0^{2\pi}\int_0^r \mu H_0 r\,\mathrm{d}r\mathrm{d}\varphi = -\mathrm{j}\omega\mu\pi r^2 H_0 \tag{4.75a}$$

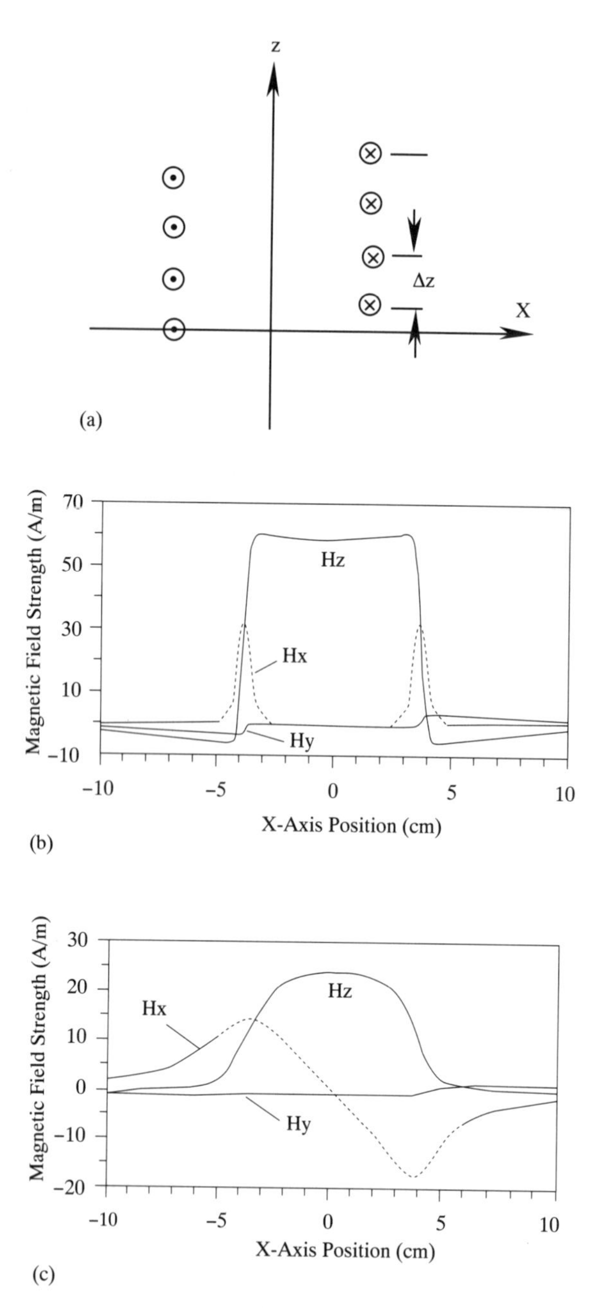

Fig. 4.19 *Outlet plane magnetic field strength, H_z component, for an 8.5 turn long helical solenoid of 3.8 cm radius, 12.7 in length. (a) Geometry for long solenoid magnetic field calculation. Coil pitch angle was determined from Δz, counter clockwise current was assumed, as shown. Calculation was executed along the x-axis at different x–y plane locations (z = constant). (b) Mid-coil plane magnetic field strength components, for an 8.5 turn helical solenoid of 3.8 cm radius, 12.7 cm long. (c) Outlet plane magnetic field strength components 1 cm below the coil, for an 8.5 turn helical solenoid of 3.8 cm radius, 12.7 cm long.*

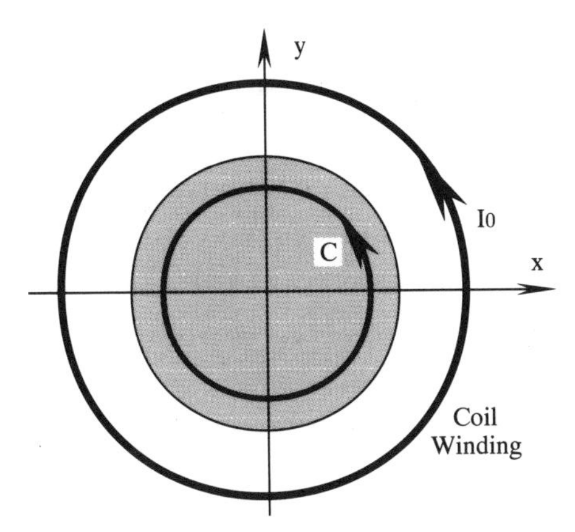

Fig. 4.20 *End view of the lossy cylinder in a solenoid coil: integration contour C encloses magnetic flux.*

and:

$$E_\varphi = \frac{-j\omega\mu r}{2} H_0. \tag{4.75b}$$

Consequently, the volumetric power density, $Q_{gen}(W/m^3)$, depends on the square of the frequency and of the radius, r, from the coil center:

$$Q_{gen} = \sigma_{eff}|\mathbf{E}|^2 = \sigma_{eff}[\omega\mu H_0 r]^2. \tag{4.76}$$

In the remainder of this discussion we will assume that all of the losses are represented by an effective electrical conductivity, and will omit the subscript.

The magnetic field in the example is co-axial to the cylinder, and either solenoid or Helmholtz coils (Figure 4.18b) would give useful approximations to this heating pattern. As can be seen in the above function, even if the magnetic flux density is constant over the volume of interest, the power generation term is parabolically weighted toward the periphery; we expect edge heating in induction geometries.

In higher electrical conductivity materials—for example, metals or graphite fiber composites—the electromagnetic fields are more complex than the simple reduction of Faraday's law in the previous example. In the high conductivity case, induction of current in the material (the source of the heating electric field, **E**) will serve to reduce the strength of the local magnetic field, **H**. So, the coupled equation set (4.77), Faraday's law and Ampere's law respectively, must be solved:

$$\oint_C \mathbf{E} \cdot d\mathbf{L} = -j\omega \left[\iint \mathbf{B} \cdot d\mathbf{S}\right] \quad \text{or} \quad \nabla \times \mathbf{E} = -j\omega\mu \mathrm{H} \tag{4.77a}$$

$$\oint_C \mathbf{H} \cdot d\mathbf{L} = \iint (\sigma + j\omega\varepsilon)\mathbf{E} \cdot d\mathbf{S} \quad \text{or} \quad \nabla \times \mathrm{H} = (\sigma + j\omega\varepsilon)\mathbf{E} \tag{4.77b}$$

In the geometry of Figure 4.21, as a first order analysis, we assume that the coil and cylinder (inner radius a, outer radius b) are infinite in length, and that the magnetic field is

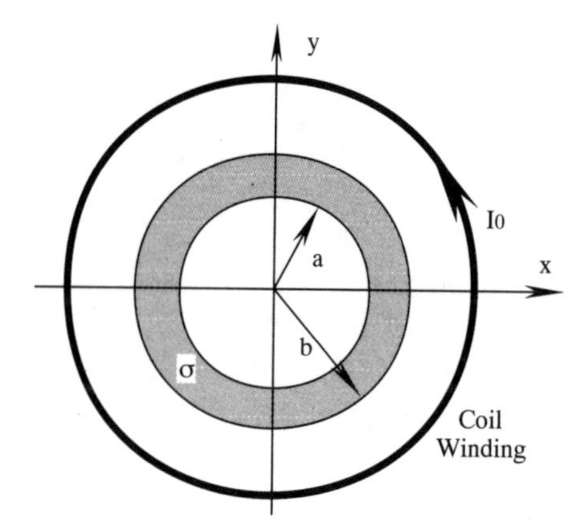

Fig. 4.21 *Conductive cylinder in uniform **H**-field.*

axial. These assumptions reduce the problem to one dimension. The point form differential equations may then be solved using the condition that the total field $H_{\mathrm{T}} = H_0 - H_i(a)$ for $r \leqslant a$; where H_0 is the coil magnetic field in the absence of the cylinder, and $H_{\mathrm{i}}(r)$ the reaction field due to the induced current density in the cylinder. The solution for $H_{\mathrm{T}}(r)$ is a zeroth order Bessel function, J_0:

$$r\frac{\mathrm{d}^2 H_{\mathrm{T}}}{\mathrm{d}r^2} + \frac{\mathrm{d}H_{\mathrm{T}}}{\mathrm{d}r} + k^2 r^2 H_{\mathrm{T}} = 0$$

gives:

$$H_{\mathrm{T}}(r) = H_{\mathrm{T}}(b)\frac{J_0(kr)}{J_0(kb)} \tag{4.78a}$$

where:

$$k^2 = -\mathrm{j}\omega\mu(\sigma + \mathrm{j}\omega\varepsilon) \cong -\mathrm{j}\omega\mu\sigma = -\gamma^2 \tag{4.78b}$$

and the approximation in the propagation coefficient, k, applies only in the case of a cylinder which is a very good conductor. It is important to note that while the boundary conditions require the fields to be quasi-static in the $z-$ and φ-directions, the high conductivity of these materials means that wave propagation must be considered in the r-direction.

Since the argument of J_0 is complex, the Bessel function will be maximum at the outer edge — its magnitude increases without bound for large k and/or r (see Figure 4.22). So, the result must be scaled by the outer edge field at $r = b$ — note that the total field $H_{\mathrm{T}}(b)$ is *not* H_0 due to the action of the induced currents in the cylinder. In this solution γ is the complex propagation constant (as discussed in Chapter 1) which contains both the skin depth, $\delta(\delta = 1/\operatorname{Re}\{k\})$ and the wavelength, $\lambda(\lambda = 2\pi/\operatorname{Im}\{k\})$. At very high conductivities the wavelength and skin depth are small, so the full time-varying solution must be used. The wavelength and phase effects occur only in the direction of propagation (the negative r-direction) so that the fields are in phase in the φ and z directions. So, we actually have a hybrid static and wave solution, similar to the uniform plane wave problem discussed in Chapter 1.

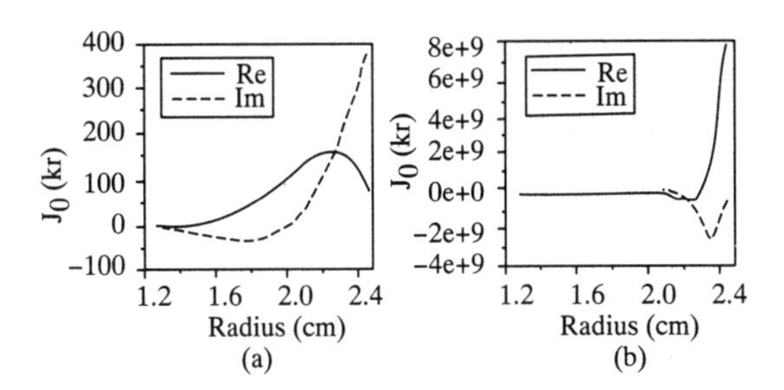

Fig. 4.22 *Complex Bessel function behavior ($x = \gamma r$) at 27 MHz for a = 1.27 cm and b = 2.47 cm for (a) $\sigma = 1000$ S/m, and (b) $\sigma = 10\,000$ S/m.*

The complexity of the field inter-relationships and the effects of the induced current in the cylinder make a numerical solution necessary. A finite difference model (FDM) was created which used 15 nodes to represent the cylinder thickness. Evaluating the boundary condition at $r = b$ is the most difficult part of the solution. Iterative approaches are not well behaved due to lack of sufficient constraints. So, the boundary field magnitude, $|H_{\mathrm{T}}(b)|$, was found from the magnitude of the interior field, $|H_{\mathrm{T}}(a)|$ by inverting the solution above — $|H_{\mathrm{T}}(b)| = |J_0(kb)||H_T(a)|/|J_0(ka)|$. H_0 was used as the phase reference for the solution. The magnitude of the interior field was estimated from an effective "shield factor" for the cylinder thickness:

$$|H_{\mathrm{T}}(a)| = H_0 e^{-\mathrm{OD}} \tag{4.79}$$

where the "optical depth", OD = $(b - a)/\delta$.

The phase of $H_T(b)$ was found by first calculating the induced electric field in the cylinder:

$$E_\varphi(r) = \frac{-\mathrm{j}\omega\mu}{r}\int_0^r H_{\mathrm{T}}(x)x\,\mathrm{d}x = \frac{-\mathrm{j}\omega\mu H_{\mathrm{T}}(b)}{\mathrm{J}_0(kb)}\int_0^r J_0(kx)x\,\mathrm{d}x \tag{4.80a}$$

and:

$$\int_0^r J_0(kx)x\,\mathrm{d}x = \frac{a^2 J_0(ka)}{2} + \int_a^r J_0(kx)x\,\mathrm{d}x \tag{4.80b}$$

where x is a dummy variable of integration which has the significance of radius. The total induced current in the cylinder, $I = \iint (\sigma + \mathrm{j}\omega\varepsilon)\mathbf{E}\cdot\mathrm{d}\mathbf{S}$, (applied to the integral form of Ampere's law) gives the phase of $H_{\mathrm{T}}(b)$ — that is, $H_{\mathrm{T}}(b)$ has the phase required so that the induced current, I, opposes H_0 (by Lenz's law):

$$H_{\mathrm{T}}(b) = H_{\mathrm{T}}(a) - I = H_{\mathrm{T}}(a) - \int_a^b (\sigma + \mathrm{j}\omega\varepsilon)E_\varphi(r)\,\mathrm{d}r. \tag{4.81}$$

In order to preserve accuracy, the Bessel functions were integrated between nodes by rectangular integration. Each Δr segment in the 15-node representation for the power density field was divided into 100 steps, and the Bessel functions integrated stepwise to determine I. This was necessary because the complex phase relationships of the Bessel functions for high values of k strongly influenced the integration results.

Subsequently, the electric field at each cylinder node was calculated from equation (4.80) using the boundary total magnetic field. In Figure (4.23) the electric field results are shown

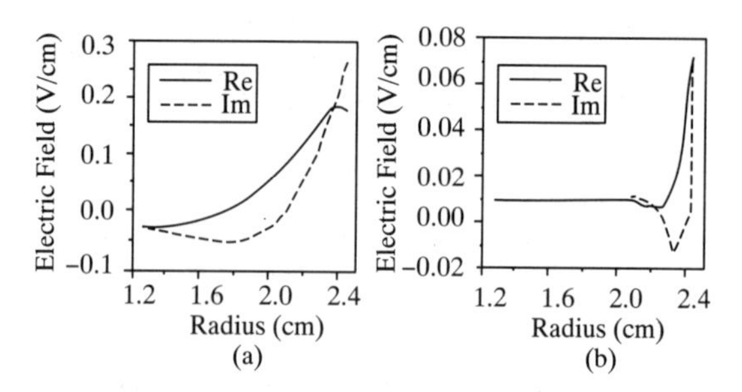

Fig. 4.23 *Electric field for the cylinder in an external field of 100 A/m at 27 MHz for (a) $\sigma =$ 1000 S/m and (b) $\sigma =$ 10 000 S/m.*

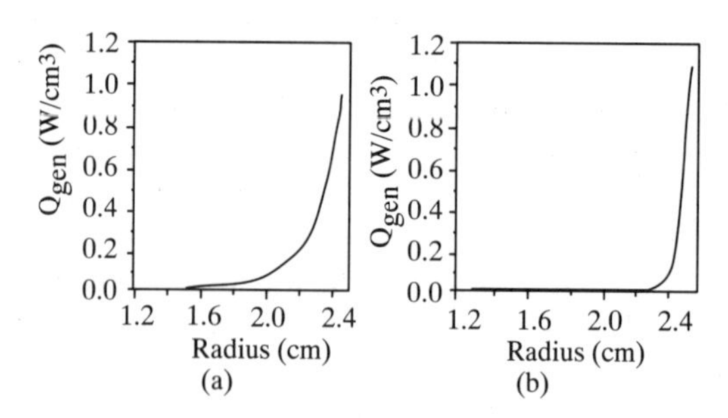

Fig. 4.24 *Resulting power density term, Q_{gen}, at 27 MHz for (a) $\sigma =$ 1000 S/m and (b) $\sigma =$ 10 000 S/m.*

for two conductivities. The power generation term, Q_{gen} was then determined from: $Q_{gen} = \sigma|\mathbf{E}(r)|^2$ calculated at each of the 15 nodes in the cylinder (Figure 4.24). The total coupled power (per unit length of cylinder) can then be determined by trapezoidal integration of $Q_{gen}(r)$.

Calculations were performed for two electrical conductivities, 10 000 S/m and 1000 S/m. The shallow penetration in the higher conductivity material is apparent in Figure (4.24). Though the surface power density is higher for the 10 000 S/m material, the total power in the cylinder is higher for the 1000 S/m material. In fact, the total coupled power at 27 MHz would be maximum if the cylinder had a conductivity of about 100 S/m, as can be seen in Figure 4.25a. The total absorbed power vs frequency for fixed conductivity and magnetic field strength is shown in Figure 4.25b. As would be expected, the total power increases with increasing frequency. These results suggest an interesting engineering trade-off between effective electrical conductivity and frequency of operation. There is a specific effective conductivity which gives maximum absorption at fixed frequencies; and the optimum frequency for power absorption decreases with increasing σ_{eff}.

As mentioned, we should expect that magnetic field heating will be non-uniform. The radial dependence has been described. One would expect some axial dependence in a finite coil length due to end fringing fields. However, in the solenoid coil geometry reflections from the ends of a finite cylinder of the same length as the coil will be minimal. This is because the field polarization gives only a tangential component in the **E**-field at the extremities.

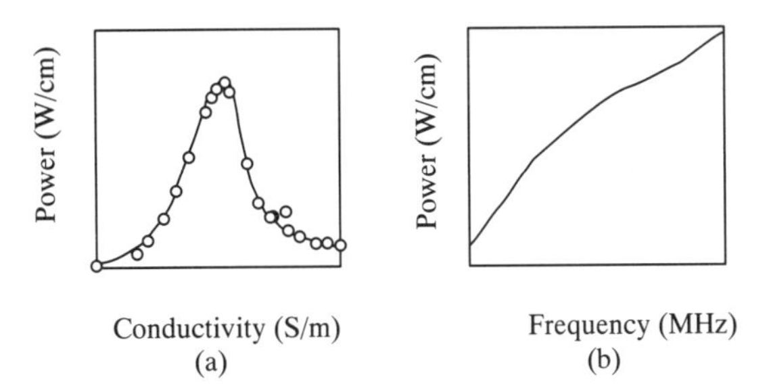

Fig. 4.25 *Total coupled power per unit length of cylinder in an external field of 47 A/m, (a) at 27 MHz as a function of electrical conductivity, and (b) as a function of frequency for $\sigma = 10\,000$ S/m.*

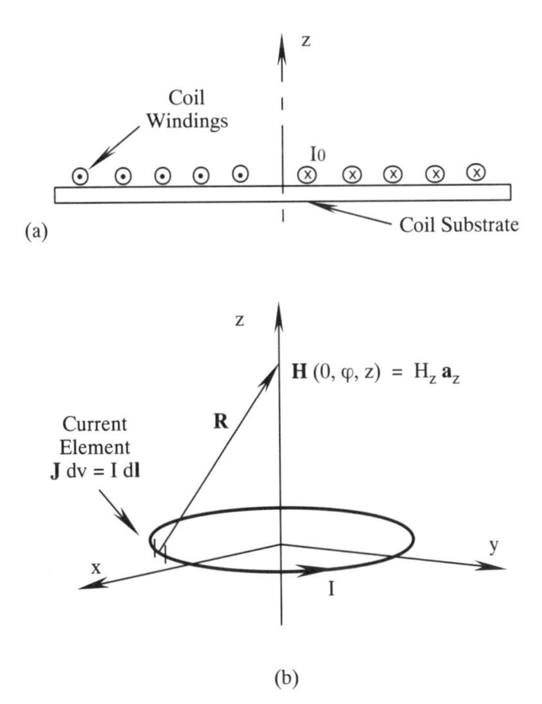

Fig. 4.26 *Two flat coil geometries. (a) The pancake coil is essentially an Archimedes spiral geometry. (b) Single ring of current.*

4.3.2 Ring and pancake coil magnetic field applicators

A pancake coil is a flattened planar winding (Figure 4.26a). The pancake coil may be approximately thought of as a collection of concentric co-planar rings of increasing radius, each with the same current. We will begin with a short discussion of the single-ring applicator (Figure 4.26b), continue with the pancake coil field distribution and present two examples of materials heated with the pancake coil.

Single current ring magnetic field

The magnetic field strength on the z-axis for a ring of current, I_0, may quickly be found from the Biot–Savart law (equation (1.7)) since symmetry dictates that there will be only a z-direction component (Figure 4.27):

$$\mathbf{H} = \frac{a^2 I_0}{2\sqrt{(a^2+z^2)^3}}\mathbf{a}_z. \tag{4.82}$$

Using the Biot–Savart law it may also be shown that far from the ring ($r \gg a$) the field behaves as though the ring were a magnetic dipole:

$$\mathbf{H} = \frac{I_0 \pi a^2}{4\pi r^3}(2\cos(\theta)\mathbf{a}_r + \sin(\theta)\mathbf{a}_\theta) \tag{4.83}$$

in which the equivalent dipole moment, $\mathbf{m}$ (Am^2) is: $\mathbf{m} = \pi a^2 I_0 \mathbf{a}_z$ for the ring. Far from the ring the magnetic vector potential, $\mathbf{A}$, also has a convenient form:

$$A = \frac{\mu(\pi a^2) I_0 \sin(\theta)}{4\pi r^2}\mathbf{a}_\phi. \tag{4.84}$$

However, near the ring (the case in which we are most interested) the calculation is more complex. The magnetic flux density, $\mathbf{B}$, may be calculated from the vector magnetic potential, $\mathbf{A}$. Because of the proximity of the ring the geometric simplification used in (4.83) and (4.84) no longer applies. Here we must use the complete expression for the scalar position, R; and for simplicity (with no loss in generality for the solution) we assume that the point at which the potential is to be determined is in the x–z plane (at $\varphi = 0$), so that:

$$\mathbf{A}(x, y, z) = \int_0^{2\pi} \frac{\mu I_0 a \, \mathrm{d}\varphi' \mathbf{a}_\phi}{4\pi\sqrt{a^2 + r^2 + z^2 - 2ar\cos(\varphi')}}. \tag{4.85}$$

The functional form for the vector potential may be manipulated to obtain an elliptic integral of the first kind. First, we substitute $\varphi' = \pi - 2\theta$, recast the integral in terms of θ, and factor the square root of $\{a^2 + r^2 + z^2 + 2ar\}$ out of the denominator. Integrating φ' from 0 to 2π

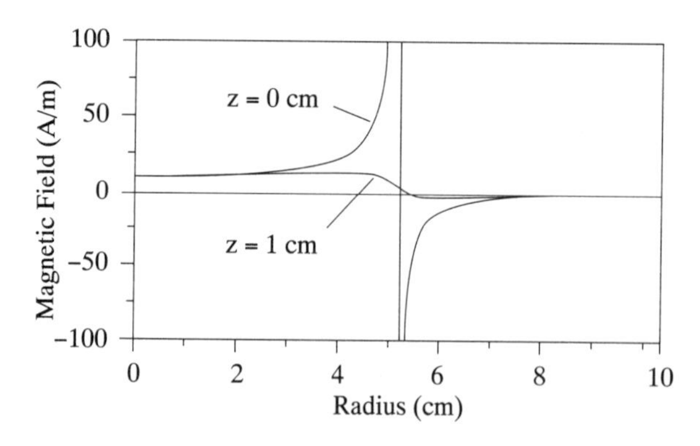

Fig. 4.27 *Magnetic field strength vs radius for a single ring of 5 cm radius at a current of 1 A calculated at z = 0 and z = 1 cm.*

is equivalent to θ from 0 to π, so the vector potential is:

$$F(\eta, k) = \int_0^{\eta} \frac{d\theta}{\sqrt{1 - k^2 \sin^2(\theta)}} \tag{4.86a}$$

$$k^2 = \frac{4ar}{(a+r)^2 + z^2} \tag{4.86b}$$

$$\mathbf{A}(r, \varphi, z) = \frac{\mu I_0}{2\pi\sqrt{(a+r)^2 + z^2}} F(\pi, k)\mathbf{a}_{\varphi} \tag{4.86c}$$

where F denotes the elliptic integral of the first kind with arguments η and k as shown; and for the full ring we use $\eta = \pi$ to find **A**. Values for the elliptic integrals are tabulated, or may be derived numerically. Now we can determine **B** or **H** from the curl of **A**. If numerical methods are used, it is actually simpler to evaluate the Biot–Savart law integral directly—one must be careful to make the calculation in Cartesian coordinates to avoid the problems created by variations in the vector directions of $\mathbf{a}_{\varphi}$ and $\mathbf{a}_r$ as φ varies. Results of a numerical calculation are shown in Figure 4.27 in which $H_z(r)$ is plotted at $z = 0$ and $z = 0.5a$. Note that the magnetic field has a singularity in the x–y plane at $r = a$ on the coil, as is expected.

Pancake coil field distribution

The pancake coil has a high axial magnetic field strength below the center as shown in Figure 4.28. The pancake coil in the Figure has an outer radius of 5 cm and consists of 5.5 coplanar turns. The normalized magnetic field strength (per Ampere of coil current) was determined by direct numerical integration of the Biot–Savart law. A highly peaked axial field with center-weighted flux density gives a bit more uniform heating than a constant axial field would. The enclosed magnetic flux (the right hand side of Faraday's law) for the pancake coil is shown in Figure 4.29a. If we assume that the induced currents are too small to significantly affect the magnetic field, the induced electric field at 27 MHz would be as shown in Figure 4.29b. For comparison purposes the linearly varying electric field for a hypothetical uniform H_z with the same total flux is also shown. Note that the inflexion in the enclosed flux (4.29a) results in an **E**-field maximum inside the radius of the pancake coil.

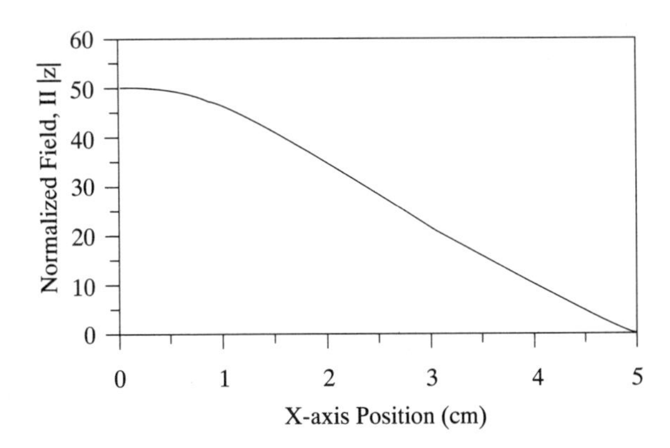

Fig. 4.28 *Pancake coil (5.5 turns, 5 cm radius) normalized magnetic field strength (A/m per ampere of coil current) calculated 1.6 cm below the coil plane along the X-axis.*

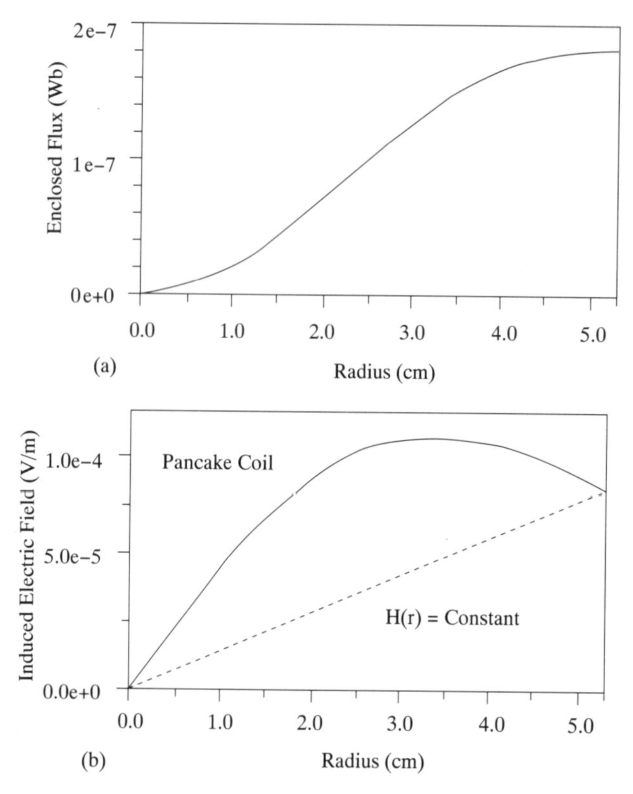

Fig. 4.29 *(a) Total magnetic flux enclosed vs radius for the example 5 cm pancake coil at a current of 1 A. (b) Induced electric field for the 5 cm pancake coil and a hypothetical constant* ***H****-field applicator with the same total flux (182 nWb).*

Thin sheet workpiece

Thin sheets of semiconducting material placed in a normal circularly symmetric magnetic field ($\mathbf{H} = H_z\mathbf{a}_z$) from either the pancake coil or solenoid coil will have very little heating in the center with increasing heating as the total enclosed flux increases. Note that by Faraday's law the only contribution to circulating currents induced in the sheet comes from the normal component of the magnetic field.

For low conductivity materials the induced currents do not significantly affect the applied external magnetic field and the electric fields can be quite high in the sheet. As the conductivity of the sheet increases the induced currents act against the external field to reduce its magnitude, just as we saw in the cylinder heating problem. Consequently, a high conductivity sheet can form an effective shield for the normal component of the magnetic field.

To see how this happens we may look at a hypothetical thin sheet of uniform electrical conductivity. Assume that a normal external field which is constant to a finite radius is applied (Figure 4.30a) to a very large thin sheet. The field will induce currents in each Δr ring (4.30b) according to the conductivity and enclosed flux. The problem is underconstrained at this point since we do not yet know how the induced currents affect the enclosed flux. By considering each Δr ring to be a source we can use the Biot–Savart law to evaluate the fields of each ring with the total ring current as the unknown, as we did

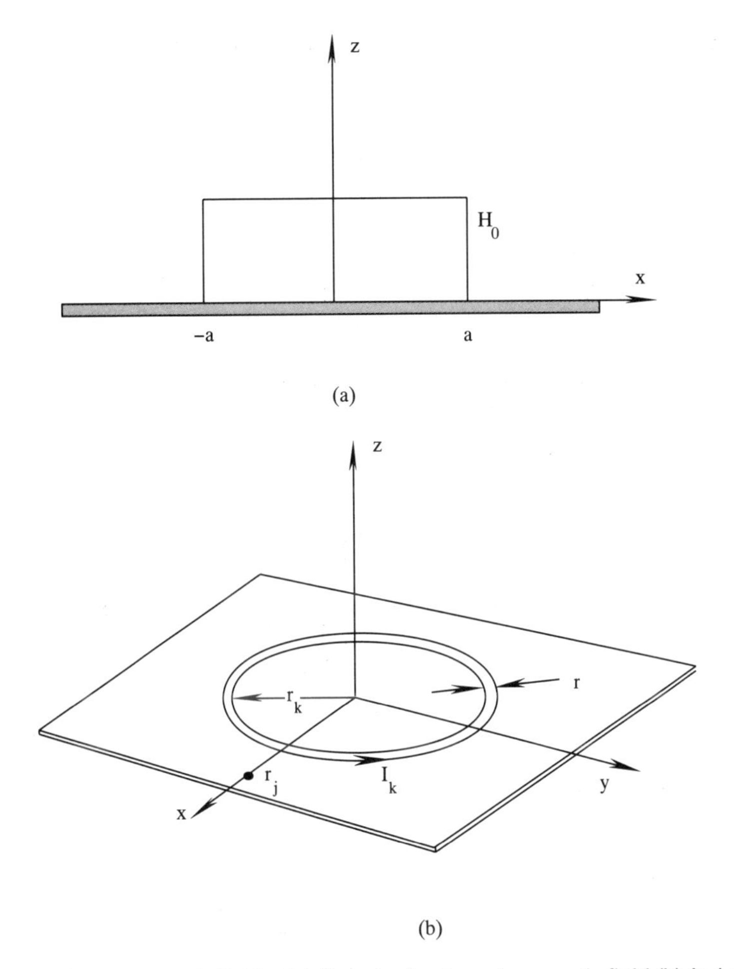

Fig. 4.30 *Thin sheet magnetic fields. (a) Constant external magnetic field (b) Induced current rings Δr thick.*

for the cylindrical shells above. At the center of the ring plane the induced magnetic field strength from N rings will be:

$$\mathbf{H}_i(0,0) = \sum_{j=1}^{N} \frac{r_j^2 I_j}{2\sqrt{(r_j^2 + z^2)^3}} \mathbf{a}_z = \sum_{j=1}^{N} \frac{I_j}{2r_j} \mathbf{a}_z \tag{4.87}$$

where now I_j represents the induced current in each ring of radius r_j and I_j is such that the external field, H_0, is diminished. Certainly, H_0 is the maximum upper bound for the magnitude of H_i since we may not have an induced field which is stronger than the external field which induces it, and as σ approaches infinity each ring must have a total enclosed flux of zero. So, we may quickly find limiting values for the I_j: first, assume that σ is low enough that no decrease in H_0 is generated (I_j will be linear for $r < a$ if H_0 is constant, as is E_ϕ in Figure 4.29b, and then decrease as $1/r$ for $r > a$). Second, assume that σ is large enough that the total enclosed flux is zero for each ring (Figure 4.30b). This results in a linear system of equations in which the total flux at each radius is zero:

$$\sum_{j=1}^{N}\sum_{k=1}^{N} C_{jk} I_k = 0 \tag{4.88a}$$

or:

$$\begin{bmatrix} C_{11} & C_{12} & \cdots & C_{1N} \\ C_{21} & C_{22} & \cdots & C_{2N} \\ \cdots & & & \\ C_{N1} & \cdots & & C_{NN} \end{bmatrix} \begin{bmatrix} I_1 \\ I_2 \\ \cdots \\ I_N \end{bmatrix} = \begin{bmatrix} 0 \\ 0 \\ 0 \\ \cdots \\ 0 \end{bmatrix} \tag{4.88b}$$

and the various coupling factors, C_{jk}, are found by integrating each normal field flux contribution from the center to radius r_j:

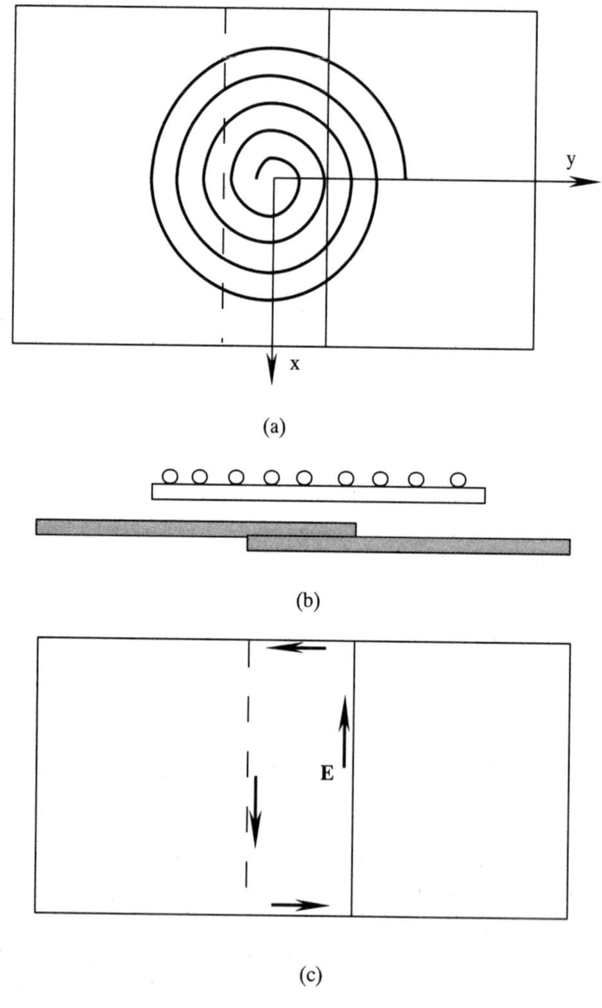

Fig. 4.31 *Lap seam weld geometry (a) Top view of pancake coil over lap seam. (b) Side view of overlap region under pancake coil. (c) Electric field confined to lap region due to air interface boundary condition.*

$$C_{jk} = \frac{\mu_0}{2} \int_0^{r_j} \left[\int_0^{2\pi} \frac{(r - r_k \cos(\varphi)) r_k \, d\varphi}{\{r^2 - 2 r r_k \cos(\varphi) + r_k^2\}^{3/2}} \right] r \, dr. \quad (4.89)$$

For simplicity we have located (r_j, φ, z) in the x–y plane at $x = r_j$ with no loss in generality due to circular symmetry.

Middling values of conductivity need an additional constraint since the magnetic field integral equations are under-constrained for this geometry. For the finite conductivity case the easiest approach is to recast the problem in terms of the electric field integral equations, as described in the Section 4.4 of this chapter.

Lap seam weld field distribution

Oftentimes the induced currents are constrained by surface **E**-field boundary conditions. This works to an advantage in the induction welding of lap seams. As an illustrative example, we inspect the case of a pancake coil used to weld a lap seam of graphite fiber thermoplastic composite material. In this instance we must heat the seam area to temperatures in the neighborhood of 350°C in order to fuse the plastic (Figure 4.31). In the figure the pancake coil induces an electric field in the x–y plane which is confined to the overlap region by boundary conditions at the edges of the top and bottom surfaces. For relatively high electrical conductivity sheets, such as the graphite fiber composites, very little induced current appears outside the overlap region since the contours of integration in Faraday's law are large. Consequently, significant heating is realized only in and very near the overlap region.

Experimental studies of lap seam welding have been performed to illustrate the feasibility of induction heating. Small sheet segments of thermoplastic graphite fiber composite 2.5 cm by 8 cm were welded with the pancake coil arrangement. The sheets were insulated top and bottom by 1 cm thick fiberglass duct insulation board. The coil consisted of 5 turns (10.5 cm outside diameter) was centered over the seam, and at a current of 2.7 A rms a weld (350°C) was obtained in approximately 4 minutes. The weld strengths were comparable to those reported for conduction heat transfer implements.

4.4 MICROWAVE APPLICATORS

Microwave applicator and load combinations must be analyzed using the full time-varying Maxwell equations. There are a number of advantageous formulations which may be employed to calculate induced fields, but of necessity even ordinary geometries must be handled by numerical means — except in the simplest of cases. We begin by re-formulating the appropriate relations and show how lossy materials may be represented in the models. We discuss standard traveling wave and slotted waveguide applicators (useful for moderate and high loss materials), resonant cavities (useful for low loss materials), and conclude with the multimode cavity, perhaps the most common MW applicator.

4.4.1 Time-varying analytical methods

In time varying solutions it is often most advantageous to recast the electric field calculation in terms of the scalar and vector potential, the so-called electric field integral equation formulation (EFIE). The EFIE method derives from the method of induced sources, and will be developed in this section. The EFIE formulation can be used with both bounded and open model spaces, whereas the magnetic field integral equation (MFIE) is only generally useful in closed spaces, as we saw in the discussion of the thin sheet induced current problem. The EFIE model may be solved in frequency space using one of several approaches (this tacitly assumes a sinusoidal source) or in the time domain using the finite difference time domain (FDTD) approach. A short description of these methods is included here.

The electric field integral equation approach expresses the unknown electric field scattered from the dielectric, $\mathbf{E}^s$, in terms of the vector magnetic potential, $\mathbf{A}$, and scalar electric potential, V:

$$-\mathbf{E}^s = j\omega\mathbf{A} + \nabla V \tag{4.90}$$

where $\mathbf{A}$ and ∇V are specifically from the current and charge induced by the incident field, and the potentials are the full retarded potentials from Chapter 1:

$$V(r,t) = \iiint_{\text{vol}} \frac{\rho_v(t - r/u)\text{d}v}{4\pi\varepsilon r} \tag{4.91a}$$

$$\mathbf{A}(r,t) = \iiint_{\text{vol}} \frac{\mu\mathbf{J}(t - r/u)\text{d}v}{4\pi r} \tag{4.91b}$$

in the time domain, and:

$$V(r,\omega) = \iiint_{\text{vol}} \frac{\rho_v(\omega)\text{e}^{-j\beta r}\text{d}v}{4\pi\varepsilon r} \tag{4.92a}$$

$$\mathbf{A}(r,\omega) = \iiint_{\text{vol}} \frac{\mu\mathbf{J}(\omega)\text{e}^{-j\beta r}\text{d}v}{4\pi r} \tag{4.92b}$$

in the frequency domain. As before, the volume integral is over the sources, ρ_v and $\mathbf{J}$.

Alternately, one may recast the (unforced or source-free) point forms of Maxwell's equations in the frequency domain in terms of the electric field or magnetic field to obtain the Helmholtz equations:

$$\nabla^2\mathbf{E} - \gamma^2\mathbf{E} = 0 \tag{4.93a}$$

$$\nabla^2\mathbf{H} - \gamma^2\mathbf{H} = 0 \tag{4.93b}$$

where γ is, again, the propagation constant:

$$\gamma = \alpha + j\beta \tag{4.93c}$$

where $\gamma^2 = -\beta^2 = -\omega^2\mu\varepsilon$ in a lossless medium, and we remember that the skin depth is $\delta = 1/\alpha$. Or, in a general region which may contain sources one may use the scalar electric potential, V, in either the frequency or time domain:

$$\nabla^2 V + \omega^2\mu\varepsilon V = \nabla^2 V + \beta^2 V = -\frac{\rho_v}{\varepsilon} \tag{4.94a}$$

$$\nabla^2 V - \mu\varepsilon\frac{\partial^2 V}{\partial t^2} = -\frac{\rho_v}{\varepsilon} \tag{4.94b}$$

or the vector magnetic potential, $\mathbf{A}$:

$$\nabla^2\mathbf{A} + \omega^2\mu\varepsilon\mathbf{A} = \nabla^2\mathbf{A} + \beta^2\mathbf{A} = -\mu\mathbf{J} \tag{4.95a}$$

$$\nabla^2\mathbf{A} - \mu\varepsilon\frac{\partial^2\mathbf{A}}{\partial t^2} = -\mu\mathbf{J}. \tag{4.95b}$$

Note that equations (4.94a) reduce to the Poisson equation as β^2V becomes small (i.e. at low frequency) and further to the Laplace equation in a source-free region. Any of the above point form (differential equation) formulations may be used to calculate local fields. The relations differ only in the forcing functions (the right hand side of the equation). So, in a source-free region (a passive load) the relations are each a form of the wave equation and all have the same general solution.

The externally applied incident fields, $\mathbf{E}^{\mathrm{i}}$ and $\mathbf{H}^{\mathrm{i}}$, induce currents in a lossy dielectric; thus the sources for the scattered fields, $\mathbf{E}^{\mathrm{s}}$ and $\mathbf{H}^{\mathrm{s}}$, are the induced current density and induced charge. The major problem reduces to estimating the induced sources so that the total electric field can be calculated, and from that the local heating.

Method of induced sources

The method of induced sources replaces a complex dielectric material with an equivalent current density, $\mathbf{J}^{\mathrm{s}}$. The scattered fields from induced currents within the material ($\mathbf{E}^{\mathrm{s}}$ and $\mathbf{H}^{\mathrm{s}}$) may then be calculated from the sources, as well as heating within the material. We begin the formulation of this technique by noting that we must write the solution in two regions: the surrounding medium (region 1) which is presumed composed of free space (ε_0 and μ_0) and which contains the source of the incident field, $\mathbf{J}^{\mathrm{i}}$, and in the material (region 2) which is assumed to be fully described by a complex electric permittivity, ε^*. Faraday's law in point form applies in both regions:

$$\nabla \times \mathbf{E} = -\mathrm{j}\omega\mu_0\mathbf{H} \tag{4.96}$$

and Ampère's law for each region is:

$$\text{Region 1}: \quad \nabla \times \mathbf{H} = \mathrm{j}\omega\varepsilon_0\mathbf{E} + \mathbf{J}^{\mathrm{i}} \tag{4.97a}$$

$$\text{Region 2}: \quad \nabla \times \mathbf{H} = \mathrm{j}\omega\varepsilon_0\mathbf{E} + \mathbf{J}^{\mathrm{s}} \tag{4.97b}$$

where $\mathbf{J}^{\mathrm{s}}$ the induced source is:

$$\mathbf{J}^{\mathrm{s}} = \mathrm{j}\omega(\varepsilon - \varepsilon_0)\mathbf{E} = \mathrm{j}\omega\varepsilon_0(\varepsilon_{\mathrm{r}} - 1)\mathbf{E} \tag{4.97c}$$

and ε_{r} is the complex relative permittivity. In equation (4.96) and equations (4.97) the electric and magnetic fields are the total fields which include both the incident and scattered contributions:

$$\mathbf{E}^{\mathrm{tot}}(x, y, z) = \mathbf{E}^{\mathrm{s}}(x, y, z) + \mathbf{E}^{\mathrm{i}}(x, y, z) \tag{4.98a}$$

$$\mathbf{H}^{\mathrm{tot}}(x, y, z) = \mathbf{H}^{\mathrm{s}}(x, y, z) + \mathbf{H}^{\mathrm{i}}(x, y, z). \tag{4.98b}$$

We must eliminate the scattered field by expressing it in terms of the total field. The formulation continues by recasting Faraday's law in terms of the vector magnetic potential, where we let $\mathbf{B}^{\mathrm{s}} = \nabla \times \mathbf{A} = \mu\mathbf{H}^{\mathrm{s}}$:

$$\nabla \times (\mathbf{E}^{\mathrm{s}} + \mathbf{j}\omega\mathbf{A}) = 0 \tag{4.99}$$

and $\mathbf{A}$ is created by $\mathbf{J}^{\mathrm{s}}$. As we have seen before, any curl-free vector (equation (4.99)) may be written as the gradient of a scalar (as in equation (4.90)), so using the Lorentz gauge ($\nabla \cdot \mathbf{A} = -\mathrm{j}\omega\mu\varepsilon V$, see Chapter 1), and equation (4.99), $\mathbf{E}^{\mathrm{s}}$ can be expressed as:

$$\mathbf{E}^{\mathrm{s}} = -(\nabla V + \mathrm{j}\omega\mathbf{A}) = \frac{\nabla(\nabla \cdot \mathbf{A}) + \gamma^2\mathbf{A}}{\mathrm{j}\omega\mu\varepsilon}. \tag{4.100}$$

The vector potential solves equation (4.95a) above where the scattered current density is the source term. We may thus rewrite equation (4.92b) in terms of the total electric field using equation (4.97c) and (4.100) and substitute the result into (4.98a) to obtain the EFIE governing equation:

$$\mathbf{E}^{\mathrm{tot}}(x, y, z) - \iiint_{\mathrm{vol}} [\text{grad div} + \beta^2](\varepsilon_{\mathrm{r}} - 1)\mathbf{E}^{\mathrm{tot}}(x, y, z)\frac{\mathrm{e}^{-\mathrm{j}\beta R}}{4\pi R}\,\mathrm{d}x'\mathrm{d}y'\mathrm{d}z' = \mathbf{E}^{\mathrm{i}}(x, y, z). \tag{4.101}$$

The electric field is determined in Cartesian coordinates at (x, y, z) from sources distributed over the (x', y', z') volume of region 2. The scalar separation distance, R, is defined in the usual way:

$$R = \sqrt{(x - x')^2 + (y - y')^2 + (z - z')^2}. \tag{4.102}$$

The resulting system of equations may be solved by the method of moments or by finite elements, as appropriate. The system of equations is expressed in Cartesian coordinates so that the unit vectors will be independent of position, as has been mentioned in earlier discussions.

Method of moments and finite element solutions

We include here a short discussion and a few comments relating to the method of moments solution to the EFIE formulation. Similar comments apply to the use of finite elements. In both methods the absorbing volume is sub-divided into small volume elements, sometimes called voxels, to perform the calculations. The governing assumption is that the electric field and material permittivity are constant over the volume. Consequently, the volume may have no dimension approaching a significant fraction of a wavelength. The usual guideline used is that no dimension may exceed approximately $\lambda/6$, though $\lambda/10$ is probably a better choice (even so, the phase error would be about 36°). The wavelength of importance is the wavelength inside the material, λ_2. For lossy materials, the ones we care most about, this may make the individual volume elements very small indeed. In distilled water at 2.45 GHz the voxel may not exceed about 2.2 mm in any dimension. The required memory for the numerical model is thus very large since full three-dimensional geometry ($3N \times 3N$ complex) and electric field ($3N$ complex) matrices result for N voxels, and very few elements are zero.

In the method of moments solution we describe the voxel source terms using rectangular "pulse" basis functions, $P_n(x', y', z')$ where:

$$P_n(x', y', z') = \begin{cases} 1 & (x', y', z') \text{ within voxel } n \\ 0 & \text{otherwise.} \end{cases} \tag{4.103}$$

The electric field for each voxel is:

$$\mathbf{E}^{\text{tot}}(x, y, z) = \sum_{n=1}^{N} P_n(x, y, z)[E_{nx}\mathbf{a}_x + E_{ny}\mathbf{a}_y + E_{nz}\mathbf{a}_z]. \tag{4.104}$$

The effects of induced electric field within a volume are lumped into an equivalent geometry matrix, G:

$$\begin{bmatrix} [E^{\text{i}}_x] \\ [E^{\text{i}}_y] \\ [E^{\text{i}}_z] \end{bmatrix} = \begin{bmatrix} [G_{xx}] & [G_{xy}] & [G_{xz}] \\ [G_{yx}] & [G_{yy}] & [G_{yz}] \\ [G_{zx}] & [G_{zy}] & [G_{zz}] \end{bmatrix} \begin{bmatrix} [E^{\text{tot}}_x] \\ [E^{\text{tot}}_y] \\ [E^{\text{tot}}_z] \end{bmatrix}. \tag{4.105}$$

There is one equation for each component of the vector. The geometric matrices have entries which describe the interaction between the field in voxel m and source in voxel n in the axial directions u and v which follow the canonical form:

$$G^{mn}_{uv} = -(\varepsilon_{\text{r}n} - 1)\frac{\partial^2 g_n}{\partial u \partial v} \tag{4.106}$$

when $u \neq v$ and $m \neq n$. Here u and v are x, y or z, as required. The term g_n includes the volume geometry from equation (4.101) and is described below. Field components along the same axes, $u = v$, for different voxels, $m \neq n$, have:

$$G^{mn}_{uu} = (\varepsilon_{\text{r}n} - 1)\left(\frac{\partial^2}{\partial u^2} + \beta^2\right) g_n. \tag{4.107}$$

Each of the g_n factors are calculated from:

$$g_n(x, y, z) = \iiint_{\text{voxel } n} \frac{\text{e}^{-\text{j}\beta R}}{4\pi R}\,\text{d}x'\text{d}y'\text{d}z'. \tag{4.108}$$

Closed form solutions for the above voxel integrals do not exist, so numerical methods are used over the cubic voxels.

It remains to inspect the self-induction terms for $m = n$. An x-direction field may not induce a non-x-direction field, so $G = 0$ when $m = n$ and $u \neq v$. For the remaining diagonal elements extreme care must be exercised when nearing the singularity in the scalar distance, R, just as we saw in the quasi-static method of moments solution. A simple alternative to numerical integration is to apply a closed form solution for a spherical volume of equivalent radius, a_n, where the voxel volume = $\frac{4}{3}\pi a_n^3$. Then, the self-induction term (when $m = n$ and $u = v$) is:

$$G^{mm}_{uu} = 1 + (\varepsilon_{\text{r}m} - 1)\left\{1 - \tfrac{2}{3}\text{e}^{-\text{j}\beta a_n}(1 + \text{j}\beta a_n)\right\}. \tag{4.109}$$

For other than cubic voxels, a cylinder may be used as an approximation instead of a sphere. Calculations show that the spherical and cylindrical approximations match to four significant figures for cubic voxels, so the cylindrical approximation may be used with equal confidence. The self-induction term formulation for right circular cylinders of height h along the z-axis and equivalent radius a_m is:

$$G^{mm}_{xx} = G^{mm}_{yy} = 1 - (\varepsilon_{\text{r}m} - 1)(S_1 + S_2 + +S_3) \tag{4.110a}$$

and:

$$G_{zz}^{mm} = 1 - (\varepsilon_{rm} - 1)(S_1 + S_4) \tag{4.110b}$$

where the S_i phase factors in equations (4.110) are:

$$S_1 = -\mathrm{j}\sin(\beta h) + \mathrm{j}\beta \int_0^h \exp(-\mathrm{j}\beta\sqrt{z^2 + a_m^2})\,\mathrm{d}z \tag{4.111a}$$

$$S_2 = \tfrac{1}{2}\int_0^h \frac{-\mathrm{j}ba_m^2\sqrt{z^2 + a_m^2}}{(\sqrt{z^2 + a_m^2})^3}\exp(-\mathrm{j}\beta\sqrt{z^2 + a_m^2})\,\mathrm{d}z \tag{4.111b}$$

$$S_3 = -\tfrac{1}{2}\int_0^h \frac{\exp(-\mathrm{j}\beta\sqrt{z^2 + a_m^2})}{\sqrt{z^2 + a_m^2}}\,\mathrm{d}z \tag{4.111c}$$

$$S_4 = \int_0^{a_m} \frac{-\mathrm{j}\beta h\sqrt{r^2 + h^2} - h}{(\sqrt{r^2 + h^2})^3}\exp(-\mathrm{j}\beta\sqrt{r^2 + h^2})_r\,\mathrm{d}r. \tag{4.111d}$$

A similarly structured finite element description may be assembled. Basis functions more accurate than the rectangular pulse bases may be then used. The improvement in accuracy may or may not be worth the increased computation time, however. The finite element method works very well in closed geometries. The FEM realization may be difficult or impossible to apply in unbounded medium problems, such as the lap-seam weld problem. Unbounded geometries are handled in FEM approaches using a so-called “infinity boundary condition”, which is sometimes under-constrained. The method of moments is robust in unbounded media, and may be used to model open geometries, such as the lap weld geometry.

Finite difference time domain method

The finite difference time domain method typically takes one of two forms: (1) direct implementation of the Helmholtz wave equations in the time domain or (2) the Yee algorithm. Working in the time domain in lossy materials is complicated by the inability to use the complex formulation for the electric permittivity. Complex permittivity formulation implies a frequency domain solution with sinusoidal harmonic signals. In each FDTD discussion we first present the formulation for lossless media and then discuss the lossy medium problem.

Helmholtz equation method The first approach uses the time domain form of the Helmholtz equations (4.93) for a source-free region:

$$\nabla^2\mathbf{E} = \mu\varepsilon\frac{\partial^2\mathbf{E}}{\partial t^2} = \frac{1}{c^2}\frac{\partial^2\mathbf{E}}{\partial t^2} \tag{4.112a}$$

$$\nabla^2\mathbf{H} = \mu\varepsilon\frac{\partial^2\mathbf{H}}{\partial t^2} = \frac{1}{c^2}\frac{\partial^2\mathbf{H}}{\partial t^2} \tag{4.112b}$$

where c is the speed of propagation and we have assumed that the medium is lossless. Approximations to the spatial and temporal second derivatives may be applied to these wave equations — there is one scalar equation for each component of the two vectors, six in all. As we have already mentioned, the frequency domain form may not be solved by finite difference methods since a singular parameter matrix results and the solution diverges. Two independent variables are involved in each equation, space and time; thus initial conditions must be specified over the entire solution space and boundary conditions for all time over the boundary. This may limit the usefulness of this approach for some problems. Nevertheless, the method works well for propagation in a homogeneous lossless medium, and has good accuracy since finite approximations to the second derivative are involved.

We present a simple lossless medium example to introduce the problem of solution stability and accuracy. Assume a one-dimensional problem (Figure 4.32a) such as a uniform plane wave propagating in the $+z$-direction where $E_y = E_z = 0$ and $E_x = E_x(z, t)$ (thus $\mathbf{H} = H_y(z, t)\mathbf{a}_y$, though we will only write the electric field). Assume a constant finite space increment, Δz, and a time step, Δt. Then:

$$E(k, n+1)-2E(k, n)+E(k, n-1) = \left(\frac{c\Delta t}{\Delta z}\right)^2 [E(k+1, n)-2E(k, n)+E(k-1, n)] \quad (4.113)$$

where k is the z-axis index and n the time index. Note that three complete node matrices for E_x are required for storage, one each at time $n-1$, n and $n+1$, in order to avoid bias

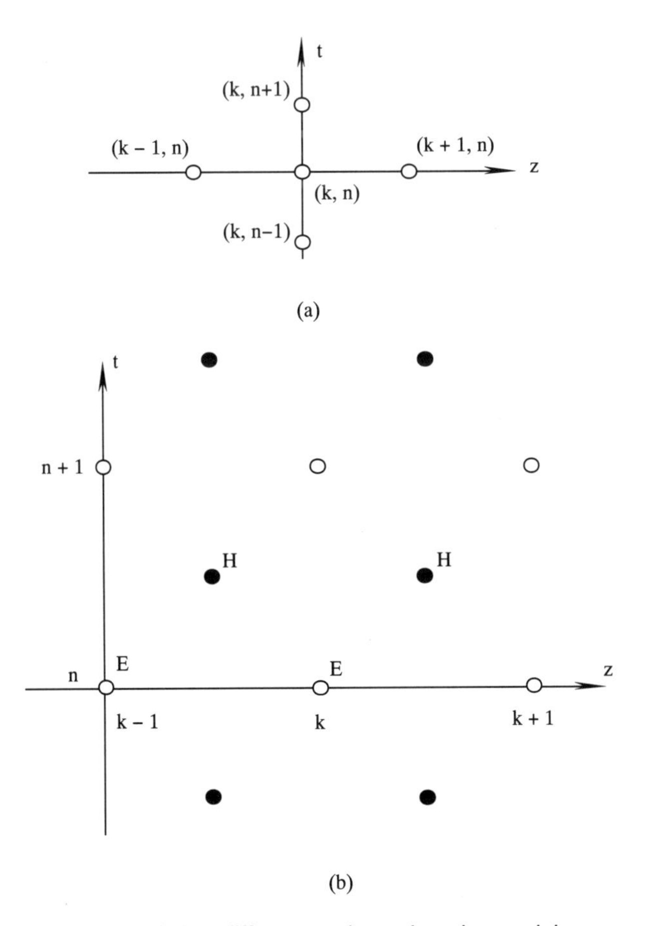

Fig. 4.32 *One-dimensional finite difference time domain model spaces. (a) Helmholtz equation geometry (b) Yee algorithm solution space.*

in the solution. For stability the factor, $\rho = c\Delta t/\Delta z$, must be less than 1. In the case of a harmonic solution we need many time steps to get a good representation of the waveform. It can be shown that if $\Delta z < 0.055\lambda$ and $\Delta t < 0.35/\omega$ an accuracy of about 1% in the approximations to the sines and cosines is obtained (these parameters give 18 points over one wavelength, a good representation of a sinusoidal wave). So, if the medium is free space and the frequency is 2.45 GHz we need a space increment less than 6.7 mm and a time increment less than 22 ps. Rewriting the relation in terms of the factor ρ we may quickly obtain the relaxation equation for $E(n, k)$:

$$E(k, n) = \frac{1}{2(1-\rho^2)}[E(k, n+1) + E(k, n-1)] - \frac{\rho^2}{2(1-\rho^2)}[E(k+1, n) + E(k-1, n)] \tag{4.114}$$

and the reason for the restriction on ρ is now obvious.

Since a second order differential equation is solved two initial conditions over the solution space and two boundary conditions for all time are required. The boundary conditions are implemented as described in the discussion of the finite difference method above. The initial conditions are new to this discussion. Ordinarily, we use the value of the function at $t = 0$ and its first derivative evaluated at $t = 0$ to form the initial conditions:

$$E_x(z, 0) = f(z) \quad \text{and} \quad \frac{\partial}{\partial t}E_x(z, 0) = g(z). \tag{4.115}$$

Implementation of the first initial condition is straightforward. We must be a little more careful in evaluating the derivative for the second condition using only a forward difference. This is because it is only of first order (Δt) accuracy and the second derivative in equation (4.113) is second order (Δt^2) accurate. It may be shown by Taylor series expansion that a better approximation to the boundary field at $t = \Delta t$ can be obtained from:

$$E_x(z, \Delta t) = (1-\rho^2)E_x(z, 0) + \Delta t g(z) + \rho^2[E_x(z-\Delta z, 0) + E_x(z+\Delta z, 0)]. \tag{4.116}$$

The above relations are easily generalized to two- or three-dimensional geometries wherein the same stability and accuracy relations apply.

When the medium is lossy we must re-derive the applicable Helmholtz equations including conductivity. We will assume that all of the loss terms, from both translational motion of charge and rotational and vibrational motion, are lumped into the conductivity:

$$\nabla^2\mathbf{E} = \mu\varepsilon\frac{\partial^2\mathbf{E}}{\partial t^2} + \mu\sigma\frac{\partial\mathbf{E}}{\partial t} = \frac{1}{c^2}\frac{\partial^2\mathbf{E}}{\partial t^2} + \mu\sigma\frac{\partial E}{\partial t} \tag{4.117a}$$

$$\nabla^2\mathbf{H} = \mu\varepsilon\frac{\partial^2\mathbf{H}}{\partial t^2} + \mu\sigma\frac{\partial\mathbf{H}}{\partial t} = \frac{1}{c^2}\frac{\partial^2\mathbf{H}}{\partial t^2} + \mu\sigma\frac{\partial H}{\partial t}. \tag{4.117b}$$

This adds a first order time derivative to the left hand side of (4.113); however, since we are using the central difference formulation to estimate the first derivatives, only two changes result:

$$E(k,n) = \frac{1}{2(1-\rho^2)}\left[\left(1+\frac{2\Delta t\sigma}{\varepsilon}\right)E(k,n+1) + \left(1-\frac{2\Delta t\sigma}{\varepsilon}\right)E(k,n-1)\right]$$
$$-\frac{\rho^2}{2(1-\rho^2)}[E(k+1,n)+E(k-1),n)] \tag{4.118}$$

The factors in front of the two time increment terms are seen to include the time constant of the medium (ε/σ). We now have an additional constraint on the stability: $2\Delta t\sigma/\varepsilon < 1$; so, $\Delta t < \min\{\varepsilon/2\sigma, 0.35/\omega\}$. High conductivity media require much shorter time steps to obtain a stable solution, as might be expected.

The Yee algorithm The second method, the Yee algorithm, formulates Maxwell's equations directly by writing out the appropriate curl functions. We can revisit the uniform plane wave example to see the differences in this approach. Maxwell's equations for an x-polarized uniform plane wave in a lossless medium reduce to:

$$\frac{\partial}{\partial z}E_x(z,t) = -\mu\frac{\partial}{\partial t}H_y(z,t) \tag{4.119a}$$

$$\frac{\partial}{\partial z}H_y(z,t) = -\varepsilon\frac{\partial}{\partial t}E_x(z,t). \tag{4.119b}$$

As should be remembered from our discussion of Laplace equation solutions, to get second order accuracy we evaluate first derivatives at the center of the interval between nodes used in the calculation, i.e. by the central difference formula. Here each relation involves two central differences, one on the space axis and the other on the time axis. The Yee algorithm makes two half-steps in space and in time alternating between the two relations, as in Figure 4.32b. For example, beginning with (4.119a) at $t = n$ in order to estimate the time derivative of H_y in the first equation we need E_x at two neighbor nodes:

$$E_x(k+1,n) - E_x(k,n) = \frac{-\mu\Delta z}{\Delta t}[H_y(k+0.5,n+0.5) - H_y(k+0.5,n-0.5)] \tag{4.120a}$$

and to estimate the time derivative of the **E**-field we use the alternative relation:

$$H_y(k+0.5,n+0.5) - H_y(k-0.5,n+0.5) = \frac{-\varepsilon\Delta z}{\Delta t}[E_x(k,n+1) - E_x(k,n)]. \tag{4.120b}$$

The solution is executed by alternating between (4.120a) and (4.120b) and by locating the nodes used to represent the **H**-field between those for the **E**-field, as shown in the figure.

For a lossy medium equation (4.119a) is unchanged but we must add a term to the right hand side of equation (4.119b):

$$\frac{\partial}{\partial z}H_y(z,t) = \sigma E_x(z,t) - \varepsilon\frac{\partial}{\partial t}E_x(z,t). \tag{4.121}$$

Consequently, the first relaxation equation (4.120a) is unchanged, but the second becomes:

$$H_y(k+0.5,n+0.5) - H_y(k-0.5,n+0.5) =$$
$$\frac{-\varepsilon\Delta z}{\Delta t}\left[E_x(k,n+1) - \left(1-\frac{\sigma\Delta t}{\varepsilon}\right)E_x(k,n)\right]. \tag{4.122}$$

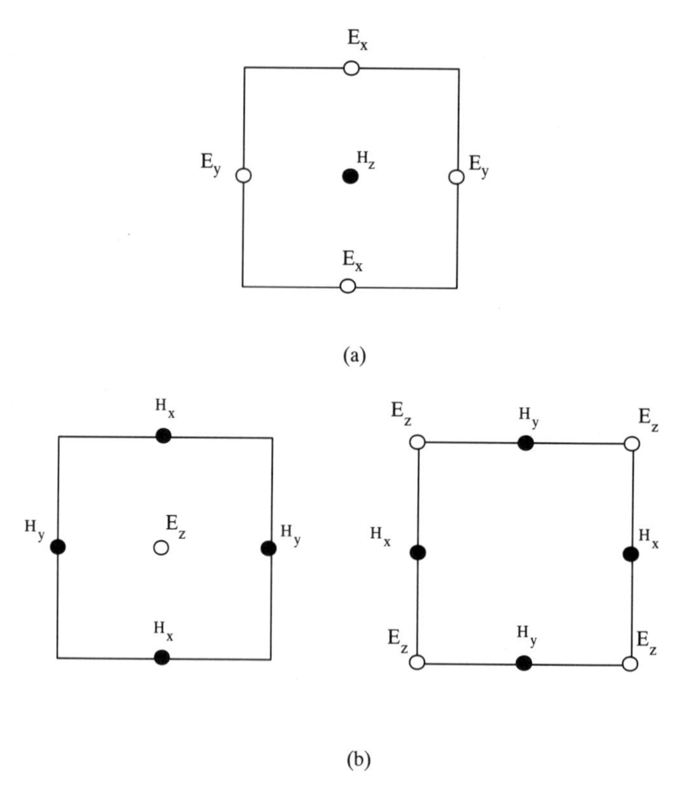

Fig. 4.33 *Two-dimensional elemental cells for the Yee FDTD algorithm. (a) Typical cell for TE mode models. (b) Two types of cells used for TM mode models.*

Here again we see the same restriction on time step: $\Delta t < \varepsilon/\sigma$. Otherwise the execution of the model is substantially similar to the lossless medium case.

The simple one-dimensional example extends to multiple dimensions. There are advantageous positions for the locations of the various nodes, however. In two-dimensions, for example, there is a preferred distribution for TE modes and a different arrangement for TM modes (Figure 4.33). Three-dimensional problems usually use a cell as depicted in Figure 4.34 in which **H**-field components are normal to faces and **E**-field parallel to edges. An excellent and extensive discussion of the Yee algorithm may be found in PIER volume 2 (Kong and Morgan, 1990).

Geometric optics

Geometric optics, or ray tracing, is a method which can be quite powerful for modeling inhomogeneous media. When inhomogeneities are sufficiently large that the internal evanescent modes do not interact, and the inhomogeneities are spaced far enough apart that the external evanescent modes do not interact, we may gain a close approximation of the field distribution by following individual rays, treating each ray as approximately representing a uniform plane wave. Since none of the evanescent modes interact, this region of operation is called the optical region. When the simplifying assumptions are not valid then full scattering solutions, for example, the method of induced sources (as above), must be applied. We take

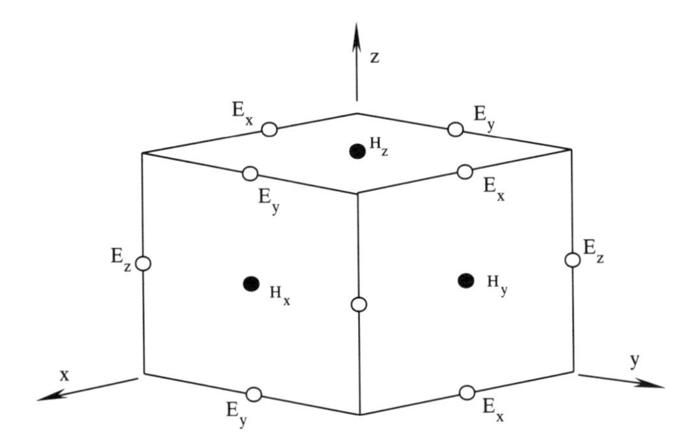

Fig. 4.34 *Three-dimensional Yee algorithm cell.*

the opportunity to present the underlying principles of the ray tracing approach since it forms the basis of a powerful modeling tool, the so-called Monte Carlo optical model method.

The basis of the geometrical optics model is obtained from the frequency domain wave equations where we let the electric permittivity vary with position. To easily relate this discussion to standard optical terminology, we remember that the refractive index of a medium, n, is the square root of the relative permittivity, ε_r:

$$n = \sqrt{\varepsilon_r}. \tag{4.123}$$

Since we have returned to frequency domain representation, we will use a complex permittivity, ε_r^*, which corresponds to a complex refractive index in a lossy medium. The wave equation (1.34) is thus modified:

$$\nabla^2 \mathbf{E} - \gamma^2 \mathbf{E} = -\nabla \left(\mathbf{E} \cdot \frac{\nabla \varepsilon}{\varepsilon} \right) \tag{4.124}$$

where γ is the propagation coefficient, $\gamma = \alpha + \mathrm{j}\beta$.

In a medium in which ε varies slowly with position its gradient will be small and the right hand side will be negligible compared to the second term on the left hand side. Formally, if:

$$\frac{|\nabla \varepsilon|}{\gamma^2 \varepsilon} \ll 1 \quad \text{or} \quad \frac{\lambda^2 |\nabla \varepsilon_r|}{4\pi^2} \ll 1 \tag{4.125}$$

the right-hand side can be neglected, where the second expression applies in a low loss medium. The dominant mechanism of geometric optics is most easily seen in media of negligible loss ($\alpha \to 0$). In that case $\gamma^2 = -\beta^2 = -n^2\beta_0^2$ (where β_0 applies in free space) and if (4.125) applies the field equation will have the approximate form:

$$\nabla^2 \mathbf{E} + n^2 \beta_0^2 \mathbf{E} \approx 0 \tag{4.126}$$

and the wave solutions apply. This is equivalent to saying that ε is constant over dimensions of at least a few wavelengths in the medium; that is, any inhomogeneities are wavelengths in dimension. Whenever this assumption applies the ray tracing method may be used with confidence and the field interaction with the inhomogeneities may be analyzed as simple reflection-refraction problems.

The solutions for any component of **E** in (4.126) (Cartesian coordinates are used because of the arbitrary geometry) will have the form:

$$\psi = \psi_0(x, y, z)e^{-j\beta L(x,y,z)}e^{j\omega t} \tag{4.127}$$

where ψ is any component of **E**, both the amplitude, ψ_0, and phase, βL, are functions of position, and L includes variations in refractive index. By substituting this solution into (4.126), computing the required gradients and canceling common exponentials we may quickly show that the governing equation becomes:

$$\psi_0[-\beta_0^2|\nabla L|^2 - j\beta_0\nabla^2 L] + \nabla^2\psi_0 - j2\beta_0\nabla L \cdot \nabla\psi_0 + n^2\beta_0^2\psi_0 = 0. \tag{4.128}$$

In this expression:

$$|\nabla L|^2 = \left(\frac{\partial L}{\partial x}\right)^2 + \left(\frac{\partial L}{\partial y}\right)^2 + \left(\frac{\partial L}{\partial z}\right)^2. \tag{4.129}$$

Separating equation (4.128) into real and imaginary parts gives two equations to solve:

$$-\psi_0\beta_0^2|\nabla L|^2 + \nabla^2\psi_0 + n^2\beta_0^2\psi_0 = 0 \tag{4.130a}$$

$$\psi_0\nabla^2 L + 2\nabla L \cdot \nabla\psi_0 = 0. \tag{4.130b}$$

We may gain some insight by rearranging and noting that $\nabla\psi_0/\psi_0 = \nabla(\ln\psi_0)$:

$$|\nabla L|^2 - \frac{\nabla^2\psi_0}{\beta_0^2\psi_0} = n^2 \tag{4.131a}$$

$$\nabla^2 L + 2\nabla L \cdot \nabla(\ln\psi_0) = 0. \tag{4.131b}$$

Further, if we apply one additional assumption, that $\beta^2\psi_0 \gg \nabla^2\psi_0$ in equation (4.131a), we obtain the approximation for the phase function, which is known as the equation of the *eikonal*:

$$|\nabla L|^2 \approx n^2. \tag{4.132}$$

This implies that regions of constant L are constant phase fronts. We may express the gradient of L in terms of a vector, **n**, whose magnitude is the refractive index, $n^2 = \varepsilon_r$, and direction is in the direction normal to L = constant, the direction of propagation:

$$\nabla L = \mathbf{n}. \tag{4.133}$$

Rays propagate normally to equi-phase fronts (parallel to **n**; see Figure 4.35) and we may complete a model by tracing out the directions of individual rays, remembering that inhomogeneities must be at least a few wavelengths in dimension. We may add the effect of a lossy medium in a simple way by letting the ray decrease in amplitude along $e^{-\alpha L}$, where α is the **E**-field attenuation coefficient. The effect of high loss materials (where electrical conductivity significantly affects the wavelength) is included in the complex expressions for α and β, equations (1.35). At any location along a ray the local volume power density term may be determined from the Poynting power theorem results. The fluence rate in the wave,

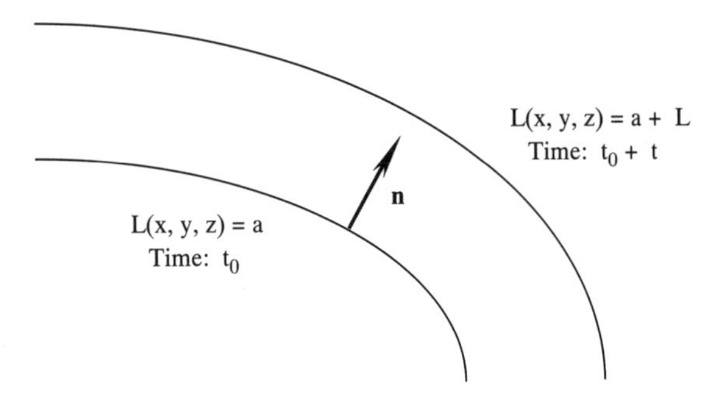

Fig. 4.35 *Wavefront propagation in geometrical optics.*

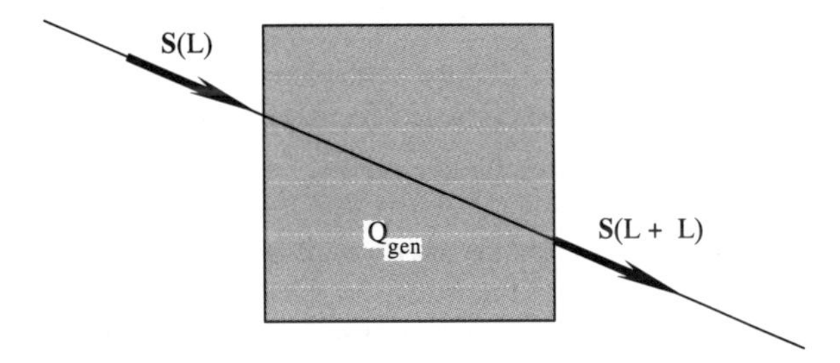

Fig. 4.36 *Volumetric heating in geometric optics model.*

S (W/m^2), is the familiar Poynting vector:

$$\mathbf{S} = \tfrac{1}{2}\{\mathbf{E} \times \mathbf{H}^*\} = \frac{|E_0|^2}{\eta} \mathrm{e}^{-2\alpha L} \mathbf{a}_n \tag{4.134}$$

where η is the characteristic impedance of the medium (complex) and E_0 is now in RMS V/m. Its divergence, $\nabla \cdot \mathbf{S}$, gives the volume power deposited in the medium, Q_{gen} (W/m^3), which is equal to the fluence rate lost from the ray (see Figure 4.36):

$$Q_{\text{gen}} = -\nabla \cdot \mathbf{S} = +2\alpha \frac{|E_0|^2}{\eta} \mathrm{e}^{-2\alpha L} \mathbf{a}_n. \tag{4.135}$$

The total power absorbed in each finite volume of a model object is the net power absorbed from the intersection of N rays launched from a source which intersect the volume:

$$Q_{\text{gen j}} = \sum_{i=1}^{N} w_{ij} \tag{4.137a}$$

$$w_{ij} = \iiint_{\text{vol } j} 2\alpha_j \frac{|E_{0i}|^2}{\eta_j} \mathrm{e}^{-2\alpha L_i} \, \mathrm{d}x_j \mathrm{d}y_j \mathrm{d}z_j \tag{4.137b}$$

where the w_{ij} are weighting coefficients which describe the intersection of ray i with model object finite volume j. The integral in (4.137b) is the "ray integral" along the propagating ray. The number N must be sufficiently large that a realistic estimate of all of the intersections can be obtained.

Scattering from a spherical load

When the conditions for geometric optic solutions ($\lambda/d \ll 1$) are not satisfied we must apply diffraction theory, or scattering calculations. Though these methods are similar in many ways they are often treated differently. In scattering calculations there are two instructive closed form solutions, which will now be discussed. Scattering from cylinders and spheres is an important consideration because the internal heating rates are usually very position dependent. Typical waveguide applicators utilize cylindrical samples and vessels, particularly in microwave chemistry experiments. Also, the behavior of droplets in a microwave field or the effect of spherical inclusions in a catalyst matrix are calculable as scattering problems. We begin with the spherical scattering case for a conducting sphere and then treat scattering from dielectric spheres. The following section will briefly treat cylinders. Additional discussion may be found in Born and Wolf (1980), Bohren and Huffman (1983), Harrington (1961), Jackson (1975) and in Kong (1986).

There are three important scattering regimes for both conducting and dielectric spheres, Rayleigh scattering, Mie scattering and optical scattering (wherein geometric optics applies). We will first consider scattering from metallic spheres assuming that they are perfectly conducting. This subject is of importance when the case of metallic particles suspended within a dielectric matrix is considered. We will analyze a single scattering center, which provides some insight into a very sparsely distributed conducting sphere mixture. If the sphere volume fraction is large then the scattered waves (i.e. the evanescent modes) from adjacent spheres will interact, creating a very difficult analytical problem which must be solved numerically.

Conducting spheres Since there is no energy absorption within a conducting sphere ($\mathbf{E} = 0$ for $r \leqslant a$) we may describe the wave interaction by looking at the magnitude of the reflected fluence rate, $\mathbf{S}_\mathrm{r} = 0.5|\mathbf{E} \times \mathbf{H}^*|$ (W/m^2) received at some receiving location located R from the scattering center (Figure 4.37). The incident wave on the scattering center has fluence rate magnitude S_i, and we consider the scatterer as a point source for the reflected wave (fluence rate magnitude S_r). The analytical expressions for this case are quite involved. We will not present this case in detail here since we are more concerned with the dielectric sphere case. A detailed discussion may be found in Bohren and Huffman (1983), Kong (1986), or Jackson (1975). The detailed solution for the metallic sphere case may be derived from the dielectric sphere case by letting ε_2 approach infinity.

We will instead describe the sphere in terms of the equivalent "scattering cross-section", s_c, an equivalent area (m^2):

$$s_\mathrm{c} = 4\pi R^2 \frac{S_\mathrm{r}}{S_\mathrm{i}}. \tag{4.138}$$

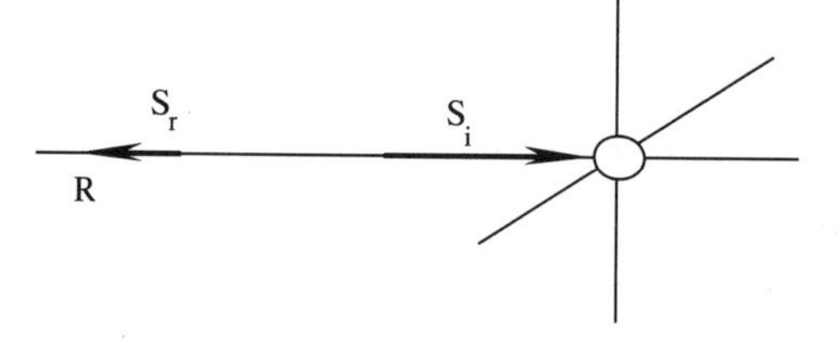

Fig. 4.37 *Metallic sphere scattering center.*

This formulation is typically used to describe a radar return signal where the received signal, S_r, is estimated by inverting the relation.

Rayleigh scattering is characterized by dominant backward scattering (diffuse reflection along the incident pathway), and applies to spheres with radius much smaller than the incident field wavelength, i.e. radius $a < 0.1\lambda_i$. Mie (or resonance) scattering applies to spheres with radius between about $0.1\lambda_i$ and $2\lambda_i$. Conducting spheres with $a > 2\lambda_i$ may be analyzed optically — that is, ray tracing may be used assuming specular reflection.

Very small spheres in the Rayleigh scattering range have solutions which are closely approximated by the quasi-static Laplace solution previously discussed. One may treat the induced surface charge (at $\theta = 0$ and $\theta = \pi$) as forming a radiating dipole, and apply standard antenna solutions for the very short dipole case (see Iskander (1992), Paris and Hurd (1969), or any of the many antenna texts). We will apply this approach in the dielectric sphere case as an example. The usual approach to the scattering cross-section is to normalize the result to sphere dimensions by comparing the scattering cross-section to the incident plane projected cross-section, πa^2. For Rayleigh scattering the normalized cross-section depends approximately on λ^{-4}:

$$\frac{s_c}{\pi a^2} = 9\left[2\pi\frac{a}{\lambda}\right]^4. \tag{4.139}$$

At higher frequencies (the Mie range) the normalized scattering cross-section oscillates about 1.0 and considerable forward scattering is observed owing to resonance interference around the sphere. Finally, in the optical region the normalized scattering cross-section is 1.0, and the scattering area is the projected cross-sectional area (Figure 4.38).

Dielectric spheres The problem of Rayleigh scattering for a small dielectric sphere makes use of the electrostatic solution derived above for a sphere in a uniform external electric field. In the Rayleigh range the particle radius is small compared to λ; so the internal field is essentially uniform and z-directed (Figure 4.11) from a z-polarized incident wave (equations (4.62)) and the external field is given by equations (4.61). The scattered waves for the Rayleigh sphere are derived by treating the sphere as a small (but finite length) dipole antenna and calculating the radiated fields. The derivation is well described by Kong (1986).

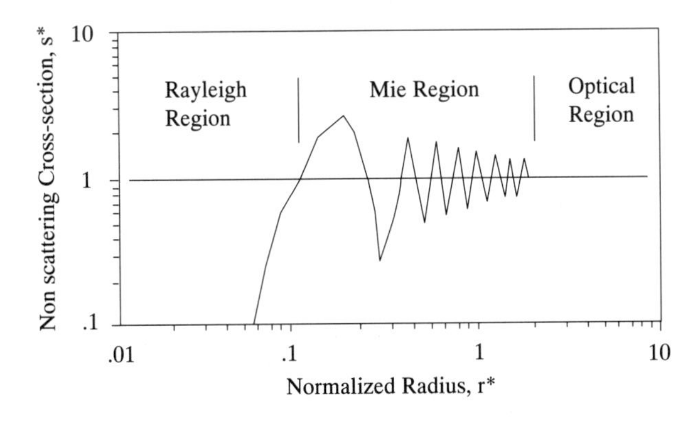

Fig. 4.38 *Normalized scattering cross-section, $s^* = s_c/\pi a^2$, as a function of normalized sphere radius, $r^* = a/\lambda$. The asymptotic limit is $s^* = 1$ far into the optical region.*

The major result is that the external electric field has radial (evanescent) and θ-components (radiating spherical wave):

$$\mathbf{E} = \frac{-\omega\mu\beta a^2}{\eta}\left(\frac{\varepsilon_2 - \varepsilon_1}{\varepsilon_2 + 2\varepsilon_1}\right) \mathrm{E}_0 \left(\frac{a}{r}\right) \mathrm{e}^{\mathrm{j}\beta r} \times \left\{\left[\left(\frac{\mathrm{j}}{\beta r}\right)^2 + \frac{\mathrm{j}}{\beta r}\right] 2\cos\theta \mathbf{a}_r + \left[\left(\frac{\mathrm{j}}{\beta r}\right)^2 + \frac{j}{\beta r} + 1\right] \sin\theta \mathbf{a}_\theta\right\} \quad (4.140)$$

where ε_1 is outside the sphere and ε_2 is inside, β is the phase coefficient in the surroundings, and η is the characteristic impedance of the surroundings. In the far field the scattered fields are:

$$E_\theta = -\beta a^2 \left(\frac{\varepsilon_2 - \varepsilon_1}{\varepsilon_2 + 2\varepsilon_1}\right) \left(\frac{a}{r}\right) E_0 \mathrm{e}^{\mathrm{j}\beta r} \sin\theta \quad (4.141a)$$

$$H_\varphi = \frac{E_\theta}{\eta} \quad (4.141b)$$

and the normalized scattering cross-section is calculated from:

$$\frac{s_c}{\pi a^2} = \frac{8}{3}\left(\frac{\varepsilon_2 - \varepsilon_1}{\varepsilon_2 + 2\varepsilon_1}\right)^2 \left(2\pi \frac{a}{\lambda}\right)^4. \quad (4.142)$$

This relation compares rather closely to equation (4.139).

The relationship between forward and backward scattering in the Mie range depends on radius as well as electric permittivity owing to the complex interference patterns within the sphere. Dielectric spheres larger than about $5\lambda_2$ are usually in the optical range and their internal heating patterns may be correctly estimated by ray tracing and geometric optics methods.

The spherical scattering problem was first solved by Mie in 1905. Imagine an x-polarized plane wave incident on a sphere along the z-axis, as in Figure 4.39. Note that the geometry is different than that used for the Rayleigh scattering calculation. A complete derivation may be found in Kong (1986) or in Bohren and Huffman (1983). We follow the notational development used by Kong here because it is the most clearly stated, of the many treatments, in terms of the sphere material properties, ε_2 and μ_2.

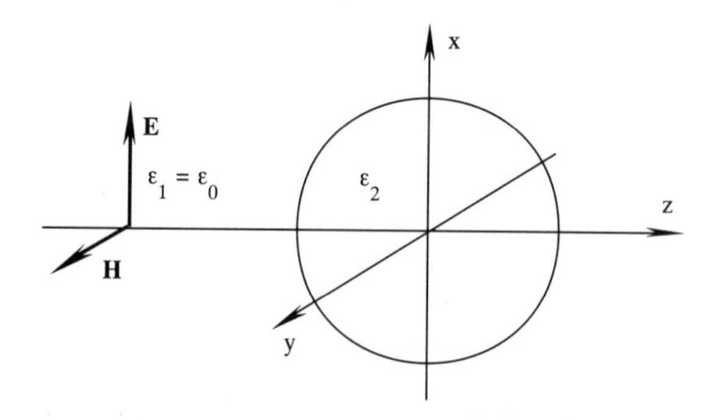

Fig. 4.39 *Geometry for Mie scattering calculation.*

The electric and magnetic field components are obtained from the respective Debye potentials, Π_e and Π_m. A Debye potential is a potential which solves the vector Helmholtz equation (equations (4.93)):

$$\nabla^2\Pi + \beta^2\Pi = 0 \tag{4.143}$$

where either Π_e or Π_m is used. The Debye potentials may be used to decompose a spherical wave into a TM to r component:

$$\mathbf{A} = \mu r \Pi_e \mathbf{a}_r \tag{4.144a}$$

$$\mathbf{H} = \frac{1}{\mu}\nabla \times \mathbf{A} = \frac{1}{\sin\theta}\frac{\partial \Pi_e}{\partial \varphi}\mathbf{a}_\theta - \frac{\partial \Pi_e}{\partial \theta}\mathbf{a}_\varphi \tag{4.144b}$$

and a TE to r component:

$$\mathbf{E} = \nabla \times (r\Pi_m \mathbf{a}_r) = \frac{1}{\sin\theta}\frac{\partial \Pi_m}{\partial \varphi}\mathbf{a}_\theta - \frac{\partial \Pi_m}{\partial \theta}\mathbf{a}_\varphi. \tag{4.145}$$

The solution of Maxwell's equations is obtained by superposition of Bessel functions, Legendre polynomials and sine waves subject to boundary conditions. For the present case they may be solved by:

$$E_r = \frac{\mathrm{j}}{\omega\varepsilon}\left[\frac{\partial^2(r\Pi_e)}{\partial r^2} + \beta^2 r\Pi_e\right] \tag{4.146a}$$

$$E_\theta = \frac{\mathrm{j}}{\omega\varepsilon}\frac{1}{r}\frac{\partial^2(r\Pi_e)}{\partial r\partial\theta} + \frac{1}{\sin\theta}\frac{\partial(\Pi_m)}{\partial\varphi} \tag{4.146b}$$

$$E_\varphi = \frac{\mathrm{j}}{\omega\varepsilon}\frac{1}{r\sin\theta}\frac{\partial^2(r\Pi_e)}{\partial r\partial\varphi} - \frac{\partial(\Pi_m)}{\partial\theta}. \tag{4.146c}$$

The relations for the magnetic field components are:

$$H_r = \frac{-\mathrm{j}}{\omega\mu}\left[\frac{\partial^2(r\Pi_m)}{\partial r^2} + \beta^2 r\Pi_m\right] \tag{4.147a}$$

$$H_\theta = \frac{-\mathrm{j}}{\omega\mu}\frac{1}{r}\frac{\partial^2(r\Pi_m)}{\partial r\partial\theta} + \frac{1}{\sin\theta}\frac{\partial(\Pi_e)}{\partial\varphi} \tag{4.147b}$$

$$H_\varphi = \frac{-\mathrm{j}}{\omega\mu}\frac{1}{r\sin\theta}\frac{\partial^2(r\Pi_m)}{\partial r\partial\varphi} - \frac{\partial(\Pi_e)}{\partial\theta}. \tag{4.147c}$$

The sphere is located in a surrounding medium with properties ε_1 and μ_1, and an x-polarized uniform plane wave impinges:

$$\mathbf{E} = E_0 \mathrm{e}^{\mathrm{j}\beta z}\mathbf{a}_x = E_0 \mathrm{e}^{\mathrm{j}\beta r\cos\theta}\mathbf{a}_x \tag{4.148a}$$

$$\mathbf{H} = \frac{E}{\eta}\mathbf{a}_y = \sqrt{\frac{\varepsilon_1}{\mu_1}}E_0 \mathrm{e}^{\mathrm{j}\beta r\cos\theta}\mathbf{a}_y. \tag{4.148b}$$

The wave is expanded in terms of spherical harmonics:

$$e^{j\beta r\cos\theta} = \sum_{n=0}^{\infty}(-j)^n(2n+1)\Psi_n(\beta r)P_n(\cos\theta) \tag{4.149a}$$

where P_n is the Legendre polynomial of the first kind and degree n, and Ψ_n is the spherical Bessel function:

$$\Psi_n(\beta r) = \sqrt{\frac{\pi\beta r}{2}}J_{n+\frac{1}{2}}(\beta r) \tag{4.149b}$$

where J is the cylindrical Bessel function.

The Debye potentials for the solution inside of the sphere are:

$$\Pi_e^i = \frac{-E_0\cos\varphi}{\omega\mu_2 r}\sum_{n=1}^{\infty}c_n\beta_2 r\Psi_n(\beta_2 r)P_n^1(\cos\theta) \tag{4.150a}$$

$$\Pi_m^i = \frac{E_0\sin\varphi}{\beta_2 r}\sum_{n=1}^{\infty}d_n\beta_2 r\Psi_n(\beta_2 r)P_n^1(\cos\theta) \tag{4.150b}$$

where P_n^1 is the Legendre function (general form $P_n^m(\cos\theta)$) with $m = 1$ and of degree n — the Legendre polynomial results when $m = 0$, P_n^0. For completeness, the Debye potentials for the scattered field (outside of the sphere) are:

$$\Pi_e^s = \frac{-E_0\cos\varphi}{\omega\mu_1 r}\sum_{n=1}^{\infty}a_n\beta_1 r\xi_n^{(1)}(\beta_2 r)P_n^1(\cos\theta) \tag{4.151a}$$

$$\Pi_m^s = \frac{E_0\sin\varphi}{\beta_1 r}\sum_{n=1}^{\infty}b_n\beta_1 r\xi_n^{(1)}(\beta_2 r)P_n^1(\cos\theta) \tag{4.151b}$$

where the $\xi_n^{(1)}$ are spherical Hankel functions of the first kind:

$$\xi_n^{(1)}(\beta r) = \sqrt{\frac{\pi\beta r}{2}}H_{n+\frac{1}{2}}(\beta r) \tag{4.151c}$$

and H denotes the cylindrical Hankel function.

By enforcing the boundary conditions the coefficients for the potentials inside of the sphere are:

$$c_n = \frac{(-j)^{-n}(2n+1)}{n(n+1)(\beta_1 a)(\beta_2 a)} \times \frac{j\sqrt{\varepsilon_2\mu_1}}{\sqrt{\varepsilon_2\mu_1}\xi_n^{(1)'}(\beta_1 a)\psi_n(\beta_2 a) - \sqrt{\varepsilon_1\mu_2}\xi_n^{(1)}(\beta_1 a)\psi_n'(\beta_2 a)} \tag{4.152a}$$

and

$$d_n = \frac{(-j)^{-n}(2n+1)}{n(n+1)(\beta_1 a)(\beta_2 a)} \times \frac{-j\sqrt{\varepsilon_1\mu_2}}{\sqrt{\varepsilon_2\mu_1}\xi_n^{(1)}(\beta_1 a)\psi_n'(\beta_2 a) - \sqrt{\varepsilon_1\mu_2}\xi_n^{(1)'}(\beta_1 a)\psi_n(\beta_2 a)}. \tag{4.152b}$$

We note that the derivatives of the spherical Hankel and Bessel functions are required in the denominator of both terms in alternate roles. Again, for completeness, we present the coefficients for the scattered field:

$$a_n = \frac{(-\mathrm{j})^{-n}(2n+1)}{n(n+1)} \times \frac{-\sqrt{\varepsilon_2\mu_1}J'_n(\beta_1 a)J_n(\beta_2 a) + \sqrt{\varepsilon_1\mu_2}J_n(\beta_1 a)J'_n(\beta_2 a)}{\sqrt{\varepsilon_2\mu_1}\xi_n^{(1)'}(\beta_1 a)\psi_n(\beta_2 a) - \sqrt{\varepsilon_1\mu_2}\xi_n^{(1)}(\beta_1 a)\psi'_n(\beta_2 a)} \quad (4.153a)$$

$$b_n = \frac{(-\mathrm{j})^{-n}(2n+1)}{n(n+1)} \times \frac{-\sqrt{\varepsilon_2\mu_1}J_n(\beta_1 a)J'_n(\beta_2 a) + \sqrt{\varepsilon_1\mu_2}J'_n(\beta_1 a)J_n(\beta_2 a)}{\sqrt{\varepsilon_2\mu_1}\xi_n^{(1)}(\beta_1 a)\psi'_n(\beta_2 a) - \sqrt{\varepsilon_1\mu_2}\xi_n^{(1)'}(\beta_1 a)\psi_n(\beta_2 a)}. \quad (4.153b)$$

As $\beta_2 a$ shrinks to a very small number these results approach the Rayleigh scattering solution.

These expressions are sufficiently complex that numerical calculations are required. We have made calculations at 2.45 GHz for a sphere of TX-151, a water-based gel medium ($\varepsilon'_r =$ 60, $\varepsilon''_r = 15$, at 2.45 GHz). The solution for radii of 2.75 cm (about $\lambda_0/4$), 4.5 cm (about $3\lambda_0/8$), and 12.2 cm (λ_0) are shown in Figure 4.40. Clearly, the internal electric field pattern is "optical" in this high loss material for radii above λ_0 and center-weighted for smaller radii. Experiments with molded spherical shapes verify the predicted heating patterns.

Scattering from a cylindrical load

Wires or conducting pipes in waveguides or multimode cavities are effectively perfectly conducting scatterers. Ordinary beakers and test tubes filled with water fall into the scattering range for dielectric cylinders at microwave frequencies. One must expect that even if the vessel is smaller in diameter than one wavelength the power density will not be uniform within the solution. This is why liquid super heating effects may be observed in beakers irradiated with microwave fields. Scattering from cylinders follows a formulation similar in many ways to the spherical case. We will assume that the cylinders are infinite in length in order to make the calculation tractable. Finite long cylinders follow similar trends even though the end effects change the field distribution.

Conducting cylinders The sphere is isotropic and so the polarization of the incident field is of no consequence. The cylinder is another matter entirely. Here we need to rewrite the incident field in terms of the cylinder geometry. A simple well behaved case will illustrate the intracies of this problem. Imagine a z-polarized wave propagating in the $+x$ direction ($E_0 e^{-\mathrm{j}\beta x}$) with respect to the cylinder (Figure 4.41). It is convenient to express the incident wave (in a lossless medium) in terms of a superposition of cylindrical waves (see Kong (1986) or Paris and Hurd (1969)):

$$e^{-\mathrm{j}\beta x} = \sum_{n=-\infty}^{\infty} \mathrm{j}^{-n} J_n(\beta r) e^{\mathrm{j}n\varphi} \quad (4.154a)$$

so that:

$$E_{\mathrm{i}z} = E_0 \sum_{n=-\infty}^{\infty} \mathrm{j}^{-n} J_n(\beta r) e^{\mathrm{j}n\varphi} \quad (4.154b)$$

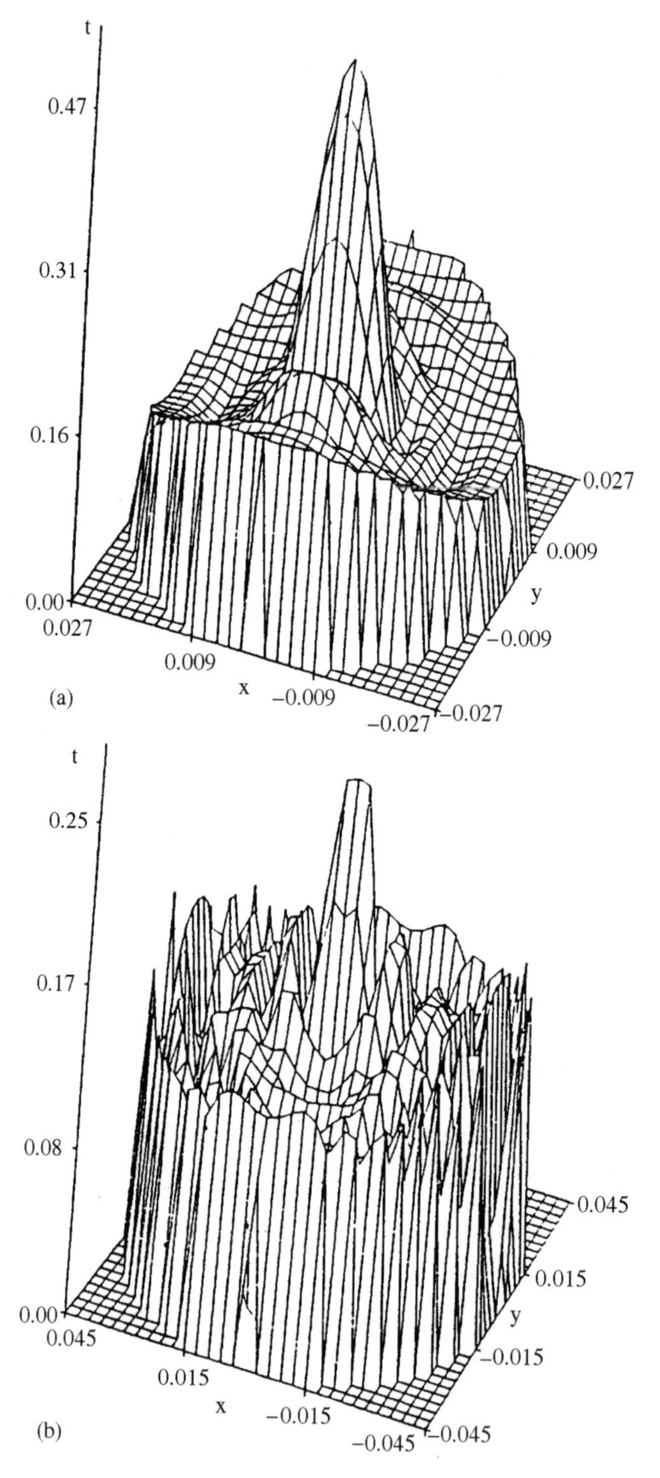

Fig. 4.40 *Mie scattering calculations for a sphere of TX-151 ($\varepsilon_r' = 60$, $\varepsilon_r'' = 15$). Electric fields in x–y plane for radii: (a) a = 2.75 cm; (b) a = 4.5 cm; (c) a = 12.2 cm.*

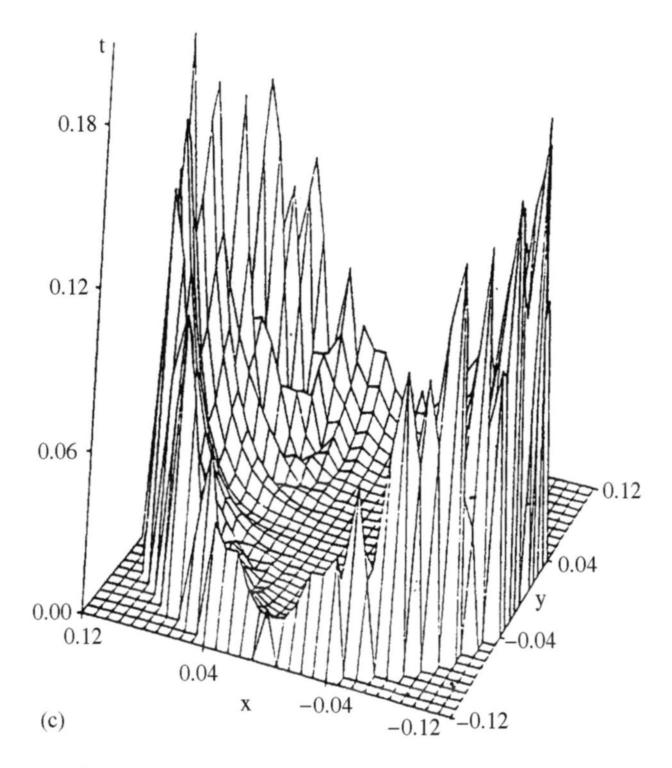

Fig. 4.40 *(continued)*

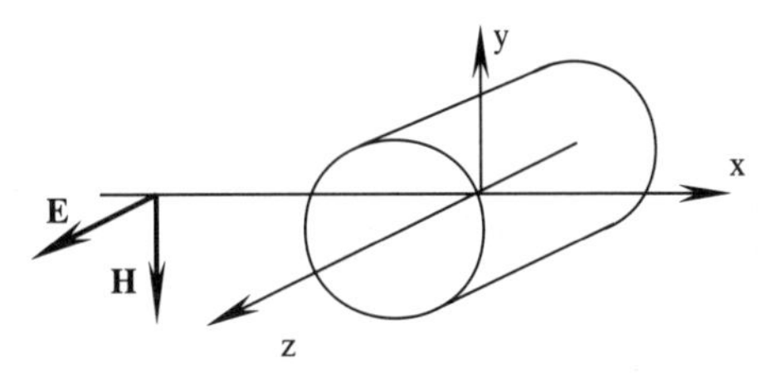

Fig. 4.41 *Conducting cylinder geometry.*

where J_n are cylindrical Bessel functions of the first kind with real arguments. The scattered field is represented with an infinite series of Hankel functions of the second kind, $H_n^{(2)}$:

$$E_z^{\mathrm{s}} = E_0 \sum_{n=-\infty}^{\infty} A_n \mathrm{j}^{-n} H_n^{(2)}(\beta r) \mathrm{e}^{\mathrm{j}n\varphi} \tag{4.155a}$$

where:

$$H_n^{(2)}(r) = J_n(r) - \mathrm{j}Y_n(r) \tag{4.155b}$$

and the Y_n are Bessel functions of the second kind and order n. The total field is then the sum of the two components.

The applicable boundary condition is that $E_z = E_{\tan} = 0$ at $r = a$ for all values of n. This gives the unknown coefficients:

$$A_n = -\frac{J_n(\beta a)}{H_n^{(2)}(\beta a)} \tag{4.156}$$

and the total electric field:

$$E_z = E_0 \sum_{n=-\infty}^{\infty} \mathrm{j}^{-n} \left[J_n(\beta r) - \frac{J_n(\beta a)}{H_n^{(2)}(\beta a)} H_n^{(2)}(\beta r) \right] \mathrm{e}^{\mathrm{j}n\varphi}. \tag{4.157}$$

The result is interesting but obscures the physical meaning of the interaction somewhat. We can gain a little more insight by inspecting the far field ($\beta r \gg 1$) expression. Here the Hankel function approaches:

$$H_n^{(2)} \approx \sqrt{\frac{2}{\pi \beta r}} \exp\left\{ -\mathrm{j}\left(\beta r - \frac{2n+1}{4}\pi \right) \right\}. \tag{4.158}$$

Thus, the scattered field becomes:

$$E_z^{\mathrm{s}} \approx E_0 \sqrt{\frac{2}{\pi \beta r}} \left[\sum_{n=-\infty}^{\infty} A_n \mathrm{j}^{-n} \exp\left\{ j\left(n\varphi + \frac{2n+1}{4}\pi \right) \right\} \right] \mathrm{e}^{\mathrm{j}(\omega t - \beta r)} \tag{4.159}$$

which may be interpreted as a cylindrical wave propagating in the $+r$-direction with uniform phase velocity $u_{\mathrm{p}} = \omega/\beta$. As should be expected, the scattering is critically dependent on the polarization of the incident wave with respect to the cylinder long axis. In fact, in experimental determinations it takes measurements at five independent polarizations to fully characterize the scattered waves.

Dielectric cylinders Infinite dielectric cylinders in the Mie scattering range have internal interference patterns similar in many ways to the dielectric sphere case. The fields may be calculated by following a development comparable to that for the spheres. The major difference is that the wave polarization is of critical importance in the cylinder case. The polarization which maximally couples to the cylinder is a z-polarized wave (Figure 4.42). It is also the most strongly scattered. A geometry similar to an infinite cylinder is seldom encountered, with the important exception of a small cylindrical object in the center of the broad wall of a TE_{10} waveguide, a common heating geometry. Therefore, we present the solution for an infinite dielectric cylinder, according to Harrington (1961), with z-polarized

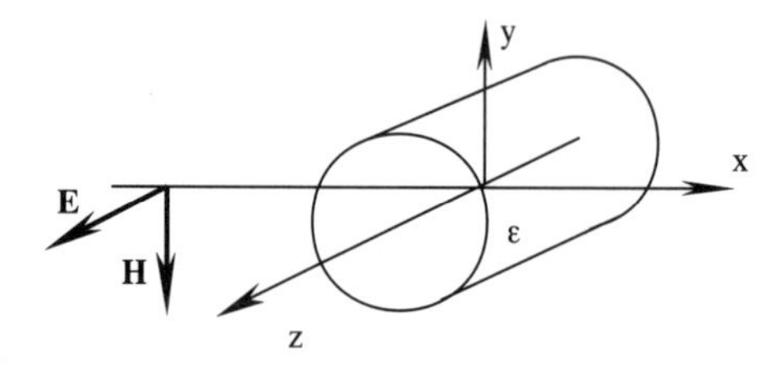

Fig. 4.42 *Polarization of assumed wave impinging on dielectric cylinder.*

incident wave:

$$\mathbf{E} = E_0 e^{-j\beta x} \mathbf{a}_z = E_0 e^{-j\beta r \cos\varphi} \mathbf{a}_z. \tag{4.160}$$

The field internal to the cylinder is given by:

$$\mathbf{E}_i = E_0 \sum_{n=-\infty}^{\infty} j^n c_n J_n(\beta_2 r) e^{jn\varphi} \mathbf{a}_z \tag{4.161a}$$

where the coefficients are:

$$c_n = \frac{J_n(\beta_1 a) + a_n H_0^{(2)}(\beta_1 a)}{J_n(\beta_2 a)}. \tag{4.161b}$$

In equation (4.161b) the a_n are the coefficients used to represent the scattered field (in region 1, outside of the cylinder):

$$a_n = \frac{-J_n(\beta_1 a)}{H_n^{(2)}(\beta_1 a)} \left\{ \frac{\dfrac{\varepsilon_2 J'_n(\beta_2 a)}{\varepsilon_1 \beta_2 a J_n(\beta_2 a)} - \dfrac{J'_n(\beta_1 a)}{\beta_1 a J_n(\beta_1 a)}}{\dfrac{\varepsilon_2 J'_n(\beta_2 a)}{\varepsilon_1 \beta_2 a J_n(\beta_2 a)} - \dfrac{H_n'^{(2)}(\beta_1 a)}{\beta_1 a H_n^{(2)}(\beta_1 a)}} \right\} \tag{4.162a}$$

and the scattered field is:

$$E^s = E_0 \sum_{n=-\infty}^{\infty} j^{-n} a_n H_n^{(2)}(\beta_1 r) e^{jn\varphi} \tag{4.162b}$$

The total external field is then the sum of (4.162b) and (4.160).

The trend in internal heating is similar to the Mie sphere results. Few infinite cylinder loads are realized other than in a waveguide. Finite cylinders (beakers etc.) are often used in multimode cavities, however, and will exhibit many of the same internal heating effects. Most ordinary beakers should be expected to have strong internal heating in a microwave field, just as we saw for the comparably-sized spheres.

Edge and corner boundary effects

We have already looked at the first-order discontinuity of a plane wave incident on the surface of a material in Chapter 1. There we saw the specular reflection coefficients which describe the incident and reflected wave interaction. The previous sections in this chapter have described scattering from smooth curved boundaries. It remains to treat edges and corners, as we did for RF heating. It might at first seem that the wave nature of the problem would make it nearly intractable. However, many of the trends we saw in our low frequency discussion in terms of the Laplace equation apply at microwave frequencies as well. That is, if one views a plane surface as a first order discontinuity, the intersection of two planes to form an edge as a second order discontinuity, and a three-plane intersection forming a corner as third order, then we might expect that higher order discontinuities would have stronger fields and thus higher heating rates. This is in fact what we observe experimentally. The reason may be traced to the requirement for simultaneously satisfying tangential and normal electric field boundary conditions in more than one direction at higher order discontinuities. We reserve discussion of this phenomenon for the section on multimode cavities, since that is where the observations have been made.

4.2.2 Traveling wave and slotted waveguide applicators

Traveling wave applicators are typically constructed of waveguides terminated by matched or nearly matched loads. They are called traveling wave devices because the location of electric field and magnetic field maxima change with time, as we saw in the discussion of waveguide propagation in Chapter 2. This is in contrast to standing wave devices (resonators), discussed in the next section, in which the field maxima always occur at the same location due to interference between the forward and reflected waves, as was introduced very briefly in Chapter 2. This section presents several commonly used traveling wave applicator configurations.

Rectangular waveguide applicators

The rectangular waveguide is almost always used in the TE_{10} mode to heat dielectric materials of various geometries. As mentioned, this is the fundamental waveguide mode and is the one least-disturbed by the difference between the permittivity of the load and surrounding air. Nevertheless, significant perturbations in the waveguide fields are often created by the load in order to satisfy the local boundary conditions. The scattered fields result in reflection in the guide and evanescent modes near the load.

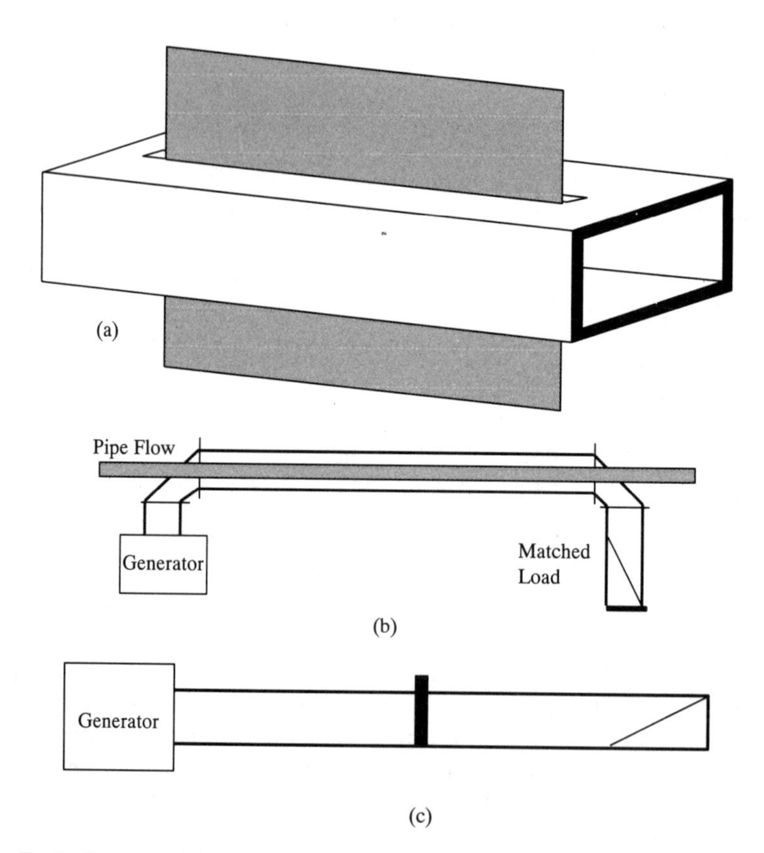

Fig. 4.43 *Typical waveguide traveling wave applicators. (a) Longitudinal slot in center of broad wall. (b) Pipe through **E**-plane bends. (c) Test tube in broad wall center.*

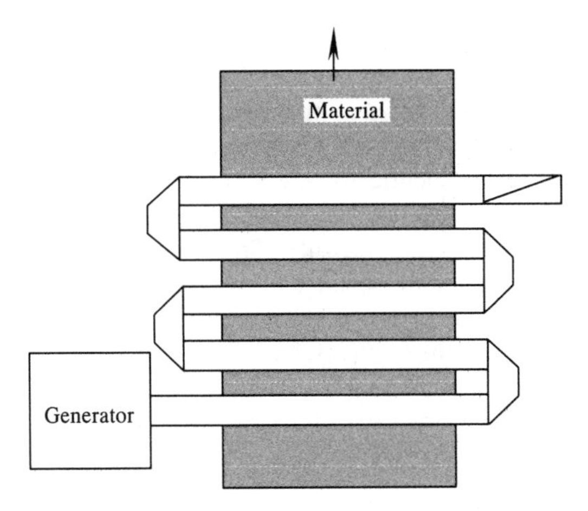

Fig. 4.44 *Meandering waveguide applicator.*

The most common form of traveling wave applicator is a longitudinal slot in the broad wall of a waveguide — the slot is non-radiating because it does not cut any wall current lines for the TE_{10} waveguide mode. Thin sheets of material to be irradiated may be passed through the longitudinal slot, as in Figure 4.43a. Another standard configuration utilizes a pipe aperture in an **E**-plane bend of a rectangular waveguide through which liquids to be heated flow in a microwave-transparent tube (made of teflon, polystyrene, etc.), Figure 4.43b. Small samples may be studied using a test tube placed through the center of the broad wall of the waveguide, Figure 4.43c.

Slow heating at relatively low power can be obtained in a thin web using a "meandering" or "serpentine" waveguide configuration, Figure 4.44. Here the waveguide makes several loops and the material travels several times through the waveguide, and will eventually absorb the majority of waveguide power so that little is dissipated in the matched load. For example, assume that 90% of the inlet power reaches the other end of a waveguide on each pass. That is, 90% reaches the inlet of the second pass, 81% the third pass, and so on. Then for, say, 10 passes, only $(0.9)^{10} = 35\%$ of the generator power arrives at the matched load, most of the rest having been absorbed by the web (there are some losses in the waveguide walls). The meandering waveguide applicators have been well studied for industrial use. The problem is to obtain uniform heating along the common path of the waveguide as a function of the intrinsic absorption of the material inside the guide. This is complicated by the tendency of the edges of the web to experience higher power densities than the central portion owing to boundary conditions. Infrared, or other, sensors may be used between waveguide passes to sense the web temperature distribution and control generator power.

Waveguide slot radiators and arrays

These systems consist of placing the material to be heated near a guide in which slots have been cut to allow irradiation of the product below the guide. The entire heating volume must be placed within a large shielded enclosure for personnel safety and to prevent unlawful radiation. Consequently, the slotted system and enclosure constitute a large multimode cavity, which will be discussed later. Ironically, in some sense the waveguide slot array then becomes another sort of aperture for introducing microwaves into a multimode cavity, but the material is in the near field of the slot rather than in the far field.

Apertures, or diffraction screens, in large thin sheets form a general class of radiators. They may be thought of as equivalent to the complementary radiator formed of the material removed from the aperture—except that the roles of the **E** and **H** fields are reversed (Figure 4.45). This is known as Babinet's principle (Jackson 1975). A thin slot one half-wavelength long behaves nearly identically to a half-wave dipole antenna. A standard half-wave dipole has a radiation impedance of $75 + j42.5\ \Omega$. By slightly shortening it, a half-wave dipole can be made to resonate, virtually eliminating its input reactance. The amount of shortening required to obtain slot resonance depends on width, waveguide wall thickness and on whether or not the slot ends are rounded. The behavior of a single linear slot depends on its length, angle and slot width (Figure 4.46a).

Because the primary driving source for the slot is a magnetic field, slots must be located so that they cut wall currents in order to radiate. We will first discuss single slots and then follow with the performance and typical architectures of slot arrays.

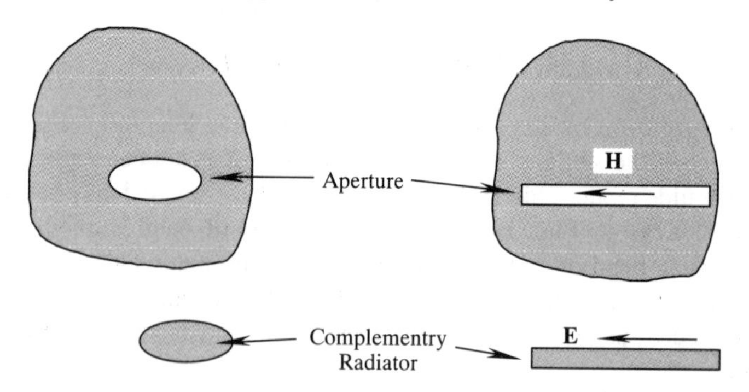

Fig. 4.45 *Illustration of Babinet's principle.*

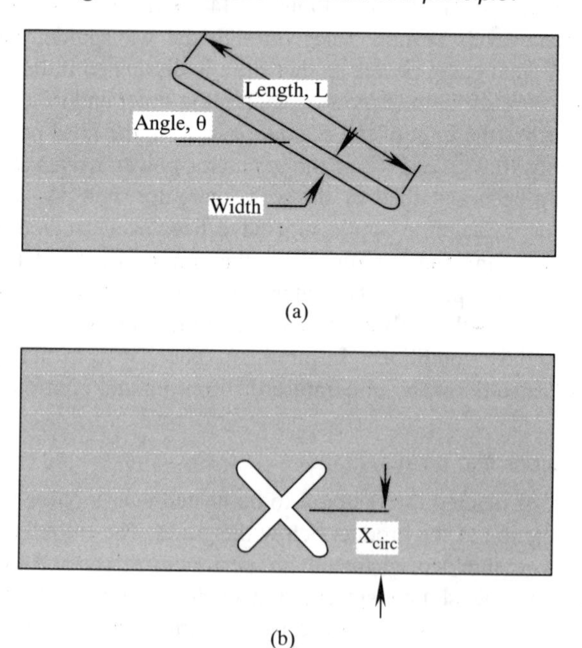

Fig. 4.46 *Single broad wall slot geometries. (a) Linear slot geometry. (b) X-slot.*

Broad wall single slots Calculationally, an inclined slot is electrically equivalent to a shunt impedance. The real part of the impedance is a measure of the radiated energy, and the imaginary part is a measure of the electric length of the equivalent propagation in the guide. It is possible to design slots which have no imaginary component in their equivalent impedance — so-called zero phase slots — by slightly shortening them. Wall thickness and slot geometry play a role in determining the resonance length. The resonance length for a slot in a typical X-band waveguide, for example, is about 2% longer than for an infinitesimally thick wall. A rounded slot end has a resonant frequency about 2% higher than that of an equal slot length with squared ends. The radiation resistance obtained is dependent on slot angle in the waveguide wall.

A circularly polarized wave may be obtained from a self-complementary X-slot (Figure 4.46b) (Simmons 1957). This configuration is in a sense self-resonant since the reflection coefficients of the opposite arms are of equal magnitude and 180° out of phase with each other, and thus cancel. As a result the X-slot may be nearly matched to the waveguide and is a very effective radiator. The X-slot will radiate a circularly polarized wave if it is located where two orthogonal components of either an electric or a magnetic field are equal in magnitude. As we discussed in Chapter 2 the TE_{10} waveguide mode consists of E_y, H_x and H_z components. According to Simmons there is one location in the broad wall where circular polarization can be obtained, at x_{circ}:

$$x_{\mathrm{circ}} = \frac{a}{\pi} \operatorname{cotan}^{-1} \left\{ \pm \sqrt{\left(\frac{2a}{\lambda_0} \right)^2 - 1} \right\}. \tag{4.163}$$

So, for WR-284 waveguide at 2.45 GHz ($a = 7.21$ cm, $\lambda_0 = 12.2$ cm) $x_{\mathrm{circ}} = 2.68$ cm and the X-slot should be 2.68 cm from either of the narrow walls.

Narrow wall single slots The wall currents in the narrow walls are oriented orthogonally to those in the broad wall. Consequently, while a centered longitudinal slot is non-radiating in a broad wall, a vertical slot is non-radiating in the narrow wall (Figure 4.47). Other characteristics of the narrow wall slot follow the same trends as for a broad wall slot. However, since the wall is narrow there is virtually no chance of working with a $\lambda/2$ slot; one is limited to very short slots indeed. Consequently, the usual wall slot applicator is found in the broad wall of the waveguide.

Arrays of slots Practical slot applicators consist of many slots in a linear array. When a series of zero-phase slots is placed at fixed distances — multiples of $\lambda_g/2$ — such that each such slot can be adjusted individually, one may obtain a nearly uniform distribution. We

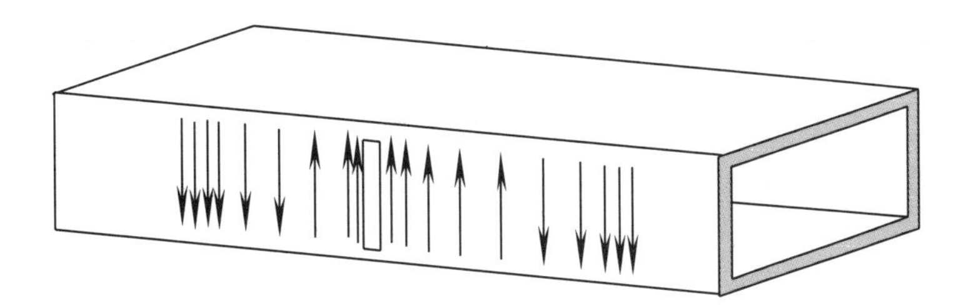

Fig. 4.47 *Narrow wall slot radiators.*

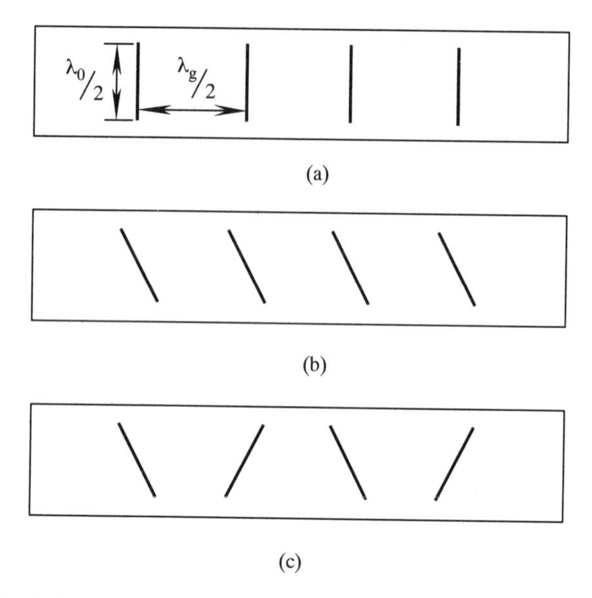

Fig. 4.48 *Broad wall linear slot arrays. (a) Transverse array ($\theta = 90^\circ$). (b) Inclined array. (c) Complementary array.*

will see, however, that a practical linear array of slots has slightly non-uniform radiation, with the first and last slots radiating more power than the center slots.

There are several typical broad wall linear array architectures (see Figure 4.48). Interestingly, the slot length is expressed in terms of the free space wavelength, λ_0, and the slot location in terms of the waveguide wavelength, λ_g. This is because the slot is radiating (approximately) into free space while its driving function is determined by the waveguide fields. Separating the slots by $\lambda_g/2$ ensures that they are driven in phase with each other creating a so-called "broadside" array. That is, the major lobe of the array factor is directed away from the waveguide into the material and the individual slot waves reinforce each other rather than combining destructively. There is, of course, a traveling wave in the waveguide, but at each instant in time the slots are driven in phase with each other.

The equivalent circuit of a linear slot is series R and jX (Oliner 1957) (Figure 4.49). Rotating the slot changes the impedance by a correction factor. The correction factor, $F(\theta)$, is:

$$F(\theta) = \left[\frac{\cos(\pi a'/2a)}{1 - a'/a)^2}\right] \frac{2}{A(\theta)\sin\theta + (\lambda_g/2a)B(\theta)\cos\theta} \tag{4.164a}$$

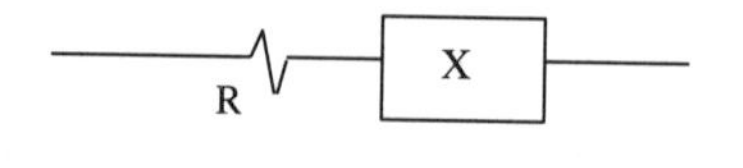

Fig. 4.49 *Slot equivalent impedance.*

where a' is the slot length, a the waveguide width and the related terms are:

$$\begin{bmatrix} A(\theta) \\ B(\theta) \end{bmatrix} = \frac{\cos(\pi\xi/2)}{1-\xi^2} \pm \frac{\cos(\pi\eta/2)}{1-\eta^2} \tag{4.164b}$$

$$\begin{bmatrix} \eta \\ \xi \end{bmatrix} = \frac{a'}{a}\sin\theta \pm \frac{2a'}{\lambda_g}\cos\theta. \tag{4.164c}$$

Though the relationship is not easy to see, inspection of equations (4.164) indicates that a slot at $90° - \alpha$ (the inclined slot of Figure 4.48b) would have an impedance factor with opposite sign to a slot at $90° + \alpha$. Thus, alternating slot angles between the two values gives a compensating impedance combination.

An X-slot array must have its elements separated by λ_g rather than $\lambda_g/2$ in order to obtain a broadside array factor because the slots themselves are self-complementary (Figure 4.50).

Some experimental results have been obtained to show the relative performance of slot arrays in WR430 waveguide at 2.45 GHz (λ_g = 14.8 cm, Z_0 = 448 Ω). Results are summarized in Table 4.3. All of the linear slots were resonant at 5.84 cm in length ($\lambda_0/2$ = 6.1 cm) and 0.16 cm in width. The transverse slots had a calculated impedance of $R + jX = -37 - j4.9$ Ω (see Oliner 1957). Decreasing the slot angle, θ, increases the impedances until at 0° the slot is non-radiating. In order to obtain a slot impedance close to the 448 Ω waveguide impedance we would need an inclination angle somewhere near $\theta = 30°$. A complementary slot array at such an angle would further cancel the load impedance presented by the slots. However, while this would reduce reflections it would mean that little power would be radiated from each slot since few current lines are cut. If a long array is to be used, a small slot angle would be a desirable design feature to reduce reflections into the generator and to even out the radiated power along the array. If a short array is necessary, the radiated power may be increased by using a large θ at the expense of more reflected power into the generator circuit.

Of the linear arrays the transverse configuration radiated the most waveguide power (41%) but was not the best match to waveguide impedance. A typical transverse array of five elements radiated up to 40% of the waveguide forward power and reflected about

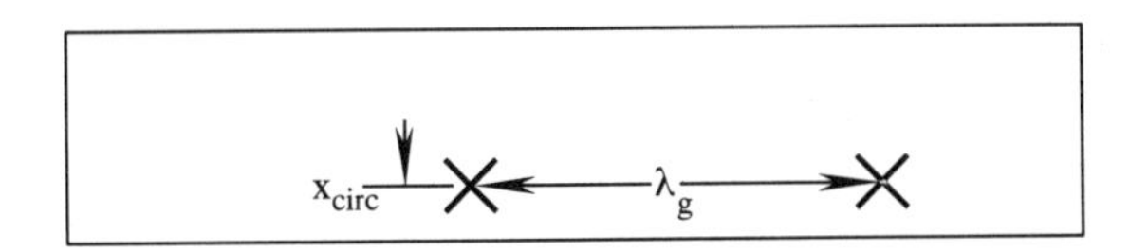

Fig. 4.50 *X-slot array.*

Table 4.3 *Five-element linear and two-element X-slot array performance.*

Array type	Power radiated P_{rad}/P_f (SD) %	Power reflected P_{ref}/P_f (SD) %
Transverse	41.4 (0.44)	24.1 (1.7)
Inclined ($\theta = 60°$)	36.4 (1.1)	20.3 (1.1)
Complementary	39.3 (1.3)	23.6 (1.3)
X-slot	57.3	5.9

24%. A five-element inclined array, with $\theta = 60^\circ$, had slots better-matched to waveguide impedance. The complementary slot array ($\theta = 60^\circ$ and $\theta = 120^\circ$) had a higher reflection coefficient than the inclined array owing to two effects: (1) an odd number of slots (five) was used in the array leaving one slot uncompensated, and (2) slot ends were separated by only 4.45 cm and some interaction between adjacent slots is likely. The radiated power of each slot was evaluated using thermographic imaging of the workpiece below the waveguide. In all of the linear arrays tested the temperature rise under the first and last elements (#1 and #5) in the array was about 60% higher than for the central (#3) element — 13°C compared to 8°C. The temperature rises for the #2 and #4 elements were between those values. So, we would expect higher heating rates near the ends of a slot array.

The two-element pair of X-slots was seen to radiate a much larger fraction of the waveguide power than the five transverse elements, 57% compared to 41%, and is much better matched than any of the linear arrays (5.9% reflection). For coupling waveguides through slot apertures, where the goal is maximum power radiated in minimum length with minimum reflection, the X-slot pattern is clearly superior. Applications such as coupling a waveguide into a vacuum chamber might fall into this category — and circular polarization might be an advantage in that case. If the application involves uniform heating of a wide thin-web workpiece the X-slots are probably much too localized to provide adequate performance. A complementary slot array is probably a better choice.

4.4.3 Single-mode cavity applicators

Traveling wave applicators work very well for medium to high loss materials. Low loss materials often require much higher electric fields to undergo the desired transformations. We may think of low loss materials, in general, as high impedance materials. To obtain adequate power absorption it is often advantageous to use the resonant properties of circuits in order to amplify the field such that adequate power is transferred to the product.

We may easily construct a resonant cavity within a rectangular waveguide in which very high field strengths can be obtained. Typical modes which can be easily excited from standard sources are TE_{10k}, where k (integer) indicates the number of maxima within the cavity. Many resonant applicators use cylindrical cavities and, for example, the TM_{011} mode. We begin with a discussion of pertinent resonance relations, then treat rectangular waveguide resonant structures and conclude with cylindrical cavities.

Resonance conditions

Obtaining well behaved fields in a resonant cavity is a delicate process. We may obtain stable resonant field modes and high field strengths as long as we do not perturb the cavity too much — that is, as long as the cavity load does not absorb too much power. The cavity Q-factor (as introduced in Chapter 3) is high when the load is not very lossy. When power absorbed by the load is high the load is strongly coupled to the cavity, its effect on the modes (and field strengths) realized within the cavity may be significant, and the Q-factor will be low. The discussion here is limited to the case of small perturbations created by the cavity load. The governing relations for cavity resonance are quite instructive and are briefly reviewed here.

Consider a large simple resistance, 1 MΩ for example. If we apply a voltage of 220 V from a source of resistance $R_0 < 1$ MΩ, the absorbed power is low, and the power absorbed

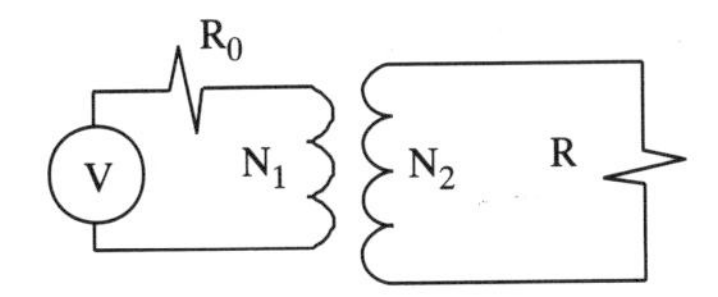

Fig. 4.51 *Transformer matching of load and source impedance.*

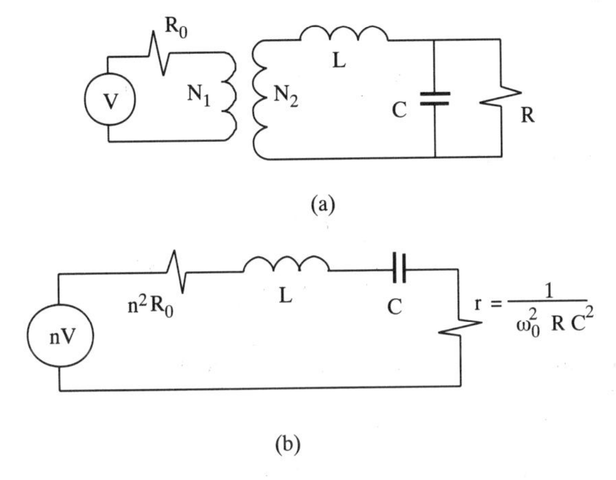

Fig. 4.52 *RF circuit transformed impedances. (a) Example RF circuit. (b) Primary representation.*

by the resistor is low. To obtain the dissipation of a large amount of power, a transformer is used to increase the voltage (Figure 4.51). At low frequencies we need only consider the turns ratio of the transformer, $n = N_1/N_2$. For maximum power transfer it will be recalled that the optimum value for n is $(R/R_0)^{0.5}$. At RF frequencies we must use the full transformer model as we did in the discussion of matching in Chapter 3.

A more realistic RF resonant circuit would include L and C elements, such as we saw previously, in which R is relatively large and simulates the low loss material (Figure 4.52a). Its "primary" representation would be as shown in Figure 4.52b where the source has been transformed as well. The relations for resonant systems are well known and we very briefly review them here. The resonance frequency is:

$$\omega_0 = \frac{1}{\sqrt{LC}}. \tag{4.165a}$$

The equivalent series load resistance is now r, which depends on the resonance frequency, ω_0:

$$r = \frac{1}{\omega_0^2 RC^2}. \tag{4.165b}$$

The Q-factor of the unloaded cavity is:

$$Q_0 = \frac{\omega_0 L}{r} = \omega_0 RC \tag{4.165c}$$

and finally, the loaded Q-factor including the source is:

$$Q_L = Q_0 \left(1 + n^2 \frac{R_0}{r}\right). \tag{4.165d}$$

When the generator, V, operates at resonance a very high voltage is generated across both L and C, and across r too. The impedance presented to the ideal generator is customarily expressed using these terms:

$$Z_L = r\left[1 + n^2 \frac{R_0}{r} + jQ_0\left(\frac{\omega}{\omega_0} - \frac{\omega_0}{\omega}\right)\right]. \tag{4.166}$$

Consequently, the load power is:

$$P_L = \frac{n^2 V^2 r}{(r + n^2 R_0)^2 + Q_0^2\left(\frac{\omega}{\omega_0} - \frac{\omega_0}{\omega}\right)^2}. \tag{4.167}$$

Recall that the maximum power available from the generator is:

$$P_{max} = \frac{V^2}{4R_0}. \tag{4.168}$$

So, the load power is maximum at the resonance frequency and when $n = (r/R_0)^{0.5}$ the power in the load can approach P_{max}, as we know from our discussion of RF generators in Chapter 3.

Rectangular waveguide resonant cavities

We now treat the case of a rectangular waveguide load. Consider a charge or load, as in the last example, which is of fairly high impedance; similar to an open circuit in a propagating transmission line, $R \gg Z_0$ (Figure 4.53). The problem will be to determine what sort of network must be placed in front of the load such that we can dissipate a large amount of power in the load by forming a resonant cavity. The process is philosophically identical to the RF impedance-matching strategies discussed in some detail in Chapter 3 (Section 3.3).

We restrict our choice of circuits to purely imaginary impedances, short circuits and coupling irises, because it is preferable that the coupling system does not absorb energy. We place a sliding short circuit d_1 behind the load and a shunt coupling reactance, normalized

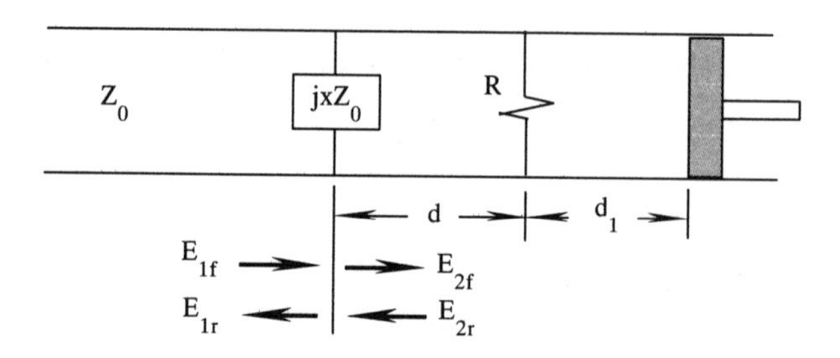

Fig. 4.53 *High impedance waveguide load circuit model.*

reactance = x, a distance d in front of it. Further, the forward waves are described by electric fields E_{1f} and E_{2f}, with respective reflected waves E_{1r} and E_{2r}, where the E_2 are inside the resonant region

The load will have a reflection coefficient (Γ_r) which would be near +1 in a traveling wave applicator because R is large. However, the equivalent reflection coefficient in a standing wave applicator has a minimum, ρ, when $d_1 = \lambda_g/4$:

$$\rho = \frac{R - Z_0}{R + Z_0}\left(1 - \frac{2Z_0}{R}\right). \tag{4.169}$$

When this equivalent reflection coefficient is transformed to the plane of the shunt reactance (at distance d toward the generator) the complex reflection coefficient is:

$$\frac{E_{2r}}{E_{2f}} = \rho e^{-j2\beta d} = \rho e^{-j4\pi d/\lambda_g}. \tag{4.170}$$

The hybrid quadrupole transfer parameter matrix which describes a single shunt reactance is:

$$\begin{bmatrix} E_{1r} \\ E_{2f} \end{bmatrix} = \frac{1}{1 + j2x}\begin{bmatrix} -1 & j2x \\ j2x & -1 \end{bmatrix}\begin{bmatrix} E_{1f} \\ E_{2r} \end{bmatrix}. \tag{4.171}$$

After some algebra we may obtain a relation for the S-parameter of the forward wave coupled into the cavity region, S_{12}:

$$S_{21} = \frac{E_{2f}}{E_{1f}} = \frac{j2x}{1 + j2x + \rho e^{-j4\pi d/\lambda_g}} \tag{4.172}$$

and a relation for the reflection S-parameter, S_{11}, the reflection coefficient toward the generator:

$$S_{11} = \frac{E_{1r}}{E_{1f}} = \frac{4x^2}{1 + j2x + \rho e^{-j4\pi d/\lambda_g}} - \frac{1}{1 + j2x} \tag{4.173}$$

Both expressions may be simplified using the common denominator:

$$D = 1 + j2x + \rho e^{-j4\pi d/\lambda_g}. \tag{4.174}$$

Our goal in this design is to select d and x to maximize E_{2f}. This is achieved when D is minimized. Remembering that the load reflection coefficient is approximately $1 - 2Z_0/R$, we must place the shunt impedance at $d = \lambda_g/4$ in front of the load to minimize D with respect to d. With that placement, we need x to be small to minimize D with respect to x; that is, as near a short circuit as is reasonable. Coupling is through a small shunt impedance such as a capacitive iris. The result is a resonant section of waveguide $\lambda_g/2$ long (or some multiple of $\lambda_g/2$) in which the load is at the center. The resonant mode is written TE_{101} for a length of $\lambda_g/2$ (one maximum in the E-field), TE_{103} for length $3\lambda_g/2$ (three maxima), etc. We will review the general representation of rectangular cavity resonant modes at the end of this section.

We may now calculate the Q-factor by letting the operating frequency, ω, vary around ω_0, where small first-order perturbations are assumed, so:

$$\omega = \omega_0(1 + \delta) \tag{4.175a}$$

$$\lambda = \lambda_0(1 - \delta) \tag{4.175b}$$

$$\lambda_g = \lambda_{g0}(1 - \delta) \tag{4.175c}$$

and we write the shunt impedance factor in polar form:

$$1 + j2x = \sqrt{1 + 4x^2}\, e^{j\phi} \tag{4.176}$$

We can now recast D into a more useful expression:

$$D = \frac{1}{\sqrt{1 + 4x^2}} \left\{ \sqrt{1 + 4x^2} - \frac{R - Z_0}{R + Z_0} \exp\left[-j\left(\phi + \frac{4\pi d}{\lambda_{g0}} + \frac{4\pi \delta d}{\lambda_{g0}} \right) \right] \right\}. \tag{4.177}$$

At resonance we will have:

$$\frac{4\pi d}{\lambda_{g0}} + \phi = \pi \tag{4.178}$$

and so we can simplify D slightly:

$$D \approx \frac{1}{\sqrt{1 + 4x^2}} \left[\sqrt{1 + 4x^2} - \frac{R - Z_0}{R + Z_0} + j\pi\delta \frac{R - Z_0}{R + Z_0} \right] \tag{4.179}$$

where we have used the first two terms of the series expansion for the exponential, which is only reasonable to do because δ is a very small number, $\delta \ll 1$. We can compare equation (4.179) with the comparable relation (4.166) for RF circuits to obtain:

$$D \approx \frac{1}{\sqrt{1 + 4x^2}} \left[\sqrt{1 + 4x^2} - \frac{R - Z_0}{R + Z_0} \right] (1 + j2Q_L\delta) \tag{4.180}$$

where the loaded Q-factor is:

$$Q_L = \frac{R - Z_0}{2(\sqrt{1 + 4x^2}(R - Z_0) - R + Z_0)} \tag{4.181}$$

and, again, represents the ratio of stored energy to the losses. When the coupling is very small ($x \to 0$) the loaded Q-factor, Q_L, approaches the unloaded Q-factor, Q_0:

$$Q_0 = \frac{R - Z_0}{4Z_0}. \tag{4.182}$$

But, of course, this is not desirable since no power would be coupled into the cavity. As x varies we may obtain an optimal coupling value which depends on the load R. When optimal coupling is attained ($1 + 4x^2 = 1/R$) we have:

$$Q_L = \frac{Q_0}{2} \tag{4.183}$$

there is no reflection toward the generator, and the generator power is completely absorbed by the load.

Using network analyzers (see Chapter 5) we may learn much about the behavior of these RF and microwave resonant circuits. The standard network analyzer display is a plot of

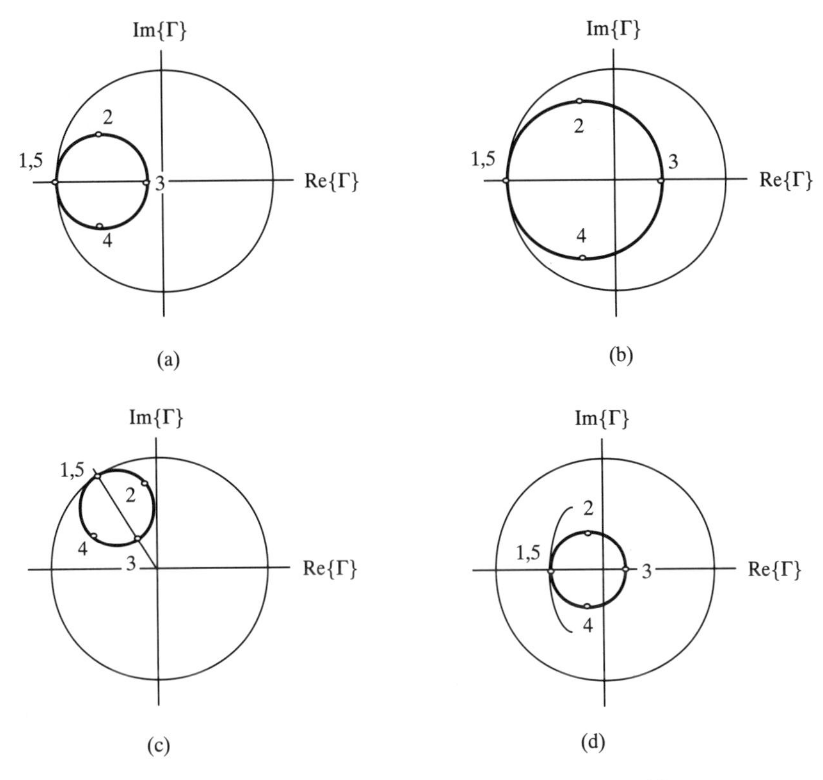

Fig. 4.54 *Example single-mode resonant cavities.*

impedance as frequency varies on a Smith chart. A short description of Smith charts was given in Chapter 2—there are also many classical tutorial texts on them. We may study a resonant applicator in detail by plotting reflection coefficient loci. A single-mode cavity will follow a circular path on the Smith chart as frequency varies (Figure 4.54). As frequency increases, the reflection coefficient will rotate clockwise on the circle. Four examples are shown in the figure. In each case the minimum reflection coefficient magnitude is at the resonance frequency, ω_3 (at the point marked 3 in the picture). The anti-resonant frequencies are ω_1 and ω_5 (marked 1, 5 with ω_1 signifying the lower of the two) and have the largest magnitude reflection coefficient. The first of the four cases (Figure 4.54a) shows a cavity which is under-coupled—the radius of the coupling circle is less than 0.5. The second case (Figure 4.54b) is an over-coupled resonant cavity and the radius of the circle is greater than 0.5. To minimize the reflection toward the generator we would like to obtain critical coupling, so that ω_3 would fall at the center of the Smith chart. No net reflection toward the generator will result from the critically-coupled circuit operated at ω_3.

The frequencies ω_2 and ω_4 are used in the calculation of the cavity Q-factor:

$$Q_L = \frac{\omega_3}{\omega_4 - \omega_2}. \tag{4.184}$$

The unloaded Q-factor can also be determined graphically (see Gintzon 1957). When the anti-resonant points (1, 5) are not at $\Gamma = -1$ (Figure 4.54c) the plane of the coupling network is not at the reference plane of the impedance measurement—it has been shifted by $\phi = 2\beta d_0$ where d_0 is the location error. When the anti-resonant points are not on the $|\Gamma| = 1$ outer circle (Figure 4.54d) the coupling network is lossy.

To complete this discussion we briefly review the resonant modes of a rectangular cavity (including a waveguide). More detail may be found in Pozar (1990) and in other classical texts. The obvious starting point is the rectangular waveguide solutions (Chapter 2) which already satisfy the boundary conditions on the walls (at $x = 0$ or a, and $y = 0$ or b), and add the conditions at the ends, $z = 0$ and d, so that the transverse electric field ($\mathbf{E}_t$) of either TE_{mn} or TM_{mn} modes is:

$$\mathbf{E}_t(x, y, z) = \mathbf{e}(x, y)[A^+ e^{-j\beta_{mn}z} + A^- e^{j\beta_{mn}z}] \tag{4.185}$$

where A^+ and A^- are the forward and reflected wave amplitudes and $\mathbf{e}$ is the transverse variation (i.e. the vector direction and sines and cosines of the mode structure). The mode structure for generalized TE modes is given in equations (2.50) (E_x and E_y components are transverse) while that of the TM modes is shown in equations (2.59). We also recall from Chapter 2 that the propagation constant in a rectangular guide is:

$$\beta_{mn} = \frac{2\pi}{\lambda_g} = \sqrt{\omega^2 \mu \varepsilon - \left(\frac{m\pi}{a}\right)^2 - \left(\frac{n\pi}{b}\right)^2}. \tag{4.186}$$

This relation defines the guide wavelength for the rectangular cavity. Using it we may set the cavity length in order to realize desired modes in the empty cavity. In order that the fields of (4.178) are totally internally reflected (at resonance) the forward and reflected waves must cancel—that is, $A^+ = -A^-$. The transverse field must be zero at the ends of the cavity to match the required tangential **E**-field boundary condition; so the cavity length, d, must be an integral multiple of $\lambda_g/2$, or:

$$\beta_{mn} d = k\pi \qquad \text{where } k = 1, 2, 3, \ldots \tag{4.187}$$

There are k maxima between $0 < z < d$, m maxima over $0 < x < a$, and n maxima over $0 < y < b$. If $b < a < d$ (as we have just assumed) then the dominant mode is the TE_{10k} mode. The dominant rectangular cavity TM mode is TM_{111}.

In terms of other devices, it is not at all easy to obtain single-mode operation in coaxial transmission lines or ridged waveguides. The principle concern for those conditionally stable devices is that the heated material must be of very small volume indeed in order to stay near the resonant frequency of the empty cavity.

Cylindrical resonant cavities

In addition to the widely used TE_{10k} rectangular waveguide resonant modes, cylindrical waveguides may form cavities in TE_{nm} or TM_{nm} modes. Note the change in sequence of the subscripts to follow the notation introduced in Chapter 2. Remember that n is the order of the Bessel function and m the root which is used. In terms of the above notation, we may express the transverse electric field as:

$$\mathbf{E}_t(r, \varphi, z) = \mathbf{e}(r, \varphi)[A^+ e^{-j\beta_{nm}z} + A^- e^{j\beta_{nm}z}]. \tag{4.188}$$

Here $\mathbf{e}(r, \varphi)$ contains the mode structure and vector sense of equations (2.71) for TE modes and equations (2.75) for TM modes. The r and φ components comprise the transverse electric field. As we saw in Chapter 2, the applicable propagation constant for each TE_{nm} mode is:

$$\beta_{nm} = \frac{2\pi}{\lambda_g} = \sqrt{\omega^2\mu\varepsilon - \left(\frac{(ha)'_{nm}}{a}\right)} \tag{4.189}$$

where, as in Chapter 2, $(ha)'_{nm}$ denotes the mth root of the derivative of the nth order Bessel function, $J'_n(ha)$, and a is the cylinder radius (equation (4.189) is a more convenient adaptation of equation (2.70)). Now, when we add the cavity ends at $z = 0$ and d we get a resonance frequency for the TE_{nmk} mode of:

$$\text{TE:} \qquad f_{nmk} = \frac{c}{2\pi\sqrt{\mu_r\varepsilon_r}}\sqrt{\left(\frac{(ha)'_{nm}}{a}\right)^2 + \left(\frac{k\pi}{d}\right)^2} \tag{4.190}$$

where the cavity may be filled with a lossless material of relative properties μ_r and ε_r. The analogous relation for a TM_{nmk} mode is:

$$\text{TM:} \qquad f_{nmk} = \frac{c}{2\pi\sqrt{\mu_r\varepsilon_r}}\sqrt{\left(\frac{(ha)_{nm}}{a}\right)^2 + \left(\frac{k\pi}{d}\right)^2} \tag{4.191}$$

and now $(ha)_{nm}$ is the mth root of the Bessel function, $J_n(ha)$.

Practical cylindrical resonant cavities are relatively simple in construction. Figure 4.55a illustrates one design of a cylindrical single mode cavity excited to energize the TM_{010} mode (radius $a = \lambda_g/4$; height = $\lambda_g/2$, where λ_g is that of the cylinder). This mode may be used to heat a low loss rod of small diameter. The TM_{010} mode has no maxima in the **H**-field within the cavity (only on the end caps). The **E**-field maximum ($\mathbf{E} = E_z\mathbf{a}_z$) occurs near the middle of the rod at $z = 0$ (Figure 4.55b) and is tangential to the rod boundary. The rod may be moved if even heating is required for a particular application. Often, however, we are attempting to weld or bond low loss cylindrical objects; in which case the seam should be placed at $z = 0$. As a function of radius within the rod, the electric field varies according to the rod diameter and electric permittivity. The calculation is identical to that for the internal fields of a scattering cylinder, which was previously described. This cavity mode is easily energized from a TE_{10} waveguide, as shown in the drawing. A circular or elliptical inductive iris has been included for impedance matching.

We note, alternatively, that we may create TE_{01k} modes in the cavity by turning the waveguide 90° on its axis (so that $\mathbf{E} = E_\varphi\mathbf{a}_\varphi$ in the cylinder when the waveguide is driven by a TE_{10} mode). This initially appears to be advantageous since one might expect that the heating would be axisymmetric. However, as we mentioned in Chapter 2, this mode is not very stable and readily slips into other modes when the cavity is loaded.

There are four additional observations of importance when attempting to use this type of applicator. First, maintaining single-mode operation may be quite difficult when a lossy load is placed in the cavity, even with TM_{01k} excitation. The reason is that the evanescent modes energized by the load boundary conditions may force higher order mode patterns than those predicted for the empty cavity. Thus, a lossy load may shift a single-mode (when empty) cylindrical cavity into a multimode cavity when loaded. Second, referring to the cylindrical waveguide discussion in Chapter 2 we see that many roots of the Bessel functions and their derivatives are located fairly close together. This means that several modes of both

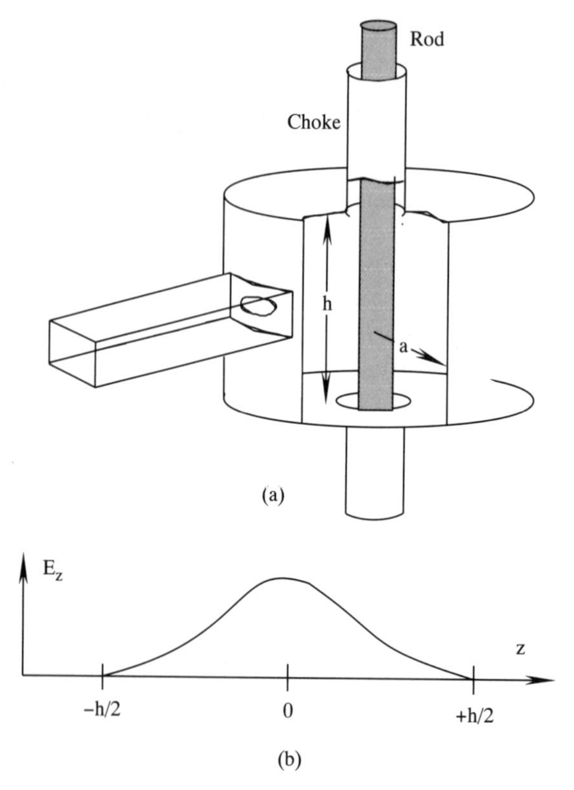

Fig. 4.55 *Cylindrical resonant cavity (a) TM010 cavity schematic view (b) Electric field z-component profile.*

TE_{nmk} and TM_{nmk} types may appear within a cylindrical cavity over a fairly narrow range of frequencies — even when only a small perturbation is created by the load. Determining exactly which mode is operative strictly from swept-frequency measurements of complex impedance or reflection coefficient may not be possible. One may, in fact, have to assess the heating patterns in the material in order to decide. We will discuss multimode operation in the next section in more detail. Third, in some practical systems where the load might not be so small or may be quite lossy, we can still achieve good matching by placing the matching network (say, an iris or multiple-stub tuner) a half-guide wavelength toward the generator. This gets the matching network out of the evanescent fields created by the aperture transition between the waveguide and the cylinder and improves performance in many cases. The heating modes may still be quite complex in the load, however. Finally, we may in fact take advantage of the tendency to shift modes in order to make the heating more uniform. If the resonant cavity has two (or more) modes which give complementary heating patterns (one center-weighted and one perimeter-weighted, for example) we may change the cavity dimensions and switch the modes alternating among the identified modes to make the time-average heating essentially uniform. This is an increasingly common technique to heat ceramic materials which are of relatively low thermal conductivity and usually require resonant cavities to achieve required field strengths.

4.4.4 Multimode cavities

When the dimensions of the applicator are large in comparison with one wavelength, the applicator is not single mode. Such an applicator is termed an oversized or multimode cavity. By far the single most common microwave applicator in use is a multimode cavity since we are often required to heat irregularly-shaped larger objects of moderate to high loss. As was true for single-mode cavities, the reflection coefficients for multimode cavities may be reduced for some values of the frequency, and it is sometimes possible to determine a matching impedance for a multimode cavity applicator. However, the cavity impedance behavior, as shown by its reflection coefficient (Figure 4.56), is much more complex than for the single-mode cavities of Figure 4.54. Each loop corresponds to a resonance mode; and as the number of permissible modes increases (i.e. as the cavity increases in size) the loops become tighter and it becomes impossible to distinguish between them. Also, when the cavity load changes slightly, we may deviate significantly from the match point. So, it is not an advantage to have a high-Q (narrow bandwidth) multimode cavity (and, in fact, that may be an oxymoron for a larger cavity).

Industrially, the typical multimode cavity is a large metallic chamber through which a conveyor belt moves the material, called a "tunnel" multimode cavity applicator (Figure 4.57). The cavity field is produced by antennas which irradiate the belt area and fill the cavity with microwaves. As one example, this design has been used for defrosting (tempering) frozen fish or meat, in which the temperature must be increased from -18°C to -3°C. Of course, it is important that the design must prevent field leakage at the conveyor belt input and output, and the design of chokes for these apertures is one of the more important considerations. The microwaves are introduced through antennas or apertures of varying design; and three examples are shown in Figure 4.58. Home microwave ovens are

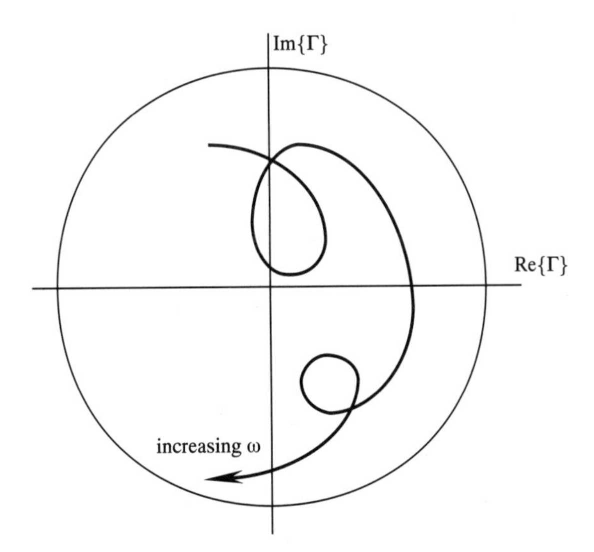

Fig. 4.56 *Typical multimode cavity reflection coefficient vs frequency.*

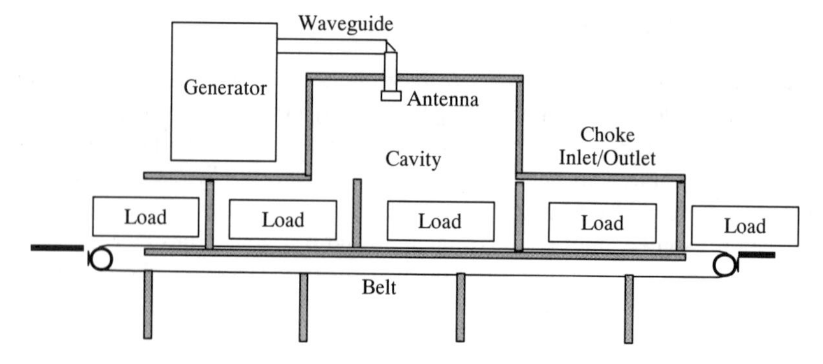

Fig. 4.57 *Typical tunnel multimode cavity.*

also multimode cavities with many mixed resonant modes. The material in the oven is often on a turntable, or a stirrer is used to aid in mode-mixing though these are usually of little help. Large cavities have, of course, been well studied (mostly empirically) and many patents have been issued for devices and applicator designs.

Mode calculation approach

As the dimensions of the cavity increase much above λ_g the number of possible modes increases, essentially without bound. For a rectangular cavity of dimensions a, b and d, one often sees the dispersion relation used to describe the number of modes:

$$\left(\frac{2\pi f}{c}\right)^2 = \left(\frac{m\pi}{a}\right)^2 + \left(\frac{n\pi}{b}\right)^2 + \left(\frac{k\pi}{d}\right)^2 \tag{4.192}$$

where f is the frequency of operation, c the propagation speed and m, n and k are integers corresponding to the permitted TE_{mnk} and TM_{mnk} modes; the similarity to equation (4.186) is obvious. Even for relatively small cavities, such as home microwave ovens, the number of modes permitted by this relation is very large indeed. However, we must bear in mind that equation (4.192) applies only to an empty cavity.

When we place the usual loads in such a cavity an entirely different set of permitted modes exists which depend on load shape and electric permittivity. The additional modes are actually the evanescent modes due to scattering fields created by the object boundary conditions. In radiation problems, such as waveguide propagation, these are termed non-radiating since they die out near the perturbation or waveguide obstacle which creates them and do not persist into the far field—except that the net result of these evanescent modes is, of course, the reflected wave. In an applicator we are almost always in the near field; certainly, the fields within the heated object are near fields. Consequently, the permitted cavity (and thus heating) modes often depend rather critically on the load position; especially if the load volume is a significant fraction of the cavity volume. This is an important consideration when one is designing a load for a home oven, for example, since the volume of a typical frozen meal may be 10% of the oven volume. Those loads are frequently located within a few wavelengths of the oven wall on all sides and their boundary conditions interact very strongly with the boundary conditions at the wall to determine the dominant heating modes.

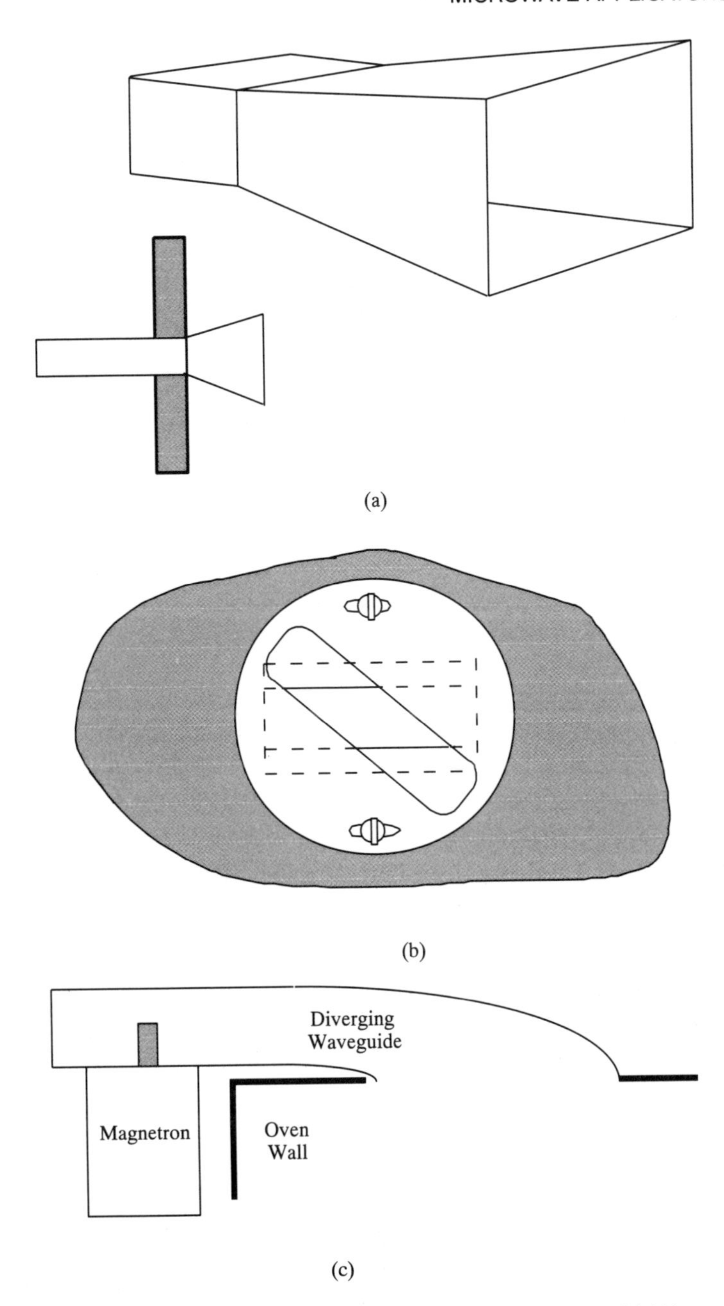

Fig. 4.58 *Multimode cavity antennas. (a) Pyramidal horn antenna. (b) Waveguide aperture with adjustable iris. (c) Typical home oven configuration.*

Mode-matching models have been applied by several groups to study heating and evanescent modes in smaller multimode cavities about the size of home ovens. The results indicate the sensitivity of the types of modes realized to the geometry and permittivity of the load. The mode-matching method utilizes series representations of the evanescent modes to determine the fields. When the scattering structure is periodic in shape the evanescent modes are described by a Floquet series. The more usual case, however, is that of a much less regular surface geometry.

A detailed description of the mode-matching method may be found in Chapter 9 of Itoh (1989). While the method has been most extensively used to model discontinuities in waveguides and microstrip transmission lines, it applies equally well to multimode cavity loads since they may also be treated as discontinuities. We will be careful to use the complete description of propagation in lossy materials. Remember that for a TEM wave (which we do not have in the waveguide case above) the propagation coefficient, γ, consists of the attenuation coefficient, α, and phase coefficient, β:

$$\gamma = \alpha + \mathrm{j}\beta = \sqrt{\mathrm{j}\omega\mu(\sigma + \mathrm{j}\omega\varepsilon)} = \sqrt{\mathrm{j}\omega\mu(\omega\varepsilon'' + \mathrm{j}\omega\varepsilon')}. \tag{4.193}$$

Here we have included all losses in the imaginary part of the permittivity, assuming that μ is real. The two parts are:

$$\alpha = \beta_0\sqrt{\varepsilon_\mathrm{r}'}\sqrt{\frac{1}{2}\left[\sqrt{1 + (\varepsilon''/\varepsilon')^2} - 1\right]} \tag{4.194a}$$

$$\beta = \beta_0\sqrt{\varepsilon_\mathrm{r}'}\sqrt{\frac{1}{2}\left[\sqrt{1 + (\varepsilon''/\varepsilon')^2} + 1\right]} \tag{4.194b}$$

where β_0 is the free space propagation coefficient, $\beta_0 = 2\pi/\lambda_0$ and the index of refraction also appears ahead of the large radical. At all points on the boundary of the cavity $\mathbf{E}_\mathrm{t} = 0$ and $\mathbf{H}_\mathrm{n} = 0$, assuming very good conductors for cavity walls—the $\mathbf{H}_\mathrm{n}$ condition comes from observations regarding induction heating, see Subsection 4.3.2 above. At all points on the interface between the air (material 1) and load (material 2), as in Figure 4.59, $E_{\mathrm{t}1} = E_{\mathrm{t}2}$ and $(\omega\varepsilon_1'' + \mathrm{j}\omega\varepsilon_1')\,E_{\mathrm{n}1} = (\omega\varepsilon_2'' + \mathrm{j}\omega\varepsilon_2')E_{n2}$. The respective fields must satisfy these conditions for all modes. The propagation coefficients are different in the waveguide cases of the previous section specifically because they do not support TEM propagation; they support either TE or TM modes but not both.

The first step is to expand the **E**- and **H**-field components in terms of modal components in the two regions. The polarization of the incident wave is of some importance in this. Remember that any wave may be decomposed into the sum of components TE and TM to

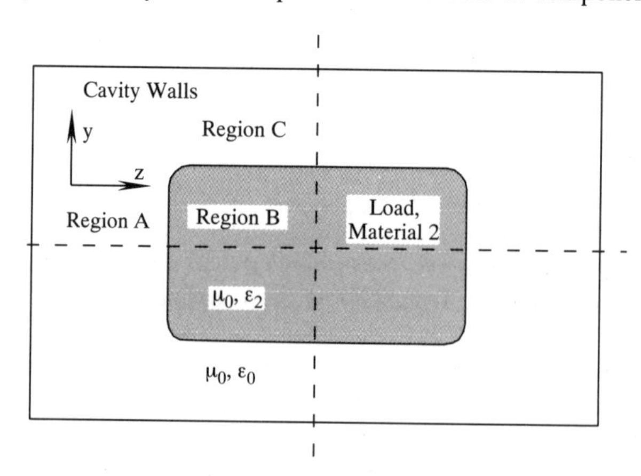

Fig. 4.59 *Mode matching example 2-D geometry.*

the surface — the TE wave has an electric field parallel to the surface (perpendicular to the surface normal) and a TM wave has a magnetic field parallel to the surface (perpendicular to the surface normal). In the example geometry (Figure 4.59) assume a TE wave with $\mathbf{E} = E_y\mathbf{a}_y$ (and we will also look at $\mathbf{H} = H_x a_x$), propagating in the $+z$-direction. Also, we will assume that the load is symmetrically located so that we only have to solve one quarter of the cavity space. The three regions of solution — A, B and C — are shown in the Figure. Decomposing into modes for each region, denoted by $\phi_{\mathrm{An}}\phi_{\mathrm{Bn}}$, and ϕ_{Cn} we would obtain a series of expressions, one for each region. For region A:

$$E_y = \sum_{n=1}^{\infty}(A_n^+ \mathrm{e}^{-\gamma_{\mathrm{an}} z} + A_n^- \mathrm{e}^{\gamma_{\mathrm{an}} z})\phi_{\mathrm{an}}(y)$$

$$H_x = \sum_{n=1}^{\infty}(A_n^+ \mathrm{e}^{-\gamma_{\mathrm{an}} z} - A_n^- \mathrm{e}^{\gamma_{\mathrm{an}} z})Y_{\mathrm{an}}\phi_{\mathrm{an}}(y) \tag{4.195a}$$

for region B:

$$E_y = \sum_{n=1}^{\infty}(B_n^+ \mathrm{e}^{-\gamma_{\mathrm{bn}} z} + B_n^- \mathrm{e}^{\gamma_{\mathrm{bn}} z})\phi_{\mathrm{bn}}(y)$$

$$H_x = \sum_{n=1}^{\infty}(B_n^+ \mathrm{e}^{-\gamma_{\mathrm{bn}} z} - B_n^- \mathrm{e}^{\gamma_{\mathrm{bn}} z})Y_{\mathrm{bn}}\phi_{\mathrm{bn}}(y) \tag{4.195b}$$

and for region C:

$$E_y = \sum_{n=1}^{\infty}(C_n^+ \mathrm{e}^{-\gamma_{\mathrm{cn}} z} + C_n^- \mathrm{e}^{\gamma_{\mathrm{cn}} z})\phi_{\mathrm{cn}}(y)$$

$$H_x = \sum_{n=1}^{\infty}(C_n^+ \mathrm{e}^{-\gamma_{\mathrm{cn}} z} - c_n^- \mathrm{e}^{\gamma_{\mathrm{cn}} z})Y_{\mathrm{cn}}\phi_{\mathrm{cn}}(y) \tag{4.195C}$$

Each region has its own propagation (γ), admittance (Y) and mode set (ϕ) because we have assumed that the load is close to the walls and we do not have a TEM wave — only TE or TM propagation is permitted. As the separation between the load and walls and the dimension of the load itself increase the propagation approaches TEM and γ_{in} and Y_{in} (where i = a, b, c) approach γ (equation (4.193)) and the material characteristic admittance, $1/\eta$, respectively.

To solve for the modes one applies the necessary boundary conditions at the respective interfaces. This generates a set of linear equations in terms of the modes which may be solved for the respective coefficients. It is outside the scope of this chapter to delve into the solution methods. One of the more instructive descriptions of this technique is that in Chapter 9 of Itoh (1989). Nevertheless, when the solution is obtained dominant modes are marked by the largest coefficients.

As a concluding observation, we should note that in many instances (especially for lossy loads) the modes realized will be completely dominated by load shape. Mode-stirring devices — irregularly shaped reflectors which are rotated, usually in the incident field (see Figure 4.60) — are often used to "mix up" or vary the oven modes and make local field

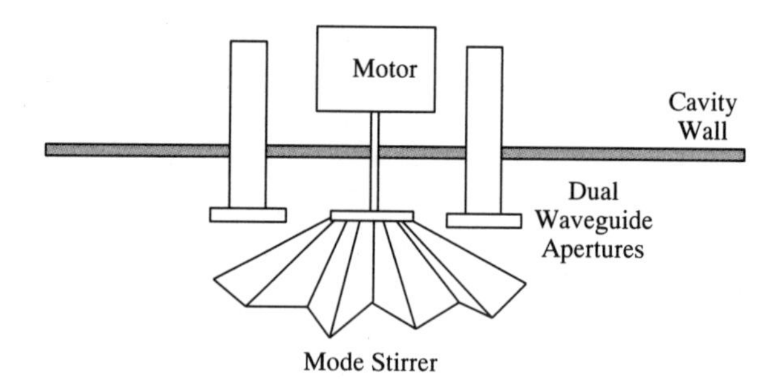

Fig. 4.60 *Typical mode-stirring device.*

interference maxima move about in the multimode cavity. However, attempts to smooth out or eliminate hot spots in a lossy load by using mode stirrers may be futile if the load geometry and boundary conditions dominate mode coefficients. That is, when the load is close to the walls of the cavity, often the interaction between load and walls determines the heating patterns in spite of the best efforts of any mode stirring devices.

As the critical dimensions become large the mode-matching method loses effectiveness because the number of modes to be matched increases rapidly. We must therefore use a different approach in large cavities. When the load is very small within a large cavity (as we will see in the next section), the heating pattern may be determined by the load geometry (in high permittivity or high conductivity materials) irrespective of whether or not a mode stirring device is active. In both the small load and large load cases, we may still achieve a relatively uniform heating pattern by using multiple frequencies within the cavity. There are commercial heating devices which use traveling wave tube amplifiers to illuminate multimode cavities with frequencies swept over about a 10–20% bandwidth (say 2.4 to 2.7 GHz or 10 to 14 GHz). By sweeping the frequency multiple modes are realized in the load which move about to obtain a more uniform time-average heating pattern. Of course, the cavity leakage radiation must be much reduced when swept frequencies are used.

Applications of geometric optics

At first glance the large multimode cavity seems an intractable problem. Yet, we may, in fact, take advantage of the extremely large number of modes in a large cavity, like the tunnel multimode applicator, to simplify the calculation to manageable proportions. It turns out that for large cavities we may return to geometric optics (or a close cousin of our own design) in order to estimate heating fields. We will take a very brief look at two of these methods to conclude the chapter.

Ray tracing Monte Carlo method First, we should look at the cavity itself. In the limit, as the load becomes a very small fraction of the cavity volume and yet has dimensions large compared to λ, we may imagine the cavity as a sort of "black body" absorber. That is, referring to Figure 4.61, a photon launched from the antenna bounces from wall to wall (in perhaps lossy reflections) until it is either absorbed in the load (ray A) or by non-ideal reflections at the wall (ray B). In this case the large cavity is operating like an "integrating sphere" as used in optical experiments. This paradigm can be used to construct a Monte Carlo model of the heating process. We launch a ray from the antenna with polarization and direction obtained from a random sequence of numbers. The probability density function

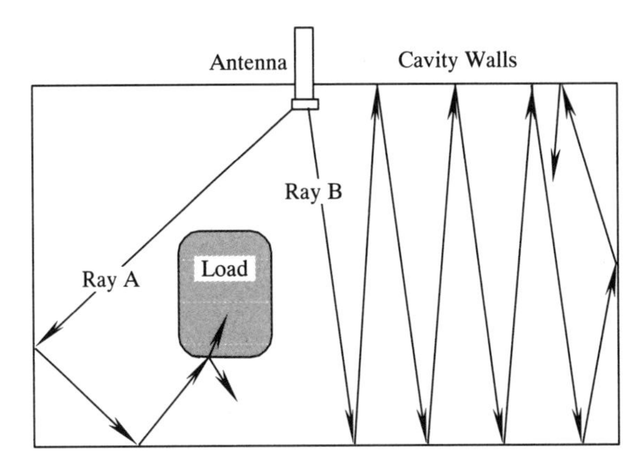

Fig. 4.61 *Large multimode cavity with small load.*

of the random sequence is formulated to mimic the illumination function of the antenna (i.e. the far field radiation pattern). The parameters which must be specified are the **E**-field magnitude, the propagation direction and the polarization (**E**-field direction). Cartesian coordinates must be used.

Once launched, we use the rules of oblique reflection for plane waves (equations (1.41)) to predict the ray direction, polarization and magnitude from wall reflections (Γ). We decompose the ray into TE and TM to the surface components in order to accomplish the calculations. When the ray intersects the load we apply, in addition, the refraction relations (equations (1.41) for T and Snell's law (1.44) for transmission angle, θ_t) and split the ray into two parts. The part transmitted is allowed to decay according to the load attenuation coefficient, α. If it reaches the back wall of the load, the same rules may be applied and the ray split further into transmitted and internally-reflected parts. We note that TM to surface rays have a critical "Brewster's" angle for total transmission through the surface, but TE rays do not. The ray within the load dissipates power as it propagates, and its contribution to each voxel (volume element) in the load is tracked. After a large number of rays have been analyzed (sufficient that the probability density function has been well represented) the model is concluded and the relative volumetric distribution of power is represented by the cumulative power in each voxel. It may take 10^6 or 10^7 rays to complete a model with satisfactory accuracy.

Method of moments It remains only to discuss an alternative approach which can be used when the object is not large compared to λ. In that case we may not trust ray tracing (geometric optics) since the internal field interactions (interference phenomena) prevent us from treating the impinging ray as a plane wave. This is the case in the Mie and Rayleigh scattering regimes. An alternate approach which has been used to estimate the heating patterns in small objects in large cavities is to apply the method of moments. We may do this by assuming that the impinging radiation can be decomposed into a sequence of plane waves of random polarization and direction of propagation. In that sense we are applying the integrating sphere concept to the cavity. Then, the internal electric field is calculated using the method of moments for each wave condition. After a sufficient number of waves have been solved in this way, the total internal power is calculated by superposition of the electric fields (since phase is important).

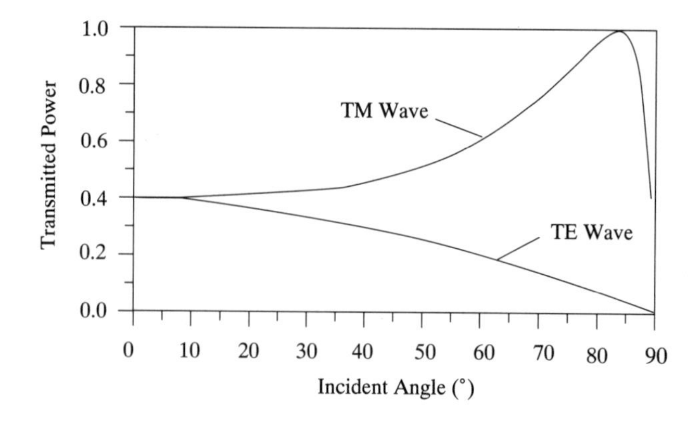

Fig. 4.62 *Plot of fraction of wave power transmitted, $P = 1 - |\Gamma|^2$, into TX-151 ($\varepsilon_r' = 60$, $\varepsilon_r'' = 20$) vs angle of incidence for TE and TM waves.*

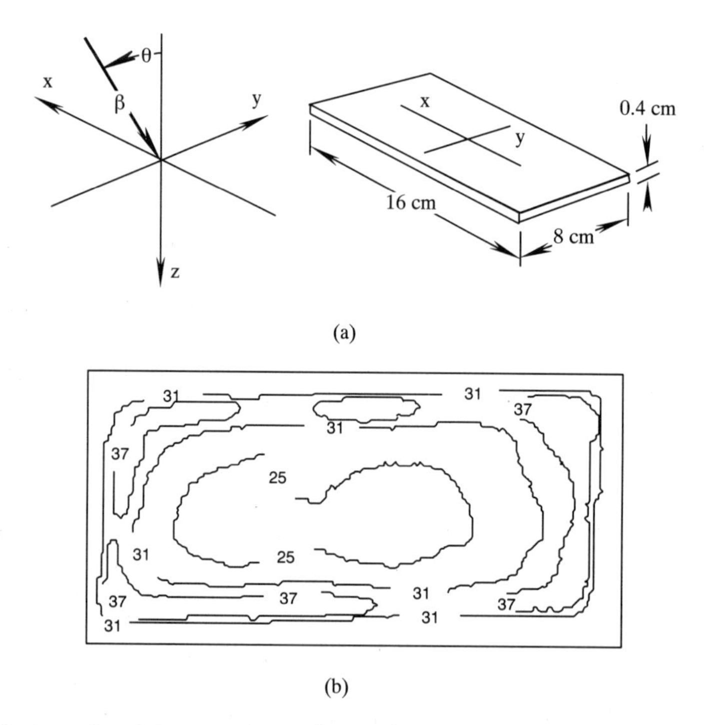

Fig. 4.63 *Rectangular slab geometry and experimental results. (a) Model space geometry and polarization reference direction. (b) Isothermal contours of experiment conducted at 200 W net forward power in a large multimode cavity for 1 minute.*

We show the results of just such a calculation for a finite rectangular slab (16 cm × 8 cm × 0.4 cm) of TX-151 gel (Pearce and Yang 1988). TX-151 has $\varepsilon_r' = 60$ and $\varepsilon_r'' = 20$ ($\beta_2 = 7.85\beta_0$) at 2.45 GHz. The Brewster's angle is about 83° and incident TM waves at this angle are totally transmitted (see Figure 4.62). The method of moments is well adapted to this problem since infinity boundary conditions are not well behaved here (thus FDM and FEM approaches are of limited usefulness). Figure 4.63a shows the problem geometry and reference angles. All polarizations are defined relative to the broad top surface normal. Note that TE to this surface is a different polarization to the edges. Figure 4.63b is a measured thermal image of the top surface which shows the corner- and edge-weighted heating patterns relative to the flat surface.

Model results for individual waves, each with electric field strength 1.0 V/cm, are summarized in Figure 4.64. The first result is for normal incidence ($\theta = 0$) in which the wave is actually TEM to the surface. Edge-weighted heating may be seen in this and all of the TE incident modes (i.e. $\mathbf{E} = 1.0\mathbf{a}_y$ with $\mathbf{H} = H_x\mathbf{a}_x + H_z\mathbf{a}_z$) with the exception of $\theta_{\text{TE}} = 90^\circ$, where significant corner heating is observed ($Q_{\text{gen}} = 0.90$ W/cm^3). In terms of total dissipated power, the TM waves contribute more to the total heating. The effect of incident angle, θ_{TM}, on the absorption pattern may be seen in Figures 4.64b through 4.64d. The most significant contribution to corner heating is observed at $\theta = 45^\circ$, with a very sharp (but very weak) contribution at 90°. Consequently, it appears that the dominant absorption is most likely from TM rather than TE modes at the corners. TE modes do make significant

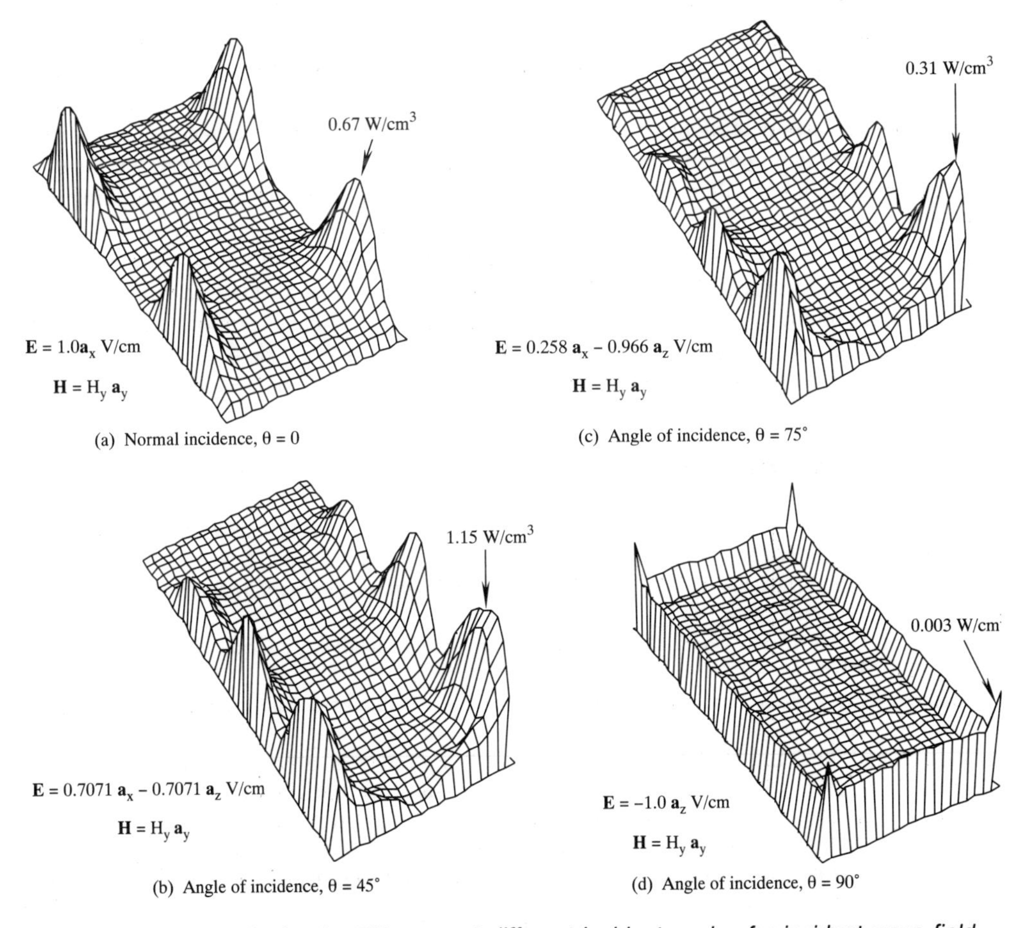

Fig. 4.64 *Model results for the TM wave at different incident angles for incident wave field strength 1 V/cm. (a) Normal incidence, $\theta = 0$ (b) Angle of incidence, $\theta = 45^\circ$ (c) Angle of incidence, $\theta = 75^\circ$ (d) Angle of incidence, $\theta = 90^\circ$.*

contributions along respective edges. Symmetry in the measured results comes from the random distribution of incident angles.

4.5 SUMMARY

This chapter has reviewed many of the common RF and MW applicator–load combinations of practical interest. In many, but not all, cases the load and applicator are tightly coupled so that changes in load properties or geometry significantly affect the power delivery system. This is an especially important consideration in resonant applicators and in RF systems. The effects are strong enough that in some designs feedback control of matching networks is required. Various analytical and numerical modeling approaches have been described. We regret that there is neither time nor space to deal with each in satisfying detail. For substantive discussions the reader is referred to several classical and modern treatments listed in the bibliography. However, we have attempted to describe the influence of the load material properties and geometry on the field distributions realized for a number of instructive practical situations. We do hope that they may prove useful. Above all, it is obvious that no single technique can be universally applied to make calculations since all have their advantages and disadvantages and the solutions are driven by the particular characteristics of each class of problem. A recurring observation is that heating patterns are almost always determined by load boundary conditions, and those need to be carefully regarded in designing an applicator for a specific process. We ought to expect that the order of the discontinuity will strongly influence heating distribution. That is, imagining bulk material as zeroth order, a plane boundary as a first order discontinuity (**E**-field boundary conditions on one axis only), an edge as a second order discontinuity and a corner as third order, we should expect that the volume power deposition would increase with increasing order of discontinuity — in both RF and MW systems. The single exception to this might be objects within the Mie scattering range where significant internal field concentrations may result from wave interference effects giving central, rather than peripheral, heating.

REFERENCES

Becker, E. B., Carey, G. F. and Oden, J. T. (1981) *Finite Elements, An Introduction* Prentice-Hall, Englewood Cliffs, NJ.

Bohren, C. F. and Huffman, D. R. (1983) *Absorption and Scattering of Light by Small Particles* Wiley-Interscience, New York.

Booker, H. G. (1946) Slot aerials and their relation to complementary wire aerials (Babinet's principle) J. IEEE **III-A**, 620–626.

Boonton, R. C. (1992) *Computational Methods for Electromagnetics and Microwaves* Wiley-Interscience, New York.

Born, M. and Wolf, E. (1980) *Principles of Optics (6th Ed.)* Pergamon Press, Oxford.

Cheng, D. K. (1989) *Field and Wave Electromagnetics (2nd Ed.)* Addison Wesley, Reading, MA.

Gintzon, E. L. (1957) *Microwave Measurements* McGraw-Hill, New York.

Harrington, R. F. (1961) *Time-Harmonic Electromagnetic Fields* McGraw-Hill, New York.

Irons, B. M. (1970) A frontal solver solution program for finite element analysis, *Int. J. Numer. Methods Engr.*, **2**, 5–32.

Iskander, M. (1992) *Electromagnetic Fields and Waves* Prentice-Hall, Englewood Cliffs, NJ.

Itoh, T. (1989) (Ed.) *Numerical Techniques for Microwave and Millimeter — Wave Passive Structures* John Wiley, New York.

Jackson, J. D. (1975) *Classical Electrodynamics (2nd Ed.)* John Wiley, New York.

Kong, J. A. (1986) *Electromagnetic Wave Theory* Wiley-Interscience, New York.

Kong, J. A. (Chief Ed.) and Morgan, M. A. (1990) (Volume Ed.) *Finite Element and Finite Difference Methods in Electromagnetics Scattering* (Progress in Electromagnetics Research, volume 2) Elsevier, New York.

Oliner, A. A. (1957) The impedance properties of narrow radiating slots on the broad face of a rectangular waveguide *Trans. IRE Antennas and Propagation*, **45**, 4–20.

Paris, D. T. and Hurd, F. K. (1969) *Basic Electromagnetic Theory* McGraw-Hill, New York.

Pearce, J. A. and Faulkner, C. M. (1992) RF processing of thermoset and thermoplastic composites, *Microwave Processing of Materials III* (Beatty, Sutton and Iskander, Eds.), *Proc. Materials Res. Soc.* **269**, 397–408.

Pearce, J. A. (1988) A research program for dielectric heating and drying of industrial materials *Proc. Material Res. Soc.* **124**, 329–334.

Pozar, D. M. (1990) *Microwave Engineering* Addison-Wesley, Reading, MA.

Sadiku, M. N. O. (1989) *Elements of Electromagnetics* Saunders, New York.

Simmons, A. J. (1957) Circularly polarized slot radiators *Trans. IRE Antennas and Propagation*, **45**, 31–36.

5 Instrumentation and Measurement Methods*

5.1 INTRODUCTION

In this chapter, we will present the most common measurement methods presently in use in radio frequency and microwave engineering. These methods combine techniques for both low and high power systems. Both are required because high power methods are often derived from low power techniques; and high power RF and microwave techniques require complementary methods for evaluating component impedances and matching needs. This chapter is a critical review of the measurement techniques.

5.2 POWER MEASUREMENT

Perhaps the most fundamental measurement is power. The determination of forward and reflected power at microwave frequencies is facilitated by the directional coupler discussed in Chapter 3. For radio frequency systems the analogous methods are often difficult to use, so load power must sometimes be determined calorimetrically. We begin with RF power estimation, then discuss microwave instrumentation and conclude with calorimetric methods which may be used in either system.

5.2.1 Radio frequency methods

A commercial device of moderate expense, similar to a directional coupler, is available for RF systems, but it is only useful when the VSWR is small (less than about 3 to 5). This device may only be used with acceptable accuracy when dynamic load matching, as described in Chapter 3, is used. For high VSWR systems relative load power may be estimated from voltage or current measurements.

Matched transmission line methods

When the RF transmission line is connected between a matched generator and load the line power can be measured with commercial field sensors such as those provided by the Bird Corporation or the Coaxial Dynamics Corporation in the US and by Sairem in Europe. These sensors are of proprietary construction. They convert the transmission line signals into a dc current which is displayed on inexpensive d'Arsonval meters specially calibrated in power units.

* This chapter has been written with the participation of Dr J. M. Thiebaut.

High VSWR methods

When a high VSWR exists on the transmission line the estimation of power is not trivial. None of the commercial devices mentioned above will give acceptable measurements. Power may be calculated from separate voltage and current measurements, but this is not as easy as it might appear since the load phase angle, θ, is likely to be near -90° and small uncertainties in θ will create large uncertainties in the power. Alternately, we can get a workable estimate of total generator power by multiplying the vacuum tube dc supply voltage (usually fixed and constant) by the anode supply current measurement. System losses other than in the load are included in this measurement, however, and the uncertainty may be high.

Voltage sensors The simplest method for reducing RF transmission line and applicator potentials to measurable levels is a voltage divider, Figure 5.1a. The divide ratio is determined by the impedances Z_1 and Z_2 plus the effect of the cable and oscilloscope input impedance. The coaxial cable adds a capacitance in parallel with the oscilloscope. Typical oscilloscope input impedances are 1 MΩ resistance in parallel with 12 pF capacitance (about $491\ \Omega\underline{/-89.97^\circ}$ at 27 MHz). The typical capacitance per unit length of RG-58A/U cable ($Z_0 = 50\ \Omega$) is about 101.3 pF/m. So, a 1 m length of that cable will drop the unterminated transmission line plus oscilloscope impedance to about $52\ \Omega\underline{/-90^\circ}$ at 27 MHz. In that case the voltage divider ratio will be determined by:

$$\frac{v_{\text{measured}}}{v_{\text{input}}} = \frac{Z_2 \parallel -j52\ \Omega}{Z_1 + (Z_2 \parallel -j52\ \Omega)}. \tag{5.1}$$

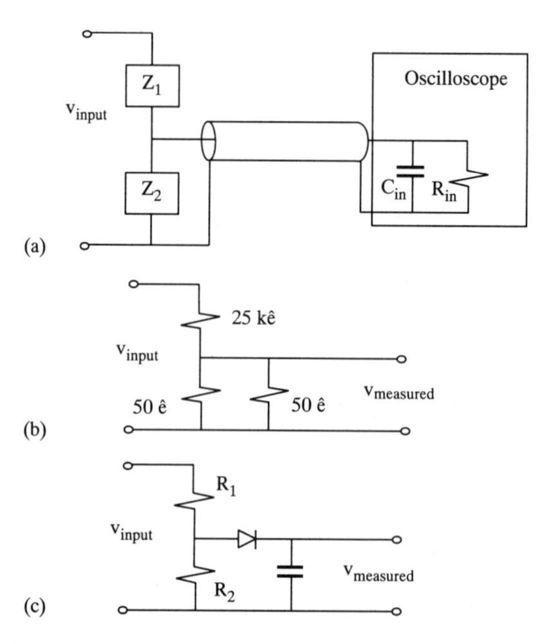

Fig. 5.1 *Voltage divider circuitry. (a) General circuit showing cable and oscilloscope. (b) Resistive voltage divider design example. (c) Resistive divider with dc conversion.*

In order that the divide ratio be dominated by the Z_1 and Z_2 impedances, Z_2 must be small compared to the combined cable load impedance, $-j52\ \Omega$. However, because the cable length is often variable, it makes considerable sense to create a standard transmission line by terminating the oscilloscope input with a 50 Ω resistor (for the 50 Ω RG-58A/U cable). Then the cable plus load comprises a 50 Ω resistive load at Z_2 irrespective of cable length. We will assume that this has been done in the following discussion.

Z_1 and Z_2 may be any impedance as long as they are of one type. The simple case, which is always discussed in elementary circuit analysis, is when Z_1 and Z_2 are both resistors, R_1 and R_2. The voltage divider acts as a source with Thevenin equivalent source impedance equal to the parallel combination: $R_{TH} = R_1 \parallel R_2$. Initially, we will make R_{TH} 50 Ω in order to match the divider to the transmission line; however, this is not a necessary design choice. Also, we want the source Thevenin equivalent voltage, v_{TH}, to be much smaller than the unknown potential, v_{input}, so we will have $R_1 \gg 50\ \Omega$ and the source impedance will be approximately R_2 which must be 50 Ω. The parallel combination of the 50 Ω cable load and R_2 is then 25 Ω, the effective divider second impedance.

For a desired divide ratio of 1000, as an example, we would have to have $R_1 = 24.975$ kΩ (nominal 25 kΩ resistor), Figure 5.1b. There are two major disadvantages to resistive voltage dividers which may now easily be seen. First, for a divide by 1000 used to measure 20 kV_{rms} the resistor R_1 must be capable of dissipating 16 kW. Some relief is realized if the divide ratio is set to 10^6, but this would require $R_1 = 25$ MΩ and 16 W would be dissipated in it. Second, these large resistance values will assuredly have significant capacitance which will all but kill the assumption of a purely resistive divider at this frequency.

An alternative design is to convert the RF signal to dc at the divider, as illustrated in Figure 5.1c. In this design the diode and small capacitor are used to provide smooth dc potentials. Since the signal is a dc signal, the length of cable and voltmeter impedance are immaterial—the cable capacitance serves only to enhance the small filter capacitor. A few commercial devices of this design are available but are typically limited to low maximum voltages by the diode characteristics.

As a practical matter making Z_1 and Z_2 inductive is not workable either for many of the same reasons. We can, however, make Z_1 and Z_2 capacitors, as in Figure 5.2a. Ignoring the effects of the terminated cable for the moment, the divide ratio for a capacitive divider is:

$$\frac{v_{measured}}{v_{input}} = \frac{Z_2}{Z_1 + Z_2} = \frac{1/j\omega C_2}{1/j\omega C_1 + 1/j\omega C_2} = \frac{C_1}{C_1 + C_2} \tag{5.2}$$

and voltage division comes from a small value of C_1 (high impedance) and a large value of C_2. It is to our advantage to have C_1 as small as possible to limit the RF current in the divider. The smallest commercial vacuum capacitor is 6.5 pF. For a divide ratio of 1000, as in the previous example, we must have $C_2 = 6.494$ nF, nominally 6.5 nF. Vacuum capacitors do not go this high, so we formed a divider by connecting six 1.1 nF transmitter capacitors in parallel. The parallel combination was used because of maximum current limitations in a larger single capacitor. Note that the divider output impedance is determined by C_2: $Z_{TH} \approx 0.9\ \Omega\angle{-90^\circ}$ at 27 MHz, very small compared to the terminated cable impedance of 50 Ω. Also at 20 kV_{rms} the current in C_1 would be 22 A_{rms}. A sketch of a practical capacitive voltage divider design is contained in Figure 5.2b.

The most practical and inexpensive voltage sensor for RF systems is actually an electric field sensor. If we imagine that the case is one of estimating the voltage at some location on a coaxial transmission line, then a small probe placed in the wall of the outer conductor will

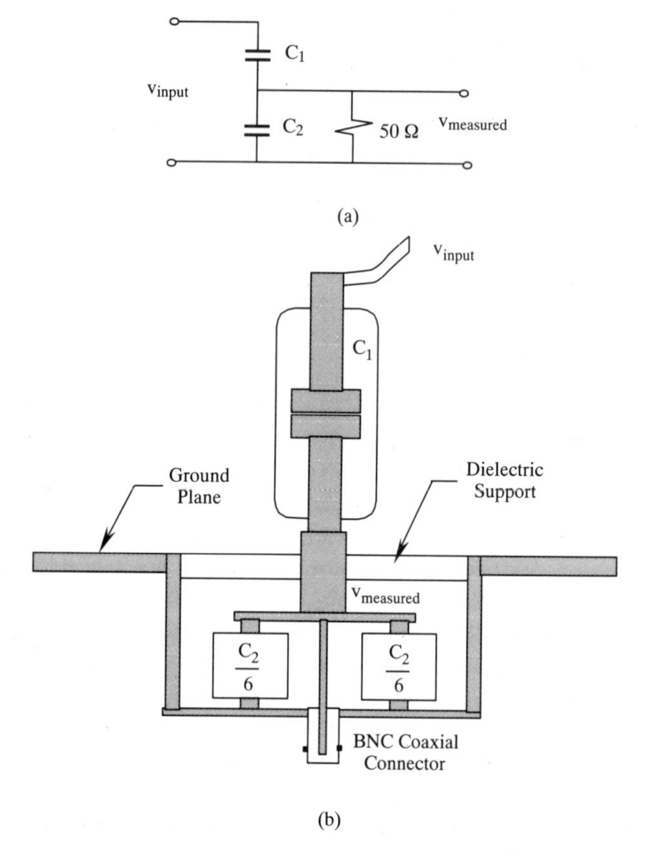

Fig. 5.2 *(a) Capacitive voltage divider. (b) Practical realization.*

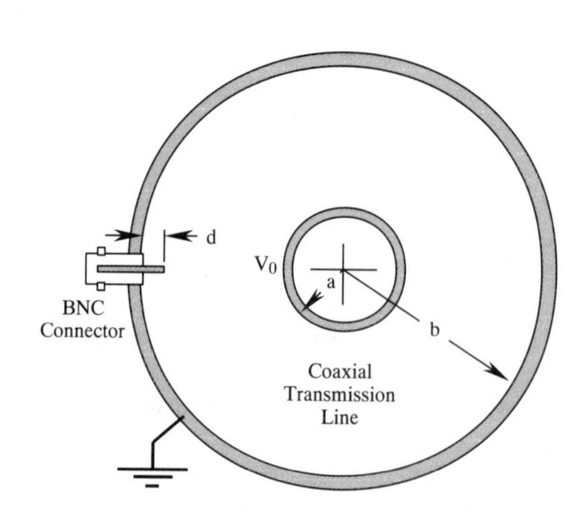

Fig. 5.3 *Voltage-sensing coaxial cable probe.*

provide the signal. Referring to Figure 5.3, we see that the probe extent into the dielectric, d, will set the "divide ratio". This is not really a divider, but is instead an electric field sensor. We may quickly solve the 1-D cylindrical Laplace equation to show that the potential between the inner conductor of radius a at potential V_0 and grounded outer conductor of radius b is:

$$V(r) = V_0 \frac{\ln(r/b)}{\ln(a/b)}. \tag{5.3}$$

Assuming that the presence of the probe does not disturb the electric field significantly, the end of the probe will have the potential at $r = b - d$. The length d can be selected to obtain the desired divide ratio, $V(b-d)/V_0$.

Current sensors A relatively simple current sensor can be built into the outer conductor of a coaxial transmission line. The measurement of the current is obtained by diverting a part of the magnetic flux associated with the outer wall current through a ferrite core to induce a voltage in a shielded-coil secondary wound around it (see Figure 5.4). With this sensor, the secondary voltage is proportional to the current, and the shield prevents contributions from the electric field of the center conductor, even in a high power transmission line. The directivity can be better than 30 dB by careful construction.

For either twin-lead or single-conductor configurations we may determine the conductor current from measurements of the magnetic field around the conductor. Perhaps the most reliable estimates of magnetic field are obtained from a Moebius loop sensor since it is relatively insensitive to electric fields. This device is described in more detail in Section 5.4. Two alternative geometries for conductor current measurement are shown in Figure 5.5. We may obtain either a dc output signal or an ac signal depending on how the Moebius loop is configured. In Figure 5.5a we see the Moebius loop between twin-lead transmission lines, and in 5.5b the Moebius loop is located below an essentially non-current-carrying section of coaxial transmission line so that the outer conductor acts as a shield to the electric field alone.

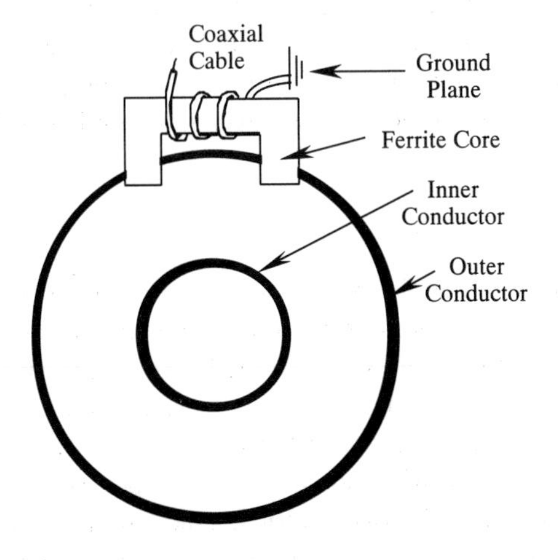

Fig. 5.4 *Current sensor for coaxial T-L.*

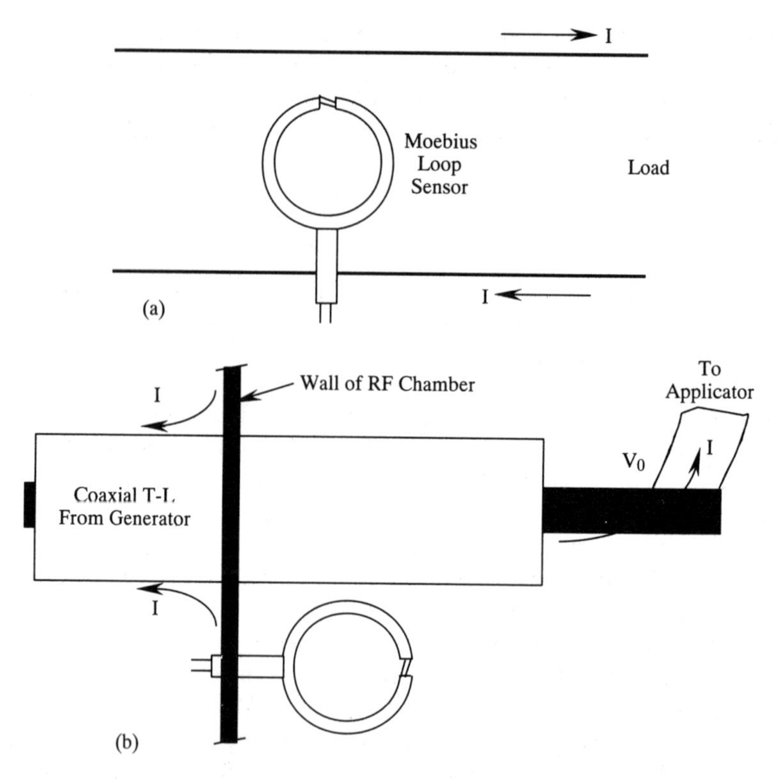

Fig. 5.5 *Moebius loop sensor used to measure conductor current: (a) between twin transmission lines; (b) suspended below* ***E****-field shield.*

5.2.2 Microwave methods

Power measurement in high power waveguide systems is always accomplished with a dual directional coupler, as was previously mentioned. The power-sensing element is basically an absorbing device consisting of a bolometer, thermistor or diode. At low power a sensor in a coaxial mounting may be used directly on a waveguide terminated in a coaxial transition, as described in Chapter 3.

Low power density waves

There are several types of commercial microwave power sensors available. They function over a wide frequency range and measure the power directly at low levels, typically in the 10 μW–1 W range. The sensor mount is well suited for working in a coaxial connector—an N-connector is the standard mount. It can also be used in a rectangular waveguide by connecting a coaxial waveguide transition to the waveguide, as illustrated in Figure 5.6.

There are two common types of sensor head. One type uses a bolometer, which is a simple metallic wire resistor of fairly low electrical conductivity which absorbs energy. The variation in the resistance due to temperature rise gives the power measurement directly. The other type of sensor head is a thermistor. The power-sensitive element is a semiconductor, rather than a metal, which has a resistance versus temperature according to:

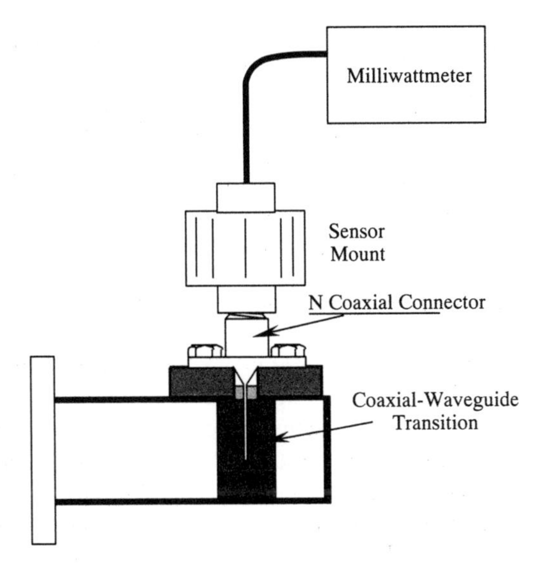

Fig. 5.6 *Typical waveguide power measurement apparatus.*

$$R(T) = R_0 \exp\left[\beta\left(\frac{1}{T} - \frac{1}{T_0}\right)\right] \tag{5.4}$$

where $R_0(\Omega)$ is measured at the reference temperature T_0 (K) and β is the sensitivity, typically 1000–4000 (K). Thermistor resistances decrease with increasing temperature. In both types of sensor the temperature increases until heat transfer out of the mount equals microwave power absorbed; so, the thermal design of the mount is critical to the power range. The mounts are designed to maintain room temperature at the rated maximum power for the sensor.

These two designs have particular advantages and disadvantages. Bolometers have rapid response times, of the order of 1 ms. Thermistors are generally slower, though response times of the order of tens of milliseconds can be obtained, but they are more sensitive than bolometers since resistance sensitivities of 1% or 2% per degree can be obtained. Both types can be used in the power range of 1 μW–10 mW, and have a linear power response since the steady state temperature is determined by the heat transfer. Many commercial units are equipped with automatic zeroing functions, to eliminate reference drift, and with internal calibration sources.

Diodes (standard p-n junction) or Schottky barrier (metal–semiconductor junction) devices are more sensitive, but their response is less linear. Most diode sensors require calibration before, and sometimes during, use.

High power density waves

When a high power is to be measured, we use a directional coupler, as has been previously described (see Figure 5.7). Assuming that the generator is connected to port 1 and the load to port 2, if we connect power-measuring devices to ports 3 and 4 we will get the desired information. Knowledge of the directional coupler coupling factor, C (from Chapter 3), is essential to this measurement.

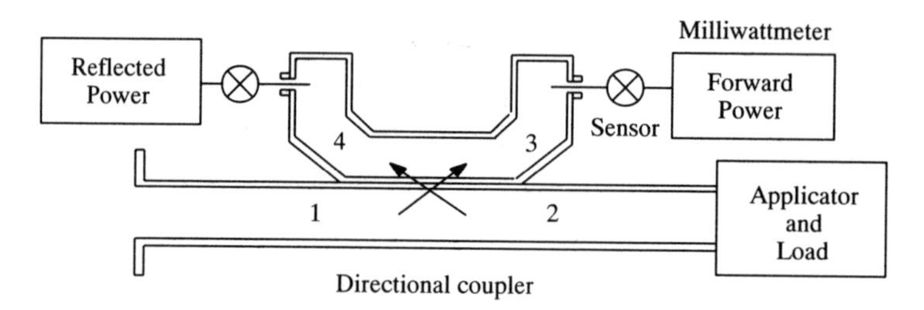

Fig. 5.7 *High power waveguide measurement.*

Since most sensors can measure up to 100 mW, and we need to measure power levels of the order of 1 kW, the calculated coupling factor of the directional coupler must be of the order of −60 dB. For very high precision and accuracy measurements, the fact that the directivity, D (Chapter 3), is not infinite must also be taken into account. Good directional couplers can be purchased commercially, with the coupling and directivity given by the manufacturer over the useful frequency range. The determination of the coupling factor requires careful technique and such measurements of attenuation are usually obtained by substitution methods.

5.2.3 Calorimetric methods

It is possible to directly measure the power level of a high power generator by the use of calorimetry. For example, a rectangular waveguide traversed by an inclined tube through which an absorbing liquid, such as water, flows (Figure 5.8) can be used to measure waveguide power. We may also use a wedge-matched load which is cooled by flowing water. The inclined tube and wedge load are well matched for reasons discussed in Chapter 3, so reflections are negligible. The temperature difference between the liquid input and output and its mass flowrate are used to determine the absorbed power. The power is then calculated from:

$$P = \frac{\partial m}{\partial t} c_p [T_2 - T_1] \tag{5.5}$$

where $\partial m/\partial t$ is the mass flow (kg/s), c_p the constant pressure specific heat (J/kg-°C) and T_1 and T_2 the inlet and outlet temperatures, respectively. As with all calorimeters, the fluid must be adiabatic between the points at which T_1 and T_2 are measured, so these measurements must be intimate with the tube exit and inlet planes. Alternatively, one may know the equivalent heat loss to the surroundings so that the apparent power can be corrected to obtain a more accurate estimate of waveguide power. This method is an absolute method based on

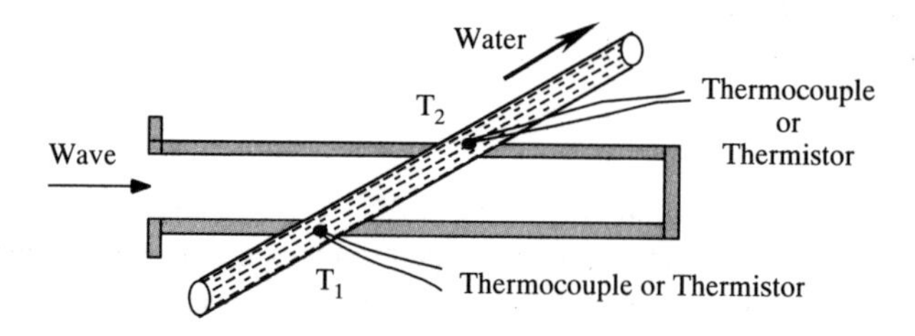

Fig. 5.8 *Calorimetric waveguide power measurement.*

a physical standard, and it has been used for the calibration of low power bolometers and other sensitive equipment.

Interestingly, this method is also applicable to determination of the local volumetric power density in a material medium. From Chapter 1 we bring forward the energy balance (equation (1.50)) assuming that local Fourier heat transfer, convection, and radiation are negligible:

$$\rho c_{\mathrm{p}} \frac{\partial T}{\partial t} = q_{\mathrm{gen}} = \omega \varepsilon'' |\mathbf{E}|^2 \tag{5.6}$$

where ρ is the density (kg/m3) and q_{gen} is the volumetric power density (W/m^3). The conditions for this relation to hold are satisfied at the initial onset of heating because for low temperature gradients heat transfer is negligible. So, if one knows the initial rate of rise and the medium properties one can calculate local power density and the magnitude of the electric field.

5.3 FREQUENCY MEASUREMENT

Frequency measurements in microwave and RF fields are now done using a counter. Counters function effectively up to 20 GHz. The modern governing principle is to impinge a wave onto a chain of bistatic transistors, open the gate for a given time and to count the impulses into which the signal has been transformed. Modern frequency counters resolve at least nine significant figures and are accurate and reliable with fairly inexpensive electronics. For both RF and microwave ranges the input signal may be derived from a simple **E**-field probe (or a Hertzian dipole antenna) as described in Section 5.4 of this chapter.

In traditional instruments, a reference oscillator and mixer were used to measure frequency, as shown in Figure 5.9. The reference oscillator supplies a constant frequency f_0 to a harmonic generator. The harmonic generator produces a signal which is at an integer multiple (nf_0, a higher frequency) for input to the mixer. A directional coupler permits the mixing of the reference wave, at unknown frequency f, with that of the generator. A mixer multiplies two signals, so two time-domain sinusoidal signals are multiplied in this case. Using the trigonometric identity for the product of two sine waves it is easy to show that the result of mixing is to generate two signals; one at the sum and the other at the difference of the two frequencies, $f + nf_0$ and $f - nf_0$. The beat (difference) frequency of the mixed waves is amplified and measured with an oscilloscope or a lower frequency counter. By varying the harmonic of the reference oscillator, two successive beat frequencies, f_1 and

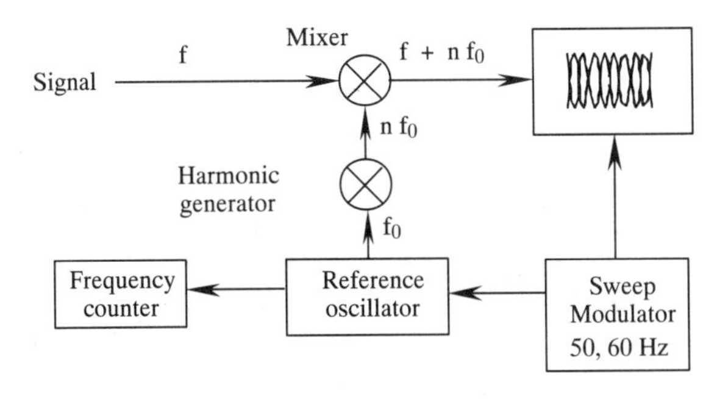

Fig. 5.9 *Frequency measurement using a mixer.*

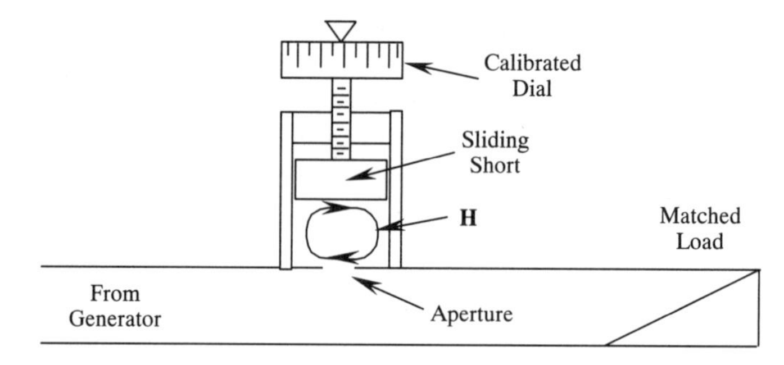

Fig. 5.10 *TE_{111} mode cylindrical cavity frequency meter connected to rectangular waveguide broad wall.*

f_2, can be observed. Then:

$$f = nf_1 \qquad f = (n+1)f_2$$

$$f = \frac{f_1 f_2}{f_1 - f_2}. \tag{5.7}$$

The advantage of this older design is that one can estimate the purity of the generated frequency by observing the beat signal as the reference oscillator is swept by a 50 or 60 Hz sweep.

As was briefly mentioned in Chapter 2, an inexpensive and moderately accurate frequency meter can be made with a cylindrical cavity similar to that shown in Figure 5.10. Cylindrical cavities were discussed in Chapter 4. The cavity sliding short is adjusted until resonance is obtained and the frequency is indicated by measuring the position of the short. In the circuit of Figure 5.10 resonance is indicated by a minimum in the reflected power. Reflections are created by the evanescent modes surrounding the aperture in the waveguide wall when the cavity is off resonance. The aperture is **H**-field coupled. We remember that the TE_{11} mode is dominant in a cylindrical waveguide since it has the lowest cutoff frequency. Typical cylindrical cavity frequency meters are operated in the TE_{111} resonant mode. Cavities are sized to cover a specific frequency range which is limited by the cutoff and mode conversion frequencies. Ultimately, the resolution and accuracy of these meters is limited by dimensional instability due to mechanical hysteresis and thermal expansion–contraction cycles. Also, a signal with strong harmonics or out-of-band components due to modulation effects is not well managed with this technique.

5.4 ELECTRIC AND MAGNETIC FIELD MEASUREMENTS

The electric and magnetic field strengths are measured by using them to induce a signal and measuring its amplitude. Sensors behave as antennas when their critical dimensions approach significant fractions of a wavelength. Typical magnetic field sensors are coils which are either very small compared to a wavelength or "resonant" in the sense that the circumference is one wavelength. Electric field sensors are almost always approximately Hertzian dipoles, so considerable mismatch between source and sensor is experienced. The design principles for both microwave and RF sensors are very similar, so we treat both types together.

5.4.1 Electric field measurement

We should note, before we begin that electric field sensing at radio frequencies differs slightly from that at microwave frequencies in that the usual sensors are so small and the transmission lines so short that one can conceptually separate the electric field from the magnetic field using quasi-static analysis. At microwave frequencies the electric and magnetic fields are inextricably linked since wave propagation principles apply, so measuring one will yield the other if the waveguide or bulk medium characteristic admittance is known.

The basic electric field sensor is a very short monopole, which is the coaxial equivalent of a very short dipole antenna. A monopole antenna (Figure 5.11) with length, h, very short compared to a wavelength (h less than about 0.1λ) has a transmitting radiation resistance given by:

$$R_{\text{rad}} = 40\pi^2 \left(\frac{h}{\lambda}\right)^2 \tag{5.8}$$

where λ is the wavelength in the medium surrounding the antenna and we have assumed radiation in free space where $\eta_0 = \{\mu_0/\varepsilon_0\}^{0.5} = 120\pi\Omega$. By the reciprocity theorem the transmitting and receiving characteristics are the same; so the source resistance is given by the same relation. If $h = 0.1\lambda$ the source resistance is about 4 Ω. A dipole antenna radiates twice as much power as a monopole, so its radiation (and also receiving) resistance is twice the figure in equation (5.8).

E-field sensors for waveguides and coaxial lines

Monopole (and dipole) sensors are isotropic in the spherical coordinate ϕ; that is, in any plane parallel to the $x - y$ plane. These sensors have a sensitivity which depends on the angle to point P in Figure 5.11: $I_{\text{rec}} \propto \sin\theta$, so the highest sensitivity is at $\theta = \pi/2$. They are most sensitive to an **E**-field which is z-polarized, and relatively insensitive to E_x and E_y components. We have already seen how they may be used in a coaxial line (Figure 5.3). A typical waveguide installation is shown in Figure 5.12. Of course, for TE_{10} propagation we get the strongest signal in the middle of the broad wall at $x = a/2$ with the monopole in the y-direction with respect to waveguide coordinates. At the N-connector we may use either a milliwattmeter and standard power sensor to indicate the square of **E**-field strength, or a diode and low frequency voltmeter. The field probe and mount must be independently calibrated (using, for example, calorimetric methods) since the source and sensor impedances will be highly mismatched, resulting in very poor coupling into the sensor.

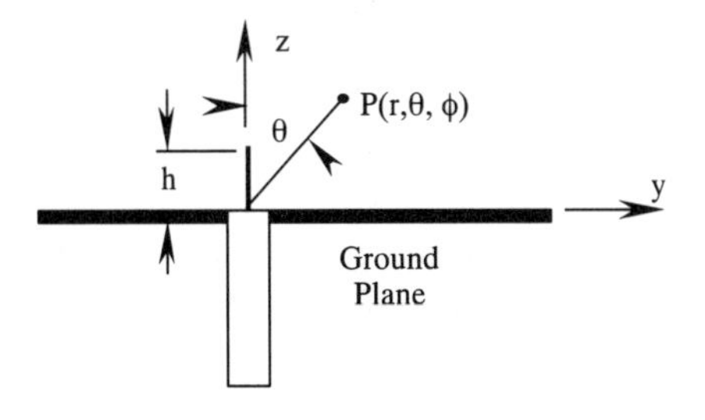

Fig. 5.11 *Short monopole* ***E****-field sensor.*

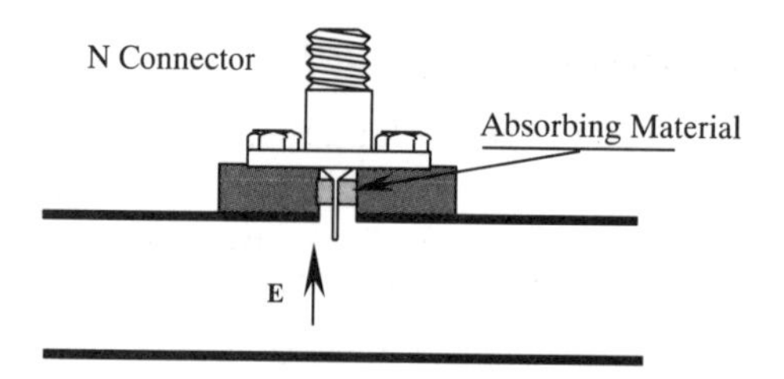

Fig. 5.12 *Monopole* ***E****-field detector in a waveguide.*

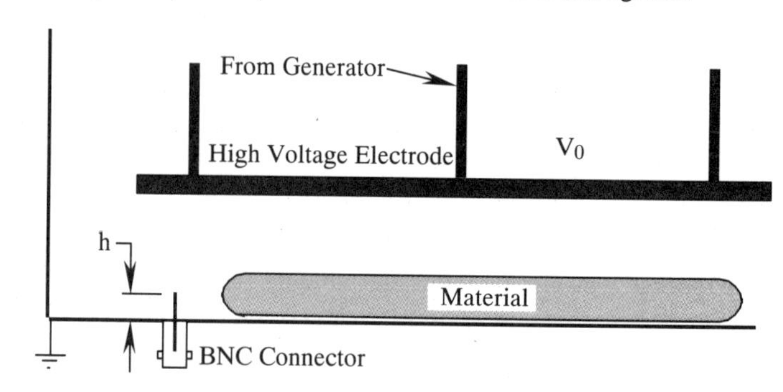

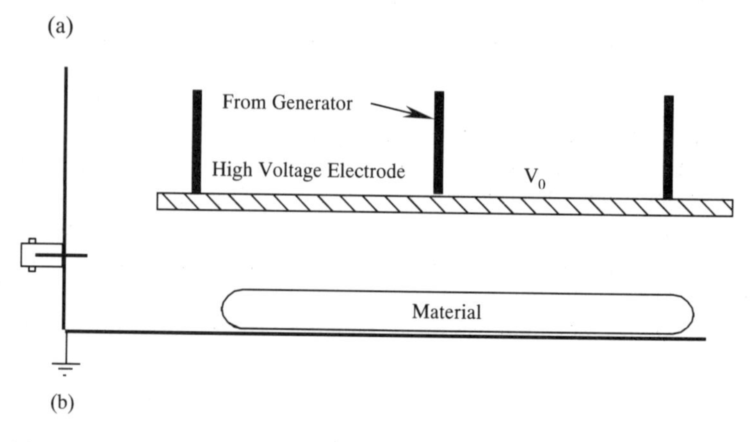

Fig. 5.13 *Monopole probe for RF enclosures. (a) Location for estimating field applied to material. (b) Side wall location for relative measurements.*

E-field sensors for large cavities

Radio frequency devices At radio frequencies the "cavity", as it were, is in fact the shielded enclosure where the material is heated. In the case where the RF applicator generates a highly localized field for welding and similar applications the field is not usually measured — one measures electrode voltage rather than the electric field. In the case of a large enclosure we may apply the monopole probe to estimate electric fields applied to the load material. An example is shown in Figure 5.13 where large flat plate electrodes are used to heat a layer of material. We can use a directional **E**-field sensor since the electric field must be normal to the conductors which comprise the enclosure. If a suitable location can be found we may obtain a reasonable estimate of the electric field applied to the material, Figure 5.13a. The sensitivity of the sensor is simply related to its length, h, and measured voltage, v_{meas}, since we have a quasi-static geometry:

$$|\mathbf{E}| = \frac{v_{\text{meas}}}{h}. \tag{5.9}$$

Also, we may mount the sensor on the side wall (Figure 5.13b) if all that is needed is a sample of the field to determine fundamental frequency, harmonic distortion, overall waveshape and grid tank circuit tuning effects. This type of sensor is much easier to apply than any of the voltage dividers described above especially at high RF power levels.

Microwave sensors In a waveguide where a single mode is propagating it is an advantage to use a directional sensor, such as a single monopole. We know *a priori* in which direction the electric field is polarized. In a large multimode cavity the electric field polarization is not known and is probably not calculable. Nevertheless, oftentimes we must have a good indication of local electric field magnitude in order to evaluate cavity performance or load susceptibility. In those cases we need to use an isotropic probe so that the total electric field may be determined. Isotropic probes are available commercially, and are often used to monitor electric field leakage from ovens and industrial microwave apparatus at low levels.

Isotropic antenna arrays may be assembled by using three orthogonal short monopoles or dipoles as illustrated in Figure 5.14. The signals are combined in a desired way (usually converted to dc with three diodes and summed) and fed through the output cable to an external readout device. However, the metallic wires in the cable will significantly disturb the field distribution in a multimode cavity as the probe is moved about. The solution is to optically couple the signal.

Perhaps the simplest form of optically coupled **E**-field sensor is a small neon bulb with its leads trimmed and shaped to form a receiving dipole, as shown in Figure 5.15. Signal strength is coupled to the readout through a fiber optic cable. The sensitivity is determined by the length of the leads since the receiving (i.e. radiation) impedance varies with the length. The device must be calibrated in a known electric field since uncertainty in the lead length and position is significant. The bulb brightness is linear with received power, and thus depends on the square of the magnitude of the electric field. This sensor has the directional characteristics of the monopole previously described. Obviously, three orthogonal sensors of

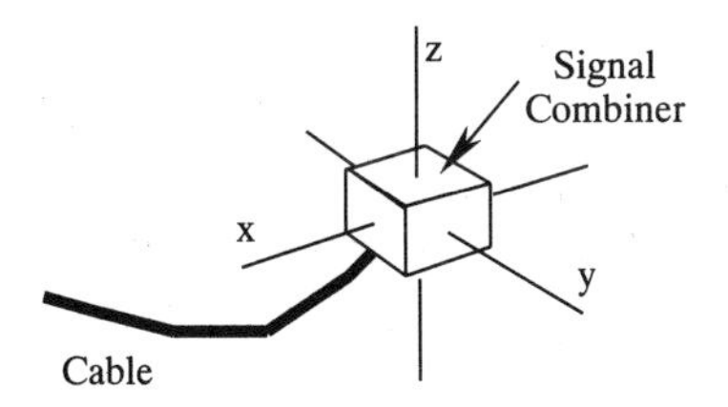

Fig. 5.14 *Isotropic dipole **E**-field sensors.*

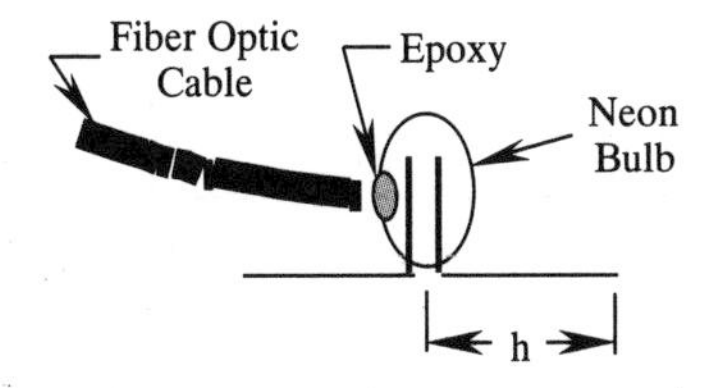

Fig. 5.15 *Optically coupled **E**-field sensor.*

this type can be placed within a small optically opaque integrating-sphere enclosure and one output fiber optic cable can be used to obtain an isotropic **E**-field measurement. The advantages are that the measurement is made very inexpensively and with minimum disturbance to the multimode cavity fields. The disadvantage of this probe is that one may only obtain an indication of the magnitude of the total electric field, not the individual vector components or their phase relationships. Of course, three fiber optic cables could be used to resolve the relative magnitudes of the components, but the phase relationships are still suppressed.

The problem of phase measurement may be addressed by modulating a small scattered field inside the cavity where the electric field must be determined. The modulated field reflects a wave in proportion to the electric field which exists where it is located. Power for the modulated wave is optically coupled to avoid interaction. Then a coherent microwave detector is used to measure the reflected wave in-phase component for simultaneously obtaining the amplitude and the phase of the electric field within the cavity.

The system design is shown in Figure 5.16. The active device consists of a phototransistor (or three for an orthogonal system) with its wires formed into a dipole antenna. It is illuminated by a 25 or 30 Hz modulated light source through an optical fiber. Light impingement from the optical fiber changes the electrical conductivity of the phototransistor semiconducting material. This, in turn, changes the local "load" placed on the electric field at the sensor location generating a reflected wave. The modulated reflected wave is proportional to the square of the local complex electric field E_{M} which exists at point M inside the applicator. A small portion of the reflected wave, e_2, propagates retrograde toward the generator:

$$e_2 = \frac{kE_{\mathrm{M}}^2}{E_{\mathrm{i}}}. \tag{5.10}$$

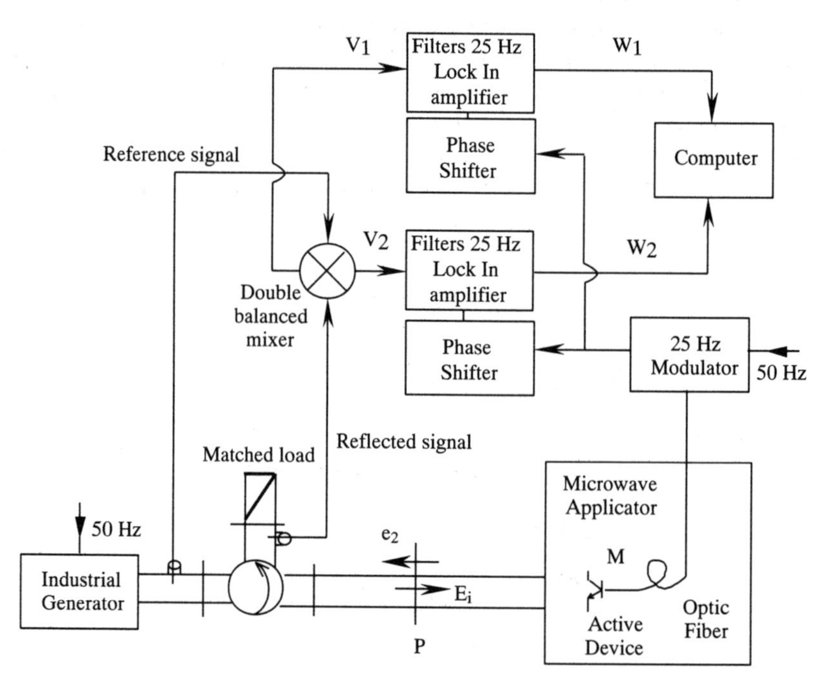

Fig. 5.16 *Synchronous detection system for complex electric field measurement in large cavities.*

This wave is deflected by a circulator into a sensing probe placed before the circulator. The reflected signal is applied to the RF input side of a double balanced mixer. The 25 Hz modulated signals V_1 and V_2 are each proportional in amplitude to the square modulus of E_M. Their relative phase difference is twice that between the microwave field E_M and the incident field E_i, the angle Φ. The lock-in amplifiers are used with signal averaging to extract the 25 Hz modulated signal from all other reflections and extraneous noise. After synchronous detection at 25 Hz using the quadrature detector:

$$W_1 = k|E_i|\left|\frac{E_M}{E_i}\right|^2 \cos(2\Phi) \tag{5.11a}$$

$$W_2 = k|E_i|\left|\frac{E_M}{E_i}\right|^2 \sin(2\Phi). \tag{5.11b}$$

Roussy *et al.* (1992a) have reported field measurement inside a slotted waveguide in front of a matched load, and in front of a short circuit with this system. Experiments reveal that the phototransistor must be very strongly illuminated in order to generate a measurable electric field intensity, e_2.

The uncertainty in the measurement of the electric field (around 40 V/m) is about 5% for the amplitude and ±5° for the phase. The achievable sensitivity is better than 10^{-2} V/m and can be improved because it is principally limited by the noise level in amplitude and frequency of the microwave generator. The system can also be used with a continuous wave (filtered dc power supply) low power generator. In that case a 1 kHz modulation signal works better; and the system can be used for experimental studies of new applicator designs in the laboratory. It also works well when the microwave generator is pulsed at 50 or 60 Hz, as industrial magnetrons usually are in industrial equipment. In that case the active device uses a 25 (or 30) Hz inner modulation. With a sensor composed of three orthogonal antennas the system is able to measure the amplitude and the phase of all the three components of the electric vector. The dynamic range is suitable for studying both loaded and empty industrial applicators.

5.4.2 Magnetic field measurement

Magnetic fields are estimated from induced potentials around enclosing loops. We recall from Faraday's law of induction that the induced potential around a contour depends on the time rate of change of the total magnetic flux enclosed by the contour. The various magnetic field transducers differ in the geometry of the enclosing contour.

Small loop transducer

An electrically small simple current loop may be used at either RF or microwave frequencies. A loop is electrically small if the circumference is much less than one wavelength, as in Figure 5.17a, where $a \ll \lambda/6$. Under this condition the magnetic field within the loop is in phase everywhere. So, we describe small loops using an equivalent magnetic dipole moment, $m = IA = \pi a^2 I$ (A-m^2). It may be shown that when a small loop in the x–y plane is used for transmitting the far fields are given by:

$$E_\varphi = \frac{\beta^2}{4\pi}\sqrt{\frac{\mu}{\varepsilon}}\frac{m \sin\theta e^{-j\beta r}}{r} \tag{5.12a}$$

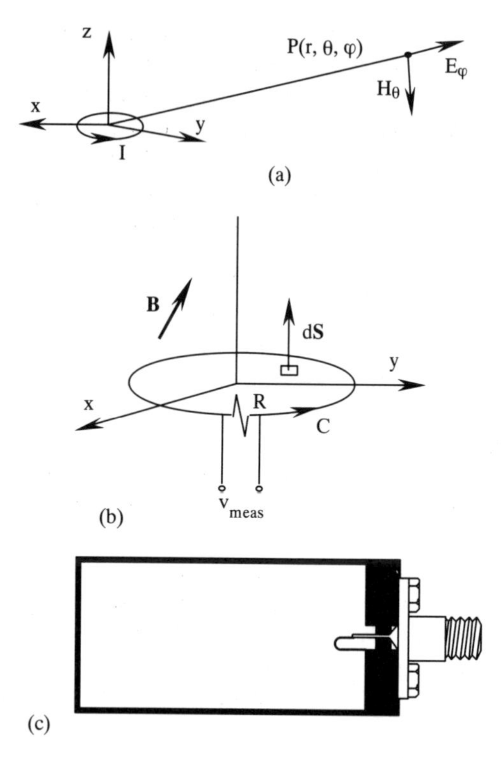

Fig. 5.17 *Small loop magnetic field transducer. (a) Far field geometry. (b) Near field geometry. (c) Small loop in waveguide narrow wall for **H**-field measurement.*

$$H_\theta = -\frac{E_\varphi}{\eta} = -\frac{\beta^2}{4\pi}\frac{m \sin\theta \mathrm{e}^{-\mathrm{j}\beta r}}{r} \tag{5.12b}$$

and the radiation resistance by:

$$R_{\mathrm{rad}} = \frac{\eta\pi}{6}(\beta a)^2. \tag{5.13}$$

For a coil radius of $a = 0.05\lambda$ in free space ($\eta = 120\pi\ \Omega$) the radiation resistance is about 1.9 Ω. By reciprocity the receiving characteristics are the same as the transmitting characteristics and the sensitivity is obtained by inverting equations (5.12) to obtain m (and thus I) as a function of E_φ and H_θ. The coil impedance is not well matched to the medium, just as for the Hertzian dipole, so the coupling into the coil will be very weak indeed. We should use these expressions when the loop is distant from a strong source which is polarized as shown, as for example in a multimode cavity. The loop should be loaded with a fairly high resistance load (say a 50 Ω terminated cable) so that the signal will be detectable and the induced currents do not interact with the local magnetic field to decrease it significantly.

For the usual case of a source in the near field of the loop we may use Faraday's law of induction (Figure 5.17b) where the loop defines the enclosing contour C and A is the loop inner area:

$$v_{\mathrm{meas}} = \oint_C \mathbf{E} \cdot \mathrm{d}\mathbf{L} = -\mathrm{j}\omega \left\{ \iint_A \mathbf{B} \cdot \mathrm{d}\mathbf{S} \right\}. \tag{5.14}$$

Then the induced voltage is a direct measure of the total enclosed magnetic flux. Note that the loop is fairly directional in the near field since only the component of the magnetic flux normal to the loop, i.e. parallel to **dS**, makes a contribution to the induced voltage. Again, the loop should have a load so that the induced current will be small. If the loop is small enough we may safely assume that B_n is constant over the loop area. A small loop is frequently used in microwave devices. For example, the magnetic field inside a standard waveguide can be measured by placing a small loop in the guide, as shown in Figure 5.17c.

Moebius loop transducer

An improved form of the small loop transducer is the Moebius loop, which is very effective for the measurement of radio frequency magnetic fields. This device has been well described by Iskander *et al.* (1981). In topology a Moebius band is a surface which folds back upon

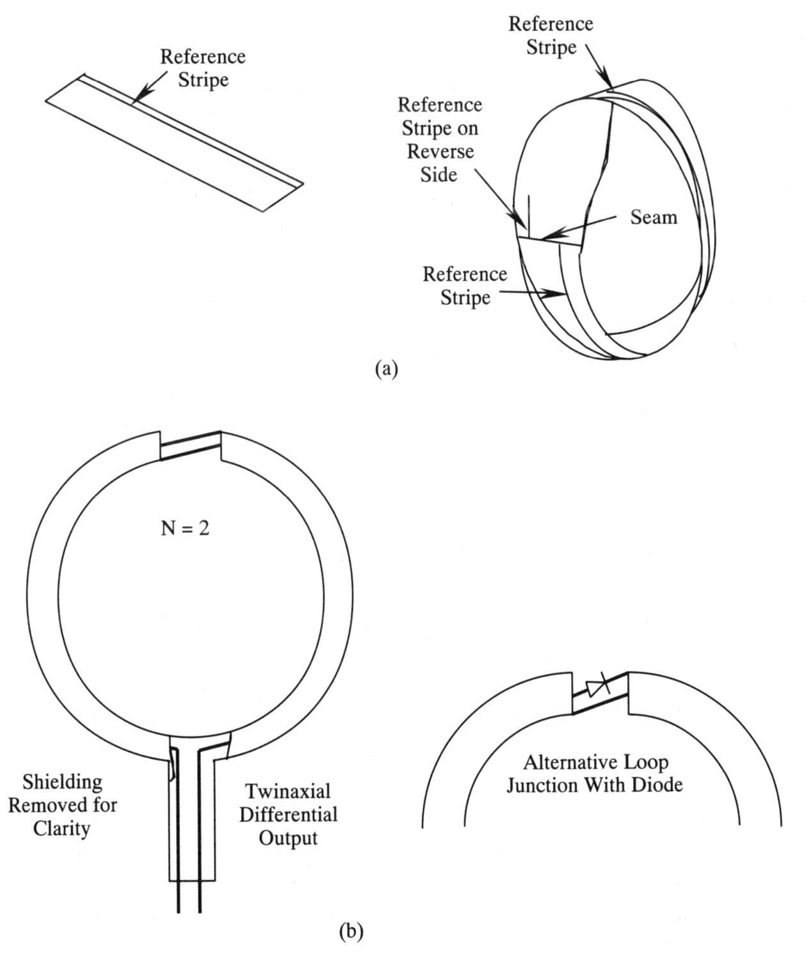

Fig. 5.18 *Principles of the Moebius loop magnetic field transducer. (a) Moebius band constructed from a strip of paper. (b) Moebius loop made from coaxial line with two types of center point junctions.*

itself so that there is only one side. One may make such a surface instead of a ring by twisting a strip of paper before gluing the ends together, Figure 5.18a. The Moebius loop magnetic field transducer is based on a similar principle but constructed from semi-rigid coaxial line. We assume in this discussion that the loop dimensions are small compared to a wavelength. The circular loop has its center conductors cross-connected so that the capacitively coupled electric field signals effectively cancel. This is the major advantage of the Moebius loop topology over a standard unshielded loop which registers both electric field and magnetic field induced signals. The Moebius loop is essentially insensitive to potential gradients in the surrounding medium and measures magnetic fields only. The loop has two turns as a result, and its output is a balanced differential line which, when shielded, is referred to as a twinaxial transmission line.

Two alternative center junction geometries are shown in Figure 5.18b. With the two coaxial center conductors simply cross-connected the output signal is a sine wave with a phase angle of -90° with respect to the total magnetic flux enclosed by the ring (from the $-j\omega$ factor in equation (5.14)). The twinaxial line feeds a so-called "balun" (for balanced–unbalanced transformer) to convert to a single-ended coaxial transmission line, Figure 5.19a. The balun should be as close as possible to the loop output. One may use this form when the wave shape needs to be observed. The standard balun as used in amateur radio transmitters matches a 75 Ω transmitter power amplifier to a 300 Ω twin lead antenna transmission line and thus has a turns ratio of 2:1. Consequently, a 50 Ω transmission line will appear as a 200 Ω load on the Moebius loop; and at the oscilloscope the voltage signal has been stepped-down by a factor of two (which just cancels the effect of having two turns in the loop). The voltage signal at the oscilloscope is then the product of the loop area, A, and the normal component of magnetic flux density, B_n, shifted in phase by -90° assuming that B_n is approximately constant over the loop.

If a diode is placed in one of the connections, as shown in the alternative form of 5.18b, then the signal is a rectified dc signal and can be read out on a standard voltmeter. It is helpful to use a small parallel capacitor (say about 1 nF ceramic) to filter the signal at the

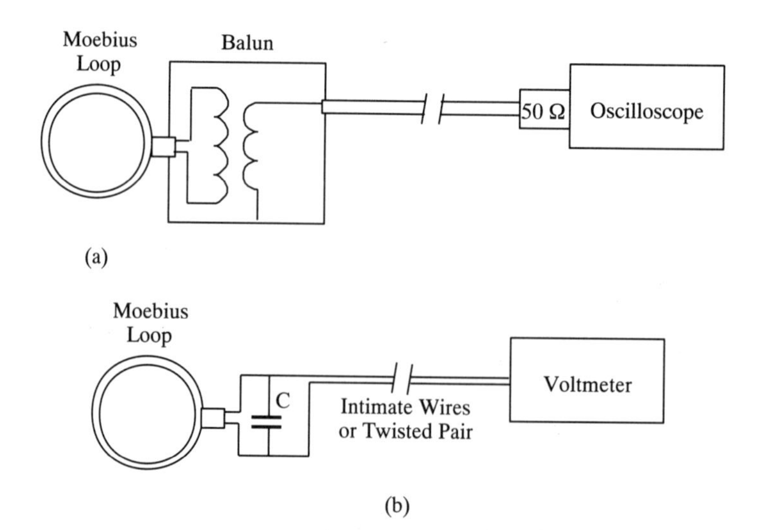

Fig. 5.19 *Alternative connection schemes for Moebius loop. (a) Twin axial balum conversion for ac Moebius loop. (b) Voltmeter readout for dc Moebius loop.*

loop output, Figure 5.19b. The voltmeter cable may be as long as necessary, but should not be grounded, and the voltmeter should either be passive or battery-operated. If a grounded transmission line or mains-powered voltmeter must be used, then the balun should be placed before the capacitor to convert to a single-ended signal.

5.5 TEMPERATURE MEASUREMENT

It is not simple to measure the temperature of a product which is irradiated by an electromagnetic field. Generally, thermocouples cannot be used because they have metallic elements which themselves are heated in, and otherwise interact with, the field. For analogous reasons, mercury thermometers also cannot be used. Alcohol thermometers are not much better because alcohol also absorbs electromagnetic energy and the thermometer fluid is itself heated. However, thermocouples can be used in waveguides when the field distribution is such that **E** is perpendicular to the thermocouple (as in Figure 5.20). In this orientation, the thermocouples do not experience significant self-heating and interact minimally with the field.

5.5.1 Infra-red methods

One of the best methods for temperature measurement of a sample in an electromagnetic field is to use infra-red radiometry or thermography. There is no contact with the sample, but such methods only give temperature information near the surface of the material; typically within about 50 μm of the surface, depending on material characteristics. Radiometric measurements use a single staring sensor which collects emitted and reflected thermal energy from the surface (assuming an opaque body). There are two overall kinds of sensors, a wide-band bolometer type which collects all wavelengths and a band-limited optical sensor. The bolometric device measures radiation heat transfer between the surface under study and the sensor itself and thus receives:

$$Q_{\text{rad}} = \sigma F_{\text{sh}}[\varepsilon_{\text{surf}} T^4_{\text{surf}} - \varepsilon_{\text{sensor}} T^4_{\text{sensor}}] \qquad \text{(W)} \tag{5.15}$$

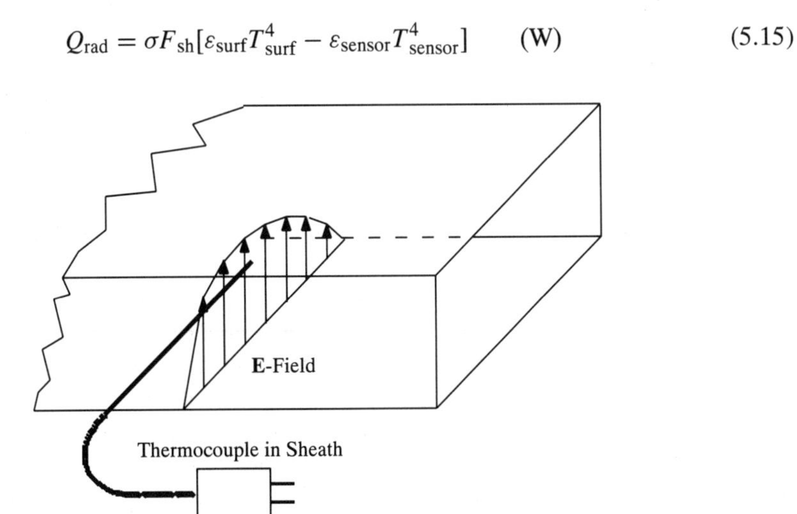

Fig. 5.20 *Thermocouple in a waveguide, acceptable orientation.*

where ε is the emissivity, T is the absolute temperature (K), F_{sh} is a shape and view factor between the two surfaces (m^2), and σ is the Stefan–Boltzmann constant (see Chapter 1). The spectral distribution of emitted radiation from an ideal black body, $W_b(\lambda, T)$ ($W/m^2/\mu m$) follows the Planck radiation law:

$$W_b(\lambda, T) = \frac{2\pi hc^2\lambda^{-5}}{\{e^{hc/\lambda kT} - 1\}} \tag{5.16}$$

where, again, T is absolute temperature (K), c is the speed of light, h is Planck's constant and k is Boltzmann's constant.

When a band-limited detector is used the received signal will depend on the spectral sensitivity of the detector. The expected function can be obtained by integrating equation (5.16) over the pass band of the detector. Equation (5.16) has a wavelength at which W_b is maximum, λ_{max}, for each temperature. It turns out that the product of λ_{max} and T is constant (this is Wien's displacement law):

$$\lambda_{max}T = 2898\ \mu\text{m K} \qquad \text{or} \qquad \lambda_{max}T = 5216\ \mu\text{m}\ ^\circ\text{R} \tag{5.17}$$

At 300 K (27°C) λ_{max} is 9.7 μm.

There are two typically used detector bands: 3–5 μm and 8–12 μm. The detector bands depend on the energy band gap (between valence and conduction bands) in the semiconductor crystal structure. All practical 8–12 μm band detectors are made of a tri-metal mixture of Hg–Cd–Te in varying proportions. The band gap (here between trap levels and the conduction band) is small enough that this material will respond to the weak photons in this wavelength band. These detectors must be operated at low temperature (typically in liquid nitrogen at 77 K) so that the photon energy will exceed thermal noise in the detector. Either Hg–Cd–Te or InSb detectors may be used in the 3–5 μm band. Both must be cooled to suppress thermal noise. Even though the peak emission lies in the middle of the 8–12 μm band, the noise process is stronger. Consequently, the signal to noise ratio is about the same for both bands.

A scanned detector is used in some devices to form a thermographic image. Thermographic imagers do not measure temperature directly; they measure surface radiosity from which we infer (or estimate) temperature. The received infra-red power again depends on surface emissivity and temperature. Errors accrue due to reflections off the surface, a finite instantaneous field of view (IFOV), and attenuation in the optical pathway. Good results are obtained as long as surface characteristics are known, the smallest thermal features to be resolved are large compared to the IFOV, and calibration sources are used. With an imaging device, as opposed to a staring sensor, reference sources may be placed in the scene for calibration. The amount of data in one image is so large (typically more than 50 000 samples) that digital analysis techniques are usually necessary.

As a final cautionary note, it should not be too surprising that the infra-red film one may buy for cameras does not indicate surface temperature. The pass band of glass optics only extends to about 1 μm, far below any significant emitted infra-red. This film measures near infra-red reflected energy, and should not be confused with the far infra-red required for temperature estimation.

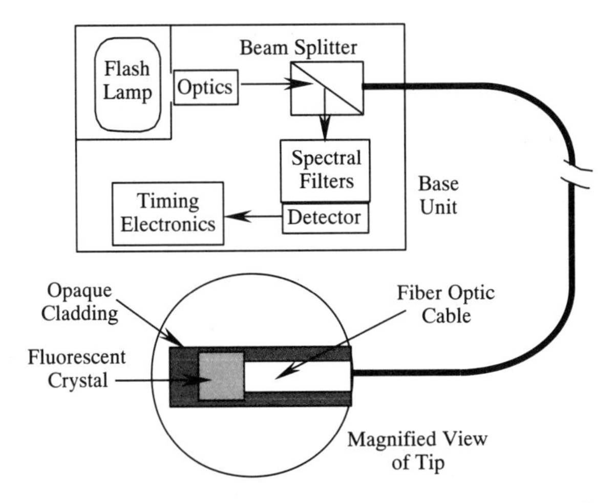

Fig. 5.21 *Optical temperature measurement system.*

5.5.2 Fluorescence methods

Within the last ten years considerable effort has been expended to develop commercial non-interfering temperature measurement devices based on fluorescence. In a standard device a small crystal of fluorescent material is located at the end of an optical fiber (Figure 5.21). The crystal is excited with a flash lamp and its return fluorescent signal (at a different known wavelength) is diverted to spectral filters and an optical detector by a beam splitter. The time constant of fluorescence decay is very temperature sensitive and independent of flash lamp intensity. By calculating the decay time constant accurate and repeatable measurements can be made up to about 300°C (limited by optical cable degradation). The measurements can be made without interference in a very strong electromagnetic field since the coupling is optical. The probe cable is quite low in thermal conductivity compared to thermocouples and other point contact devices.

5.6 IMPEDANCE MEASUREMENTS

The measurement of impedance is the basic measurement in circuit design at both microwave and radio frequencies. We have already seen how to use load and applicator impedance to design matching networks. It remains to review the fundamentals of these measurements.

5.6.1 Traditional methods

Slotted lines

One of the oldest and most common techniques involves the use of a slotted line, either a rigid air dielectric coaxial cable or a waveguide. A slotted line is a long thin non-radiating aperture (parallel to current lines) equipped with a movable monopole probe, as previously described. The realization in the broad wall of a rectangular waveguide is shown in Figure 5.22. The slot could just as easily be made in the outer wall of a coaxial line. For either RF or microwave measurements a diode may be used to rectify the signal so that a simple dc voltmeter can be used to make the measurement.

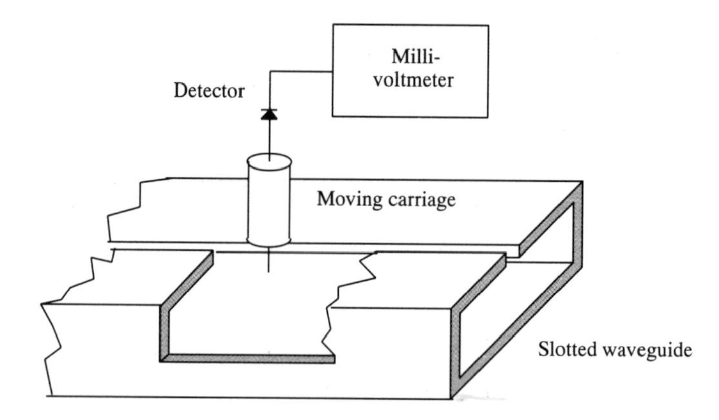

Fig. 5.22 *Slotted line for VSWR measurement.*

The signal obtained is directly related to the square of the local electric field in the line, if the diode has an approximately quadratic response curve. For this case, the standing wave ratio, SWR, is directly related to the square root of the ratio of the measured voltage maximum and minimum:

$$\mathrm{SWR} = \sqrt{\frac{V_{\mathrm{max}}}{V_{\mathrm{min}}}} \tag{5.18}$$

and the magnitude of the load reflection coefficient, $|\Gamma|$, is given by:

$$|\Gamma| = \frac{\mathrm{SWR} - 1}{\mathrm{SWR} + 1}. \tag{5.19}$$

The equivalent phase of the stationary wave is given by the position of the field maximum inside the slot region. The relative position of the load may be found by replacing the load with a short circuit and finding the location of the voltage minimum. This point is an integral number of half-wavelengths from the load plane and is the reference plane for the measurement. The difference between the location of the voltage minimum when the load is at the end of the line and the reference plane is the phase angle of Γ, $\angle\Gamma$. The normalized load impedance, $z = r + \mathrm{j}x$, may be found by rotating through an angle $\angle\Gamma$ from the maximum (at a voltage maximum $r_{\mathrm{eq}} = \mathrm{SWR}$) in the "wtg" direction at constant $|\Gamma|$. Then, for a lossless line, $Z_{\mathrm{L}} = zZ_0$. This method has been well studied at low power, and is equally useful for high power measurements. Because this method is time consuming and sensitive to nonlinearities in the detector, however, the measurements are now performed using automated systems for information retrieval.

When the SWR is large, that is when the load impedance is close to a short circuit or an open circuit, the difference between the voltage minimum and maximum is also large. The error is limited by the dynamics of the crystal detector. Therefore it is best to estimate the SWR from two measurements near a minimum. This is done, as shown in the example of Figure 5.23, by measuring the distance $(d_2 - d_1)$ for which the amplitude of the signal is twice that of the minimum. The voltage variation is given by:

$$V(d) = V_{\mathrm{min}} + V_{\mathrm{max}} \sin^2\left(\frac{4\pi d}{\lambda_{\mathrm{g}}}\right) \tag{5.20}$$

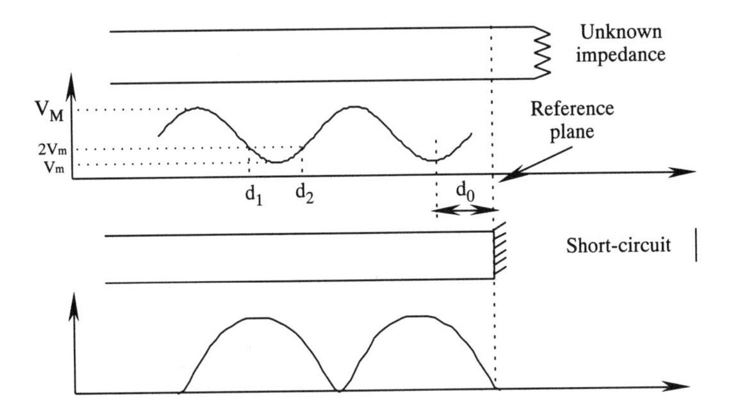

Fig. 5.23 *Determination of load impedance and location by slotted line.* $\angle\Gamma = \angle 4\pi d_0/\lambda_g$.

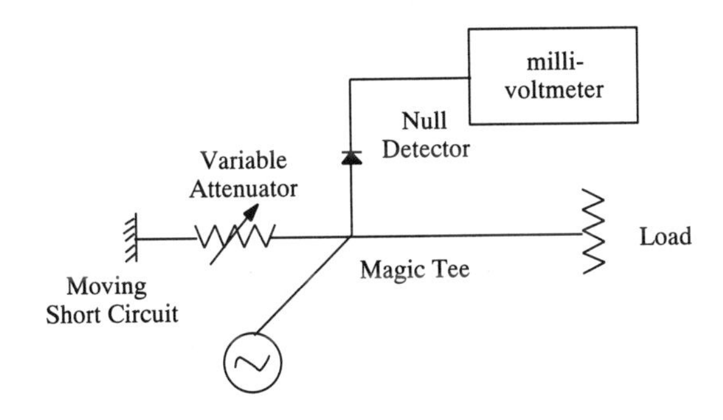

Fig. 5.24 *Microwave impedance bridge using a "magic tee".*

from which one can determine the SWR:

$$\text{SWR} = (d_2 - d_1)\frac{2\pi}{\lambda_g}. \tag{5.21}$$

Impedance bridges

Impedance measurements determined using a bridge (such as a Wheastone bridge or a Sharing bridge) are well known at low frequencies, but an equivalent technique also exists for the RF and low microwave frequency ranges. The principle is to compare the unknown impedance to a reference impedance and then to detect the difference in the reflected signal. Circuits such as these can be employed at frequencies up to 200 MHz. At microwave frequencies a "magic tee" or a 3 dB directional coupler can be used as a bridge to obtain impedance measurements. Figure 5.24 shows a "magic tee" in this configuration where the variable attenuator takes the role of the reference impedance.

5.6.2 Modern methods

Scalar reflectometry

At microwave frequencies a reflectometer consists of a dual directional coupler; one measures incident power and the other measures the reflected power. An unknown load

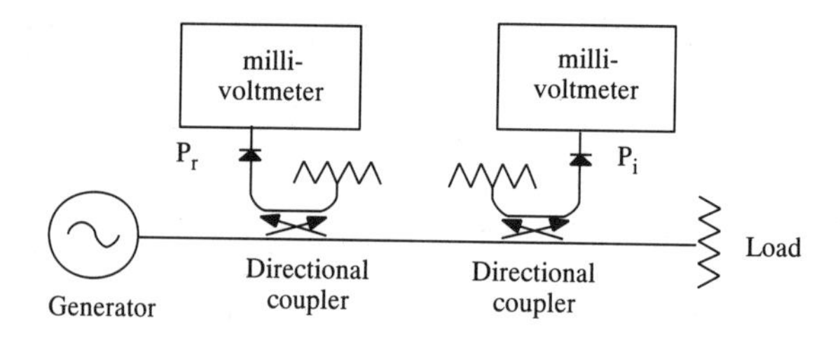

Fig. 5.25 *Double dual directional coupler scalar reflectometer for* $|\Gamma|$ *measurement.*

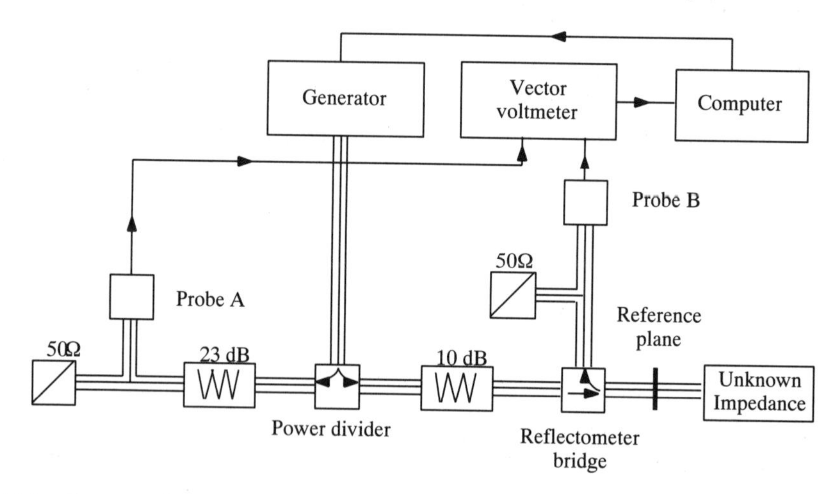

Fig. 5.26 *Vector reflectometer composed of power splitter, reflection bridge and vector voltmeter.*

impedance at the output (see Figure 5.25) will induce two voltages from which $|z|$ or $|\Gamma|$ may be directly calculated. Scalar reflectometers must be compensated for small defects in the system, such as a finite directivity in the directional couplers. The errors inherent in scalar reflectometers have been analyzed in many classical texts on microwave techniques.

Vector reflectometry

We may now make vector reflectance measurements, similar to scalar reflectometry, but automatically determining the amplitude and phase of the wave at the same time. The detection involves the mixing of the incident wave with a reference signal so that the phase between them can be estimated. A practical system is shown in Figure 5.26. Mixing is accomplished in the vector voltmeter between the reference signal and the reflected signal. The accuracy of vectorial impedance reflectometers is limited by parasitic reflections inside the lines which perturb the measurement of the reflection coefficient.

Complete compensation for line irregularities can be obtained by calibrating the meter at each intended frequency with three values of impedance at the reference plane: (1) a short circuit (Γ_{sc}); (2) an open circuit (Γ_{oc}); (3) a matched resistive load (Γ_r from $Z_L = 50\Omega$). The method of calibration is based on the calculation of the coefficients of the homographic compensation function, χ, by knowing the transformation function for the reflection coefficient at the three calibration points, as was introduced in Chapter 3. We repeat the result for impedance here for the reader's convenience. The normalized transformed impedance

z'_L is a homographic function of the actual load impedance z_L:

$$z'_L = \chi(z_L) = \frac{az_L + b}{cz_L + 1}. \tag{5.22a}$$

The coefficients a, b, c are determined by applying:

(i) $z_L = 0$, a short circuit, yields $z'_L = b$
(ii) $z_L = \infty$, an open circuit, yields $z'_L = a/c$
(iii) $z_L = 1$, a matched load, yields $z'_L = (a + b)/(c + 1)$.

Consequently, the correction formula for reflection coefficient, $\Gamma_{(\text{measured})} = \Gamma_m$ transformed to $\Gamma_{(\text{corrected})} = \Gamma_c$, is:

$$\Gamma_c = \frac{(\Gamma_m - \Gamma_r)(\Gamma_{sc} - \Gamma_{oc})}{\Gamma_m(2\Gamma_r - \Gamma_{oc} - \Gamma_{sc}) + 2\Gamma_{oc}\Gamma_{sc} - \Gamma_r\Gamma_{oc} - \Gamma_r\Gamma_{sc}} \tag{5.22b}$$

Six-port reflectometry

About ten years ago, a new method for the measurement of impedance was originally developed by the US National Bureau of Standards (now the National Institute of Standards and Technology, NIST). The apparatus consists of a measurement line with a six ports, with circuit as shown given in Figure 5.27. Port 1 is connected to the source, port 2 is connected to the unknown impedance Z_x, and ports 3, 4, 5 and 6 allow access for power detectors, which measure the amplitude of the signals. The system may be calibrated using many reference standard impedances by one of the several methods given in the literature. The success of this design is due to the recent advances in available inexpensive computing power. This design is now used for automated circuit analyzers; that is, network analyzers.

The six-port measurement system is described by the respective coefficient relations:

$$b_3 = M_3 a_2 + N_3 b_2 \tag{5.23a}$$

$$b_4 = M_4 a_2 + N_4 b_2 \tag{5.23b}$$

$$b_5 = M_5 a_2 + N_5 b_2 \tag{5.23c}$$

$$b_6 = M_6 a_2 + N_6 b_2 \tag{5.23d}$$

in which the complex load reflection coefficient is $\Gamma = b_2/a_2$ is to be measured. In general, the output wave b_3 is never zero for any value of Γ, and we force $N_3 \cong 0$ by placing a

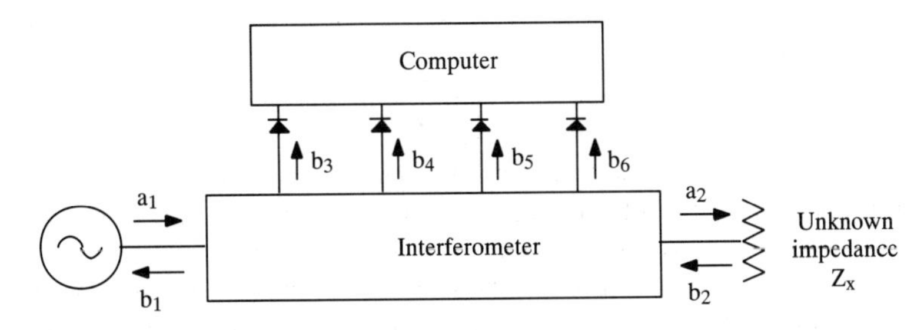

Fig. 5.27 *Six-port impedance measurement of complex impedance.*

directional coupler between the generator and port 3. If at each measurement port there is a quadratic detector, then the signals from ports 4, 5, and 6 divided by that from port 3 give:

$$\left|\frac{b_4}{b_3}\right|^2 = \left|\frac{N_4}{N_3}\right|^2 \left|\frac{(1+\Gamma M_4/N_4)}{(1+\Gamma M_3/N_3)}\right|^2 \tag{5.24a}$$

$$\left|\frac{b_5}{b_3}\right|^2 = \left|\frac{N_5}{N_3}\right|^2 \left|\frac{(1+\Gamma M_5/N_5)}{(1+\Gamma M_3/N_3)}\right|^2 \tag{5.24b}$$

$$\left|\frac{b_6}{b_3}\right|^2 = \left|\frac{N_6}{N_3}\right|^2 \left|\frac{(1+\Gamma M_6/N_6)}{(1+\Gamma M_3/N_3)}\right|^2 \tag{5.24c}$$

To determine the 11 calibration parameters, M_i and N_i, of the system, we must measure the signals at the measuring port by successively placing at least four standards at the impedance measurement plane. The system has at least 12 equations with 11 unknown quantities and is redundant and nonlinear. Parameters may be estimated using one of a number of optimizing schemes such as the Gauss–Newton method, the steepest gradient method, or the Levenberg–Marquardt method.

Once the system is calibrated, the complex reflexion factor, Γ, of the unknown impedance is then obtained by measuring the signals from the four ports and by solving the system of equations in (5.24) with the two unknown quantities $|\Gamma|$ and $\theta = \angle\Gamma$. This system is still redundant, and one can take advantage of the extra equation to verify the estimates.
Note that equations (5.24) can be linearized by substituting:

$$x = |\Gamma|\cos(\theta) \qquad \text{and} \qquad y = |\Gamma|\sin(\theta). \tag{5.25}$$

A simpler system consisting of four antennas fixed in a waveguide, which is derived from the six-port circuit, can also be used for impedance measurements. Each probe yields a signal which is linearly related to the total stationary field and is a function of the magnitude and phase of the reflection coefficient that is induced by the load impedance. This system uses a linear system of equations to determine $1+\Gamma^2$, $|\Gamma|\cos\theta$, and $|\Gamma|\sin\theta$ as a function of the voltage measured by the probes. Each probe has a signal:

$$V_i = |b_i|^2 = k_i|1 + A_i\Gamma|^2. \qquad (i = 1\text{–}4) \tag{5.26}$$

where k_i is the probe sensitivity. Expressing $\Gamma = |\Gamma|(\cos\theta + \mathrm{j}\sin\theta)$, allows V_i to be rewritten:

$$V_i = a_i(1 + |\Gamma|^2) + b_i|\Gamma|\cos(\theta) + c_i|\Gamma|\sin(\theta) + d_i. \tag{5.27}$$

The matrix equation is then:

$$\begin{bmatrix} V_1 \\ V_2 \\ V_3 \\ V_4 \end{bmatrix} = A \cdot \begin{bmatrix} 1+|\Gamma|^2 \\ |\Gamma|\cos(\theta) \\ |\Gamma|\sin(\theta) \\ 1 \end{bmatrix}. \tag{5.28}$$

Direct inversion is not possible because the matrix is singular. However, the coefficients which relate the unknown and known parameters can be obtained by reference to standard impedances. The calibration method has been proven and discussed in the literature.

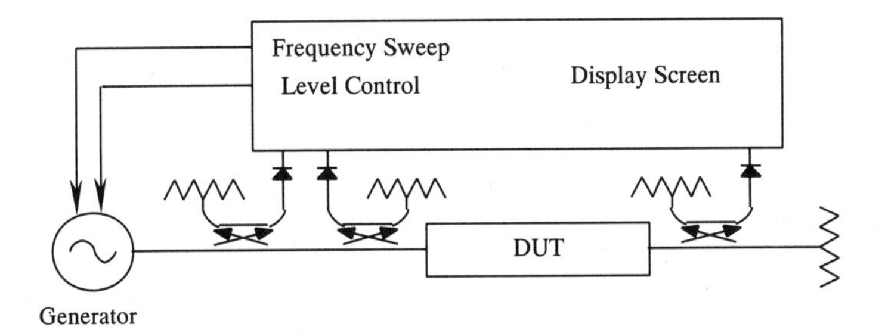

Fig. 5.28 *Scalar network analyzer.*

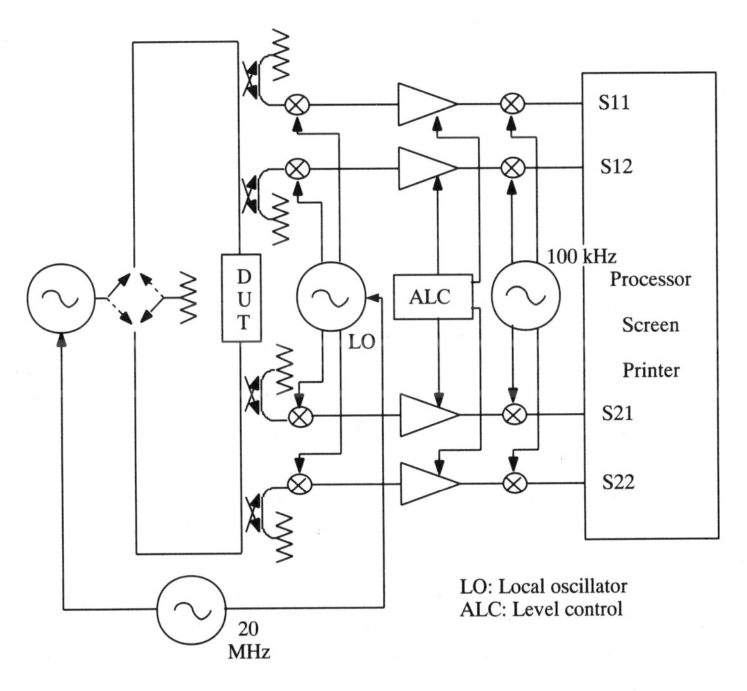

Fig. 5.29 *Fully automated vector network analyzer.*

Network analyzers

Finally, there are fully automated devices, network analyzers, which apply the six-port method. Typical schematic diagrams are suggested in Figures 5.28 and 5.29. Network analyzers may be configured to measure impedances, reflection and transmission or scattering parameters over a wide range of frequencies. We have shown the microwave realization, but both RF and microwave devices are available. Using built-in calibration procedures and the impedance correction methods outlined in equation (5.22a) the measurement reference plane may be placed at the end of a particular transmission line, impedance probe or waveguide section.

5.7 MEASUREMENT OF PERMITTIVITY

The measurement of electric permittivity is a very old and important problem. From the point of view of the theoretical aspects, it is the permittivity, as we will see, which indicates how an electric field interacts with matter on both macroscopic and microscopic scales. The electric permittivity is also important since it indicates how the field is perturbed by the material, as we have seen.

There are many methods for the measurement of dielectric properties. The best method depends upon the frequency range and often on the value of the permittivity. The conductivity of the material is also an important consideration in the choice of a method. There are special methods, termed cavity methods, for very low loss materials. Transmission line methods must be used for intermediate and high loss materials. Mention must be made that optical properties and methods can also be used to measure dielectric properties. These are employed when the material is large and no contact is required. We will restrict the discussion to some of the most practical or basic methods.

5.7.1 Radio frequency methods

Here we will discuss two applicators typically used in the RF range: a coaxial chamber, or open coaxial line, and a parallel plate capacitor.

Coaxial line methods

The open-ended coaxial line is very effective for liquids and powders. One form of this apparatus is a cell in which the central conductor has been cut, as seen in Figure 5.30. It may be used for pelletized materials as long as the major dimension of the pellets does not exceed about 20% of the chamber inner radius. Larger pellets must be measured in a larger chamber which will limit the maximum frequency to lower values than that in the figure. The temperature of the material may be varied by circulating water or other liquid through the external jacket. Contact electric resistance heaters may also be used to control chamber wall temperatures. The reference plane may be established by the three-impedance method described above.

At very low frequencies, this open coaxial cell is equivalent to a capacitor. When the capacitor is filled with a material the capacitance varies as a function of ε_r^*. The dielectric loss of the material induces a parallel resistance in its equivalent electrical circuit, such that from a measurement of the total impedance of the cell containing the material we can determine both parts of the permittivity. This is described by equation (5.29), which gives

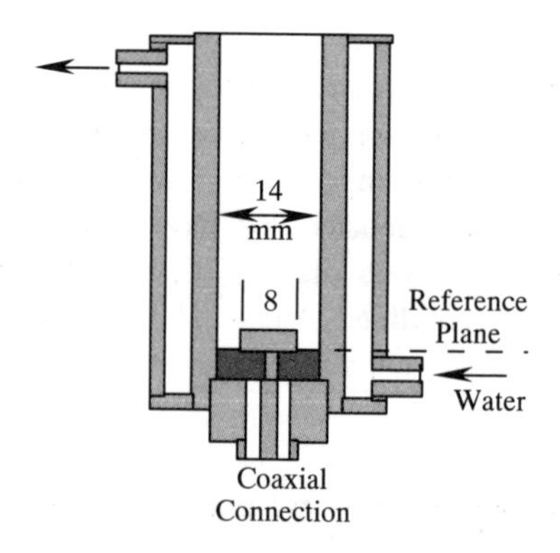

Fig. 5.30 *Open-ended coaxial chamber for RF permittivity measurement useful to about 5 GHz.*

the admittance of the cell in the reference plane of Figure 5.30:

$$Y = Y_c \sqrt{\varepsilon_r^*} \tanh \left(j \frac{\omega \sqrt{\varepsilon_r^*} d}{c} \right). \tag{5.29}$$

From this formula we see that the cell is characterized by two parameters Y_c and d where Y_c is the characteristic admittance of the cell cross-section and d is the equivalent length of the line (i.e. the stub in the cell and its fringing field in the unloaded case), which must be determined by calibration. At low frequencies where the line length is not important (that is, where the phase angle does not change appreciably within the chamber) Y is approximately given by:

$$Y = j\omega C_0 \varepsilon_r^* \tag{5.30}$$

where C_0 is the capacitance of the empty chamber. The admittance is modified when filled with a material of complex *relative* permittivity ε_r^*. At higher frequencies, it is necessary to solve the transcendental equation (5.29) using numerical methods and a computer to obtain the permittivity of a material from impedance measurements.

Parallel plate capacitor

A parallel plate capacitor arrangement, such as shown in Figure 5.31, is useful for measuring the permittivity of solid materials. A cylindrical geometry avoids the corner effects associated with rectangular plates. Essential measurements include the thickness and radius of the sample material. It is good practice to make the material sample smaller in radius than the capacitor disks. In that case the material will not participate in the fringing fields near the plate radius (thus the effective capacitance of the fringe fields is not changed by the material). Also, the electric field will be parallel to the material–air boundary and thus the same inside as outside of the material. Two measurements are made: one with the material in place and the other, C_0, with the material removed at precisely the same top plate position. The measuring micrometer is used to ensure that C_0 is as accurate as possible. Note that

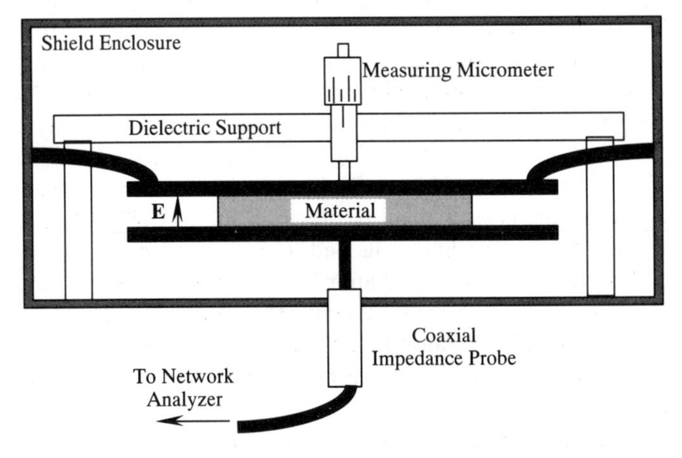

Fig. 5.31 *RF parallel plate capacitor apparatus for measuring permittivity.*

C_0 includes the plate fringing field capacitance, C_f:

$$C_0 = C_f + \frac{\varepsilon_0 A_p}{d} \tag{5.31}$$

where $A_p = \pi R_p^2$ (m^2) is the plate area, d (m) is the plate separation, and for extremely low electric permittivity materials one should actually use the permittivity of dry air, $1.0006\varepsilon_0$. The relative permittivity of the material under test can then be determined by modifying equation (5.31) to include the proportional area occupied by the material:

$$\varepsilon_r^* = \frac{\frac{d}{\varepsilon_0}[C_m - C_f] + A_m - A_p}{A_m} \tag{5.32}$$

where C_m is the measured capacitance and $A_m \pi R_m^2$ (m^2) is the area of the material. The relative permittivity is complex if and only if C_m is complex, indicating losses in the material.

5.7.2 *Microwave methods*

At microwave frequencies alternative methods are used in addition to variations on the RF approaches described. For example, a coaxial cell similar to that in Figure 5.30 may be used at industrial microwave frequencies, but the cell diameter and length are restricted to small dimensions, so the maximum pellet or particle diameter is also restricted. Further, at microwave frequencies we may make use of the wave propagation characteristics of materials to obtain sensitive measurements in low loss and low permittivity materials. The disadvantage is that often the type of applicator best suited to the measurement is dictated by the permittivity itself; thus the freedom of choice is restricted by electrical as well as structural properties.

Open coaxial lines

Perhaps the simplest apparatus for measurement of complex permittivity at microwave frequencies is the blunt coaxial transmission line, Figure 5.32, which has been carefully described by Stuchly and Stuchly (1980). This device is especially well suited for thick samples of very lossy materials of high permittivity because the governing assumption is that radiation into the medium is negligible and that the evanescent modes near the coaxial line dominate absorption. Consequently, the determination of permittivity is made in the fringing field at the end of the line. Extending the plane of the outer conductor along the surface (Figure 5.32b) controls the fringe fields and yields more reliable measurements. An equivalent circuit for the measurement is shown in Figure 5.32c. Here C_f is the fringe capacitance just inside the coaxial line at its end, C is the capacitance of the near field in the medium and G represents far field radiation. For lossy materials far field radiation can be neglected and C dominates the measurement. The medium determines the equivalent load impedance at the end of the transmission line, so the electrical length of the line is important, especially when measurements are made with swept frequency devices such as network analyzers.

The best approach is to establish the measurement reference plane at the end of the coaxial line using a short circuit, an open and a matched load and the relations of equation (5.22a)

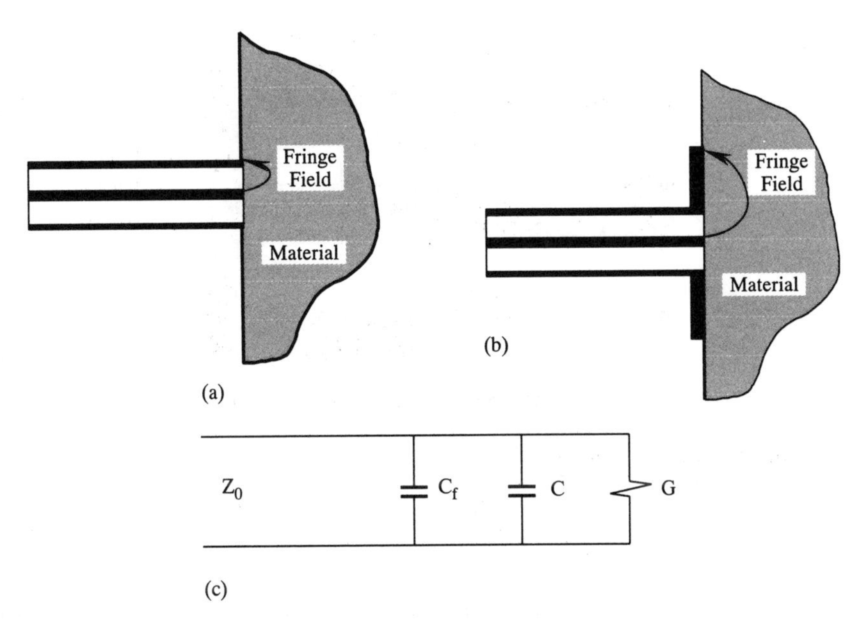

Fig. 5.32 *Open-ended coaxial line for permittivity measurement. (a) Blunt coaxial line device. (b) Extended ground plane controls fringe field. (c) Equivalent circuit.*

above. Often, however, it is not reasonable to do this because of the physical size of the probe or for other reasons. As an alternative approach, the reference plane may be placed at the probe coaxial connector. Then the line length may be estimated by placing a short circuit at the measurement plane if the connector impedance is not too much different from the transmission line impedance (a low insertion loss connector). A frequency sweep is then performed and the electrical length at each frequency, d where $0 \leqslant d \leqslant \lambda/2$, is determined from:

$$Z_m = Z_0 \left[\frac{Z_L \cos(\beta d) - jZ_0 \sin(\beta d)}{Z_0 \cos(\beta d) - jZ_L \sin(\beta d)}\right] = -jZ_0 \tan(\beta d) \tag{5.33a}$$

$$d = \frac{\lambda}{2\pi} \tan^{-1}\left(j\frac{Z_m}{Z_0}\right) \tag{5.33b}$$

where the load, Z_L, is a short circuit, Z_m is the measured impedance (imaginary), and Z_0 the characteristic impedance of the transmission line (real). All subsequent impedance measurements are corrected for line length. Note that we need not know the transmission line velocity, u, or wavelength, λ, in order to make the required calculations, but we must know Z_0. The measurement is repeated with the coaxial line immersed in a calibration liquid (such as water) and exposed to the air to determine the air load capacitance, C_0, where $\varepsilon_r^* = \varepsilon_0$, and the fringe capacitance, C_f (usually smaller than C_0). Then the probe is placed on the material surface and the measurement repeated to determine the load impedance in the medium, as represented by the complex reflection coefficient, $\Gamma = |\Gamma|\angle\phi$. The line load capacitance, C, is complex, $C = C_0\varepsilon_r^*$, and the real and imaginary parts of the relative

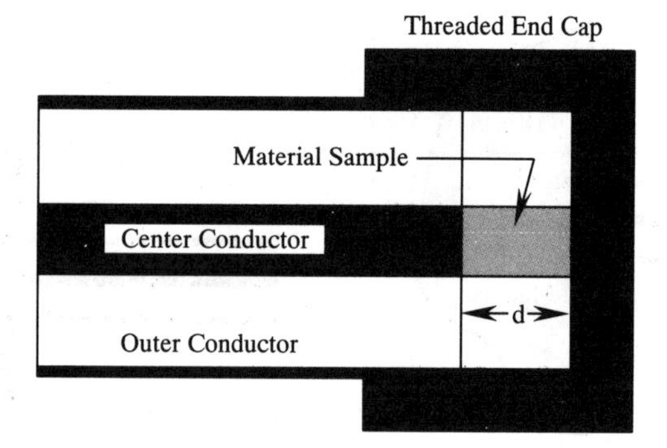

Fig. 5.33 *Variation on open coaxial line.*

permittivity are then obtained from:

$$\varepsilon_r' = \frac{2|\Gamma|\sin(-\phi)}{\omega Z_0 C_0(1 + 2|\Gamma|\cos(\phi) + |\Gamma|^2)} - \frac{C_f}{C_0} \tag{5.34a}$$

$$\varepsilon_r'' = \frac{1 - |\Gamma|^2}{\omega Z_0 C_0(1 + 2|\Gamma|\cos(\phi) + |\Gamma|^2)}. \tag{5.34b}$$

A variation on this configuration, shown in Figure 5.33, has been used to measure the permittivity of ceramics at high temperature (Iskander *et al.* 1981). The sample is placed in a threaded end cap which is screwed onto the end of the coaxial line. The end space may be heated to high temperatures for the measurement. The accuracy suffers somewhat due to small differences between the capacitance of the empty and sample-filled end cap for low permittivity materials. Nevertheless, in ceramics work one is most interested in the behavior of the permittivity around 500°C since many ceramic materials become very lossy at elevated temperatures. So, the values of permittivity may include significant uncertainty; but the temperature-dependent trends are quite well represented.

Waveguide measurement methods

Waveguide-based methods offer several advantages over open apparatus since the fields are confined to known geometries. Network analyzers can be readily applied to determine the impedance of samples within the guide and relations which describe the effect of the sample geometry used to infer the electric permittivity of the material.

Solid samples can be analyzed in the form of a thin sheet at a waveguide flange or a rectangular solid sample which is placed within the guide. In the most sensitive approach the impedance is measured by moving a short circuit piston behind the sample (see Figure 5.34). The theory of propagating fields within and behind the sample has been well described in the literature. The impedance measured at the reference plane, Z^*, is given by:

$$Z^* = \frac{Z_1}{Z_0}\frac{Z_0 \tanh(\gamma_0 d) + Z_1 \tanh(\gamma_1 d_1)}{Z_1 + Z_0 \tanh(\gamma_0 d)\tanh(\gamma_1 d_1)} \tag{5.35a}$$

where Z_1 is the wave impedance in the dielectric loaded waveguide section:

$$Z_1 = Z_0 \frac{\gamma_0 \mu_1}{\gamma_1 \mu_0} \tag{5.35b}$$

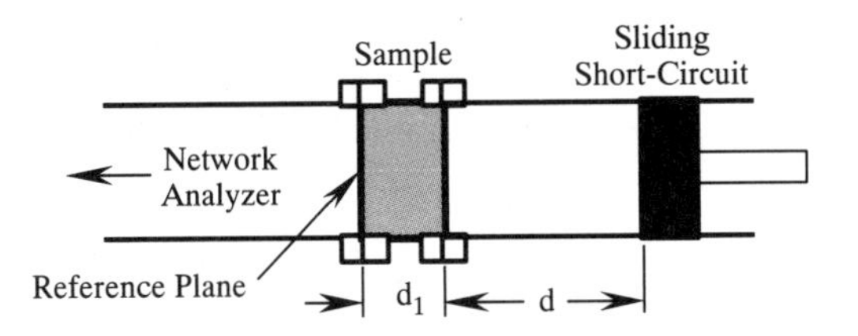

Fig. 5.34 *Finite solid sample of unknown permittivity in waveguide.*

and Z_0 is the wave impedance in the empty waveguide. The complex propagation coefficients γ_1 and γ_0 are for the TE_{10} mode in the two waveguide sections; and in this configuration we expect that $\gamma_0 = \beta_0$ for an essentially lossless guide. Note that the permittivity may not be so high that the waveguide section containing the sample supports modes other than the TE_{10} mode. Thus there is a limit on the materials which may be analyzed with this method. The experimental data are curve fitted by the method of least squares to obtain estimates of ε_r^* from Z^* as d varies.

This method is also well suited to the determination of the magnetic permeability (μ_1) of thin sheets (see Roussy *et al.* 1990) because the sheet is placed successively in a region where the electric field strength is high and the magnetic field strength low; and then in a region of low electric field and high magnetic field. The sliding short circuit method to find μ_1 works well for intermediate loss materials having $\varepsilon'' > 0.05$.

By recognizing that the admittance of any electromagnetic circuit which contains a dielectric is an analytic complex function of the permittivity of the dielectric we have derived general formulas which can be used for calibrating any dielectric measurement cell in the RF and microwave regions. (Roussy *et al.* 1990) The transformation function is:

$$\frac{Y^* - Y_0^*}{Y_0^*} = \frac{A(\varepsilon_r^* - 1)[1 + C(\varepsilon_r^* - 1)][1 + E(\varepsilon_r^* - 1)]\cdots}{[1 + B(\varepsilon_r^* - 1)][1 + D(\varepsilon_r^* - 1)]\cdots} \tag{5.36}$$

where Y^* and Y_0^* are respectively the measured admittances of the loaded and empty cell, and ε_r^* is the complex permittivity of the material. The cell coefficients A, B, C, D are real and associated with the zeros and the poles of the rational function. They can be determined in practice from a set of experimental measurements of the normalized admittance when the cell contains several materials of known permittivity. The degree of accuracy obtained depends entirely on the calibration substances, however. In general the coefficients are monotonic, with $B > C > D > E\ldots$

Cavity resonator for low loss materials

Low loss materials, such as plastics, require special attention. The imaginary part of the permittivity of low loss materials is difficult to estimate by placing the sample in a propagating line because the wave is not significantly attenuated. It is necessary to increase the apparent interaction between the waves and the material to induce measurable attenuation. This can be done by placing the material in a resonant cavity where the electric field is amplified by the quality factor, Q. When the sample size is small enough that conventional perturbation theory can be applied, then the resonant frequency and Q-factor shifts are linearly related to the permittivity of the material. That is, f_0 shifts to f' and Q_0 shifts to

Q' in the loaded cavity, and:

$$\frac{f' - f_0}{f_0} = k(\varepsilon_r' - 1) \tag{5.37}$$

$$\left(\frac{1}{Q'} - \frac{1}{Q_0}\right) = 2k\varepsilon_r''. \tag{5.38}$$

The real coefficient k which appears equations (5.37) and (5.38) depends upon the relative volume of the sample and also upon the intensity of the electric field where the sample is situated. It can be calculated for simple geometries. For a sheet of width d placed centrally in a rectangular cavity, resonating in TE_{101} mode:

$$k = \frac{2d}{\lambda_g}. \tag{5.39}$$

For a cylindrical rod sample, also centered in a TE_{101} rectangular cavity:

$$k = \frac{\pi d^2}{a\lambda_g}. \tag{5.40}$$

The same experimental design can also be used with larger volumes when the distribution of fields inside the cavity is not significantly changed by the sample. It is more accurate to determine k by calibration using reference materials. One must take into account the fringing field effects, especially when a hole has been drilled in the metallic wall of the cavity to admit a sample holder.

Measurements at high power density

A simple method for the dielectric measurements of solids is to place a small cylindrical sample in the center of a guide, parallel to the TE_{10} electric field. The impedance that this dielectric obstacle represents is then measured. Knowing the diameter of the sample, we can calculate the complex permittivity of the material. The waveguide experiment may be conducted with a small quantity of the material, and is of particular interest because it may be performed at high power. Secondly, when a material is to be transformed by microwaves, it is often necessary to measure its electric permittivity while the material is undergoing transformation. Thirdly, we may want to determine ε_r^* directly, as a function of the intensity of the field which causes the transformation. This method permits measurement of the dielectric behavior of the material as a function of the temperature, so that we may anticipate the behavior of the material in an industrial oven of higher capacity. Lastly, the measurement is useful to construct complete models of the electromagnetic effects on the material.

A typical experimental set-up is shown in Figure 5.35. The sample is cylindrical and placed parallel to the electric field at the center of the waveguide (at $x = a/2$) where the electric field is maximum. Of course, the generator is protected by circulators to prevent reflected power from damaging the magnetron, and forward and reflected power are measured with a directional coupler, neither of which are illustrated in the figure. The field is approximately homogeneous in the sample provided the diameter of the sample is

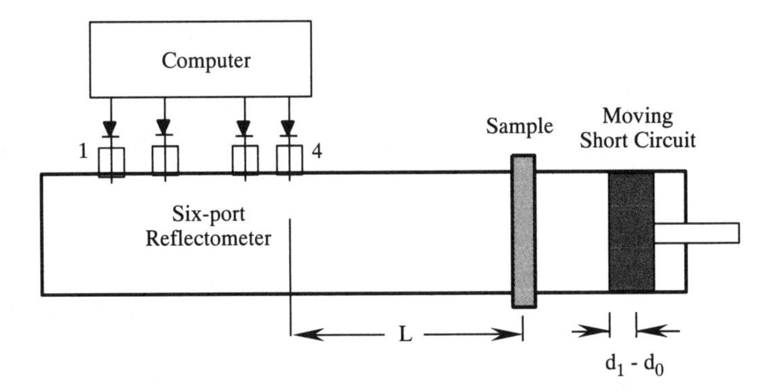

Fig. 5.35 *Small cylindrical sample in a waveguide.*

small enough. The sample constitutes a parallel impedance Z_s:

$$\frac{1}{Z_s} = Y_s = G_s + jB_s. \tag{5.41}$$

The sample temperature may be measured by a fiber optic probe or a thermocouple placed so that it does not interfere with the electric field, as was previously explained. In front of the sample is a four-probe sensor (constituting a six-port reflectometer), placed so that one of the probes (probe #4, for example) is at an electrical distance of $L = (2n + 1)\lambda_g/4$ from the axial plane of the sample. A minimum signal is maintained on that probe by moving the sliding short circuit. Under these conditions, the imaginary part of the admittance is provided by:

$$B_s = Y_0 \tan\left(\frac{2\pi(d_1 - d_0)}{\lambda_g}\right) \tag{5.42}$$

where d_0 is the position of the short circuit piston with the sample removed and d_1 is the position of the short circuit when the sample is in place.

The real part of the admittance may be deduced from the measurement of the signal P_{s2} (which should be linearized) from probe 2 placed at a maximum of the electric field, and of the signal P_{s4} from probe 4 (at the minimum):

$$\mathrm{G} = \sqrt{\frac{P_{s4}}{P_{s2}}}. \tag{5.43}$$

The fractional absorbed power is provided by:

$$P_a = \frac{4G}{G + 1}. \tag{5.44}$$

The relative electric permittivity is calculated from the relation:

$$\varepsilon_r^* = 1 + \frac{\lambda a}{\pi d^2 \lambda_g}(G + jB). \tag{5.45}$$

In the arrangement functioning in our laboratory at the Universite de Nancy, a 200 W generator supplies the microwave power. The incident power, P_i, is measured by a directional coupler of high directivity (36 dB) and a power meter. Between the generator and the reactor, a mechanical switch and two circulators in series have been placed to obtain isolation of at least 40 dB. A computer collects data in real time and calculates the expected electric values. Selectable constant incident power levels can be maintained by controlling the generator power with a software feedback algorithm. The control variables can be the electric field in the sample, the temperature of the sample, or the power absorbed by it. The algorithm can also force the selected variable to follow a specified profile (constant ramp or other specified time-dependent curve). The time between measurements is about 4 seconds under the usual operating conditions. To improve the accuracy of the measurements and stability of control, it is possible to average multiple data points.

5.8. SUMMARY

In this chapter we have collected together the diverse measurement methods which we have found most useful in microwave and RF system characterization. There are, of course, many other measurement approaches which have advantages under certain conditions. The major consideration in selecting a method for a particular measurement is that it should minimally perturb the electromagnetic field. This is where optically coupled devices make a significant contribution, though one may not always have the economic resources required. The measurement requirements in a laboratory setting are naturally more accurate than in a production setting. We find that laboratory scale experiment results can be readily applied to larger scale processes if one focuses on the proper variable. In evaluating electromagnetic heating applications, that boils down to estimating the local electric and magnetic field strengths and their positional distribution. As we expect from the discussion in Chapter 4, the field strengths will not be uniform, in general. Direct measurements of complex permittivity and permeability are important to process design, but often one only needs to know heating rates resulting from applied fields, and the loss factor alone may be sufficient. The search for new approaches to old instrumentation questions is never ending, however.

REFERENCES

Further information can be found in the following texts.

Ginzton, E. L. (1957) *Microwave Measurements* McGraw-Hill, New York.

Harvey, A. F. (1963) *Microwave Engineering* Academic Press, New York, Chap. 7.

Montgomery, C. G. (1948) *Techniques of Microwave Measurements* (*Radiation Laboratory Series M.T.T.*) McGraw-Hill, New York.

Okress, E. C. (1968) *Microwave Power Engineering* Vol 1 and 2, Academic Press, New York.

Sucher, M. and Fox, J. (1963) *Handbook of Microwave Measurements* Vol. I, II and III, Polytechnic Press, Wiley Interscience, New York.

and in the following papers:

Andrade, O. M., Iskander, M. F., Bringhurst, S. (1992) High temperature broadband dielectric properties measurement techniques *Microwave Processing of Materials III* (Beatty, Sutton and Iskander, Eds.), Proc. Materials Res. Soc., **269**, 527–539.

Iskander, M. F., Massoudi, H., Durney, C. H. and Allen, S. J. (1981) Measurements of the RF power absorption in spheroidal human and animal phantoms exposed to the near field of a dipole source *IEEE Trans. Biomed. Engr.*, **BME-28**, 258–264.

Roussy, G., Agbossou, K., Dichtel, B. and Thiebaut, J. M. (1992a) High electromagnetic field measurements in industrial applicators by using an optically modulator sensor. *J. Microwave Power Elec. Energy*, **27**, 164–170.

Roussy, G., Agbossou, K. and Thiebaut, J. M. (1992b) Improved modeling of permittivity measurement cells *IEEE Trans. Instrum. Meas.*, **IM-41** 366–370.

Roussy, G., Ghanem, H., Thiebaut, J. M., Dichtel, B. (1990) Six-port waveguide used for simultaneously measuring permittivity and permeability of solid materials in microwave region *J. Microwave Power Elec. Energy*, **25**, 67–75.

Stuchly, M. A. and Stuchly, S. S. (1980) Coaxial line reflection methods for measuring dielectric properties of biological substances at radio and microwave frequencies — a review, *IEEE Trans. on Instr. and Meas.*, **IM-29** (3), 176–183.

Thiebaut, J. M., and Roussy, G. (1991) Extension of the six-port circuit theory for using a practical directional Coupler in measuring RF impedances and RF power, *Meas. Sci. Technol.*, **2** 836.

PART 2

Material Aspects

The second part of this book, Chapters 6, 7 and 8, will consider the topic of dielectric materials in considerable detail. This is a dominant topic in the overall consideration of high power applications of electromagnetic fields since, ultimately, the dielectric properties determine the effects of electric fields on a material. Also, many of the physical effects of electric fields on matter are not included in standard electrical engineering treatments of dielectric properties, so some additional discussion is warranted. The goals of the electrical engineer, who is interested primarily in the integrity and effectiveness of insulating materials, differ significantly from the goals of the chemist or physicist, who is interested in molecular and atomic structure, and these two groups seem to have little communication with each other. To be sure, insulating materials are dielectrics; but the classification is, in fact, very much broader than that. And, in our study of high power applications at varying frequency, it is important to relate (at least conceptually) the molecular arrangement of a material to its response to an applied external electric field. We will begin the second part of the book with a more detailed look at the behavior of dielectric materials under static fields in Chapter 6, including semiconductors, since we may describe the response of many materials to radio frequency fields using static and quasi-static concepts. We will then progress to consideration of the dynamic response in Chapter 7, and finally we will discuss a universal law of dielectrics which covers all known materials over at least eight decades of frequency derived by Johnscher. We will also approach the topic of discretely inhomogeneous materials in Chapter 8.

6 Introduction to the Macroscopic Theory of Dielectrics

In this chapter we examine some fundamental concepts used to describe the behavior of matter under the influence of an electric field, which can be of low or high intensity, static or slowly time-varying (quasi-static). The discussion in this chapter is confined to conventional solids and liquids so we will avoid consideration of discharge phenomena (ionization and arc formation), and spontaneously electrified materials with specific properties. The macroscopic theory of dielectrics, which is the topic of this chapter, supposes that the material is, in the main, continuous; even though we fully realize that matter is composed of atoms which are organized into molecules and molecules assembled with some order on a larger scale than atoms to form matter. The distinction between continuous and discontinuous descriptions is currently under discussion in many areas of physics, and is of considerable interest here as well because the scale of electric phenomena is determined by the wavelength.

The most serious criticism of the macroscopic theory of matter — which dates from the 1920s — is that it limits one to classical mechanical descriptions of the individual electrical species and associated phenomena. A rigorous approach would be to apply quantum mechanics and statistical thermodynamics to obtain these descriptions. This, however, would be an enormous task (and intractable, in the wide sense), so we will confine the discussion to a classical framework in this chapter. The reader is advised that while this approach cannot explain all dielectric phenomena, it does explain the vast majority of experimental observations. This is because the correspondence principle of quantum mechanics usually applies — that is, in the limit the quantum results converge to the classical results when either the dimension of the sample or the time scale of the observations is large.

6.1 THE ORIGIN OF THE ELECTRIC POLARIZATION OF MATTER

The distinguishing feature of dielectric materials is that they are polarizable. That is, indigenous charge in the material can move in response to an external electric field. Many dielectrics are also semiconducting and have motion of both bound charge (polarization, or displacement, current density) and free charge (conduction current density). An ideal conductor — for which metals such as gold, silver and copper are good examples — has little, if any, polarization effect since an external electric field generates current density (or surface charge in the static case) in the virtual absence of an internal electric field owing to the high electrical conductivity. So these materials may usually be treated as ideal conductors with the electric permittivity of free space, rather than as dielectrics. Ideal dielectrics consist of bound charge alone, no free charge, and in addition have mobilities of zero for any free charge applied to their surface—and thus are insulators. Semiconducting materials are dielectrics between these two extremes (including, for example, ionic solutions) and exhibit both polarization and conduction current density in response to an applied electric field.

6.1.1 The polarization vector field

We have already seen that the capacitance of a free-space capacitor increases when the space in the capacitor is replaced by a dielectric material. For an arbitrary quasi-static electric field distribution the capacitance changes from C_0 to $C' > C_0$:

$$C' = \varepsilon_r C_0 = \varepsilon_r \left\{ \frac{\oint_\Sigma \varepsilon_0 \mathbf{E} \cdot d\mathbf{S}}{-\int_b^a \mathbf{E} \cdot d\mathbf{L}} \right\} \tag{6.1}$$

where ε_r is the relative permittivity of the material (unitless), ε_0 is the electric permittivity of free space (8.854×10^{-12} F/m and $\varepsilon = \varepsilon_r\varepsilon_0$), Σ completely encloses the charge source of $\mathbf{E}$, the direction of $d\mathbf{S}$ is parallel to $d\mathbf{L}$, and the potential integration is from the reference point b to point a, as we saw in Chapter 2. We have been careful to define the concept of capacitance in the quasi-static framework because if the quasi-static assumption does not apply lumped parameter descriptors such as capacitance, resistance and inductance make no sense and must be replaced by consideration of wave propagation (see Chapter 3).

From the macroscopic point of view, the increase in capacitance is usually explained by introducing the concept of a polarization vector field in the material, $\mathbf{P}$ (units of C-m/m^3 = C/m^2) which forms in response to an externally applied electric field, $\mathbf{E}$:

$$\mathbf{P} = \lim_{\Delta v \to 0} \left\{ \frac{\sum_{\Delta v} \mathbf{p}_i}{\Delta v} \right\} = \varepsilon_0 \chi \mathbf{E} \tag{6.2}$$

where $\mathbf{p}_i$ represents individual (local) dipole moments (C-m), Δv is the volume under consideration and χ is the susceptibility (unitless). Polarization in this form is the net electric dipole moment—that is, the dipole moment per unit volume or the vector sum of all of the individual dipoles in the material—which forms in response to the external electric field. We will see in future discussion that the dipole contribution to polarization represents an important portion of the total; but that polarization actually also includes higher moments, such as quadrupole and octupole moments, as well. The additional electric moment represented by $\mathbf{P}$ is added to its source electric field to give the total electric flux density, $\mathbf{D}$:

$$\mathbf{D} = \varepsilon_0 \mathbf{E} + \mathbf{P} = (1 + \chi)\varepsilon_0 \mathbf{E}. \tag{6.3}$$

The electric flux density, $\mathbf{D}$, is affected by local polarization of the material medium because its spatial derivative (actually, its divergence) depends on local charge independently of the

Table 6.1 *Relationship of electric flux density,* **D**, *and electric field,* **E**.

Charge Distribution	Electric Flux Density **D**	Electric Field **E**
Point charge $\rho_V\, dv = q$	$\frac{q}{4\pi r^2}\mathbf{a}_r$	$\frac{q}{4\pi\varepsilon r^2}\mathbf{a}_r$
Line charge $\rho_V\, dv = \rho_L\, dL$	$\frac{\rho_L}{2\pi r}\mathbf{a}_r$	$\frac{\rho_L}{2\pi\varepsilon r}\mathbf{a}_r$
Surface charge $\rho_V\, dv = \rho_S\, dS$	$\frac{\rho_S}{2}\mathbf{a}_n$	$\frac{\rho_S}{2\varepsilon}\mathbf{a}_n$

medium property — by Gauss' electric law. The polarization vector field represents the existence of bound charge in the material. Bound charge is "locked" into the matrix of the dielectric and is capable only of restricted motion under an external electric field. We will use ρ_b (C/m^3) to represent bound charge, while ρ_v represents free charge (charge which can move without restriction in the material), as we have used previously. From equation (6.3) and the constitutive relation it is easy to see that, in an ideal dielectric, the permittivity, ε, is simply related to the susceptibility, χ:

$$\varepsilon = (1 + \chi)\varepsilon_0. \tag{6.4}$$

The susceptibility, χ, represents the generation of the equivalent of net charge by the orientation of the electric dipole moments of bound charge in the medium. A non-zero polarization vector field represents energy absorbed from the external field (i.e. work done in orienting or creating local dipole moments), which is why all of the energy (or power) absorption terms in a dielectric rely on the product of **E** and **D**, as was described in Chapter 1.

The difference between **D** and **E** is illustrated by the representative expressions in Table 6.1. Note that the electric flux density, **D**, depends only on the charge distribution, while the force field, **E**, depends on the medium electric permittivity as well. Consequently, for a given charge distribution — say, for a spherically symmetric charge of q (C) — the electric flux density at every point is independent of the material, while in a dielectric ($\varepsilon > \varepsilon_0$) the force on a test charge would be reduced over that of the same charge geometry in free space: the polarization contribution acts to *reduce* the electric field in the material.

6.1.2 Microscopic view of polarization

Consider a simple hydrogen molecule (Figure 6.1a). When an electric field is applied to the molecule, the electron density around the two nuclei is perturbed due to the electric force exerted by the external field. The centroid of the electron distribution (charge $= -2e$) will be different than that of the pair of nuclei ($+2e$), creating a net dipole moment. The dipole moment associated with this difference in the centroids of positive and negative electric charges will be proportional to the applied electric field. As a practical matter the hydrogen molecule is difficult to depict since it is nearly always a gas and the large number of possible

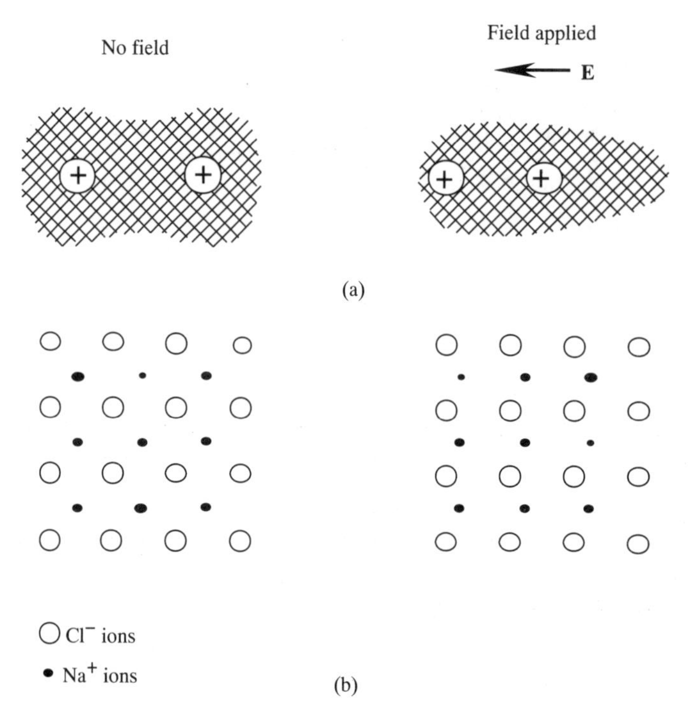

Fig. 6.1 *(a) Electronic polarization. (b) Atomic polarization (NaCl crystal lattice).*

states makes an accurate sketch intractable. It is easier to view the atomic polarization effect in a crystal lattice, such as NaCl shown in Figure 6.1b. This representation provides a simple model of electronic polarization:

$$\mathbf{p}_e = \varepsilon_0 \alpha \mathbf{E} \tag{6.5}$$

where $\mathbf{p}_e$ is the net electric dipole moment (in C-m) of the charge distribution and the constant of proportionality, α, is the polarizability (in m^3 at the microscopic level), a function of the charge geometry induced by the external electric field, **E**. We have used SI (i.e. MKS) units here and the 4π term common in CGS unit descriptions (as in most classical texts on electrodynamics, such as Jackson (1975)) does not appear. The dipole moment representation is somewhat simplistic, but we will see next that a mathematical approach provides a more suitable image of the dipole moment in terms of the unit volume polarization. The polarization term combines vectorially with the electric field which is its source — remember that **p** is defined as directed from the negative charge to the positive charge in the dipole; while **E** originates on positive charges and terminates on negative charges, so **p** tends to align with **E**; and when **p** is non-zero (for a fixed charge geometry) the net polarization contributes a field in opposition to **E**.

Electric polarization at the microscopic level

The origin of polarization and of the higher order moment contributions can be seen in a detailed look at a typical molecular structure. We may consider a small sample of material as an ensemble of point-charges — nuclei and electrons constituting atoms and molecules — sparsely distributed and interacting in a vacuum. For a given configuration

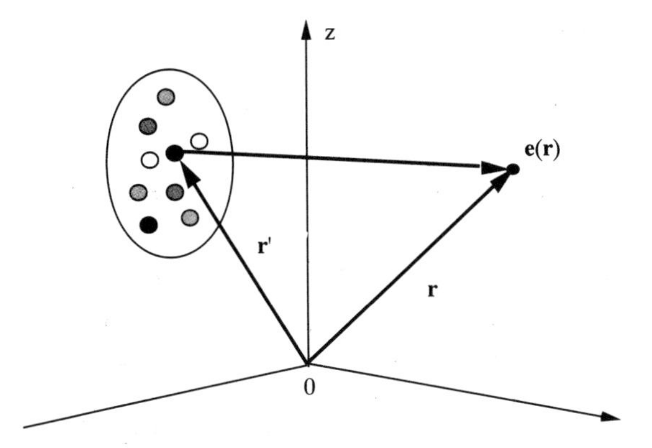

Fig. 6.2 *Calculation of incremental electric field,* **e**, *at location* **r** *due to microscopic point charges located at* **r**′.

over a lifetime on the order of 10^{-20} s, all of these charges may be considered localized as in a static distribution, see Figure 6.2. The local (microscopic) electric field, **e**, due only to this charge distribution is the vector sum of the point charge fields, and is conservative:

$$\nabla \times \mathbf{e} = 0; \qquad \text{and} \qquad \nabla \cdot \mathbf{e} = \frac{\rho_v}{\varepsilon_0}$$

so:

$$\mathbf{e}(\mathbf{r}) = \int_{\text{vol}} \frac{\rho_v(\mathbf{r}')(\mathbf{r} - \mathbf{r}')}{4\pi\varepsilon_0 \left(\sqrt{(\mathbf{r} - \mathbf{r}')^2}\right)^3} \, \mathrm{d}v' \tag{6.6}$$

where the lower case **e** represents an instantaneous electric field vector calculated over times short compared to atomic relaxation times in free-space at position **r**, and $\rho_v(\mathbf{r}')$ is actually a highly localized volume density of charge, $\rho_{\text{local}}(\mathbf{r}')$. The charges are considered discrete and located at positions $\mathbf{R}_{ij}$:

$$\rho_v(\mathbf{r}')\,\mathrm{d}v' = \rho_{\text{local}}(\mathbf{r}')\,\mathrm{d}v' = \sum_i^M \sum_i^N q_{ij}\delta(\mathbf{R}_{ij} - \mathbf{r}') \tag{6.7}$$

where there are N charges in each group (nuclei, electrons), M groups (atoms, ions, molecules) of discrete charge, q_{ij} (C), and δ is the Dirac delta function.

The electric field described by macroscopic electromagnetic theory, **E**, is the time-average value of the instantaneous random electric field, **e**, calculated on a macroscopic scale:

$$\mathbf{E}(\mathbf{r}) = \int_{-\infty}^{+\infty} \mathbf{e}(\mathbf{r}) f_{\mathbf{e}}(\mathbf{r}, t)\,\mathrm{d}t = \langle \mathbf{e} \rangle \tag{6.8}$$

where $f_{\mathbf{e}}$ is the probability density function of the random electric field, and the limits on the integration mean simply that the integration time is long compared to atomic motion times. In general, the time-average and space-differentiation operations may be commuted so that, from equation (6.6), $\mathbf{E}(\mathbf{r})$ satisfies:

$$\nabla \times \mathbf{E}(\mathbf{r}) = 0 \qquad \text{and} \qquad \nabla \cdot \mathbf{E}(\mathbf{r}) = \frac{\langle \rho_{\text{local}}(\mathbf{r}) \rangle}{\varepsilon_0} \tag{6.9}$$

which are, of course, Maxwell's equations for a static electric field.

Relation between polarizability and macroscopic polarization

We may recalculate the local charge density, ρ_{local}, by supposing that over a practical time scale (long compared to 10^{-20} s) the mean value may be used. If so, then some particles will occupy their mean positions and we can relabel the charge using a discrete representation. The position of the charge cluster i is $\mathbf{R}_i$ — there are j charges in molecule i — and the internal coordinates of other charges in the ith subset of charges are given by:

$$\mathbf{R}_{ij} = \mathbf{R}_i + \mathbf{r}_{ij} \tag{6.10}$$

where the distances $\mathbf{r}_{ij}$ are small compared to $\mathbf{R}_{ij}$, as shown in Figure 6.3. The localized volume charge distribution can then be recast using a Taylor series:

$$f(x + \Delta x) = f(x) + \Delta x \frac{\partial f}{\partial x} + \frac{\Delta x^2}{2!} \frac{\partial^2 f}{\partial x^2} + \cdots \tag{6.11}$$

where the partial derivatives are evaluated at x. In this case position vectors are involved, so the derivatives are replaced by the respective gradient operators:

$$\rho_{\text{local}}(\mathbf{r})\, \mathrm{d}v = \sum_i \sum_j q_{ij} \delta(\mathbf{R}_{ij} - \mathbf{r})$$

$$= \sum_i \sum_j q_{ij} \left\{ \delta(\mathbf{R}_i - \mathbf{r}) + \mathbf{r}_{ij} \nabla \delta(\mathbf{R}_i - \mathbf{r}) + \tfrac{1}{2} (\mathbf{r}_{ij} \nabla^2)^2 \delta(\mathbf{R}_i - \mathbf{r}) + \cdots \right\}. \tag{6.12}$$

This result may be rewritten in terms of the monopole moment (total charge) in group i, the dipole moment of the group, and the quadrupole tensor moment of the group:

$$\rho_{\text{local}}(\mathbf{r})\, \mathrm{d}v = \sum_i \left\{ q_i - \mathbf{p}_i \cdot \nabla + \mathbf{Q}_i \nabla^2 - \cdots \right\} \delta(\mathbf{R}_i - \mathbf{r}) \tag{6.13}$$

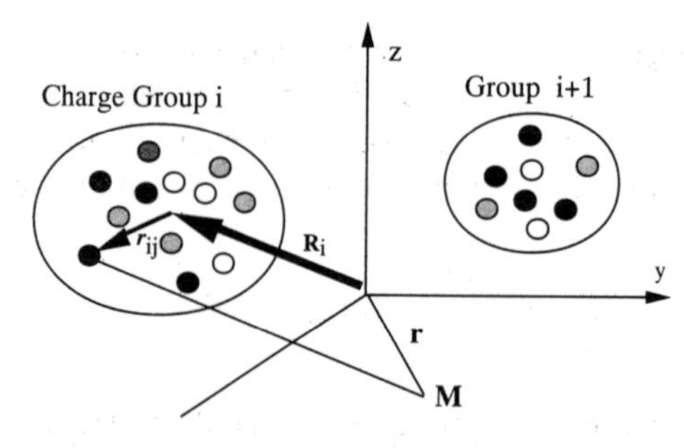

Fig. 6.3 *Net moment, **M** calculated at location **r** due to the j charges which comprise molecule i.*

where the respective moments are given by:

charge: $q_i = \sum_j q_{ij}$

dipole moment: $\mathbf{p}_i = \sum_j q_{ij} r_{ij}$

quadrupole moment: $(\mathbf{Q}_i)_{xy} = \frac{1}{2} \sum_j q_{ij} (3\mathbf{r}_{ij}\mathbf{r}_{ij} - I r_{ij}^2)$ (6.14)

and where I represents the unit dyadic (i.e. $\mathbf{a}_x\mathbf{a}_x + \mathbf{a}_y\mathbf{a}_y + \mathbf{a}_z\mathbf{a}_z$).

If we neglect higher terms (third order and above) in the series, the mean local charge density is the total electric charge (free charge plus bound charge) concentrated at the point **r** minus the divergence of the polarization vector, **P**, which is the net electric moment of order 1 at the point:

$$\langle \rho_{\text{local}}(\mathbf{r}) \rangle = \rho_{\text{v}}(\mathbf{r}) - \nabla \cdot \mathbf{P} \tag{6.15}$$

where $\rho_{\text{v}}(\mathbf{r})$ is the time-average free charge at the point **r**:

$$\rho_{\text{v}}(\mathbf{r}) = \left\langle \sum_i q_i \delta(\mathbf{R}_i - \mathbf{r}) \right\rangle = \lim_{\Delta v \to 0} \left\{ \frac{\Delta q}{\Delta v} \right\}_{\mathbf{r}} \tag{6.16}$$

and the negative of the divergence of what we have called the polarization vector field (6.15) is, in fact, the bound charge, $\rho_{\text{b}}(\mathbf{r})$.

In the Coulomb force theorem, the divergence of the polarization vector field (i.e. ρ_{b}) will be added to the divergence of the electric field (including ε_0). Remember that the polarization vector field actually represents the mean electric moment (of all kinds) per unit volume of the matter (C-m/m^3) and thus is added directly to the electric field-ε_0 product. The bound charge has many terms in addition to the dipole term when higher order moments (quadrupole, octupole, etc.) are included:

$$\rho_{\text{b}}(\mathbf{r}) = \left\langle \sum_i \mathbf{p}_i \delta(\mathbf{R}_i - \mathbf{r}) \right\rangle - \nabla \left\{ \langle \sum_i \mathbf{Q}_i \delta(\mathbf{R}_i - \mathbf{r}) \rangle \right\} + \cdots \tag{6.17}$$

Note that the dipole moment does not depend upon the origin chosen for calculation. This property does not hold for the other terms, especially for the quadrupole moment. When the material is placed in an external electric field, this term is not zero; it is added to (i.e. induced by) the electric field. The electrostatic Coulomb theorem takes the form:

$$\nabla \cdot (\varepsilon_0 \mathbf{E}) = \rho_{\text{v}} - \nabla \cdot \mathbf{P}. \tag{6.18}$$

In conclusion, the polarization appears as a complement to be added to the electric field to account for the presence of oscillating charge. It is an intrinsic property of the material which is modified by an externally applied electric field. This leads to the standard definition of the electric flux density, **D**, which we have used without detailed explanation up to this point. The electric flux density is also called the induction field or electric displacement:

$$\mathbf{D} = \varepsilon_0 \mathbf{E} + \mathbf{P} = (1 + \chi)\varepsilon_0 \mathbf{E} = \varepsilon_{\text{r}} \varepsilon_0 \mathbf{E} \tag{6.19}$$

with the electrostatic properties:

$$\nabla \cdot \mathbf{D} = \rho_v \qquad \text{and} \qquad \nabla \times \mathbf{D} = 0. \tag{6.20}$$

6.1.3 Discussion

In dielectric materials there are many sources for the polarization vector field. One is electronic polarization, $\mathbf{P}_e$, which concerns the displacement of the electrons in relation to their associated nuclei (as mentioned above). The displacement of nuclei relative to other nuclei in a molecule results in atomic polarization, $\mathbf{P}_a$. These two types of polarization are similar in that they are nearly independent of the material temperature and frequency of the field — they are both instantaneous responses, and are always in phase with the external field. These two types of polarization combine to generate what is collectively termed induced polarization, $\mathbf{P}_\alpha$:

$$\mathbf{P}_\alpha = \mathbf{P}_e + \mathbf{P}_a. \tag{6.21}$$

Induced polarization occurs in molecules which do not have a permanent electric moment.

Polar molecules, molecules with a permanent electric moment, will rotate under the influence of an electric field so that their dipole moments tend to align with the field (recall the sign convention on $\mathbf{p}$). They thus create an orientation polarization, $\mathbf{P}_u$, which is both temperature and frequency dependent. The temperature and frequency dependence arises primarily from agitation of molecules and exchange of energy between dipoles. The total polarization is then the vector sum of contributions:

$$\mathbf{P} = \mathbf{P}_\alpha + \mathbf{P}_u. \tag{6.22}$$

Another kind of polarization is due to chemical bond shifting in molecules — which actually may be included as a form of electronic polarization, but is often treated separately. For example, we think of the CO_2 molecule as a simple balanced double bond, O=C=O, which is accurate in a time-average sense. In fact, the molecule oscillates (at frequencies in the neighborhood of 10^{17}–10^{20} Hz) between two states: O≡C–O→O=C=O→O–C≡O (Glasstone 1942). The net result is that the molecule creates a transient dipole; and there are many interesting phenomena associated with transient dipole-induced dipole interactions: CO_2 forms a molecular solid (lattice sites occupied by molecules rather than atoms) which is held together by relatively weak van der Waals forces.

For a field of high intensity, 1000 kV/m and above, matter can be modified by the field which leads to the observation of other phenomena. For example, non-polar molecules may generate induced dipoles or the chemical equilibrium between polar molecules can shift to favor more highly polar species. These cases are examples of dielectric saturation, avalanche multiplication or breakdown and will not be treated in this text.

6.2 DIFFERENT TYPES OF DIELECTRIC MATERIALS

6.2.1 Non-electret materials

We begin by discussing materials which do not have a permanent net polarization. Excluding electret and ferroelectric materials is not as restrictive as it may at first appear, since even the polar materials described above will not have a *net* dipole moment until an external field has been applied. In fact, the vast majority of polar materials falls into this category. For non-electret materials the relation between the vector polarization and the electric field has been described macroscopically using many different formulations.

Isotropic and anisotropic descriptions

As may quickly be seen from equation (6.19), the polarization vector in a simple material is linearly related to the susceptibility, χ. For χ to be a scalar constant the polarization must be in a linear isotropic dielectric. For a linear anisotropic dielectric χ is a second order tensor:

$$\mathbf{P} = \begin{bmatrix} P_x \\ P_y \\ P_z \end{bmatrix} = \varepsilon_0 \begin{bmatrix} \chi_{xx} & \chi_{xy} & \chi_{xz} \\ \chi_{yx} & \chi_{yy} & \chi_{yz} \\ \chi_{zx} & \chi_{zy} & \chi_{zz} \end{bmatrix} \begin{bmatrix} E_x \\ E_y \\ E_z \end{bmatrix} \tag{6.23}$$

where a Cartesian coordinate description has been used for convenience. Unlike the electrical conductivity, it is not clear, *a priori*, that any of the off-diagonal terms will be zero for the susceptibility tensor. However, since we are not describing electrets, in order for off-diagonal terms to be non-zero the *net* polarization (in the macroscopic sense) must not be in line with the external field which induces it—a physically unrealistic situation. Therefore, we may safely treat the off-diagonal terms as negligible for non-electret materials.

Nonlinear dielectrics

Some materials exhibit hyperpolarizability phenomena. In those substances we introduce higher order terms into the constitutive relation (6.19):

$$\mathbf{P} = \varepsilon_0(\chi + \xi|\mathbf{E}|^2 + \cdots)\mathbf{E} \tag{6.24}$$

where the constant $\xi < 0$. We would ordinarily expect to see a term sensitive to the first power of the magnitude of $\mathbf{E}$ in (6.24). However, it may be shown that the first order term depends on $\cos(\theta)$, the second on $\cos^2(\theta)$, etc.; and $\langle\cos(\theta)\rangle = 0$, so this term does not survive the expectation operation while the higher order terms do. We will show some examples of nonlinear materials in later chapters.

Materials with memory

The electric polarization may consist of residual polarization due to a previously applied electric field. Such a material exhibits "memory" and the polarization at any time, t, depends on the complete electric field history:

$$\mathbf{P}(t) = \int_{-\infty}^{t} \varepsilon_0 f(t-\tau)\mathbf{E}(\tau)\,\mathrm{d}\tau \tag{6.25}$$

where f is a scalar function which describes the memory in an isotropic material. We expect that, due to randomizing influences, when an electric field is applied and then removed the material will eventually relax back to a state where no net polarization is observed, in accordance with the second law of thermodynamics. This is termed dielectric relaxation. In the Debye model (and in experimental data for most liquids) the relaxation function, f, is exponential. Solids which have been studied have a relaxation function which follows two power laws, t^{-n} at very small t, and $t^{-(m+1)}$ for large t (Johnscher 1983). So, for solid materials it is important to use a log–log plot which will show this relationship rather than a semi-log plot which obscures it. We will discuss these phenomena in considerable detail in future chapters.

6.2.2 Electrets and mechanical polarization

Electrets, and/or piezoelectric compounds, have a polarization independent of an external electric field. Electret materials have bound charge distributions which give rise to a permanent static electric field in the absence of free charge. As a practical matter, electrets are created by transforming the material under an electric field so that it has a very long memory. For example, plexiglas or a mixture of carnuba wax and beeswax may be made into an electret by subjecting them to high temperature in a strong electric field. When cooled the material has an extremely long memory, and for all practical purposes may be considered permanently polarized. Electrets have a permanent static electric field, analogous to the magnetic field of a permanent magnet. Ferroelectric compounds also have a very large polarization even when the electric field is quite small. These materials are highly a anisotropic so the full tensor description is required to determine the polarization:

$$\mathbf{P}(t) = \varepsilon_0 \int_{-\infty}^{t} \mathbf{f}(t-\tau)\mathbf{E}(\tau)\,\mathrm{d}\tau \tag{6.26}$$

where $\mathbf{f}(t)$ is now a second order tensor and none of the off-diagonal terms may be considered zero due to residual polarization.

Polarization due to applied mechanical forces, or piezoelectricity, is described by:

$$\mathbf{P} = \mathbf{P}(\mathbf{F}) \tag{6.27}$$

where $\mathbf{F}$ is the applied force. In piezoelectric materials deformation of the crystal structure (strain) in response to the applied force produces unbalanced net polarization. In practical piezoelectrics the polarization relaxes to zero if the polarization charge is drained by current flow. When connected to an external circuit for measurement the voltage induced by the strain decreases with a time constant proportional to the circuit current. For this reason extremely high input impedance amplifiers (charge amplifiers) are used to measure slowly varying piezoelectric strain in motion transducers and electret microphones. In most of the important piezoelectric substances the relationship is reciprocal so that an applied electric field can be used to create deformation of the piezoelectric crystal. Ultrasound transducers (both transmitter and receiver) are usually made from piezoelectrics.

6.3 IDEAL ISOTROPIC DIELECTRICS

We will return, now, to the fundamental equations of a perfect isotropic dielectric, for which:

$$\mathbf{P} = \varepsilon_0 \chi \mathbf{E} \tag{6.28}$$

and χ is a scalar constant. The relationship between the susceptibility (unitless), the electric permittivity, ε (F/m) and the electric flux density, $\mathbf{D}$ (C/m^2) is summarized by:

$$\mathbf{D} = \varepsilon_0(1 + \chi)\mathbf{E} = \varepsilon_r\varepsilon_0\mathbf{E} = \varepsilon\mathbf{E} \tag{6.29}$$

where ε_r is the relative permittivity of the material (unitless), and ε_0 is the electric permittivity of free space (F/m).

6.3.1 Estimating the permittivity

Static dielectric permittivity is generally a thermodynamic state of the substance: it is a function of the internal pressure and of the temperature. For gases, ε_r is very close to 1.0, but for liquids, especially polar substances like water, the permittivity can be much higher than ε_0. The definition of the permittivity which we have derived through the use of Maxwell's equations is, in some materials, only a theoretical construct. This is because the electric field which is under consideration in the above relations is that *inside* the material; this field often cannot be directly measured. So, the determination of the permittivity remains, in many cases, extremely difficult, especially at high frequency. Nevertheless, experimental estimation of the internal electric field can be accomplished by the placement of a small sample of the material in a capacitor of suitable geometry. Of course, whenever concepts of lumped impedance, such as capacitance, are used the quasi-static assumption must apply. The value of the capacitance will vary with the shape of the sample according to equation (6.1). By consideration of the variation of capacitance with the shape, the permittivity can be estimated. This is a well studied problem in electromagnetics, and we present some useful and important results. At high frequency we may look at interface reflection phenomena to infer dielectric properties, but scattering processes in inhomogeneous materials can completely frustrate this approach. Resonant cavity perturbation methods are most often used to estimate high frequency properties, and these will be treated in later chapters as well as Chapter 5.

Results for a spherical sample in a uniform external field

When a linear isotropic homogeneous spherical sample is placed in a uniform external electric field, $\mathbf{E}_0 = \mathrm{E}_0\mathbf{a}_z$ (far away from the sphere), the electric field within the sample is uniform with the same orientation which would be observed if the sample were not there (see Figure 6.4); this was discussed in the macroscopic solution of a dielectric sphere in a uniform field in Chapter 4. In order to match the necessary boundary conditions the electric field in the surrounding medium (air for this example) is modified. Since there is no free charge in an ideal dielectric, there is no surface charge on the sphere, and at all points on the surface:

$$E_{t1} = E_{t2} \qquad \text{and} \qquad \varepsilon_1 E_{n1} = \varepsilon_2 E_{n2}. \tag{6.30}$$

We showed in Chapter 4 that the formal solution for the potential inside the sphere (from the Laplace equation) is given by:

$$V_1 = \frac{-3\varepsilon_0}{2\varepsilon_0 + \varepsilon_1}\mathrm{E}_0 r\cos(\theta) \tag{6.31}$$

if we assume free-space properties around the sphere. From this solution the internal electric field is:

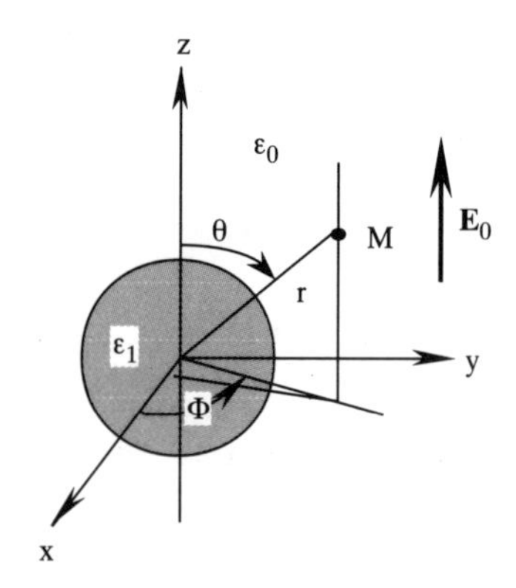

Fig. 6.4 *Sphere in a uniform external field.*

$$\mathbf{E}_1 = -\nabla V_1 = \frac{3\varepsilon_0 E_0}{2\varepsilon_0 + \varepsilon_1}\left[\cos(\theta)\mathbf{a}_r - \sin(\theta)\mathbf{a}_\theta\right]$$

$$\mathbf{E}_1 = \frac{3\varepsilon_0 E_0}{2\varepsilon_0 + \varepsilon_1}\mathbf{a}_z = \frac{3\varepsilon_0}{2\varepsilon_0 + \varepsilon_1}\mathbf{E}_0. \tag{6.32}$$

So, for well behaved materials the sample electric field can easily be written in terms of the electric permittivity.

Flat disc normal to the **E***-field*

If the shape of the sample is in the form of a flat disc perpendicular to the same electric field (Figure 6.5), the electric field inside of the material is dominated by the normal field boundary conditions near the center:

$$\mathbf{E}_1(0, z) = \frac{1}{\varepsilon_{r1}}\mathbf{E}_0 \tag{6.33}$$

where ε_{r1} is the relative permittivity (unitless), and the external field is bent to satisfy both conditions at the edge. Here we must be careful about the corners of the disc, as a non-uniform electric field will exist there, even in the static case, as we discussed in Chapter 4.

Long cylinder parallel to **E**

If an infinitely long cylinder is placed with its axis parallel to the applied electric field (Figure 6.6), the internal electric field is the same as if there were no material in the field

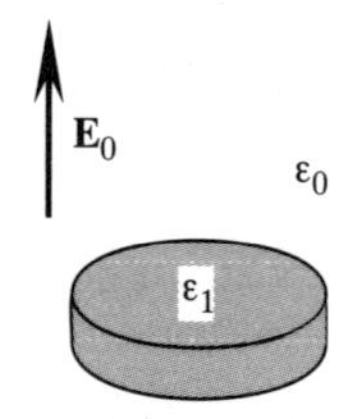

Fig. 6.5 *Disc in a uniform external* **E***-field.*

Fig. 6.6 *Cylinder in a uniform external **E**-field.*

because the surface of the cylinder must satisfy the tangential boundary conditions alone:

$$\mathbf{E}_1 = \mathbf{E}_0. \tag{6.34}$$

Of course, infinite cylinders are rare indeed, but the same solution applies for a finite cylinder placed between parallel plates (RF case) or between the broad walls of a waveguide in the TE_{10} mode. In fact, the latter case is preferred for estimating dielectric properties at high field strength in a waveguide as long as the cylinder diameter is small compared to the internal wavelength, λ_1, as discussed in Chapter 5.

6.3.2 Application of formal solutions of the Laplace equation

We will modify the formal solution of the Laplace equation for the dielectric sphere in a uniform electric field originally discussed in Chapter 4 to apply to the microscopic case. We illustrate its use in the estimation of permittivity. The classical solution for a dipole placed within a hollow spherical cavity is then derived. The two solutions are combined to model nonpolar but polarizable materials.

As always when the Laplace equation is used the unstated assumptions are that: (1) materials in the region of solution are source-free (no net charge), (2) the materials are linear homogeneous and isotropic with the exception of the interfaces, at which boundary conditions apply, and (3) the solution space field is due only to charges at the boundaries.

Dielectric sphere within a dielectric medium

The first case involves the determination of the field in the center of a spherical dielectric of radius a situated in a dielectric of different permittivity, both under the influence of an external electric field, $\mathbf{E}_0$, which is uniform far away from the dielectric sphere, as in Figure 6.7. The pertinent result was mentioned in equation (6.32) above. In the solution space ε_2 is the permittivity of the sphere, ε_1 is the permittivity of the external medium, and the applied field, $\mathbf{E}_0$, is along the z-axis: $\mathbf{E}_0 = \mathbf{E}_0\mathbf{a}_z$. We identify the potentials in regions 1 and 2 by V_1 and V_2, respectively. The potential is axisymmetric in each region (independent of φ) and the general solution is given by (r, θ, φ are the spherical coordinates):

$$V_1 = \sum_{k=0}^{+\infty} \left(A_k r^k + \frac{B_k}{r^{k+1}} \right) P_k(\cos\theta) \tag{6.35}$$

$$V_2 = \sum_{i=0}^{+\infty} \left(C_i r^i + \frac{D_i}{r^{i+1}} \right) P_i(\cos\theta) \tag{6.36}$$

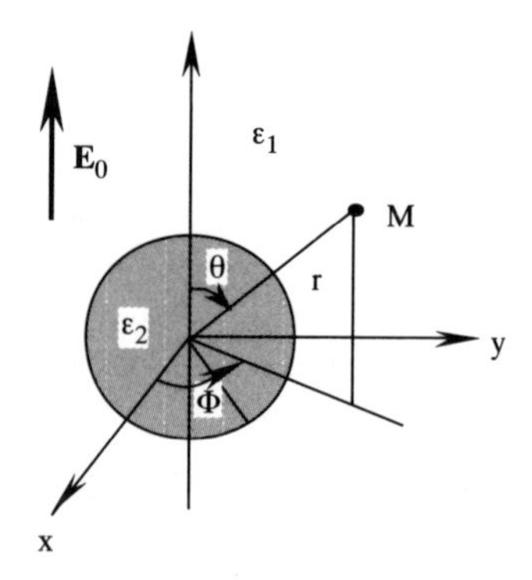

Fig. 6.7 *Field within a dielectric sphere in a dielectric medium.*

where $P_j(\cos\theta)$ are the Legendre polynomials of the first kind and degree j (see Chapter 4). The coefficients A_k, B_k, C_i, D_i are calculated to satisfy the boundary conditions. By way of review, far from the embedded sphere:

$$\lim_{r\to+\infty} V_1 = -E_0 z = -E_0 r\cos\theta \tag{6.37}$$

the potential is continuous:

$$\lim_{r\to a+} V_1 = \lim_{r\to a-} V_2 \tag{6.38}$$

and the normal component of the vector **D**-field is continuous:

$$\varepsilon_1 \lim_{r\to a+} \frac{\partial V_1}{\partial r} = \varepsilon_2 \lim_{r\to a-} \frac{\partial V_2}{\partial r}. \tag{6.39}$$

Then, we choose the reference surface such that both V_1 and V_2 are zero on the $x-y$ plane. It follows that:

$$V_1 = \left(\frac{\varepsilon_2-\varepsilon_1}{\varepsilon_2+2\varepsilon_1}\frac{a^3}{r^3} - 1\right)E_0 z \tag{6.40}$$

$$V_2 = \frac{-3\varepsilon_1}{\varepsilon_2+2\varepsilon_1}E_0 z \tag{6.41}$$

and the field inside the sphere is:

$$\mathbf{E}_2 = \frac{3\varepsilon_1}{2\varepsilon_1+\varepsilon_2}\mathbf{E}_0. \tag{6.42}$$

The field is uniform, and is proportional to the externally applied electric field. If the middle of the sphere, which contains material 2, is evacuated, the field for a cavity, E_0', is:

$$\mathbf{E}_0' = \frac{3\varepsilon_1}{2\varepsilon_1+\varepsilon_0}\mathbf{E}_0. \tag{6.43}$$

Dipole placed within a cavity

The second example calculation involves the case of a dipole, **p**, situated within the hollow cavity of region 2, and is similar to the previous solution, but in the absence of the uniform external field. We may still use the Laplace equation because we have not added any *net* charge to the system. This case is shown in Figure 6.8, in which **p** represents the dipole, oriented along the z-axis, and 1 represents the material surrounding the free-space cavity. The center of the cavity is the origin of coordinates and again $V = 0$ on the x–y plane. The same potential equations apply (equations (6.35) and (6.36)) and the conditions on **E** in equation (6.39) are also applicable (so that the coefficients k and i which are greater than one still yield zero values) and:

$$V_1 = \left(Ar + \frac{B}{r^2}\right)\cos\theta \qquad \text{for } r > a \tag{6.44}$$

$$V_0 = \left(Cr + \frac{D}{r^2}\right)\cos\theta \qquad \text{for } r < a. \tag{6.45}$$

At infinity, the electric field is zero, and so is V_1, so it follows that $A = O$. The potential very near the dipole should be that of the dipole itself:

$$\lim_{r\to 0} V_0 = \frac{p\cos\theta}{4\pi\varepsilon_0 r^2}. \tag{6.46}$$

Therefore $D = p$. By the application of the limit conditions given by equations (6.38) and (6.39) it follows that:

$$\frac{B}{a^2} = aC + \frac{D}{a^2} \tag{6.47}$$

and:

$$\lim_{r=a} \varepsilon_1 \frac{dV_1}{dr} = \lim_{r=a} \varepsilon_0 \frac{dV_0}{dr} \tag{6.48}$$

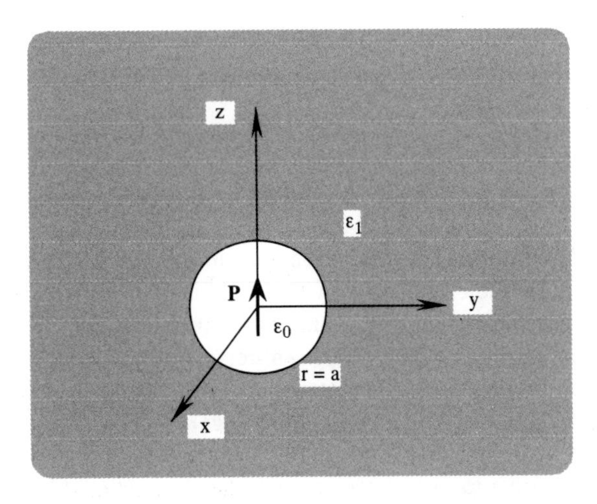

Fig. 6.8 *Dipole within a cavity (no external field applied).*

By combining (6.47) and (6.48), we obtain:

$$-2\varepsilon_1 \frac{B}{a^3} = \left| C - \frac{2D}{a^3} \right| \varepsilon_0 \tag{6.49}$$

from which expressions for B, C, and V_0 are found:

$$B = \frac{3\varepsilon_0 D}{\varepsilon_0 + 2\varepsilon_1} = \frac{3\varepsilon_0 D}{\varepsilon_0 + 2\varepsilon_1} \frac{p}{4\pi\varepsilon_0} \tag{6.50}$$

$$C = -\frac{2D(\varepsilon_1 - \varepsilon_0)}{a^3(\varepsilon_0 + 2\varepsilon_1)} = -\frac{2(\varepsilon_1 - \varepsilon_0)}{a^3(\varepsilon_0 + 2\varepsilon_1)} \frac{p}{4\pi\varepsilon_0} \tag{6.51}$$

and

$$V_0 = \left[\frac{2(\varepsilon_1 - \varepsilon_0)}{a^3(\varepsilon_0 + 2\varepsilon_1)} r + \frac{1}{r^2} \right] \frac{p\cos\theta}{4\pi\varepsilon_0}. \tag{6.52}$$

The potential V_0 in the center of the cavity occupied by the dipole is the superposition of two potentials. One potential is that of the dipole itself and the other is the reaction of the field of the dipole under the influence of the dielectric (material 1). This term will be:

$$V = -\frac{2p(\varepsilon_1 - \varepsilon_0)}{a^3(\varepsilon_0 + 2\varepsilon_1)} \frac{r\cos\theta}{4\pi\varepsilon_0}. \tag{6.53}$$

The total field, which is calculated from $\mathbf{E} = -\nabla V$, is therefore the sum of two terms, the field of the dipole and the field of the reaction:

$$\mathbf{E}_\mathrm{R} = \frac{2(\varepsilon_1 - \varepsilon_0)}{a^3(\varepsilon_0 + 2\varepsilon_1)} \frac{\mathbf{p}}{4\pi\varepsilon_0}. \tag{6.54}$$

So, the reaction field, $\mathbf{E}_R$, is proportional to the polarization, with constant of proportionality $\kappa(\mathrm{m}^{-3})$:

$$\mathbf{E}_\mathrm{R} = \kappa \frac{\mathbf{p}}{\varepsilon_0}. \tag{6.55}$$

The results contained in equations (6.43) and (6.55) were obtained assuming electrostatics, but apply equally well to quasi-static solutions. They may be used whenever the quasi-static assumption is valid.

6.3.3 Calculus of nonpolar but polarizable molecular materials

The material considered in this section is nonpolar in the absence of an external field and consists entirely of polarizable molecules with a polarization isotrope, α (m^3). That is, the molecules acquire a dipole moment $\alpha\varepsilon_0\mathbf{E}_0$ under a local electric field, $\mathbf{E}_0$, in the material and it has an electric permittivity, ε, which is due entirely to the polarizable molecules. In order to study the dielectric properties of this material we will perform a *gedanken* experiment in which we remove (without disturbing the surroundings) a very small spherical cavity of the material (radius a, containing a single molecule) and calculate the electric field components which we must add in order to restore the cavity to the same state it had before the material was removed. So, the cavity must have a dipole moment, $\mathbf{p}$, equivalent to that which would exist if the material was in place, and the sum of all fields must give the local field, $\mathbf{E}_{\text{eff}} = \mathbf{E}_0$, as shown in Figure 6.9a. The effect of creating the cavity (internal permittivity ε_0) is to generate a cavity field, $\mathbf{E}_C$, different from $\mathbf{E}_0$, as we saw in the previous section. The dipole moment will create a reaction field, $\mathbf{E}_R$, due to the dielectric interface, as we have seen. So the effective electric field inside the cavity has two terms:

$$\mathbf{E}_{\text{eff}} = \mathbf{E}_C + \mathbf{E}_R \tag{6.56}$$

where $\mathbf{E}_C$ represents the field in a spherical cavity (resulting from an external field) and $\mathbf{E}_R$ is the reaction field (from induced dipoles which would have been in the cavity). The cavity field exists at the level of a supposed spherical molecule but in the absence of the molecule (the cavity thus has ε_0), and was summarized in equation (6.43):

$$\mathbf{E}_C = \frac{3\varepsilon}{2\varepsilon + \varepsilon_0}\mathbf{E}_0 \tag{6.57}$$

where $\mathbf{E}_0$ is the external field, i.e. the field in the bulk undisturbed medium.

The reaction field, $\mathbf{E}_R$, is due to the induced polarization throughout the molecule represented by its net dipole moment, $\alpha\varepsilon_0\mathbf{E}_{\text{eff}}$. This field may be described by equation (6.55) with $\mathbf{E}_R$ proportional to $\alpha\mathbf{E}_{\text{eff}}$:

$$\mathbf{E}_R = \kappa\alpha\mathbf{E}_{\text{eff}} = 2\frac{\varepsilon - \varepsilon_0}{2\varepsilon + \varepsilon_0}\frac{\alpha\mathbf{E}_{\text{eff}}}{a^3} \tag{6.58}$$

where κ is the proportionality constant for each molecule. If N is the number of polarizable molecules per unit volume, $\mathbf{E}_R$ may be expressed as:

$$\mathbf{E}_R = \frac{2(\varepsilon - \varepsilon_0)}{2\varepsilon + \varepsilon_0}\frac{N\alpha}{3}\mathbf{E}_{\text{eff}}. \tag{6.59}$$

The cavity field term, $\mathbf{E}_C$, can also be rewritten:

$$\mathbf{E}_C = \mathbf{E}_{\text{eff}} - \mathbf{E}_R = \mathbf{E}_{\text{eff}}\left[1 - \frac{2(\varepsilon - \varepsilon_0)}{2\varepsilon + \varepsilon_0}\frac{N\alpha}{3}\right]. \tag{6.60}$$

The effective field, $\mathbf{E}_{\text{eff}}$, can then be written in terms of the applied field, $\mathbf{E}_0$:

$$\mathbf{E}_{\text{eff}} = \frac{3\varepsilon}{2\varepsilon + \varepsilon_0 - \dfrac{2N\alpha}{3}(\varepsilon - \varepsilon_0)}\mathbf{E}_0 \tag{6.61}$$

from which the induced polarization is:

$$\mathbf{P} = N\alpha\varepsilon_0\mathbf{E}_{\text{eff}} = \frac{3N\alpha\varepsilon\varepsilon_0}{2\varepsilon + \varepsilon_0 - \frac{2}{3}N\alpha(\varepsilon - \varepsilon_0)}\mathbf{E}_0 \tag{6.62}$$

where $\mathbf{P}$ is defined by $(\varepsilon - \varepsilon_0)\mathbf{E}_0$.

One can calculate ε as a function of α and write this in the form:

$$\frac{\varepsilon - \varepsilon_0}{\varepsilon + 2\varepsilon_0} = \frac{N\alpha}{3} \tag{6.63}$$

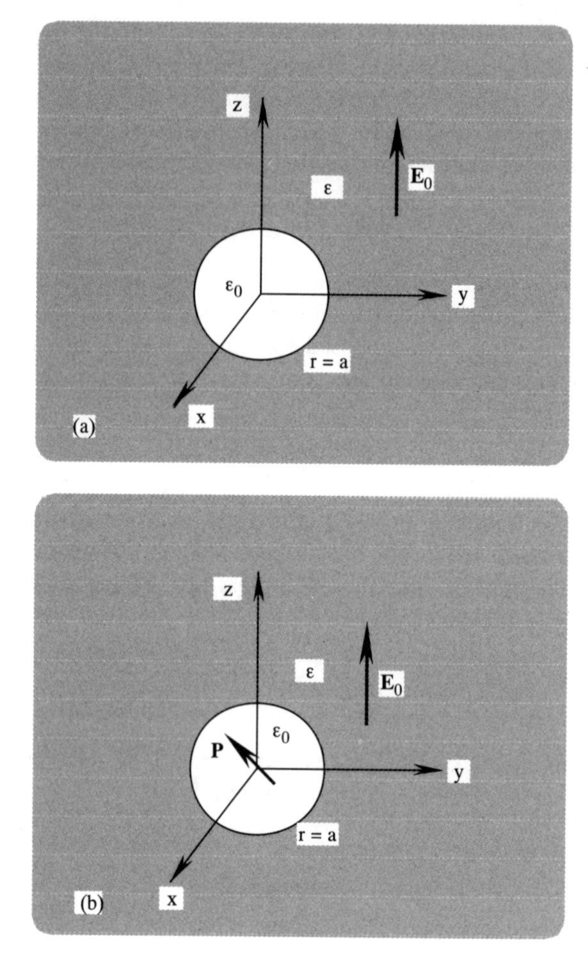

Fig. 6.9 *Calculus of permittivity of a material. (a) Nonpolar but polarizable molecular materials. (b) Coexisting polar and polarizable molecules.*

which is the well known Clausius–Mossotti formula. Inversely, the measurement of ε in nonpolar surroundings allows the calculation of the polarizability of the molecule—$N\alpha\varepsilon_0$ is the polarizability per unit volume.

6.3.4 Modeling the permittivity of a material with polar and polarizable molecules

Many interesting materials consist of molecules which are both polar and polarizable. The previous result can be extended to these materials by adding a term to the polarization vector field (6.62) which accounts for the polar part of the molecules. Again, we assume that α is the polarizability isotrope, add $\mathbf{p}$, the dipole moment in the surroundings due to the polar molecules, and create a cavity in the medium; however, the Onsager model (Figure 6.9b) differs from the Clausius–Mossotti model (Figure 6.9a) in that the perturbation in the external field created by the cavity is included; and the cavity is of the order of a single molecule in size. Here the calculation of the internal field is delicate because the dipole is randomly oriented in the cavity and one must take into account the fluctuations in its orientation due to temperature. At each moment in time, the internal field includes a term which is collinear with the instantaneous orientation of the dipole. If the reaction field, $\mathbf{E}_\mathrm{R}$ is known then the total reaction consists of a permanent dipole and induced dipole term:

$$\mathbf{E}_\mathrm{R} = \kappa\left(\frac{\mathbf{p}}{\varepsilon_0} + \alpha\mathbf{E}_\mathrm{R}\right) \tag{6.64}$$

where the terms are as previously described, and from which:

$$\mathbf{E}_\mathrm{R} = \frac{\kappa}{1-\kappa\alpha}\frac{\mathbf{p}}{\varepsilon_0}. \tag{6.65}$$

The reaction field makes no contribution to the orientation of the dipole since it is derived from it. The field which induces the orientation is the modified cavity field, $\mathbf{E}_\sigma$:

$$\mathbf{E}_\sigma = \frac{3\varepsilon_\mathrm{r}\mathbf{E}_0}{2\varepsilon_\mathrm{r}+1} + \kappa\alpha\mathbf{E}_\sigma \tag{6.66}$$

where ε_r is now the relative dielectric constant of the surrounding medium. It follows then that:

$$\mathbf{E}_\sigma = \frac{3\varepsilon_\mathrm{r}}{(2\varepsilon_\mathrm{r}+1)(1-\kappa\alpha)}\mathbf{E}_0. \tag{6.67}$$

In this formula, one recognizes that to obtain the field in an evacuated cavity, one adds the induced field to the polarization term for the molecules—that is, the dielectric polarization term is added to the dipole moment but only the component of $\mathbf{E}_\sigma$ parallel to $\mathbf{p}$ contributes to the stored energy in the dipole, $\mathbf{W}$:

$$\mathbf{W} = -\mathbf{p}\cdot\mathbf{E}_\sigma = -\mathrm{p}E_\sigma\cos(\theta). \tag{6.68}$$

The average orientation of the dipole parallel to $\mathbf{E}_\sigma$ can then be estimated from Boltzmann statistics:

$$\langle\mathbf{p}\rangle = \langle p\cos(\theta)\rangle\mathbf{a}_{\mathrm{E}_\sigma} = p\frac{\displaystyle\int_0^{2\pi}\mathrm{d}\varphi\int_0^{\pi}\cos(\theta)\exp\left(-\frac{W}{kT}\right)\sin(\theta)\,\mathrm{d}\theta}{\displaystyle\int_0^{2\pi}\mathrm{d}\varphi\int_0^{\pi}\exp-\left(\frac{W}{kT}\right)\sin(\theta)\,\mathrm{d}(\theta)}\mathbf{a}_{\mathrm{E}_\sigma}$$

$$\langle\mathbf{p}\rangle = \frac{p^2}{3\varepsilon_0 kT}\mathbf{E}_\sigma \tag{6.69}$$

where the expectation is independent of φ for obvious reasons. It is to be expected that average orientation of a dipole would be in the direction of the field which orients it, so the result is quite reasonable.

Now we are in a position to calculate the permittivity. The result of equation (6.69) permits calculation of the average value of the reaction field, $\mathbf{E}_\mathrm{R}$:

$$\mathbf{E}_\mathrm{R} = \frac{\kappa}{1-\kappa\alpha}\frac{\mathbf{p}}{\varepsilon_0} = \frac{\kappa}{1-\kappa\alpha}\frac{p^2}{3\varepsilon_0 kT}\mathbf{E}_\sigma. \tag{6.70}$$

The effective internal electric field is the vector sum of $\mathbf{E}_R$ and $\mathbf{E}_\sigma$:

$$\mathbf{E}_{\text{eff}} = \mathbf{E}_{\text{R}} + \mathbf{E}_\sigma = \left[1 + \frac{\kappa}{1-\kappa\alpha}\frac{p^2}{3\varepsilon_0 kT}\right]\mathbf{E}_\sigma. \tag{6.71}$$

Thus, for N molecules per unit volume, the time-average net polarization can be found from:

$$\langle\mathbf{P}\rangle = N\alpha\varepsilon_0\mathbf{E}_{\text{eff}} + \frac{Np^2}{3kT}\mathbf{E}_\sigma \tag{6.72}$$

where, as expected, the polarization has the first term to represent the induced polarization and the second to represent the indigenous randomly polarized molecules.

Of course, equations (6.28) and (6.29) apply, so the relative permittivity is calculable from $\mathbf{P}$ by:

$$\mathbf{P} = (\varepsilon_{\text{r}} - 1)\varepsilon_0\mathbf{E}_0$$

so that:

$$\varepsilon_{\text{r}} = \frac{\mathbf{P}}{\varepsilon_0\mathbf{E}_0} + 1. \tag{6.73}$$

The polarization due to the bulk medium electric field, $\mathbf{E}_0$, is given in a standard form known as the Onsager relation:

$$\frac{(\varepsilon_{\text{r}} - \varepsilon_{\text{r}\infty})(2\varepsilon_{\text{r}} + \varepsilon_{\text{r}\infty})}{\varepsilon_{\text{r}}(\varepsilon_{\text{r}\infty} + 2)^2} = \frac{Np^2}{9\varepsilon_0 kT} \tag{6.74}$$

where $\varepsilon_{\text{r}\infty}$ is the value of the relative permittivity of the material if the surroundings are nonpolar (as in the Clausius–Mossotti relation):

$$\frac{\varepsilon_{\text{r}\infty} - 1}{\varepsilon_{\text{r}\infty} + 2} = \frac{N\alpha}{3}. \tag{6.75}$$

As above, the Onsager relation permits the determination of the dipole moments while measuring the permittivity. The model works surprisingly well for most dielectric materials, and the rigorous results obtained from quantum mechanics add what turn out to be relatively insignificant corrections. In the next chapter we will show how to modify these formulae to include time-dependent electric fields.

6.4 ELECTRICAL CONDUCTIVITY

Electrical conductivity refers specifically to the net translational motion of free charge under the influence of an electric field. The flux density of moving charge is the conduction current density, $\mathbf{J}$, as we have previously seen. It may not be clear at this point exactly why a discussion of charge in motion belongs in a chapter on static phenomena. The reason is that at low frequency in a conducting or semiconducting material the total charge is conserved since we are neither creating nor storing charge within the medium — the right hand side of equation (1.8) (Gauss' electric law) is zero and charge is conserved. For example, in conduction in a copper wire (Figure 6.10) the injection of an electron into one end of the wire unbalances the copper atom lattice so that another electron (presumably in one of

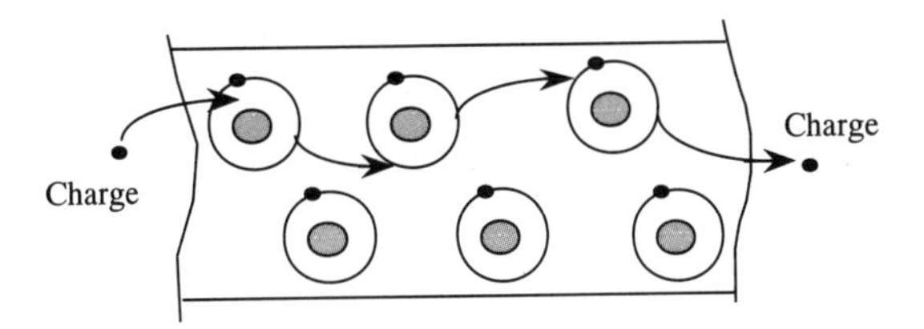

Fig. 6.10 *Conduction in a wire.*

the outer shells) is ejected from that atom and impinges on a neighbor atom destabilizing it, and so on until the other end of the wire is reached. The disturbance created by the original electron injection propagates very quickly in the wire—at speeds approaching the speed of light—even though the velocities of the individual electrons are extremely low by comparison. So the medium is sort of "stationary" even though charges are being knocked around a bit. It will turn out that the discussion of conductivity is general enough that the concepts apply even when the quasi-static assumption fails. We begin with a consideration of unhindered electron acceleration and progress to charge flux in a material medium.

6.4.1 Electron in free space

A single electron injected into free space in which there is an electric field will be accelerated according to the electrostatic Coulomb force law:

$$\mathbf{a} = \frac{\mathbf{F}}{m} = \frac{q}{m}\mathbf{E} \tag{6.76}$$

where **a** is the acceleration as long as Newtonian mechanics is applicable and q/m is the charge to mass ratio for the electron (< 0). The limits on electron velocity are relativistic and the resulting electron velocities can be quite high. This principle is used in cathode ray tubes to obtain a low-mass writing implement from an electron beam, as sketched in Figure 6.11a.

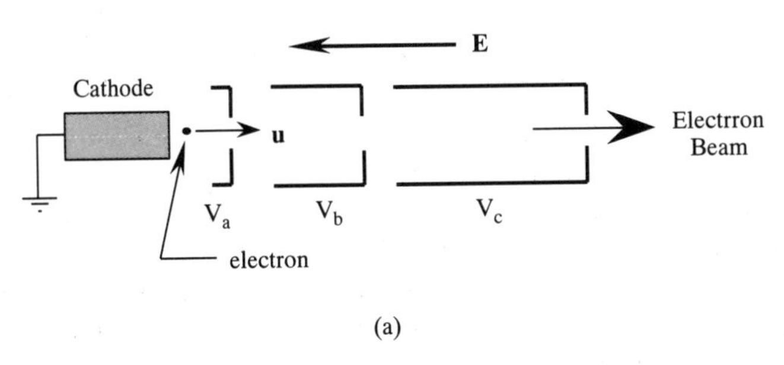

Fig. 6.11 *Electron gun in a cathode ray tube. (a) Electron beam acceleration: the heated cathode is the electron source; electrons are accelerated by successively higher potentials $V_c > V_b > V_a > 0$ to produce the electron beam. (b) Electric field deflection: deflection electric fields E_x and E_y between plates alter electron beam velocity **u** giving the desired (x, y) position on the screen. (c) Magnetic field deflection: magnetic x-deflection, $\mathbf{H}_x$, and y-deflection, $\mathbf{H}_y$, fields created by coils in the "yoke" deflect the beam velocity **u** as shown.*

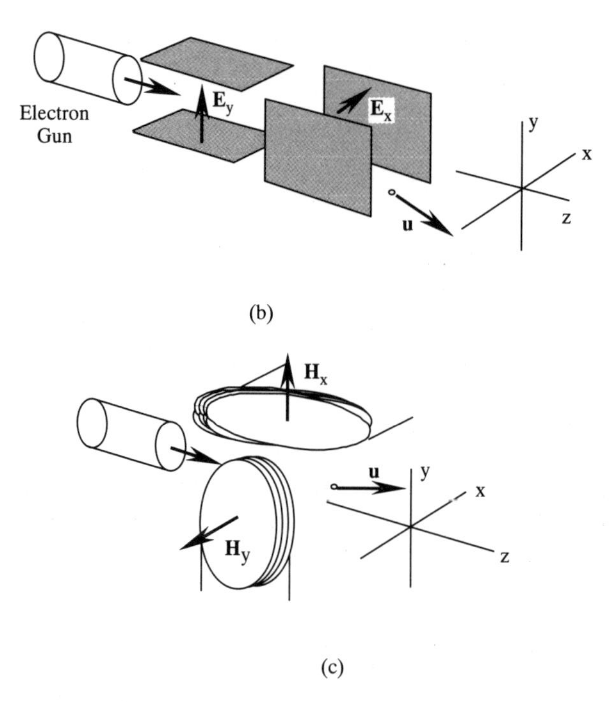

Fig. 6.11 *(continued)*

Of course, a beam of electrons comprises a current (with associated magnetic fields), so the full form of the Coulomb force law must be used to account for the interactions in the beam. The position of the beam on the display screen can be altered by either an electric field applied between plate electrodes (as is done in oscilloscope tubes to obtain highly linear beam positioning, Figure 6.11b) or by an external magnetic field (as is done in television sets and video display terminals, Figure 6.11c). Televisions use magnetic field beam deflection to obtain short tubes (wide deflection angles) and correspondingly smaller cabinets.

6.4.2 Semiconductors

Materials containing free charge — ions dispersed in solution or in lattices — may also have that charge accelerated by the electric field, **E** (V/m). The same holds for metal atoms and charge carriers in a crystal lattice even if they do not strictly have "free charge". That is, in a time-average sense atomic and molecular solids have no net free charge as all valence positions sum to a zero monopole moment. When we inject a charge carrier, as in the wire discussed above, the charge neutrality is upset and this disturbance propagates throughout the medium at a velocity determined by the lattice (or liquid) geometry. The term "charge carrier" includes electrons, electron donors (n-type semiconductor atoms or molecules) and electron acceptors (p-type semiconductor atoms or molecules).

The charge carriers do not accelerate without hindrance, however. In the lattice the charge carrier is accelerated by the electric field until it collides with atoms or molecules at neighboring lattice sites. An analogous sequence of events takes place in liquids and electrolytic solutions. There momentum transfer processes govern the collision mechanics.

The result is that, in a time- and space-average sense, the charge carriers are accelerated up to a limiting average velocity, the drift velocity $\mathbf{u}_d$ (m/s), which is limited by the mean free path between collisions:

$$\mathbf{u}_d = \mu_q \mathbf{E} \tag{6.77}$$

where μ_q is the mobility for the charge carrier (m^2/V s) in the material medium. The mobility, a measurable parameter, is determined by the mean free path between collisions, and is therefore a sensitive function of molecular geometry and lattice (or liquid) temperature. The conduction current density resulting from the charge motion is the product of the volume charge density and the drift velocity, so a simple relation for the electrical conductivity is:

$$\mathbf{J} = \rho_v \mathbf{u}_d = \rho_v \mu_q \mathbf{E} = \sigma \mathbf{E}. \tag{6.78}$$

The collision processes are inherently lossy, so conduction current density always results in heating for finite electrical conductivity (Joule or resistive heating = $\mathbf{E} \cdot \mathbf{J}$, as we have seen from the Poynting power theorem).

Typical semiconductors (both solid state crystals and ionic solutions) have both electron donors and electron acceptors and, since the respective charge carriers are likely to be of differing mass and geometry, the two species are likely to have differing mobilities. Also, electrons and electron donors (net negative charge) are accelerated antiparallel to $\mathbf{E}$ and electron acceptors (net positive charge) parallel to $\mathbf{E}$; thus, their mobilities are of opposite sign. By the definition we have used for $\mathbf{J}$ (net motion of positive charge) charge carriers of both signs contribute positive $\mathbf{J}$ by opposite motion. Thus the electric conductivity is always positive and both species contribute:

$$\sigma = p q_p \mu_p + n q_n \mu_n \tag{6.79}$$

where p is the number density of electron acceptors, n the number density of electron donors (number/m^3), and q_p and q_n are the respective valences (in Coulombs per carrier), μ the mobilities, and the $q\mu$ product is always positive.

6.4.3 Ionic solutions

Ionic solutions are especially interesting semiconducting dielectrics in high power applications. The electrical properties of many of the most interesting materials are dominated by ionic conductivity, especially in RF heating cases. We will confine our remarks in this section to electric fields in the bulk solution since the physics of metal–semiconductor interfaces is quite complex.

Concentration effects

Equation (6.79) could just as easily be expressed on a molal concentration basis for electrolytic solutions:

$$\sigma = \mu n_i N_A q_e [C] = 9.65 \times 10^7 \mu n_i [C] \tag{6.80}$$

where μ is the average mobility, n_i the number of ions in a molecule (2 for NaCl), N_A is Avogadro's number and q_e the elementary charge, and $[C]$ is the concentration (moles/l) — $\rho_v = 9.65 \times 10^7 [C]$. The mobility of an electrolyte decreases very slightly with increasing concentration. For sodium chloride electrolytic solutions between 0.01 and 1 molal, for example, the mobility ranges from 51.8×10^{-9} to $40.3 \times 10^{-9} \mathrm{m}^2/\mathrm{V\,s}$ (respectively) at 20°C.

Temperature effects

Elevations in temperature raise the mobility of ions in solution which results in an increase in the electrical conductivity. Mobility changes are such that, for moderate temperature rises, the electrical conductivity of a typical electrolytic solution increases approximately 1.5–2% per Celsius degree; that is:

$$\frac{\partial \sigma}{\partial T} \cong 0.02(T - T_0)$$

so that:

$$\sigma(T) \cong \sigma(T_0)\mathrm{e}^{0.02(T-T_0)} \tag{6.81}$$

where T_0 is the reference temperature (°C). The effect of temperature rise on electrical conductivity is, consequently, quite significant and cannot be ignored, especially in heating applications. Note that conductivity increases in a heated material may generate a positive feedback loop in which a temperature increase creates a local power density increase ($\mathbf{E}\cdot\mathbf{J} = \sigma\mathbf{E}^2$) with resulting temperature rise increase leading, eventually, to thermal runaway. We will investigate thermal runaway in a subsequent discussion.

The power generation term in materials containing electrolytic solutions is always dominated by conduction current density losses (Joule heating) at radio frequencies. We will show that even at microwave frequencies the mobility of electrolytes is such that Joule heating may dominate other terms.

6.5 SUMMARY

Our purpose in this chapter was to look closely at the behavior of dielectric materials (including semiconductors) under static electric fields. Many of the results obtained this way apply equally well to very slowly time-varying fields and some apply as long as the quasi-static assumption is valid. We have extended the usual description of polarization to include higher order electric moments and illustrated the estimation of permittivity in simple molecular structures. Because the results were obtained with static analysis they have limited validity: notably, none of the polarization processes analyzed was considered lossy, although semiconduction certainly is. We will see in the next chapter that the materials which require a more careful description of their time-dependent behavior are the materials with memory and those whose polarization process is lossy — the vast majority of interesting materials.

REFERENCES

Fröhlich, H. (1955) *Theory of Dielectrics* Oxford University Press, Oxford.

Glasstone, S. (1942) *An Introduction to Electrochemistry* van Nostrand, Toronto.

Jackson, J. D. (1975) *Classical Electrodynamics (2nd Edition)* John Wiley, New York.

Johnscher, A. K. (1983) *Dielectric Relaxation in Solids* Chelsea Dielectrics Press, London.

Von Hippel, A. R. (1954) *Dielectric Materials and Applications* John Wiley, New York.

7 Dynamic Aspects

7.1 NORMALIZED COMPLEX PERMITTIVITY

The properties of a material under an electromagnetic field which varies with time are explained, generally, by the same macroscopic variables which we used for a time-invariant electromagnetic field. When these quantities are time dependent, the electric flux density and current density are functionals of the externally applied electric field, $\mathbf{E}_0$. The magnetic flux density depends on the external magnetic field, $\mathbf{H}_0$. At any point in the material they are related to the local field strengths, $\mathbf{E}$ and $\mathbf{H}$, by the constitutive relations:

$$\mathbf{D} = f(\mathbf{E}(t)) = \varepsilon(t)\mathbf{E}(t) \tag{7.1}$$

$$\mathbf{J} = g(\mathbf{E}(t)) = \sigma(t)\mathbf{E}(t) \tag{7.2}$$

$$\mathbf{B} = h(\mathbf{H}(t)) = \mu(t)\mathbf{H}(t) \tag{7.3}$$

where the material has been assumed linear and isotropic so that the properties are scalars, but still may be time-dependent.

7.1.1 Phasor notation

In previous analyses the properties have been constants since time-dependence was neglected. The constitutive relations completed Maxwell's equations, as we discussed in Chapter 1. The possible time-dependent nature of the properties was briefly mentioned in the discussion of polarizable materials with memory in the last chapter. We are concerned with electromagnetic fields from sinusoidal generators. So mathematically the problem can be much more easily treated using complex (i.e. phasor) notation:

$$\begin{aligned}\mathbf{X}(x, y, x, t) &= \mathrm{Re}\{\mathbf{X}(x, y, z)e^{j\phi(x,y,z)}e^{j\omega t}\} \\ &= \mathbf{X}(x, y, z)\cos(\omega t + \phi(x, y, z))\end{aligned} \tag{7.4}$$

where the vector direction and magnitude are represented in $\mathbf{X}(x, y, z)$, and the vector has phase angle $\phi(x, y, z)$ with respect to some phase reference. In this notation we are only interested in the real part of the solution, but we construct a complex sinusoidal generator — $V(\omega, t, \phi) = V_0[\cos(\omega t + \phi) + \mathrm{j}\sin(\omega t + \phi)]$ — superposing the real and imaginary potentials and fields in order to simplify the solutions. Using phasor notation solves Maxwell's equations in the frequency domain; the variables are the complex conjugates of their Fourier transforms (but we are only interested in the real part, so it is in fact equivalent). This transform reduces the differential equations to algebraic equations which simplifies their solution. Of course one cannot buy a complex sinusoidal generator, so after solution we accept only the real part as the result. Whenever phasor notation is used we have made the tacit assumption that the system under study is linear (so that superposition applies). To review, the frequency domain Maxwell equations are:

$$\begin{aligned} \nabla \times \mathbf{E} &= -\mathrm{j}\omega\mathbf{B} = -\mathrm{j}\omega\mu\mathbf{H} \\ \nabla \times \mathbf{H} &= \mathbf{J} + \mathrm{j}\omega\mathbf{D} = (\sigma + \mathrm{j}\omega\varepsilon)\mathbf{E} \end{aligned} \tag{7.5}$$

where the properties are as complex as is required to complete the constitutive relations, the resulting multiplications can include tensors as required, and all fields are expressed in the frequency domain.

When complex notation is used it is assumed that all of the major variables are sinusoidal at a single frequency, ω, which is understood. Therefore, we ordinarily keep track of only the magnitude and phase angle of each variable. If the time-dependence of the variables is periodic but not sinusoidal (the usual case in real RF systems), the waveform is distorted by higher order harmonics, and may be represented by a Fourier series. The system response at each frequency, obtained from phasor analysis, allows one to re-create the global time-dependent response by the summation (superposition) of all the elementary single-frequency responses for each of the frequency components. It is possible to study any linear system by considering only responses over a range of frequencies, ω.

7.1.2 Complex properties of the medium

If the complex electric flux density, $\mathbf{D}$, is in phase with $\mathbf{E}$ then ε is real. In general, they are not in phase, and $\mathbf{D} = \varepsilon^*\mathbf{E}$ where ε^* is at least a complex number—and may be a complex tensor. For a linear isotropic material ε^* is a complex scalar constant. It follows, then, that conductivity, σ, and permeability, μ, are also, in general, complex:

$$\begin{aligned} \varepsilon^* &= \varepsilon' - \mathrm{j}\varepsilon'' \\ \mu^* &= \mu' - \mathrm{j}\mu'' \\ \sigma^* &= \sigma' + \mathrm{j}\sigma'' \end{aligned} \tag{7.6}$$

where the negative imaginary parts have that sign so that absorbed power (the Poynting power theorem in point form) will be positive. In fact, it turns out that the above generalization contains redundancy. Ampère's law may be recast in two convenient ways in order to see the redundancy we may create an equivalent conductivity:

$$\nabla \times \mathbf{H} = (\sigma^* + \mathrm{j}\omega\varepsilon^*)\mathbf{E} = ((\sigma' + \omega\varepsilon'') + \mathrm{j}(\sigma'' + \omega\varepsilon'))\mathbf{E} \tag{7.7}$$

and an equivalent permittivity:

$$\nabla \times \mathbf{H} = \mathrm{j}\omega\left(\varepsilon^* - \mathrm{j}\frac{\sigma^*}{\omega}\right)\mathbf{E} = \mathrm{j}\omega\left[\left(\varepsilon' + \frac{\sigma''}{\omega}\right) - \mathrm{j}\left(\varepsilon'' + \frac{\sigma'}{\omega}\right)\right]\mathbf{E}. \tag{7.8}$$

Without specifying the molecular processes which are the origin of permittivity and conductivity, the equivalent of dielectric dispersion is determined in either form by the sum of ε' and σ''/ω. Likewise, ε'' and σ'/ω are also interchangeable and, in addition, collectively describe the losses.

No experiment will be able to separate losses due to σ' and ε'', though they are due to two different phenomena: translational motion of free charge and rotational motion of bound charge. We will see, however, that the frequency dependence induced by the σ'/ω term can be identified in a plot of $\varepsilon''_{\text{eff}}(\omega)$, so it is indeed worthwhile to maintain separate identities for those two parameters. The σ''/ω term is another matter entirely. In view of the previous discussion it may appear at first glance that a non-zero σ'' implies that motion of free charge limited by collisions can exist in the absence of any losses, but that is not, in fact, what this term implies. Materials with that behavior constitute superconductors and can be more conveniently managed with no loss in generality by presuming an infinite real σ (so that $\mathbf{E} = 0$), as we did in Chapter 6. In fact, the physical significance of σ'' is that it represents energy stored in the kinetic energy of charge carriers in translational motion—σ'' implies that translational motion of free charge makes a contribution to reactive power (contributes to wave propagation) just as ε' does. It is easy to pick an example where this model might be used—the motion of an electron beam in free space (the example in Chapter 6) might be described in this way. However, there are more convenient descriptions of the motion of a space charge in a vacuum, so this is far from a necessary construct. We are not interested in space charge motion; and we should write the system of equations with the minimum number of essential parameters. Therefore, we will use $\sigma^* = \sigma' = \sigma$ (real only) in the analysis of electromagnetic fields. In some cases, it may be convenient to lump all of the losses into the electrical conductivity, as we did in Chapter 1. In those cases, then, $\sigma_{\text{eff}} = \sigma + \omega\varepsilon''$ and the conduction current density is imagined to include the lossy part of the displacement current density, even though we are cognizant of the difference. In other cases we may lump all of the losses into ε'' so that $\varepsilon''_{\text{eff}}(\omega) = \varepsilon''(\omega) + \sigma/\omega$.

7.1.3 Normalized susceptibility

For coherence we may define the polarization vector field as the total excess of the electric flux density (above the $\mathbf{D}$ realized in free-space by the same charge) without specifying the molecular processes which result in ε and σ:

$$\mathbf{P}(t) = \mathbf{D}(t) - \varepsilon_0 \mathbf{E}(t). \tag{7.9}$$

By application of an electric field, $\mathbf{E}(t)$, one can determine all of the properties of the dielectric from the total electric polarization, $\mathbf{P}(t)$. We need a formulation for $\mathbf{P}(t)$ in terms of $\mathbf{E}(t)$ which will accommodate the most general case, that of a material with memory.

We can characterize the time-dependence of the dielectric response using an elementary polarization, $f(t)$, which is the equivalent of an impulse response in linear system analysis. The elementary polarization is the time response obtained when the electric field is a Dirac δ-function (see Figure 7.1a):

$$\mathbf{E}(t) = \delta(t)\mathbf{a}_{\text{E}}$$

where:

$$\delta(t) = \lim_{T\to 0}\left\{\int_0^T \frac{1}{T}\,\mathrm{d}t\right\} \qquad \text{and} \qquad \int_{-\infty}^{\infty} \delta(t)\,\mathrm{d}t = 1. \tag{7.10}$$

In Figure 7.1b a typical Debye liquid has been subjected to an impulsive electric field and displays an exponential response with time constant, τ:

$$f(t) = \frac{1}{\varepsilon_0}\mathcal{P}\{\delta(t)\mathbf{a}_{\text{E}}\} = \text{A}e^{-t/\tau} \qquad t \geqslant 0 \tag{7.11}$$

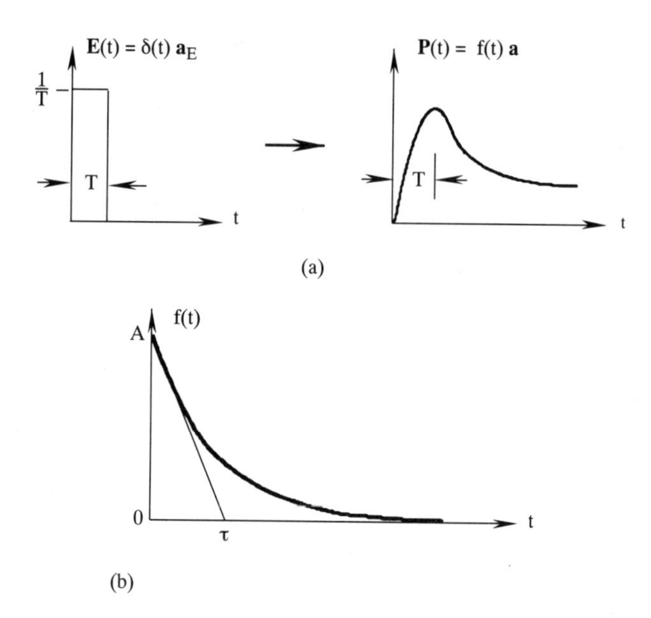

Fig. 7.1 *(a) Polarization time response when* **E** *is a Dirac δ-functions. (b) Typical polarization impulse response of a polar liquid.*

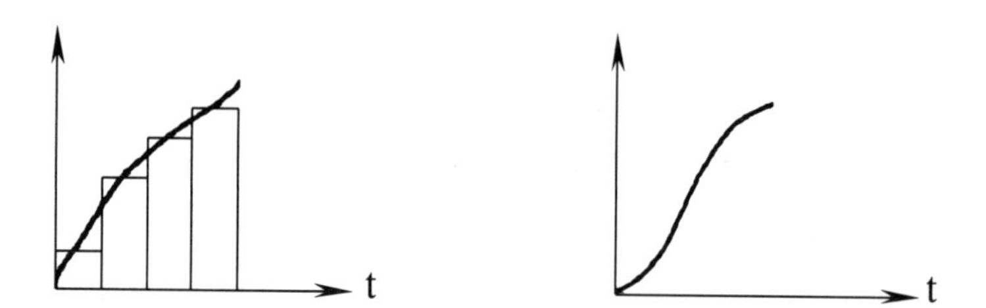

Fig. 7.2 **E***-field decomposed into Dirac delta functions and corresponding* **P***(t).*

where A is the amplitude of the response to a unit impulse and the polarization operator, $\mathcal{P}$, is a linear vector operator with the impulse electric field as input. The impulse response, $f(t)$, then, is the result of the polarization operator operating on an impulse input.

Knowledge of the impulse response allows one to calculate any polarization when the field is of a more complicated time-dependent form. In a calculus fashion, the electric field is reduced to a sequence of Dirac elements (Figure 7.2), and all of the contributions of $f(t)$ corresponding to the impulses are summed in a convolution integral:

$$\begin{aligned}\mathbf{P}(t) &= \varepsilon_0 \int_{-\infty}^{\infty} \mathbf{E}(u) f(t-u)\,\mathrm{d}u = \varepsilon_0 \int_{-\infty}^{t} \mathbf{E}(u) f(t-u)\,\mathrm{d}u \\ &= \varepsilon_0 \int_{-\infty}^{\infty} \mathbf{E}(t-v) f(v)\,\mathrm{d}v = \varepsilon_0 \int_{-\infty}^{t} \mathbf{E}(t-v) f(v)\,\mathrm{d}v \end{aligned} \tag{7.12}$$

where u and v are dummy variables of integration. As reflected in the second forms of the integral any real dielectric must be causal — that is, it may not anticipate the application of an electric field — and $f(t) \equiv 0$ for all $t < 0$. The impulse response must be finite for physical reasons — the polarization cannot increase without bound — and, as a consequence, $f(t)$ is square integrable so its Fourier transform exists. Also, $f(t)$ is a real function of t; however, its Fourier transform, $F(\omega)$, may be complex. At any time t, the total polarization is the result of the contributions of all reactions to the electric field which have occurred previous to the time t — the system has "memory" of all of the events of the field which have occurred.

Taking the Fourier transform of $\mathbf{P}(t)$ we may interchange the sequence of integration because both operators are linear:

$$\begin{aligned}\mathcal{F}\{\mathbf{P}(t)\} &= \varepsilon_0 \int_{-\infty}^{\infty} \mathbf{P}(t)\mathrm{e}^{-\mathrm{j}\omega t}\,\mathrm{d}t = \varepsilon_0 \int_{-\infty}^{\infty} \left[\int_{-\infty}^{\infty} f(u)\mathbf{E}(t-u)\,\mathrm{d}u\right] \mathrm{e}^{-\mathrm{j}\omega t}\,\mathrm{d}t \\ &= \varepsilon_0 \int_{-\infty}^{\infty} f(u)\left[\int_{-\infty}^{\infty} \mathbf{E}(t-u)\mathrm{e}^{-\mathrm{j}\omega t}\,\mathrm{d}t\right]\mathrm{d}u. \end{aligned} \tag{7.13}$$

By multiplying by $\exp(-\mathrm{j}\omega u)\exp(+\mathrm{j}\omega u) = 1$ and using the substitution $s = t - u$ we obtain:

$$\begin{aligned}\mathcal{F}\{\mathbf{P}(t)\} = \mathbf{P}(\omega) &= \varepsilon_0 \int_{-\infty}^{\infty} f(u)\mathrm{e}^{-\mathrm{j}\omega u}\left[\int_{-\infty}^{\infty} \mathbf{E}(t-u)\mathrm{e}^{-\mathrm{j}\omega(t-u)}\,\mathrm{d}t\right]\mathrm{d}u \\ &= \varepsilon_0 \left[\int_{-\infty}^{\infty} f(u)\mathrm{e}^{-\mathrm{j}\omega u}\,\mathrm{d}u\right]\left[\int_{-\infty}^{\infty} \mathbf{E}(s)\mathrm{e}^{-\mathrm{j}\omega s}\,\mathrm{d}s\right] \\ &= \varepsilon_0 \mathcal{F}\{f(t)\}\mathcal{F}\{\mathbf{E}(t)\} = \varepsilon_0 F(\omega)\mathbf{E}(\omega) \end{aligned} \tag{7.14}$$

Note that the vector nature of the fields is independent of their time dependence. Equations such as (7.12), (7.13) and (7.14) are actually three separate expressions since they hold for each component of the vector.

From the convolution property of the Fourier transform, in which convolution in the time domain (7.12) is equivalent to the product in frequency space (7.14), it may be seen that, by the definitions in use, $F(\omega)$ is the frequency domain expression for the complex conjugate form of the susceptibility, $\chi^*(\omega)$:

$$\mathbf{P}(\omega) = \varepsilon_0 F(\omega)\mathbf{E}(\omega) = \varepsilon_0 \chi^*(\omega)\mathbf{E}(\omega) \tag{7.15}$$

where the total complex susceptibility $\chi^*(\omega)$ and $f(t)$ form a Fourier transform pair. The susceptibility is complex, $F(\omega) = \chi^*(\omega) = \chi'(\omega) - \mathrm{j}\chi''(\omega) = \varepsilon^*(\omega)/\varepsilon_0$, with real and imaginary parts (assuming that $\sigma = 0$, also remember the sign convention on ε^*). Because $f(t)$ is necessarily real, the real and imaginary parts of χ^* can be obtained from the separate parts of the Fourier transform:

$$\chi'(\omega) = \int_{-\infty}^{\infty} f(t)\cos(\omega t)\,\mathrm{d}t = \int_{0}^{\infty} f(t)\cos(\omega t)\,\mathrm{d}t \tag{7.16}$$

and:

$$\chi''(\omega) = \int_{-\infty}^{\infty} f(t)\sin(\omega t)\,\mathrm{d}t = \int_{0}^{\infty} f(t)\sin(\omega t)\,\mathrm{d}t. \tag{7.17}$$

The limit of integration can be changed from $-\infty$ to 0 since $f(t)$ is causal. Also, because $f(t)$ is real $\chi(\omega)$ has an even real part and an odd imaginary part, that is, $\mathrm{Re}\{\chi(\omega)\} = \mathrm{Re}\{\chi(-\omega)\}$ and $\mathrm{Im}\{\chi(\omega)\} = -\mathrm{Im}\{\chi(-\omega)\}$.

The elementary polarization, $f(t)$ completely characterizes the dielectric response, and if continuity in f is strictly enforced (we do not allow instantaneous changes in **P**) the real and imaginary parts of the susceptibility χ' and χ'' can be calculated. Because of the special properties of $f(t)$ (it is real, causal and square-integrable), when the complex susceptibility, $\chi^*(\omega)$, is known, the elementary polarization can be calculated by taking the inverse sine or cosine transform—the derivation of these relations was nicely described by Morse and Feshback (1968):

$$f(t) = \frac{2}{\pi}\int_0^{+\infty} \chi'(\omega)\cos(\omega t)\,\mathrm{d}\omega \tag{7.18}$$

$$f(t) = \frac{2}{\pi}\int_0^{+\infty} \chi''(\omega)\sin(\omega t)\,\mathrm{d}\omega \tag{7.19}$$

One can also eliminate the function $f(t)$ between equations (7.18) and (7.19) and calculate $\chi'(\omega)$ as a function of $\chi''(\omega)$ and vice versa. Thus there exists a relationship between the real and imaginary parts of the complex susceptibility of any material, the Kramers-Krönig relations. The actual derivation of this general relation is delicate from a mathematical viewpoint; but the results are:

$$\chi'(\omega) = \chi'(\infty) + \frac{2}{\pi}\int_0^{\infty} \frac{u\chi''(u) - \omega\chi''(\omega)}{u^2 - \omega^2}\,\mathrm{d}u \tag{7.20}$$

and:

$$\chi''(\omega) = -\frac{2}{\pi}\int_0^{\infty} \frac{\chi'(u) - \chi'(\omega)}{u^2 - \omega^2}\,\mathrm{d}u \tag{7.21}$$

where $\chi'(\infty)$ represents the high frequency asymptote of the real part of the susceptibility. The Kramers–Krönig relations are used in many domains of physics, from x-ray absorption to optics, to describe absorption of all forms of electromagnetic radiation—including cases where the matter is described by quantum concepts and the radiation is described either classically or in quantum terms. These relations specify, in fact, that the absorption process is causal, as we have assumed.

If the imaginary part of the dielectric susceptibility $\chi''(\omega)$ is known over the entire range of the frequency, the real part can be calculated. Reciprocally, knowledge of the real part permits the calculation of the imaginary part over the same (very wide) range. Evidently, when the frequency is zero, the imaginary part of the susceptibility must be zero (i.e. **P** cannot be out of phase with **E** since **E** = constant). Then, the real part of the susceptibility is a simple integral of the imaginary part multiplied by $2/\pi$ and added to the value at infinity. However:

$$\int_0^{\infty} \frac{\mathrm{d}u}{u^2 - \omega^2} = 0. \tag{7.22}$$

If an arbitrary constant is added to $\chi'(\omega)$ the value of $\chi''(\omega)$ is unchanged. Likewise, the addition of a constant of the form k/ω to $\chi''(\omega)$ causes no perturbation in $\chi'(\omega)$. One is led, therefore, to use the value $\chi'(\infty)$ as the absolute origin of the real part of the susceptibility. Similarly, the origin of the imaginary part will be taken as the limit of $\omega\chi''(\omega)$ as ω

approaches zero. Then, the definition of the generalized electric permittivity is:

$$\varepsilon^*(\omega) = \varepsilon'(\infty) + \varepsilon_0\chi'(\omega) - \mathrm{j}\varepsilon_0\chi''(\omega) - \mathrm{j}\frac{\sigma}{\omega} \tag{7.23}$$

where $\varepsilon'(\infty)$ is, again, the high frequency asymptote and we have placed the conduction current density losses in the imaginary part of ε^*.

In conclusion, to satisfy the Kramers-Krönig relations as a simple formula, it follows that the real and imaginary parts of the total susceptibility are defined on an absolute basis. With these conventions, the susceptibility is normalized, and the Kramers-Krönig equations will not overlap other terms:

$$\chi'_{\mathrm{n}}(\omega) = \frac{2}{\pi}\int_0^{\infty} \frac{u\chi''_{\mathrm{n}}(\omega) - \omega\chi''_{\mathrm{n}}(\omega)}{u^2 - \omega^2}\,\mathrm{d}u \tag{7.24}$$

and:

$$\chi''_{\mathrm{n}}(\omega) = -\frac{2}{\pi}\int_0^{\infty} \frac{\chi'_{\mathrm{n}}(u) - \chi'_{\mathrm{n}}(\omega)}{u^2 - \omega^2}\,\mathrm{d}u. \tag{7.25}$$

The complex permittivity is then as described by equation (7.23).

In the previous view of dielectric phenomena, we were not concerned with making a detailed analysis of the electricity within the irradiated matter. The consequences of causality were of primary concern. Dielectric polarization phenomena can have two origins: polarization with restricted rotation of molecules, or electric charge whose translational movement is restricted. Both of these can induce relaxation delays. Each of the relaxation phenomena have the same representation as a function of frequency, which is more clearly seen on logarithmic axes. The real part of permittivity is a curve with many step decreases with increasing frequency (Figure 7.3). Corresponding to each step decrease, the total loss presents a local maximum, with a background determined by the conductivity of the material, σ/ω, which decreases in importance as ω becomes large. The global permittivity can be then decomposed into many relaxation processes, with each individual relaxation process described by:

$$\int_{\omega_1}^{\omega_2} \frac{\chi''_n(\omega)}{\omega}\,\mathrm{d}\omega = \varepsilon'(\omega < \omega_1) - \varepsilon'(\omega > \omega_2). \tag{7.26}$$

Now we may describe electrical conductivity by the limit (real and positive) of the total loss term, $\varepsilon''(\omega)$, as ω approaches 0:

$$\sigma = \lim_{\omega\to 0}\{\omega\varepsilon''(\omega)\} = \lim_{\omega\to 0}\{\mathrm{j}\omega\varepsilon^*(\omega)\}. \tag{7.27}$$

This formulation describes the free motion of discrete electric charge in translation in a stationary electric field. The complex susceptibility includes the possible translational motion of charge restricted by other charges or dipoles. The relaxation contribution to the complex susceptibility also satisfies the Kramers–Krönig relations. Formal description of the relaxation processes will be treated in the next section.

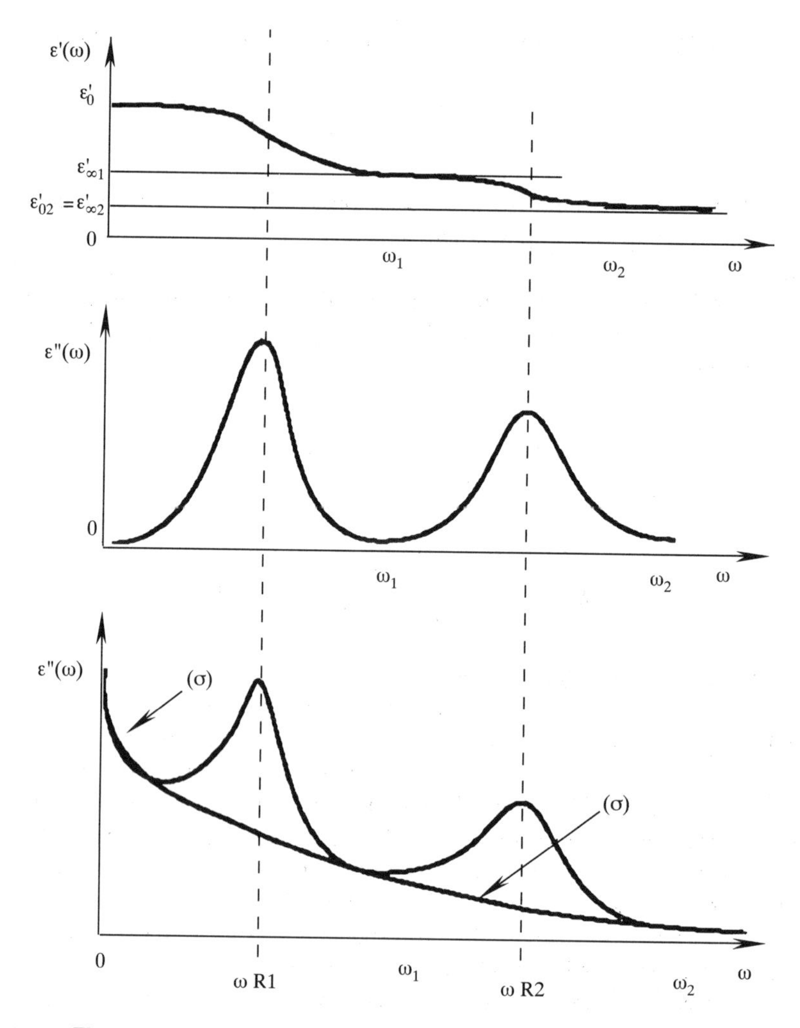

Fig. 7.3 *Generalized frequency dependence of permittivity.*

7.2 DIELECTRIC RELAXATION

We will now discuss four classical, but simple, cases of representative dielectric relaxation processes using the impulse response approach: an ideal dielectric liquid, a Debye liquid, an ideal solid dielectric and a harmonic oscillator. The ideal dielectric liquid will be treated first. Debye formulated a theory for polar liquids and Fröhlich an equivalent model for solids which will follow. A third model has been elaborated from the classical harmonic oscillator theory and modified to yield a simple relaxation formula.

7.2.1 Ideal dielectric liquid

An ideal dielectric liquid is devoid of free charge; $\sigma = 0$. In ideal liquids, the polarization follows the electromagnetic field with a simple delay (Figure 7.4). That is, the polarization

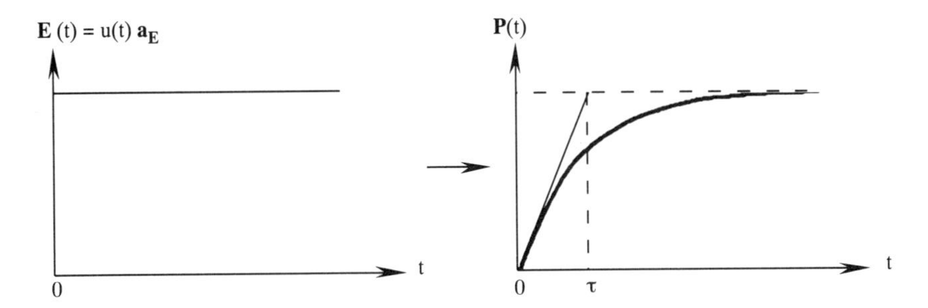

Fig. 7.4 *Step function and step response.*

is given as the solution of a first order differential equation whose coefficients are constant, with response time constant, τ:

$$\frac{\mathrm{d}\mathbf{P}(t)}{\mathrm{d}t} + \frac{1}{\tau}\mathbf{P}(t) = \alpha\varepsilon_0\mathbf{E}(t) = \alpha\varepsilon_0 u(t)\mathbf{a}_\mathrm{E} \tag{7.28}$$

where a unit step electric field has been assumed, α is, again, the polarizability, and $u(t)$ is the step function, which is a discontinuous function defined by its limit behavior:

$$u(t): \qquad \lim_{t\to 0+}\{u(t)\} = 1 \qquad \text{and} \qquad \lim_{t\to 0-}\{u(t)\} = 0. \tag{7.29}$$

The unit step function has two attractive properties over the impulse function: it is approximately physically realizable, and the response to a Dirac delta input can be derived from the derivative of the step response:

$$u(t) = \int_{-\infty}^{t} \delta(\xi)\,\mathrm{d}\xi \qquad \text{so} \qquad f(t) = \frac{1}{\varepsilon_0}\frac{\partial}{\partial t}[\mathcal{P}\{u(t)\mathbf{a}_\mathrm{E}\}] \tag{7.30}$$

where the notation is as previously described. Equation (7.28) is a forced Bernoulli differential equation with the solution shown in Figure 7.4:

$$\mathbf{P}(t) = \frac{\alpha}{\varepsilon_0}[1 - \mathrm{e}^{-t/\tau}]u(t)\mathbf{a}_\mathrm{E} \tag{7.31}$$

where we have used the unit step function to indicate that the solution is only valid for $t > 0$ to satisfy causality. From this solution, the impulse response is:

$$f(t) = \frac{1}{\varepsilon_0}\frac{\mathrm{d}P(t)}{\mathrm{d}t} = \frac{\alpha}{\tau}\mathrm{e}^{-t/\tau}u(t) + 0\delta(t) \tag{7.32}$$

a simple exponential decay. The impulse function on the right hand side resulting from the differentiation process has (fortunately) zero strength at t = 0 (where it occurs) and therefore makes no contribution to the impulse response.

Since $f(t)$ is an energy signal we may calculate the susceptibility by direct integration of the Fourier transform:

$$F(\omega) = \mathcal{F}\{f(t)\} = \int_0^{\infty} \frac{\alpha}{\tau}(\mathrm{e}^{-t/\tau})(\mathrm{e}^{-\mathrm{j}\omega t})\,\mathrm{d}t = \frac{\alpha}{1 + \mathrm{j}\omega\tau}. \tag{7.33}$$

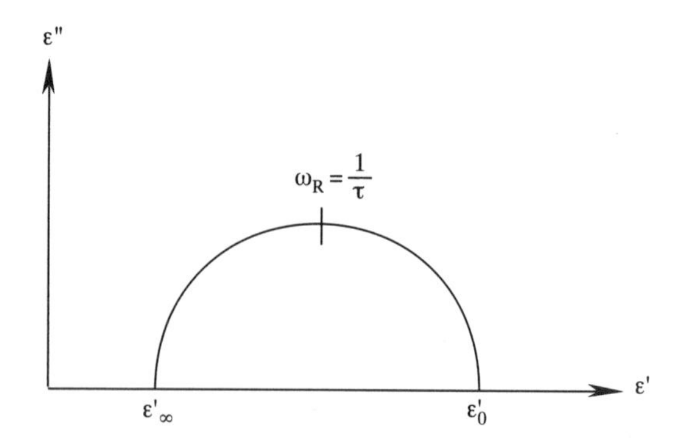

Fig. 7.5 *Ideal Debye liquid.*

The real and imaginary parts of the susceptibility — $F(\omega) = \chi'(\omega) - j\chi''(\omega)$ — and electric permittivity are then given by:

$$\chi'(\omega) = \frac{\alpha}{1 + \omega^2\tau^2} \quad \text{and} \quad \chi''(\omega) = \frac{\alpha\omega\tau}{1 + \omega^2\tau^2} \tag{7.34}$$

so that,

$$\varepsilon'(\omega) = \varepsilon'(\infty) + \frac{k}{1 + \omega^2\tau^2} \quad \text{and} \quad \varepsilon''(\omega) = \frac{k\omega\tau}{1 + \omega^2\tau^2} \tag{7.35}$$

from the general description of ε^* (equation (7.23)), where $k = \varepsilon'(0) - \varepsilon'(\infty)$, the amplitude of the relaxation response.

In the complex plane, the curve represented by equations (7.35) is a semi-circle which has its center point situated on the real axis, known as a Cole–Cole plot (Figure 7.5). The frequency for which the losses are at a maximum is the relaxation frequency, $\omega_R = 1/\tau$, so the maximum loss frequency gives τ from experimental data. Many ideal polar liquids, such as pure water, have dielectric behavior which closely follows this simple model. The simple formula has been interpreted from the point of view of microscopic behavior of molecules or ions in the material.

7.2.2 Debye description of an ideal liquid

The Debye description can be used to derive the form of the time constant and polarizability which were used to characterize the frequency response in the previous discussion. In an ideal liquid composed of polar molecules, we may focus attention upon a single isolated molecule. This molecule is rotating in the liquid, and it consequently experiences a variable electromagnetic field. First, we suppose that the molecule is nearly spherical, and then define a zone of influence of the equivalent dipole, which is limited to the volume of the molecule — i.e. we ignore interaction between adjacent dipoles. The behavior of the molecule is assumed to be that of a small electric dipole in the center of the spherical volume, as we used in Chapter 6 in a static field. The oscillating electric field will cause the dipole to rotate (Figure 7.6) in precessional motion. The balance of forces which are applied

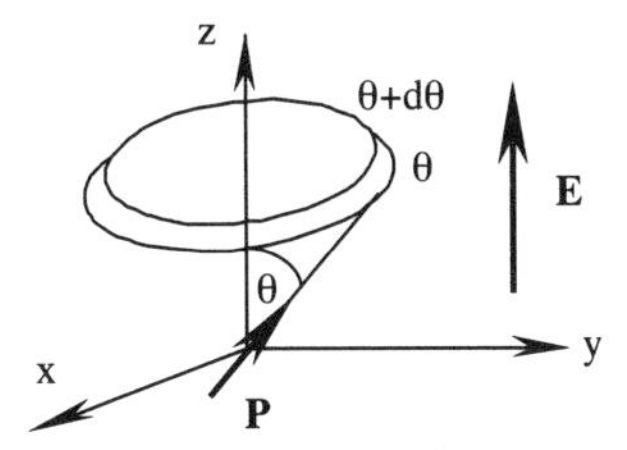

Fig. 7.6 *Net orientation of dipoles, P, in external field, E.*

to the dipole will yield an equation describing its movement from classical mechanics:

$$I\frac{\partial^2\theta}{\partial t^2} + B\frac{\partial\theta}{\partial t} + pE\sin(\theta) = 0 \tag{7.36}$$

where θ is the angle between $\mathbf{E}$ and $\mathbf{p}$, I is the moment of inertia of the dipole, the second derivative of θ is the angular acceleration, B is the microscopic viscous damping due to other molecules, and the first derivative of θ is the angular velocity. The magnitude of the torque on the dipole is $pE\sin\theta$. The angular inertia term does not make a significant contribution to the equation of motion when the relaxation time is short, and can be neglected for most materials.

The equation of motion for individual molecules must include the statistics of space orientation of the dipoles. The distribution function for dipoles is $N(\theta, t)$; so the probability density function gives the number of molecules per unit volume having an orientation between the solid angles Ω and $\Omega + \mathrm{d}\Omega$ or between the angles $\theta, \theta + \mathrm{d}\theta$:

$$\mathrm{d}N = N(\theta, t)\frac{\mathrm{d}\Omega}{4\pi} = \frac{1}{2}N(\theta, t)\sin(\theta)\,\mathrm{d}\theta. \tag{7.37}$$

If $B\partial\theta/\partial t + pE\sin\theta \approx 0$, the mean algebraic number of dipoles with orientation between θ and $\theta + \Delta\theta$ has a variation ΔN in a small time Δt given by:

$$\Delta N = -\frac{pE}{2B}N(\theta, t)\sin^2(\theta)\Delta t = \frac{pE}{B}\frac{\partial}{\partial\theta}\{N\sin^2(\theta)\}\,\mathrm{d}\theta\Delta t. \tag{7.38}$$

Consequently, the partial differential equation which describes the re-orientation of a dipole in time and space, $N(\theta, t)$, under the influence of an electromagnetic field is:

$$\frac{\partial N(\theta, t)}{\partial t} = \frac{pE}{B}\frac{\partial}{\partial\theta}\{N(\theta, t)\sin^2(\theta)\}. \tag{7.39}$$

This relation may be used for calculating the mean dipole moment of a single molecule isolated within this sphere. The spatial mean of the dipole moment is:

$$\langle p\rangle = E\{p\} = \int_0^{\pi} p\cos(\theta)\frac{N(\theta, t)}{2}\,\mathrm{d}\theta \tag{7.40}$$

and, using (7.39), the time-dependence of the mean is described by:

$$\begin{aligned}\frac{\mathrm{d}}{\mathrm{d}t}\{\langle p\rangle\} &= \int_0^{\pi} p\cos(\theta)\frac{\partial N}{\partial t}\frac{\sin^2(\theta)}{2}\,\mathrm{d}\theta \\ &= \frac{p^2E}{2B}\int_0^{\pi}\cos(\theta)\frac{\partial}{\partial\theta}\{N\sin^2(\theta)\}\,\mathrm{d}\theta\end{aligned} \tag{7.41}$$

from which it follows that:

$$\frac{\mathrm{d}}{\mathrm{d}t}\{\langle p\rangle\} = \frac{p^2 E}{2B}\int_0^{\pi} 2N\cos^2(\theta)\sin(\theta)\,\mathrm{d}\theta = \frac{p^2 E}{B}\frac{2N_0}{3} \tag{7.42}$$

where N_0 is the density of dipoles (assumed uniform) before the field is applied.

Thermal agitation is accounted for by assuming that the mean dipole moment approaches zero when the applied field is zero, and the rate of variation of the mean dipole moment is proportional to the zero-field dipole moment at any time:

$$\frac{\mathrm{d}}{\mathrm{d}t}\{\langle p\rangle\} + \frac{\langle p\rangle}{\tau} = \frac{2}{3}\frac{N_0 p^2 E(t)}{B}. \tag{7.43}$$

The mean dipole moment, $\langle p\rangle$, is thus given by a first order differential equation with time constant, τ, and a forcing function directly proportional to the applied electric field. This equation is equivalent to equation (7.27) in the introductory discussion.

By analogy to the theory of the macroscopic viscosity of small spheres of radius a in liquids, the damping constant is:

$$B = 8\pi\eta a^3 \tag{7.44}$$

where η is the viscosity. On the other side, for a constant electric field (7.40) gives:

$$\langle p\rangle = \frac{p^2 E}{3kT}. \tag{7.45}$$

So, the time constant, τ, is an inverse function of the temperature:

$$\tau = \frac{4\pi\eta a^3}{kT}. \tag{7.46}$$

As in the previous section we can obtain the frequency dependence using the Laplace transform to solve the differential equation:

$$\langle p\rangle = \frac{N_0 p^2}{3kT}\frac{E}{1+\mathrm{j}\omega\tau}. \tag{7.47}$$

The complex susceptibility, χ^*, is then proportional to the mean dipole moment:

$$\chi^*(\omega) = \frac{N_0 p^2}{3kT}\frac{1}{1+\mathrm{j}\omega\tau} \tag{7.48}$$

and the electric permittivity has the form:

$$\varepsilon^*(\omega) = \varepsilon_\infty + \frac{\varepsilon_0 - \varepsilon_\infty}{1+\mathrm{j}\omega\tau} \tag{7.49}$$

where:

$$\tau = \frac{4\pi\eta a^3}{kT} \quad \text{and} \quad \varepsilon_0 - \varepsilon_\infty = \frac{4\pi N_0 p^2}{3kT}. \tag{7.50}$$

This simple picture is a good model for the description of the process experienced by individual molecules in a liquid using a statistical framework. However, the assumptions which underlie these formulae must not be forgotten. It has been supposed that the molecule is a point dipole moment without dimension, from the electrical standpoint, and electrical interaction with the surroundings has been neglected. The finite spherical volume has been defined in order to consider the viscous influence of the surroundings, only. The interaction between molecules has been assumed to be equivalent to the viscous resistance of a liquid on the surface of the spherical volume element, which is of course questionable.

7.2.3 Extension of the Debye model to ideal solids

Let us now consider a dielectric model for solids which leads to the same relatively simple equations to describe the molecular interaction. Given a molecular solid in which a particular polar species is dispersed, assume that the orientation and position of this species are known. For simplification of the model without any adverse consequences, suppose that this dipolar species can exist in two equivalent stable positions: oriented up ($\uparrow$) or oriented down ($\downarrow$) with respect to an arbitrary coordinate frame. When no external electromagnetic field is applied, the potential energy of each position is assumed equal. For the transition of a site from one orientation to the other a certain excess energy, ΔH^* (J/molecule), must be supplied to the system. The expected number of molecules which jump the ΔH^* enthalpy barrier in the absence of an applied external field is $\exp(-\Delta H^*/kT)$, k is Boltzmann's constant; and the number jumping the barrier is very small if $\Delta H^* \gg kT$. This situation is described by the curve in Figure 7.7, in which site 1 and site 2 are at an equivalent potential (dashed curve).

If one applies an electric field, the potential of the two orientations is different. Molecules which are oriented with dipole moments parallel to the field (μ_1) are stabilized against a transition since the barrier now includes the external field as well as ΔH^*; the other molecules (μ_2) are destabilized since they are oriented against the field and the transition barrier is lowered by the external field. The dynamic equilibrium is upset and the equilibrium population of molecules which occupy states 1 and 2 changes. The rates of change of the

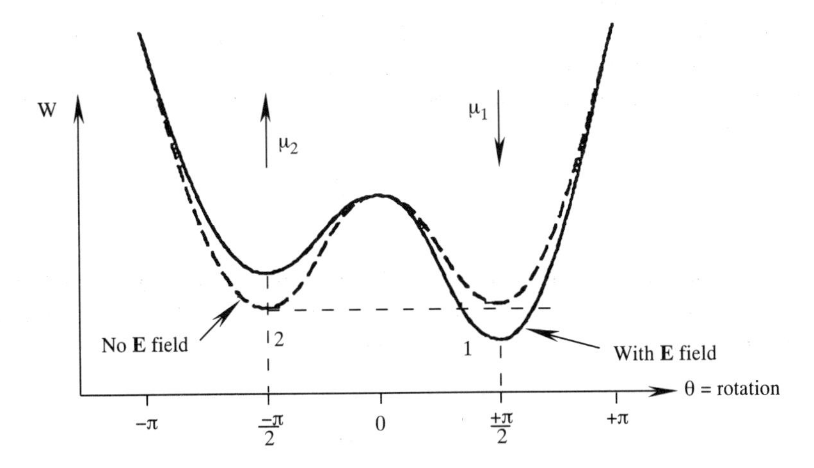

Fig. 7.7 *Potential energy of a dipole in a solid.*

populations of the two states are given by two rate equations:

$$\frac{\mathrm{d}N_1}{\mathrm{d}t} = -N_1 k_{12} + N_2 k_{21} \tag{7.51}$$

and

$$\frac{\mathrm{d}N_2}{\mathrm{d}t} = -N_2 k_{21} + N_1 k_{12} \tag{7.52}$$

where N_1 and N_2 are the populations of state 1 and 2, respectively, the total population is constant, $N_1 + N_2 = N$, and k_{12} and k_{21} are the respective rate coefficients. The transition rate coefficients (reaction velocities) k_{12} and k_{21} are the classical probabilities per second of conversion from one state to the other:

$$k_{12} = k_0 \exp\left[\frac{-\Delta H^* + \mathbf{p}\cdot\mathbf{E}}{kT}\right] \tag{7.53}$$

and;

$$k_{21} = k_0 \exp\left[\frac{-\Delta H^* - \mathbf{p}\cdot\mathbf{E}}{kT}\right]. \tag{7.54}$$

The k_{12} barrier is raised by $\mathbf{E}$ while the k_{21} barrier is lowered. We may use the entropy of the barrier, ΔS^* (J/mole K) to express the proportionality constant, k_0:

$$k_0 = \left(\frac{Nh}{kT}\right) \mathrm{e}^{[\Delta S^*/R]}. \tag{7.55}$$

The difference in the populations of states 1 and 2 can be described by the first order differential equation obtained from (7.51) and (7.52):

$$\frac{\mathrm{d}}{\mathrm{d}t}\{N_2 - N_1\} = -2k_{21}[N_2 - N_1] + k_{21}\frac{\mathbf{p}\cdot\mathbf{E}N}{kT} \tag{7.56}$$

where the exponential terms have been approximated by the first two terms of the power series representation: $\exp\{\pm x\} = 1 \pm x + \cdots$. It should be noted that this approximation for the exponential is only valid when x is very small (less than 1).

The difference in population between the two states is proportional to the electric field, and we can solve (7.56) to obtain:

$$N_2 - N_1 = \frac{\mathbf{p}\cdot\mathbf{E}}{2kT}N. \tag{7.57}$$

Whenever the polarization due to the external electric field is directly proportional to the difference $N_2 - N_1$, equation (7.57) is justified. It follows, then, that the forcing function for the polarization vector field is $[N_2 - N_1]$, and so:

$$\frac{\mathrm{d}P}{\mathrm{d}t} + \frac{P}{\tau} = \frac{N\mathbf{p}\cdot\mathbf{E}}{2kT} \tag{7.58}$$

which gives the same relaxation relationship in the idealized solid as for the Debye liquid.

7.2.4 Harmonic oscillator model

Another idealized model of an arbitrary material is a harmonic oscillator analogous to a forced spring–mass–dashpot mechanical system. Consider an electric charge q, of mass m, which is constrained to a particular position ($y = 0$) by a linear force of intensity λy (Figure 7.8). It is supposed that this charge can oscillate, and can experience dissipative viscous forces in interactions with its surroundings which are proportional to the speed at which it moves. The equation describing the position of the charge about its equilibrium position is given by

$$m\frac{\partial^2 y}{\partial t^2} + s\frac{\partial y}{\partial t} + \lambda y = q\mathrm{E}\mathrm{e}^{\mathrm{j}\omega t} \tag{7.59}$$

where the applied field is collinear with the y-axis and of frequency ω. We may quickly solve the differential equation in the frequency domain by using the Fourier transform — or, equivalently, set $y(\omega) = A\exp[\mathrm{j}(\omega t + \phi)]$:

$$y(\omega) = \frac{\dfrac{q}{m}E}{\dfrac{\lambda}{m} + \mathrm{j}\dfrac{s}{m}\omega - \omega^2}. \tag{7.60}$$

The oscillator has a natural frequency, ω_0, and a damping coefficient, α:

$$\omega_0 = \sqrt{\frac{\lambda}{m}} \qquad \alpha = \frac{s}{2m}. \tag{7.61}$$

The value of y is the position of the charge. It will be more convenient if we factor out the spring constant, λ, and rewrite the position in terms of the normalized frequency, $x = \omega/\omega_0$ with $k = s/m\omega_0$:

$$y(x) = \frac{qE}{\lambda(1 + \mathrm{j}kx - x^2)}. \tag{7.62}$$

The equivalent polarization for N identical oscillators is:

$$\mathbf{P} = qN\mathbf{y} = \frac{q^2 N\mathbf{E}}{\lambda(1 + \mathrm{j}kx - x^2)}. \tag{7.63}$$

and from this we may obtain the susceptibility:

$$\chi^*(\omega) = \frac{\chi(0)}{1 + \mathrm{j}kx - x^2} \tag{7.64}$$

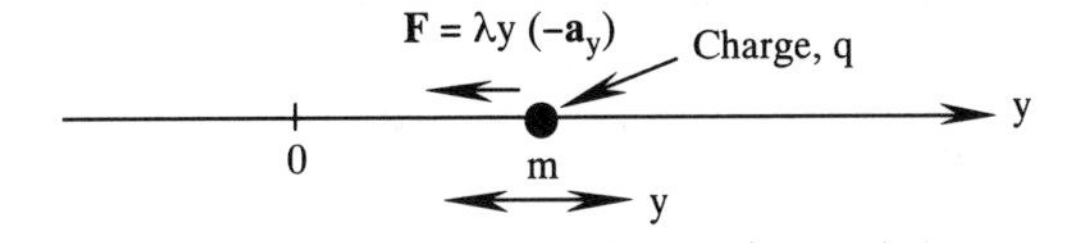

Fig. 7.8 *Harmonic oscillator model.*

where:

$$\chi(0) = \frac{q^2 N}{\varepsilon_0 m \omega_0^2}. \tag{7.65}$$

The curves of the two parts of the susceptibility reveal a number of interesting features of the behavior of the material. The real and imaginary parts are given by:

$$\chi'(\omega) = \chi(0)\frac{1-x}{(1-x)^2 + k^2x^2} \tag{7.66}$$

$$\chi''(\omega) = \chi(0)\frac{kx}{(1-x)^2 + k^2x^2}. \tag{7.67}$$

If k is small (very little damping) then oscillations persist and both $\chi'(\omega)$ and $\chi''(\omega)$ have a pronounced peak due to the inertial effects. As k increases to 1, the damping causes a spread in both $\chi'(\omega)$ and $\chi''(\omega)$ with a corresponding elimination of the peak (Figure 7.9a). Finally, for $k = 10$ (Figure 7.9a) the predicted response is more typical of experimentally observed frequency dependence for χ' and χ''. The corresponding Cole–Cole plot is shown in Figure 7.9b.

The elementary polarization, $f(t)$, can also be obtained from the differential equation by applying an electric field in the form of a Dirac delta function:

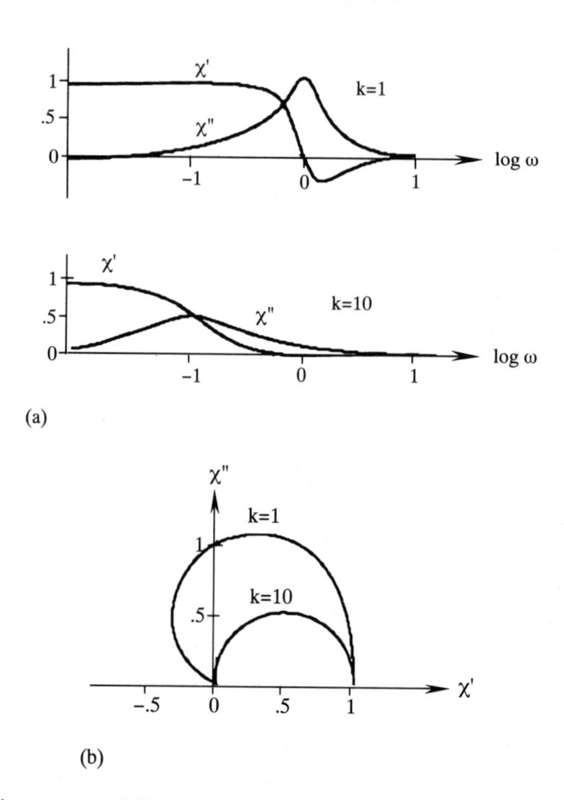

Fig. 7.9 *(a) Complex susceptibility for a damped harmonic oscillator and (b) its Cole–Cole representation.*

$$f(t) \cong e^{-\frac{1}{2}smt} \sin(\omega_0 t) \tag{7.68}$$

The special case where the inertial effects are weaker than the combined forces of the electric field and the viscosity leads to the Debye theory:

$$s\frac{\partial y}{\partial t} + \lambda y = qE e^{j\omega t} \tag{7.69}$$

for which

$$y(\omega) = \frac{qE}{1 + js\omega} e^{j\omega t}. \tag{7.70}$$

The resulting expression for the electric permittivity is recognizable:

$$\chi^*(\omega) = \frac{q^2 N}{\varepsilon_0 \lambda \left(1 + j\frac{s}{\lambda}\omega\right)} \tag{7.71}$$

which has a time constant and dc susceptibility:

$$\tau = \frac{s}{\lambda} \qquad \chi(0) = \frac{q^2 N}{\varepsilon_0 \lambda} \tag{7.72}$$

and resulting impulse response:

$$f(t) = \frac{\alpha}{\tau} e^{-t/\tau} \tag{7.73}$$

7.3 SUMMARY

Unfortunately, this simple representation is only idealized, and is not precisely verified by experiment. The following chapter describes how to modify these models to more closely match them with the experimental data.

A more elaborate theory was derived to describe the relationship between macroscopic and molecular dielectric relaxation behavior around 1960 (Glarum (1973); see Böttcher and Bordewijk (1978), Chapter 10 for a discussion). Glarum's theory was based on Kubo's general theory of statistical mechanics for linear dissipative systems. The mathematical framework uses correlation functions of fluctuating dipole moments in a system at equilibrium and a system in which the dipoles are evolving in time due to a driving electric field. The effect of the forcing field was expressed in terms an appropriate Hamiltonian operator in a quantum mechanical description of the system. While the theory is effective, it can only be successfully applied in the simpler cases, such as pure liquids composed of non-rigid molecules. Non-rigid molecules are molecules which may take many stable configurations through partial rotation or inversion. For many applications to industrial materials the Glarum theory is too complex to use.

Many simpler models have been suggested to describe the permittivity of mixtures and other interesting industrial materials. The next chapter will discuss several of the most effective models which have been used to describe experimental permittivity data.

REFERENCES

Barriol, J. (1957) *Les Moments Dipolaires* Gauthier-Villars, Paris.

Böttcher, C. J. F. and Bordewijk, P. (1978) *Theory of Electric Polarization* Elsevier, Amsterdam.

Digest of Literature on Dielectrics vol 1-29, NAS — NRC, Washington DC, 1950-1969.

Fröhlich, H. (1955) Oxford University Press, Oxford *Theory of Dielectrics*.

Glarum, S. H. *Mol. Phys.*, **27**, 241, 1973.

Johnscher, A. K. *Dielectric Relaxation in Solids* Chelsea Dielectrics Press, London, 1983.

Morse, P. M. and Feshbach, H. (1968) *Methods of Theoretical Physics* Vols 1 and 2, Academic Press, New York.

Priou, A. (1992) *Dielectric Properties of Heterogeneous Materials, Progress in Electromagnetic Research*, Volume 6 (J.A. Kong, Ed.), Elsevier, Amsterdam.

8 Generalization of Dielectric Relaxation in Real Materials

The dynamic nature of the interaction between materials and electric fields is by far the most important consideration in this section of the book since we are concerned with high intensity and high frequency fields. Also, the theoretical descriptions presented in Chapter 7 are extremely useful, in that they help to explain important phenomena. However, much of the experimental data differs from the theoretical curves. The nature of the deviations and their causes, discussed in this chapter, will complete our discussion of dielectric properties.

8.1 INTRODUCTION

We may extend the Debye formula for ideal polar liquids to include lossy solutions (equation (7.50)):

$$\varepsilon^*(\omega) = \varepsilon'(\infty) + \frac{\varepsilon'(0) - \varepsilon'(\infty)}{1 + \mathrm{j}\omega\tau} + \frac{\sigma}{\mathrm{j}\omega} \tag{8.1}$$

where $\varepsilon'(0)$ and $\varepsilon'(\infty)$ are the low and high frequency asymptotes for ε' and we have assumed a single relaxation time, τ. A Debye material has a Cole–Cole plot which is semi-circular with center on the ε' axis (i.e. the abscissa) located at $\{\varepsilon'(\infty) + \varepsilon'(0)\}/2$. The imaginary part, ε'', has a maximum at $\omega_{\mathrm{m}} = 1/\tau$ when $\sigma = 0$.

The theory is very important because it delineates the processes of dielectric relaxation. Some generalized conclusions can be drawn from this theory. First, absorption may be explained as arising from the rotational mobility of the dipolar molecules or from the hindered translation of isolated electric charges. Second, the equation is sensitive to microscopic viscosity forces; thus, one is obliged to carefully specify the internal electric forces which stabilize a liquid or the structure of solid. Third, the temperature is an important parameter due to thermal agitation of the electric species (dipoles and free charges); we will use a relaxation time, τ, which has an assumed first order temperature dependence as a model:

$$\tau = A\mathrm{e}^{U/kT} \tag{8.2}$$

where A is a proportionality constant and U is the internal energy of rotation of the dipoles. So, the frequency and temperature variations of ε^* are inseparable. However, in many cases the Debye relationship does not fit experimental data particularly well, even for fairly well behaved liquids. In most cases where Debye theory is not applicable, equation (8.2) can still be applied to model the temperature dependence.

8.2 IMPROVEMENTS TO THE DEBYE AND FRÖHLICH THEORIES

There have been many attempts to improve the Debye relations to fit experimental data. Much of the literature before 1950 contains results which are fit to empirically derived formulae; so, before we present the universal expressions we will review the improvements to the Debye relations. The formulae will be useful in the interpretation of early publications, particularly when the reported results contain only the fit parameters. There are three common improvement relations which have been applied to the Debye relations: (1) the Cole–Cole formula, (2) the Cole–Davidson formula, and (3) the Havriliak–Negami equation. Each is equivalent to assuming a continuous distribution of relaxation times. We will briefly review each and then present the associated relaxation time distributions and resulting impulse responses.

8.2.1 Cole–Cole formula

The 1941 formula of Cole and Cole is based on the observation that deviations from Debye behavior can be thought of as due to the existence of several relaxation times clustered around τ. Using a fit parameter, α, they were able in many cases to fit the measured complex susceptibility by:

$$\chi^*(\omega) = \frac{1}{1 + (j\omega\tau)^{(1-\alpha)}} \tag{8.3}$$

where $0 < \alpha < 1$. The real part of χ is:

$$\chi'(\omega) = \frac{1}{2}\left\{1 - \frac{\sinh[(1-\alpha)\ln(\omega\tau)]}{\cosh[(1-\alpha)\ln(\omega\tau)] + \sin(\alpha\pi/2)}\right\} \tag{8.4}$$

and the imaginary part is:

$$\chi''(\omega) = \frac{\cos(\alpha\pi/2)}{\cosh[(1-\alpha)\ln(\omega\tau)] + \sin(\alpha\pi/2)}. \tag{8.5}$$

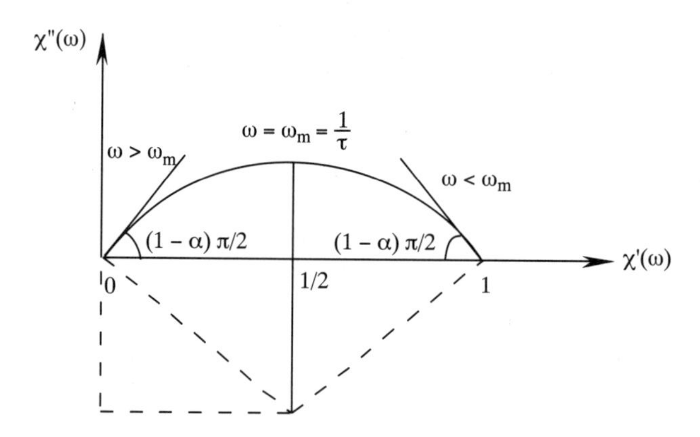

Fig. 8.1 *Cole–Cole formula.*

The Cole–Cole plot for the susceptibility is also a circular arc, but the center of the circle is below the real axis (Figure 8.1) located at (x, y):

$$x = \tfrac{1}{2} \qquad y = \tfrac{1}{2}\tan\left\{(1-\alpha)\frac{\pi}{2}\right\} \tag{8.6}$$

where $y < 0$, as expected. The loss factor again has a maximum at $\omega_m = 1/\tau$:

$$\varepsilon''(\max) = \frac{\varepsilon'(0) - \varepsilon'(\infty)}{2}\,\frac{\cos(\alpha\pi/2)}{1 + \sin(\alpha\pi/2)} \tag{8.7}$$

It is notable that when $\alpha = 0$, the Cole–Cole formula reduces to the Debye formula; thus α represents a measure of deviation from Debye behavior. The Cole–Cole formula satisfies the Kramers–Krönig relations, as well.

8.2.2 Cole–Davidson formula

Later, in 1950, Cole and Davidson proposed another formula which uses an exponent β applied to the entire denominator rather than $1 - \alpha$ applied only to the frequency term:

$$\chi^*(\omega) = \frac{1}{(1 + j\omega\tau)^\beta} \tag{8.8}$$

which has a real part of:

$$\chi'(\omega) = \cos(\beta\phi)[\cos(\phi)]^\beta \tag{8.9}$$

and an imaginary part of:

$$\chi''(\omega) = \sin(\beta\phi)[\cos(\phi)]^\beta \tag{8.10}$$

where the separation variable, $\phi = \arctan\{\omega\tau\}$.

When $\beta = 1$, the formula again reduces to the Debye relation. The rationale for this approach is that high frequency behavior often does not fit the semi-circular arc form. The plot of this curve at low frequencies is similar to the Debye plot (the tangent at the $\chi'(0)$ intercept is $\pi/2$), but at high frequencies, the tangent to the curve at $\chi'(\infty)$ is $\beta\pi/2$ rather than $\pi/2$ (Figure 8.2). Note that the maximum in χ'' is not at $1/\tau$ in this case.

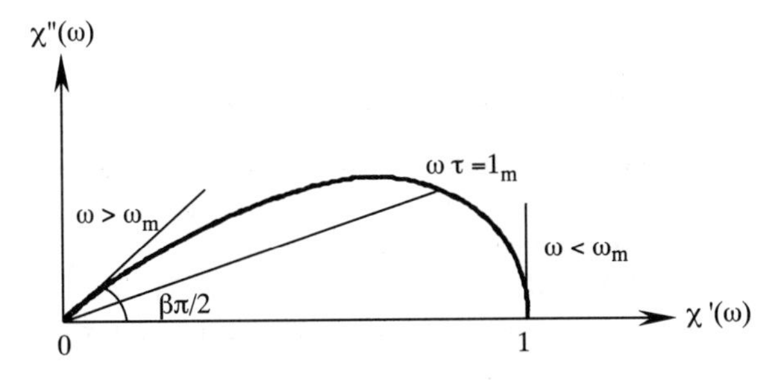

Fig. 8.2 *Cole–Davidson formula.*

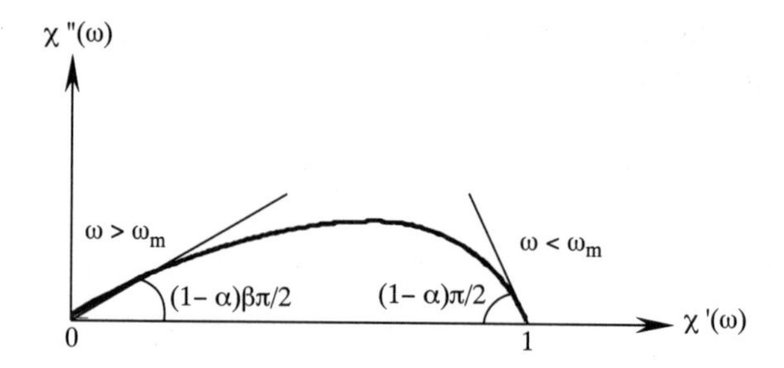

Fig. 8.3 *Havriliak–Nagami equation.*

8.2.3 Havriliak–Negami equation

The Havriliak–Negami relation combines the Cole–Cole and Cole–Davidson improvements. Power law coefficients β and $1-\alpha$ are both applied:

$$\chi^*(\omega) = \frac{1}{[1+(\mathrm{j}\omega\tau)^{(1-\alpha)}]^{\beta}}. \tag{8.11}$$

Separation of the real and imaginary parts gives complicated expressions. However, they may be easily be solved on a small computer.

$$\chi'(\omega) = \frac{\cos(\beta\phi)}{[1+2(\omega\tau)^{(1-\alpha)}\sin(\alpha\pi/2)+(\omega\tau)^{2(1-\alpha)}]^{\beta/2}} \tag{8.12}$$

$$\chi''(\omega) = \frac{\sin(\beta\phi)}{[1+2(\omega\tau)^{(1-\alpha)}\sin(\alpha\pi/2)+(\omega\tau)^{2(1-\alpha)}]^{\beta/2}} \tag{8.13}$$

where the parameter ϕ is now defined by:

$$\phi = \arctan\left[\frac{(\omega\tau)^{(1-\alpha)}\cos(\alpha\pi/2)}{1+(\omega\tau)^{(1-\alpha)}\sin(\alpha\pi/2)}\right]. \tag{8.14}$$

The Cole–Cole plot for the Havriliak–Negami equation is now an asymmetric curve at both high and low frequency intercepts. The plot intercepts the real axis at the angles: $(1-\alpha)\pi/2$ at low frequency and $(1-\alpha)\beta\pi/2$ at high frequency (Figure 8.3).

There are other formulae, but they are not often used to model the frequency dependence of experimental determinations of ε' and ε''.

8.2.4 Distribution of relaxation times

We have mentioned that the improvements in the Debye relation were associated with a distribution of relaxation times. In this section we will relate each of the formulae to their relaxation time distribution and impulse response.

In the discussion of Fröhlich theory of solids (Chapter 7), it was supposed that only two states existed for the moving molecules. However, in a more realistic sense, one must consider cases in which there are many states. The relaxation of molecules among the states is evidently much more complex since the mobility of molecules then depends upon the distribution of the states. A simple method for the description of such a situation as this is to consider a continuous distribution of relaxation times. The associated probability density function, $g(\tau)$, is the probability of having molecules in states with relaxation times between τ and $\tau + \mathrm{d}\tau$. For mathematical convenience, $g(\tau)$ will be expressed in normalized form using a reference relaxation time τ_0:

$$\underset{(0,+\infty)}{\tau} \rightarrow \underset{(-\infty,+\infty)}{\ln\{\tau\}} \quad \text{and} \quad g(\tau) \rightarrow G(\tau/\tau_0). \tag{8.15}$$

This change of variables does not introduce any alteration in the sense of the equations. Then, the complex permittivity will be given by a simple generalization taking into account that the total distribution of relaxation is normalized:

$$\chi^*(\omega) = \int_{\tau=0}^{\tau=\infty} \frac{G(\tau/\tau_0)\,\mathrm{d}(\ln(\tau))}{1 + \mathrm{j}\omega\tau} \tag{8.16}$$

where

$$\int_{\tau=0}^{\tau=\infty} G(\tau/\tau_0)\,\mathrm{d}\{\ln(\tau)\} = 1. \tag{8.17}$$

This method for describing the susceptibility is equivalent to using the impulse response, $f(t)$. The function $f(t)$ has been defined as the inverse cosine transform of $\chi'(\omega)$ and also as the inverse sine transform of $\chi''(\omega)$, which is valid because of the special properties of $f(t)$ (Chapter 7). We may also calculate $f(t)$ as a function of $G\{\tau/\tau_0\}$, and reciprocally $G\{\tau/\tau_0\}$ as a function of $f(t)$ with the one-sided Laplace transform:

$$f(t) = \int_{\tau=0}^{\tau=\infty} G\left(\frac{\tau}{\tau_0}\right)\frac{1}{\tau}\mathrm{e}^{-(t/\tau)}\,\mathrm{d}\{\ln(\tau)\}. \tag{8.18}$$

Cole–Cole formula distribution function

The Cole–Cole formula has a distribution of relaxation times of this type. Böttcher and Bordewijk (1978) present the derivation of $G\{\tau/\tau_0\}$ from $\varepsilon''(\omega)$ using a limit approximation. Briefly, the angle of intersection at the point $[\varepsilon'(0), 0]$ is $(1-\alpha)\pi/2$, as mentioned above. Though often difficult to do with experimental data, it is possible to obtain $G\{\tau/\tau_0\}$ from singularities in $\varepsilon^*(\omega)$ (see Böttcher and Bordewijk 1978):

$$G\left\{\frac{\tau}{\tau_0}\right\} = \frac{1}{2\pi}\lim_{\omega\to 0}\left[\chi^*\left\{\left(-\omega + \frac{\mathrm{j}}{\tau}\right)\tau_0\right\} - \chi^*\left\{\left(\omega + \frac{\mathrm{j}}{\tau}\right)\tau_0\right\}\right] \tag{8.19}$$

and for the Cole–Cole relation:

$$G\left\{\frac{\tau}{\tau_0}\right\} = \frac{1}{2\pi}\frac{\sin(\pi\alpha)}{\cosh\{(1-\alpha)\ln(\tau_0/\tau)\} - \cos(\pi\alpha)}. \tag{8.20}$$

When $\alpha = 0$ a single relaxation time at $\tau = \tau_0$ is realized, corresponding to the Debye model. As α increases the amplitude of the peak decreases and the peak broadens (Figure 8.4a).

In addition to this result we can obtain the impulse response of the material once a good fit has been made with experimental data using equation (8.18). In the present case,

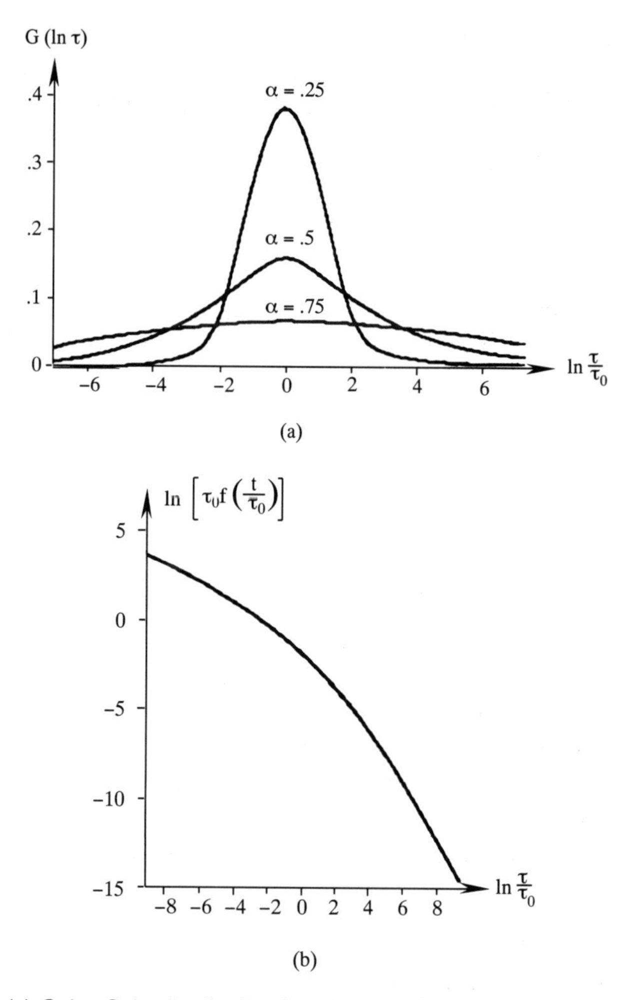

Fig. 8.4 *(a) Cole–Cole distribution function. (b) Cole–Cole impulse response.*

however, the transform is not calculable in a direct form. It is possible to apply a power series approach in order to obtain asymptotic behavior at short times ($t \ll \tau_0$):

$$f(t) \cong \frac{1}{\tau_0} \sum_{n=0}^{\infty} \frac{(-1)^n}{\Gamma\{(n+1)(1-\alpha)\}} \left(\frac{t}{\tau_0}\right)^{n(1-\alpha)-\alpha} \tag{8.21}$$

and at long times ($t \gg \tau_0$):

$$f(t) \cong \frac{1}{\tau_0} \sum_{n=1}^{\infty} \frac{(-1)^n}{\Gamma\{n(\alpha-1)\}} \left(\frac{t}{\tau_0}\right)^{-n(1-\alpha)-1} \tag{8.22}$$

with some cross-over behavior in between (Figure 8.4b).

Cole–Davidson relaxation distribution

Applying the same approach as was used for the Cole–Cole formula, the distribution function can be obtained for the Cole–Davidson case (Figure 8.5a):

$$G\left\{\frac{\tau}{\tau_0}\right\} = \begin{cases} 0 & \text{for } \tau > \tau_0 \\ \dfrac{1}{\pi}\left(\dfrac{\tau}{\tau_0 - \tau}\right)^{\beta} \sin(\pi\beta) & \text{for } \tau < \tau_0 \end{cases}. \tag{8.23}$$

The singularity at $\tau = \tau_0$ is integrable, the Laplace transform can be calculated directly in this case, and the impulse response originally presented by Davidson is (see Figure 8.5b):

$$f(t) = \frac{1}{\tau_0 \Gamma(\beta)}\left(\frac{t}{\tau_0}\right)^{\beta-1} e^{-t/\tau_0}. \tag{8.24}$$

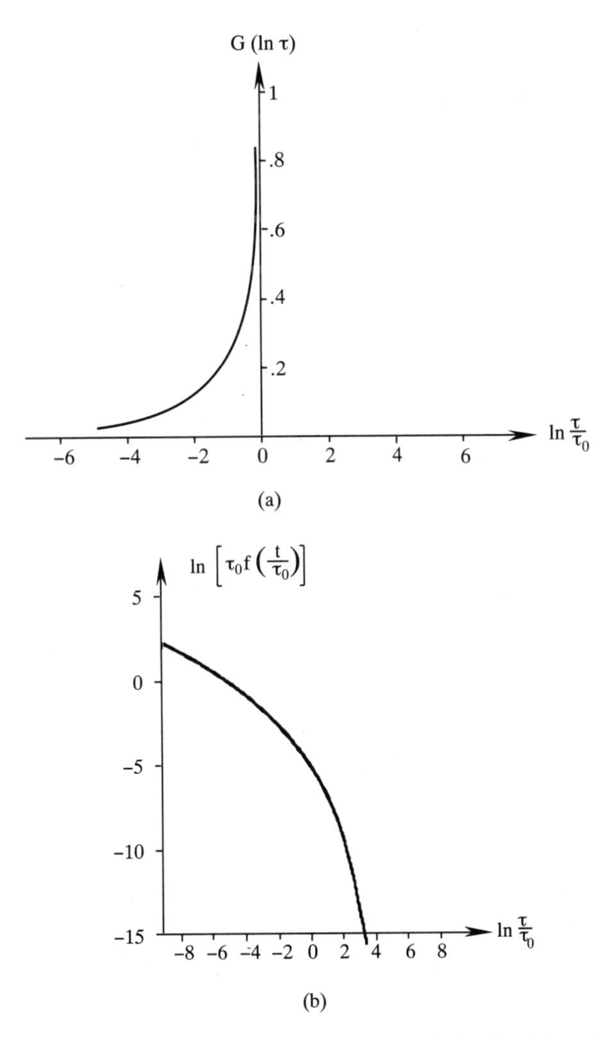

Fig. 8.5 *(a) Cole–Davidson distribution function. (b) Cole–Davidson impulse response.*

Havriliak–Negami distribution function

The complete distribution function was computed by Havriliak and Negami:

$$G\left\{\frac{\tau}{\tau_0}\right\} = \frac{1}{\pi}\frac{(\tau/\tau_0)^{\beta(1-\alpha)}\sin(\beta\theta)}{[(\tau/\tau_0)^{2(1-\alpha)} + 2(\tau/\tau_0)^{(1-\alpha)}\cos(\pi(1-\alpha)) + 1]^{\beta/2}} \tag{8.25}$$

where the simplification parameter, θ, is:

$$\theta = \arctan\left\{\frac{\sin((1-\alpha)\pi)}{[(\tau/\tau_0) + \cos((1-\alpha)\pi)]}\right\}. \tag{8.26}$$

Gaussian distribution function

In many cases, the above formulae are not sufficient to describe more complicated dielectric relaxation systems. One type of distribution which has often been used is a Gaussian relaxation time distribution. Applying the Gaussian distribution function is reasonable since by the Central Limit Theorem as the number of interacting random processes becomes large the distribution function approaches a Gaussian distribution:

$$G\{\ln(\tau)\} = \frac{a}{\sqrt{\pi}}\exp\left[-a^2\left(\ln\left(\frac{\tau}{\tau_0}\right)\right)^2\right]. \tag{8.27}$$

Here the Gaussian fit parameter, a, is the inverse of the usual standard deviation for the function, $a = 1/(\sigma\sqrt{2})$. Then, the real and imaginary parts of the susceptibility can be calculated from the following integral formulae:

$$\chi'(\omega) = \frac{a}{\sqrt{\pi}}\mathrm{e}^{-a^2x^2}\int_0^{+\infty}\exp\left[-a^2u^2\frac{\cosh((2a^2x-1)u)}{\cosh(u)}\right]\mathrm{d}u \tag{8.28}$$

for the real part, and:

$$\chi''(\omega) = \frac{a}{\sqrt{\pi}}\mathrm{e}^{-a^2x^2}\int_0^{+\infty}\exp\left[-a^2u^2\frac{\cosh(2a^2xu)}{\cosh(u)}\right]\mathrm{d}u \tag{8.29}$$

for the imaginary part, where in both expressions the simplification parameter contains the frequency dependence: $x = \ln\{\omega\tau_0\}$. Equations (8.28) and (8.29) must be calculated numerically since closed form expressions are only obtainable in very special cases (i.e. when $2a^2x$ is an even integer for ε' and an odd integer for ε'').

Other distribution functions

Many other distribution functions have been used in obtaining descriptions of susceptibility from experimental data. A complete list is not warranted; however, several of the more effective ones are given below.

Kirkwood–Fuoss distribution derived for long polymer molecules:

$$G\left\{\frac{\tau}{\tau_0}\right\} = \frac{1}{2\cosh\{\ln(\tau/\tau_0)\} + 2}. \tag{8.30}$$

Fröhlich distribution (simplified band-limited model):

$$G\left\{\frac{\tau}{\tau_0}\right\} = \begin{cases} \dfrac{1}{\ln(a_1/a_2)} & \text{for } a_1\tau_0 < \tau < a_2\tau_0 \\ 0 & \text{elsewhere.} \end{cases} \tag{8.31}$$

Matsumoto–Higasi distribution function:

$$G\left\{\frac{\tau}{\tau_0}\right\} = \begin{cases} \dfrac{p\tau^p}{(a_1^p - a_2^p)\tau_0^p} & \text{for } a_1\tau_0 < \tau < a_2\tau_0 \\ 0 & \text{elsewhere} \end{cases} \tag{8.32}$$

where p is any non-zero real number.

All of the above distribution functions satisfy equation (8.17).

8.3 TEMPERATURE EFFECTS

As the single-relaxation time (Debye) model illustrates, the temperature dependence of τ ensures that the temperature dependence and frequency dependence of χ^* are inextricably linked. The incorporation of temperature dependence into the relaxation distributions is especially worthwhile in our study of high intensity EM fields. The consequence of the frequency–temperature link is that in some situations we may encounter a positive feedback mechanism which results in thermal runaway—we will discuss runaway in more detail in Chapter 10.

In this discussion we will also separate electrical conductivity effects from susceptibility using the general formulation:

$$\varepsilon^*(\omega) = \varepsilon'(\infty) + \frac{\sigma}{j\omega} + \varepsilon_0\chi^*(\omega) \tag{8.33}$$

where $\varepsilon'(\infty)$ is again the high frequency limiting value which is essentially insensitive to temperature variation. The temperature dependence of electrical conductivity has been discussed in terms of electrolytic solutions; however, other relationships may govern different materials. The temperature dependence of susceptibility will be treated in this section.

First, reconsider the Debye model with relaxation time temperature dependence given by the Arrhenius formula (8.2):

$$\varepsilon'(\omega) = \varepsilon'(\infty) + \frac{\varepsilon'(0) - \varepsilon'(\infty)}{1 + \omega^2\tau^2} \tag{8.34}$$

$$\varepsilon''(\omega) = (\varepsilon'(0) - \varepsilon'(\infty))\frac{\omega\tau}{1 + \omega^2\tau^2} \tag{8.35}$$

and

$$\tau = Ae^{U/RT} \tag{8.36}$$

where now U is the energy per mole in a macroscopic model so that R is the universal gas constant and has replaced Boltzmann's constant in equation (8.2).

When the dependence of the variation of both the real and imaginary parts of the dielectric constants is included in the relations, the duality between temperature and frequency behavior is easily seen:

$$\varepsilon'(\omega, T) = \varepsilon'(\infty) + \frac{\varepsilon'(0) - \varepsilon'(\infty)}{2}\left[1 - \tanh\left\{\frac{U}{RT} + \ln(\omega A)\right\}\right] \tag{8.37}$$

$$\varepsilon''(\omega, T) = \frac{\varepsilon'(0) - \varepsilon'(\infty)}{2} \frac{1}{\cosh\left\{\dfrac{U}{RT} + \ln(\omega A)\right\}}. \tag{8.38}$$

For a given frequency, one can consider the curves of ε' and ε'' as a function of temperature. Generally, the quantities $\varepsilon'(\infty)$ and $\varepsilon'(0)$ are largely invariant in comparison with the other parameters. It is possible to define a temperature T_{m} as a value of the temperature for which ε'' is maximum at a known value of the frequency; similar to the way we used $1/\tau$ as the frequency which corresponded to the maximum absorption at a given temperature. When the description of the dielectric behavior follows the Debye formula, the value T_{m} is given by:

$$T_{\mathrm{m}} = -\frac{U}{R\ln(\omega A)}. \tag{8.39}$$

It is interesting that T_{m} is frequency dependent. This expression can be used to recast equations (8.37) and (8.38) in terms of T alone:

$$\varepsilon'(\omega, T) = \varepsilon'(T) = \varepsilon'(\infty) + \frac{\varepsilon'(0) - \varepsilon'(\infty)}{2}\left[1 - \tanh\left\{\frac{U}{R}\left(\frac{1}{T} - \frac{1}{T_{\mathrm{m}}}\right)\right\}\right] \tag{8.40}$$

$$\varepsilon''(\omega, T) = \varepsilon''(T) = \frac{\varepsilon'(0) - \varepsilon'(\infty)}{2\cosh\left\{\dfrac{U}{R}\left(\dfrac{1}{T} - \dfrac{1}{T_{\mathrm{m}}}\right)\right\}}. \tag{8.41}$$

Experimental data in which the electric permittivity is determined as a function of temperature can be presented on a plot similar to the Cole–Cole diagram. Figure 8.6 is a temperature plot on Cole–Cole axes for glycerol triacetate at 100 kHz.

The results in equations (8.40) and (8.41) were derived specifically for an ideal Debye material. When the Debye formula is not sufficiently accurate a distribution function, similar

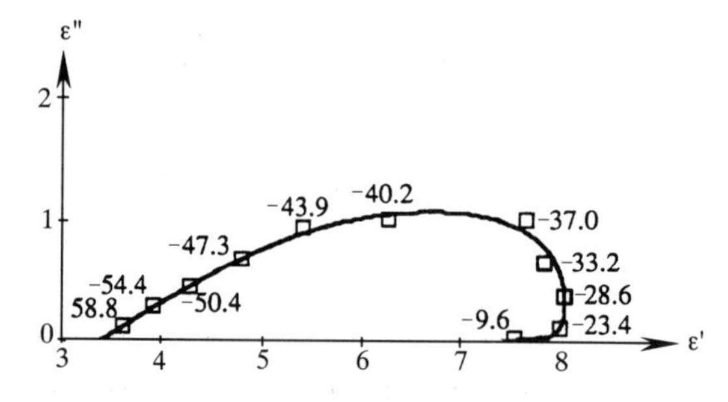

Fig. 8.6 *Cole–Cole permittivity of glycerol triacetate vs temperature at 100 kHz.*

to that used for τ above, may be applied to the pre-exponential factor, A, to obtain a more accurate representation:

$$\varepsilon''(\omega, T) = (\varepsilon'(0) - \varepsilon'(\infty)) \int_{A=0}^{A=\infty} \frac{G(A)\,\mathrm{d}\{\ln(A)\}}{\cosh\left\{\dfrac{U_A}{RT} + \ln(\omega A)\right\}}. \tag{8.42}$$

8.4 THE UNIVERSAL LAW OF DIELECTRIC PHENOMENA

In more modern approaches to the behavior of dielectric materials it has become apparent that all materials follow a similar pattern providing that the available data are sufficient to show the dependence. Consequently, the various improvements to the Debye theory are of historical interest and not at all necessary to the proper interpretation of measurements. In this section we will discuss the recent observations by Jonscher and by Dissado and Hill which support this view. Interestingly, the so-called universal law fits data collected over about eight decades of frequency, far more than any of the other approaches.

8.4.1 Jonscher formulation

After examining the results of many experiments from diverse sources, Jonscher (1983) proposed that all dielectric materials follow universal relations. He discovered that for the entirety of published data $\ln(\chi''(\omega))/\ln(\omega)$ is linear at low frequencies and also at high frequencies. A typical plot of this function is shown in Figure 8.7. Consequently, the real part of the electric susceptibility is correlated to the same indices (by the Kramers–Krönig relations, equation (7.20)) so that for the approximate formula the imaginary part and the real part have this same behavior. Jonscher justifies the empirical fit for both parts by means of the Kramers–Krönig relations:

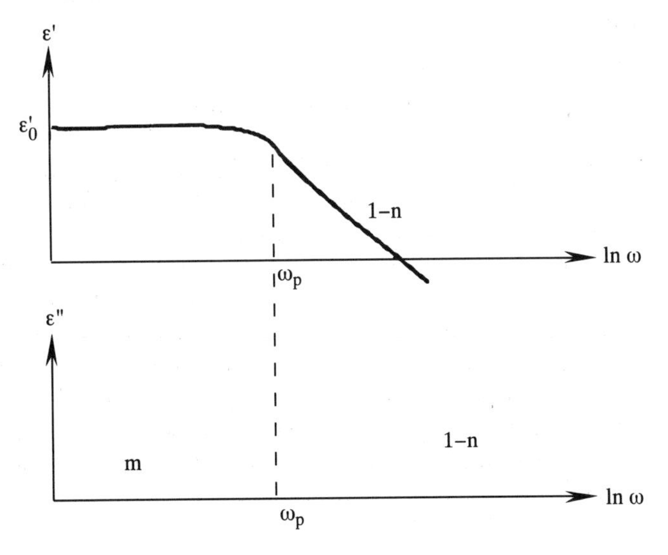

Fig. 8.7 *Log–log Jonscher plot for a representative material.*

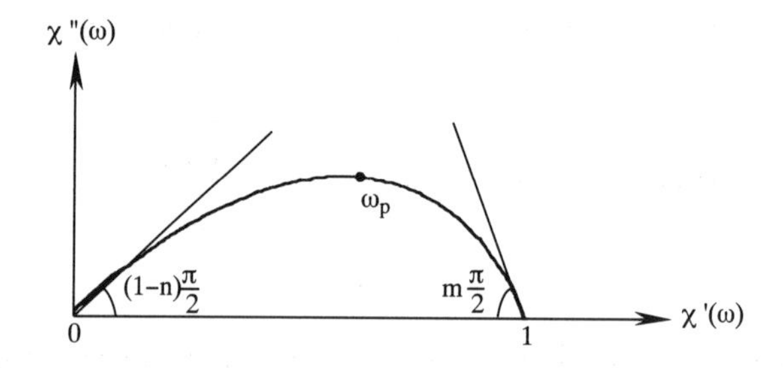

Fig. 8.8 *Cole–Cole plot for the universal electric response.*

$$\text{if} \quad \omega < \omega_\text{p}, \quad \text{then} \quad \chi''(\omega) \approx \omega^m \quad \text{and} \quad \chi'(\omega) \approx 1 - \chi''(\omega) \tag{8.43}$$

$$\text{if} \quad \omega > \omega_\text{p}, \quad \text{then} \quad \chi''(\omega) \approx \omega^{n-1} \quad \text{and} \quad \chi'(\omega) \approx \chi''(\omega) \approx \omega^{n-1} \tag{8.44}$$

where ω_p is a critical frequency near which the losses are maximum, and m and $1 - n$ are slopes derived from the log–log plot of Figure 8.8. He also verifies from the data that ω_p is temperature dependent and follows an Arrhenius function:

$$\omega_\text{p} = A\text{e}^{-W/kT}. \tag{8.45}$$

Based on these observations, the appropriate functional form for $\chi''(\omega)$ obtained by Jonscher is:

$$\chi''(\omega) = \frac{A}{(\omega/\omega_\text{p})^{1-n} + (\omega_\text{p}/\omega)^m}. \tag{8.46}$$

The values of $\chi'(\omega)$ can then be determined numerically from the Kramers–Krönig relations. Although most experimental data are acceptably fitted by this approach, there is no clear theoretical foundation for it.

8.4.2 Dissado and Hill cluster analysis

An early attempt to derive a universal theoretical description of regression to equilibrium by Dissado, Nigmatullin and Hill employed a rigorous quantum mechanical/statistical thermodynamics approach (Dissado and *et al.* 1985). Later, Dissado and Hill (1989) used a simplified cluster model to obtain the same result. They succeeded in theoretically describing the observed behavior by studying how a system perturbed away from equilibrium by an electric field returns to equilibrium as the fluctuations regress. They found that the impulse response function, $f(t)$, is governed by a second order differential equation, the coefficients of which are linearly dependent on time — a nonlinear response. It turns out that $f(t)$ is the product of three functions: (1) a decreasing exponential, $\exp\{-\omega_\text{p}t\}$, (2) a negative power function, $(\omega_\text{p}t)^{-n}$, and (3) a confluent hypergeometric function which also contains the indices $1 - n$ and m:

$$f(t) \approx a(\omega_\text{p}t)^{-n}\text{e}^{-\omega_\text{p}t}F(1 - m, 2 - n; \omega_\text{p}t) \tag{8.47}$$

where a is a constant of proportionality. They then proved, using the Fourier transform, that the complex susceptibility takes the form:

$$\chi^*(\omega) \approx \left[\frac{1}{1+\mathrm{j}\omega/\omega_\mathrm{p}}\right]^{1-n} \frac{F\left(1-n, 1-m, 2-n; \dfrac{1}{1+\mathrm{j}\omega/\omega_\mathrm{p}}\right)}{F(1-n, 1-m, 2-n; 1)} \tag{8.48}$$

where the frequency has been normalized with respect to ω_p. F is now a Gaussian hypergeometric function whose arguments redefine the indices m and n introduced by Jonscher; and the temperature dependence is included in ω_p.

It can be shown that for $m = 0$ the hypergeometric function reduces to:

$$\chi^*(\omega) = \frac{1}{\left(1+\mathrm{j}\omega/\omega_\mathrm{p}\right)^{1-n}}. \tag{8.49}$$

Likewise, if $1 - n = m = \frac{1}{2}$ then the hypergeometric function reduces to an arcsin:

$$\chi^*(\omega) = \frac{2}{\pi}\arcsin\left(\sqrt{\frac{1}{1+\mathrm{j}\omega/\omega_\mathrm{p}}}\right). \tag{8.50}$$

Definition of cluster analysis

Fundamentally, the Dissado–Hill theory examines fluctuations in clusters. Previously, dielectric relaxation was interpreted by trying to adjust the microscopic properties of the system such that these microscopic properties yielded the macroscopic properties. This approach is difficult: the primary difficulty is that there is no way to calculate all of the properties of a liquid, such as water, from the properties of the individual molecules. Some liquid properties do not depend directly upon the molecular properties, but are governed by molecular interaction. Dissado and Hill eliminated this difficulty by considering the material to be a collection of "clusters". A cluster is the smallest volume of material which has the macroscopic property under consideration. Here it is the fluctuating movement of the charge (electrical particles) connected to fluctuations in the electric field which constitutes the material. So, cluster analysis is an important analytical technique and considerable discussion is warranted.

The definition of a cluster depends upon the properties which are considered. For example, in wet paper a cluster is the arrangement of macroscopic molecules, which represent the solid paper, upon which at least one molecule of water is fixed. A cluster of rubber is at least three or more polymeric molecules connected by a molecule which includes a sulfide atom if vulcanization is due to an S-bridge. When a new substance is introduced into the rubber, the cluster must contain at least one molecule of the new substance. Consequently, the size of a cluster can change with changes in the composition of the material.

Application to a system of clusters

Fluctuation of orientation and motion of particles may be considered to include the notion of clusters in a material. Since the cluster itself is composed of different molecules the internal and external (relative to other clusters) fluctuations must be defined. If the applied electric field is of very high frequency, that is $\omega \gg \omega_\mathrm{p}$, it will induce new fluctuations which are only at high frequency. The fluctuations remain internal to the cluster, so the system can be reduced to a single cluster and other clusters (induced cluster–cluster interactions) can be ignored in the analysis. Low frequency fields, $\omega \ll \omega_{\mathrm{p}}$, on the other hand, may be thought of as inducing no modification of the individual clusters, but only changing their interaction dynamics, and thus their relative movement. The boundary frequency, ω_{p}, which separates the effects depends on the electrical size of the cluster.

The description of fluctuations was examined by quantum mechanical and statistical thermodynamic means. Dissado and Hill consider the fluctuations in motion induced by the electric field and how they regress back to an equilibrium distribution of states — that is, how they interact to mix to create the incoherent ensemble of clusters which existed prior to the application of the **E** field. The theory is similar for rapid and slow fluctuations, but the size of the elemental object in each formulation is very different: the first case is written for particles within a cluster and the second for the clusters themselves. The two different indices in equations (8.47) and (8.48) represent the differing regimes: $1 - n$ describes the effects of changes within a cluster, and m describes the cluster–cluster interactions. The m index corresponds to low frequency fluctuations in the orientation and vibrational motion between clusters. The detailed theoretical derivation of (8.47) and (8.48) is beyond the scope of this book; however, an inspection of the primary results is of interest.

Cluster analysis explains the origin of the analytical expression of the impulse response, $f(t)$. The impulse response may be approximately represented by two power laws:

$$\text{for } t \ll \frac{1}{\omega_{\mathrm{p}}} \qquad f(t) \approx \alpha t^{-n} \tag{8.51}$$

$$\text{for } t \gg \frac{1}{\omega_{\mathrm{p}}} \qquad f(t) \approx \beta t^{-m-1} \tag{8.52}$$

and near the critical frequency the material can be approximated as a Debye substance:

$$\text{for } t \approx \frac{1}{\omega_{\mathrm{p}}} \qquad f(t) \approx \gamma e^{-\omega_{\mathrm{p}} t} \tag{8.53}$$

where α, β and γ are arbitrary constants of proportionality.

The powers m and n contain significant information regarding the structure of the material. The power n is a measure of the structural order of the cluster: as n approaches zero (in the limit case) the cluster dipoles become disentangled with the surroundings and the material reverts to Debye behavior. Similarly, m is a measure of the extent to which one cluster can affect others in its environment, and can be thought of as the efficiency with which dipolar displacement can be exchanged between clusters. So m is related to the isolation of clusters. For values of m near 1 the material distributes a disturbance nearly homogeneously as it approaches equilibrium. The m and $1 - n$ indices are nearly constant under varying temperature conditions in physically stable materials.

In general, materials can contain many levels of organizational structure, similar to societies: family $\Rightarrow$ town $\Rightarrow$ county (department) $\Rightarrow$ state (province) $\Rightarrow$ country. A particular frequency range around any ω_{p} will be related to each level of organization. The dielectric susceptibility spectrum will then contain many separate relaxation domains. When these domains overlap it is not necessary to define separate clusters.

The Dissado and Hill formulation is the most comprehensive to date. It provides an incisive description of the micro-mechanics behind the experimental results reported in the literature. The use of hypergeometric functions, however, can make the results inordinately clumsy to apply. We have shown (Roussy *et al.* 1992) that the calculation of the hypergeometric function, F, can be replaced by an equivalent, but much simpler, integral form:

$$\chi'\left(\frac{\omega}{\omega_p}\right) = 1 - A\int_0^{\tan^{-1}(\omega/\omega_p)} \frac{\cos\{(1-n+m)u - m\pi/2\}}{\cos^n(u)\sin^{1-m}(u)}\,du \tag{8.54}$$

$$\chi''\left(\frac{\omega}{\omega_p}\right) = A\int_0^{\tan^{-1}(\omega/\omega_p)} \frac{\sin\{m\pi/2 - (1-n+m)u\}}{\cos^n(u)\sin^{1-m}(u)}\,du \tag{8.55}$$

where the normalization factor is given by:

$$\frac{1}{A} = \int_0^{\pi/2} \frac{\cos\{(1-n+m)u - m\pi/2\}}{\cos^n(u)\sin^{1-m}(u)}\,du. \tag{8.56}$$

These relations avoid the calculation of the Gaussian hypergeometric function, and are especially useful when the series implementation for F converges slowly. Also, these relations lead simultaneously to:

$$\frac{d\{\chi''(\omega)\}}{d\{\chi'(\omega)\}} = \tan\left\{(1-n+m)\tan^{-1}\left(\frac{\omega}{\omega_p}\right) - \frac{m\pi}{2}\right\} \tag{8.57}$$

from which the determination of $1-n$, m, and ω_p are straightforward if data at both very low and very high frequencies are available.

The expression for the distribution function of relaxation times associated with equation (8.57) has also been determined. In integral form it is:

$$\text{for } 0 < \frac{\tau}{\tau_0} < 1 \qquad G\left(1-n, m; \frac{\tau}{\tau_0}\right) = C\sin((1-n)\pi)\int_0^{\tau/\tau_0} \frac{du}{u^n(1-u)^{1-n+m}} \tag{8.58a}$$

$$\text{for } \frac{\tau}{\tau_0} > 1 \qquad G\left(1-n, m; \frac{\tau}{\tau_0}\right) = C\sin(m\pi)\int_{\tau/\tau_0}^{\infty} \frac{du}{u^n(u-1)^{1-n+m}} \tag{8.58b}$$

and in both expressions the constant C is:

$$C = \frac{1-n}{\pi F(1-n, 1-m, 2-n; 1)} = \frac{\Gamma(1-n)\Gamma(m)}{\pi\Gamma(1-n+m)}. \tag{8.58c}$$

Low frequency dispersion

At frequencies below about 1 Hz in most physical systems and about 1 kHz in biological systems a second type of dispersion has been observed experimentally. While these effects are outside the frequency range of immediate interest, we present them here very briefly as a matter of completeness. This dispersion effect occurs in systems in which some of the charges are weakly bound and partially free to move. In these cases the loss component, χ'', increases with decreasing frequency as does the real part of χ^*. This low frequency dispersion cannot be described by, nor included within, the $\sigma/j\omega$ term because of the connectedness between the real and imaginary part. It must be a part of the susceptibility since it follows Kramers–Krönig behavior if a parameter, p, is introduced:

$$\text{for } \omega \ll \omega_p \qquad \chi^*(\omega) \approx \alpha \left(\frac{j\omega}{\omega_p}\right)^{-p} \tag{8.59}$$

and α is again a constant of proportionality. It has been shown that this type of response can be explained as transport of charges between clusters and that a relationship comparable to that in equation (8.48) can be obtained:

$$\chi^*(\omega) \approx \left[\frac{1}{1 + j\dfrac{\omega}{\omega_p}}\right]^{1-n} \frac{F\left(1-n, 1+p, 2-n; \dfrac{1}{1 + j\omega/\omega_p}\right)}{F(1-n, 1+p, 2-n; 1)}. \tag{8.60}$$

In this case the index p is a measure of the efficiency of charge transfer between clusters. As this efficiency approaches 1 the ac electrical conductivity contained in $\omega\chi''(\omega)$ approaches a constant value such as that characteristic of a dc conductivity. Hence, p is also a measure of the long-range inhomogeneity.

While this consideration is of little direct interest in RF and MW field analysis, it is instructive to note that in the case of particles jumping from site to site on a "lattice" ω_p will be associated with the time of a standard jump (the time required for a charge to escape the cluster) and $\chi(0)$ will be the charge displacement. Then the value p defines the geometry of the jump path and $1 - n$ defines the geometry of the inter-cluster path. Consequently, inhomogeneity on a large scale and on a local microscopic scale can be distinguished and separately analyzed.

8.4.3 Dielectric diagnostics

The discussion so far has illuminated how the response to an electric field at varying frequency can be a probe to investigate not only the static structure of a material but also the mobility of charges. Microwave or dielectric spectroscopy studies of condensed materials gives many valuable insights into the fine structure and can be used to study material transformations. A full description of the band concept interpretation of dielectric response is outside the scope of this book. However, analysis of the electric permittivity as a function of frequency complements other assessments of material transformation under high power density fields at a fixed frequency.

8.5 COMPLEX PERMITTIVITY OF HETEROGENEOUS MIXTURES

So far we have only considered homogeneous materials. The discussion will describe the more common problem of heterogeneous materials. There exist very important classes of heterogeneous materials, such as plastics, which are very poor absorbers of microwave energy. We would like to identify additives which will enhance the absorption of electromagnetic energy such that dielectric heating becomes more attractive for industrial use. Microwave and radio frequency techniques would then be applicable for the fabrication of reinforced plastics. The problem is to control the total absorption by predicting the absorption as a function of the composition of absorbing and non-absorbing species. This is a tedious problem, even from the theoretical point of view, since it is not at all evident that ε^* depends solely upon the proportions of the composition.

Consider a large parallel plate capacitor with surface area S. Assuming an approximately uniform electric field distribution (i.e. that $S \gg d^2$), its capacitance when filled with air is:

$$C_0 = \varepsilon_0 \frac{S}{d} \tag{8.61}$$

where d is the separation distance between the plates. If two materials, 1 and 2, constitute the dielectric such that each occupies half of the volume of the capacitor, the capacitance will depend upon how the two materials are distributed. If material 1 is placed beside material 2 in parallel, each occupying half the capacitor area, as in Figure 8.9a, the equivalent capacity would be the sum of the halves — two capacitances in parallel — since the electric field is the same in each medium (tangential **E**-field boundary condition). The total capacitance is then:

$$C = \varepsilon_1^* \frac{S}{2d} + \varepsilon_2^* \frac{S}{2d} = \varepsilon_{\text{eff}}^* \frac{S}{d} \tag{8.62}$$

where the effective value of the permittivity is half of the sum of ε_1^* and ε_2^*:

$$\varepsilon_{\text{eff}}^* = \frac{\varepsilon_1^* + \varepsilon_2^*}{2}. \tag{8.63}$$

If the two materials are placed in series as two sheets with each sheet $d/2$ thick (Figure 8.9b), then the interface between the two materials is an equipotential surface and the normal **E**-field boundary condition applies. The total capacitance of this configuration is:

$$\frac{1}{C} = \frac{1}{\varepsilon_1^* 2S/\text{d}} + \frac{1}{\varepsilon_2^* 2S/\text{d}} \tag{8.64a}$$

where:

$$\varepsilon_{\text{eff}}^* = \frac{2\varepsilon_1^* \varepsilon_2^*}{\varepsilon_1^* + \varepsilon_2^*}. \tag{8.64b}$$

In like manner, the electric permittivity of a two-component heterogeneous medium depends critically upon the geometrical distribution of both constituents. The general formulation to obtain an analytical expression for the volumetric mean permittivity as a function of both media in a heterogeneous system has not yet been solved, even when the geometrical

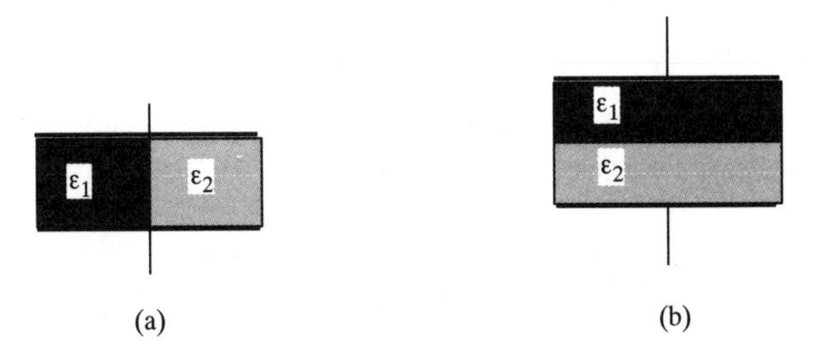

Fig. 8.9 *(a) Parallel dielectric plate capacitor. (b) Series dielectric plate capacitor.*

distribution is known. There exist, nevertheless, special cases for which useful approximate formulae have been obtained.

In the examination of these special cases, the specific types of heterogeneous media must be distinguished. The first class of heterogeneous media is described by a homogeneous material with constant ε_1 in which small particles of a second homogeneous material, with ε_2, are widely dispersed. These particles can be dispersed regularly or irregularly. In the simple case, material 2 is considered to be in discrete spheres, but could also be of asymmetric form. The second case is a heterogeneous mixture described by "layers" of two materials, which can be evenly distributed or interwoven, such that there exists a continuous path in each medium from one point to another point. The third class of heterogeneous mixture will be isolated parts of materials 1 and 2, which are independent of other parts of the same material, perhaps with vacuum interstices. These three classes are called dispersed in continuous phase, continuous–continuous medium, and dispersed–dispersed media, respectively.

There are many formulae for the representation of the permittivity of a heterogeneous mixture as a function of the permittivity of each component. Three of the most important formulae will be presented here.

8.5.1 Rayleigh formula for dispersed in continuous phase mixtures

The Rayleigh formula is for a two-component medium where material 1 is a continuum in which material 2, composed of small spherical particles, is widely dispersed. The formula was derived by inspection of the electric fields around the spheres, as we saw in Chapter 6, and is:

$$\frac{\varepsilon_{\mathrm{m}}^* - \varepsilon_1^*}{2\varepsilon_1^* + \varepsilon_{\mathrm{m}}^*} = \Phi \frac{\varepsilon_2^* - \varepsilon_1^*}{2\varepsilon_1^* + \varepsilon_2^*} \tag{8.65}$$

where Φ is the volume fraction of material 2 in 1, ε_1^* and ε_2^* are the respective permittivities for the two materials and the m subscript refers to the mixture. This formula is also valid for liquids.

8.5.2 Böttcher formula

The Böttcher formula also describes a distribution of particles dispersed in a homogeneous phase. The special characteristic of this formula is that it is symmetric over the indices for the two materials:

$$\frac{\varepsilon_{\mathrm{m}}^* - \varepsilon_1^*}{3\varepsilon_{\mathrm{m}}^*} = \Phi \frac{\varepsilon_2^* - \varepsilon_1^*}{2\varepsilon_{\mathrm{m}}^* + \varepsilon_2^*}. \tag{8.66}$$

Equation (8.66) can be expanded to include a three-component system with particles of 1 and 2 suspended in a large volume of solvent (subscript s):

$$\frac{\varepsilon_{\mathrm{m}}^* - \varepsilon_{\mathrm{s}}^*}{3\varepsilon_{\mathrm{m}}^*} = \Phi_1 \frac{\varepsilon_1^* - \varepsilon_{\mathrm{s}}^*}{2\varepsilon_{\mathrm{m}}^* + \varepsilon_1^*} + \Phi_2 \frac{\varepsilon_2^* - \varepsilon_{\mathrm{s}}^*}{2\varepsilon_{\mathrm{m}}^* + \varepsilon_2^*} \tag{8.67}$$

where Φ_1 and Φ_2 are the respective volume fractions of the two dispersed media.

8.5.3 Bruggeman–Hanaï formula

The mixture formula proposed by Bruggeman and Hanaï is:

$$\left(\frac{\varepsilon_{\mathrm{m}}^* - \varepsilon_1^*}{\varepsilon_2^* - \varepsilon_1^*}\right)^3 + (1 - \Phi)\frac{\varepsilon_{\mathrm{m}}^*}{\varepsilon_1^*} = 1. \tag{8.68}$$

This is a nonlinear formula in which the effective permittivity of the heterogeneous medium is asymmetric for ε_1^* with ε_2^* dispersed within it. The formula easily permits the calculation of the permittivity of the dispersed phase, ε_2^* in medium 1, where Φ is the again the volume fraction. Unfortunately, the determination of $\varepsilon_{\mathrm{m}}^*$ requires the solution of a complex cubic equation with three possible roots. Although inconvenient to use, the Bruggeman–Hanaï relation is valid for many mixtures, emulsions, dye baths, floatation baths, etc.

8.5.4 Looyenga equation

The Looyenga equation is a power law approximation which has been successfully used in two forms (Nelson 1992):

$$(\varepsilon_{\mathrm{m}}^*)^{1/3} = \Phi_1(\varepsilon_1^*)^{1/3} + \Phi_2(\varepsilon_2^*)^{1/3} \tag{8.69}$$

and in a more general expression:

$$(\varepsilon_{\mathrm{m}}^*)^k = \Phi_1(\varepsilon_1^*)^k + \Phi_2(\varepsilon_2^*)^k \tag{8.70}$$

where m refers to the mixture and Φ_1 and Φ_2 are the respective volume fractions ($\Phi_1 + \Phi_2 = 1$), and k is constrained to $0 < k < 1$.

8.5.5 Lichtnecker formula

Lichtnecker proposed a power law mixture formula for a single component (ε_2) suspended in a host medium (ε_1) with volume fraction of suspended material Φ:

$$\varepsilon_{\mathrm{m}}^* = \varepsilon_1^*\left(\frac{\varepsilon_2^*}{\varepsilon_1^*}\right) = (\varepsilon_1^*)^{(1-\Phi)}(\varepsilon_2^*)^{\Phi} \tag{8.71}$$

where $\varepsilon_{\mathrm{m}}^*$ is the mixture permittivity. This relation is easily extended to multiple components dispersed in a single host medium:

$$\varepsilon_{\mathrm{m}}^* = \prod_{i=2}^{\mathrm{N}}(\varepsilon_i^*)^{\Phi_i} \tag{8.72}$$

where: $\Sigma\Phi_i = 1$.

None of the above formulae have a strong theoretical basis, although most of them give acceptably good approximations at low concentration. The underlying assumption is that the particles are relatively small and dispersed in a relatively large volume of known electric permittivity (air or solvent). For intermediate or extreme particle concentrations their behavior is highly variable and thus questionable. The relative applicability of the relations has been reviewed many times with mixed results (Priou (1992)). Nelson *et al.* have applied this type of mixture formula to the electric permittivity of volumes of various grains.

8.5.6 Maxwell–Wagner effect

A special discussion of the Maxwell–Wagner effect is appropriate at this point since it is a prime example of the inversion which occurs when a small volume of semiconductor is suspended in a dielectric matrix of negligible electrical conductivity. For example, we may encounter isolated pockets of electrolyte or semiconducting impurities suspended in a dielectric matrix. When isolated regions of semiconducting material are suspended in a non-conducting medium the application of an external electric field generates a non-uniform charge distribution.

Conducting and semiconducting sphere fields

The case of conducting and semiconducting inclusions within a dielectric substance is more complex than it first appears. For example, a conducting sphere in a uniform electric field will have a surface charge induced on it so that the internal electric field is zero. If the sphere is a perfect conductor the internal electric field will be zero at all frequencies. We may see this clearly by reviewing the solution of the Laplace equation for this situation which was discussed in Chapter 4. Of course, the underlying assumptions are that the particles are spherical and their radius, a, is very small with respect to a wavelength at the frequency of operation — so the quasi-static assumption applies.

For perfectly conducting metallic spheres, the electric field near the sphere is:

$$\mathbf{E} = E_0 \left\{ \left[1 - \left(\frac{a}{r}\right)^3\right] \sin(\theta)\mathbf{a}_\theta - \left[1 + 2\left(\frac{a}{r}\right)^3\right] \cos(\theta)\mathbf{a}_r \right\} \tag{8.73a}$$

and at the surface of the metallic sphere ($r = a^+$) the electric field has only a normal component:

$$\mathbf{E}(a^+, \theta) = -3E_0 \cos(\theta)\mathbf{a}_r \tag{8.73b}$$

A surface charge, ρ_s, cancels the electric field so that the field inside the sphere is zero:

$$\rho_s = -3\varepsilon'_1 E_0 \cos(\theta) \tag{8.73c}$$

where ε'_1 is the permittivity of the surroundings. This results in a net negative charge on the upper hemisphere ($0 < \theta < \pi/2$) exactly balanced by a net negative charge on the lower hemisphere ($\pi/2 < \theta < \pi$). In that sense the particle is "polarized" because of the redistribution of free charge induced by the external field and a dipole results.

The perfectly conducting sphere produces a surface charge immediately since its surface charge time constant, $\tau = \varepsilon/\sigma \approx 0$ (see Chapter 1). For a semiconducting spherical inclusion (Figure 8.10) the internal field is a function of the dimensions of the sphere and the frequency of the external electric field, in addition to the electric properties. To study what happens we may imagine a semiconducting sphere suspended in a perfect dielectric exposed to an instantaneous step external electric field. At $t = 0^+$ (the instant after the field is applied) the surface charge will be zero everywhere on the sphere since charge has not had time to

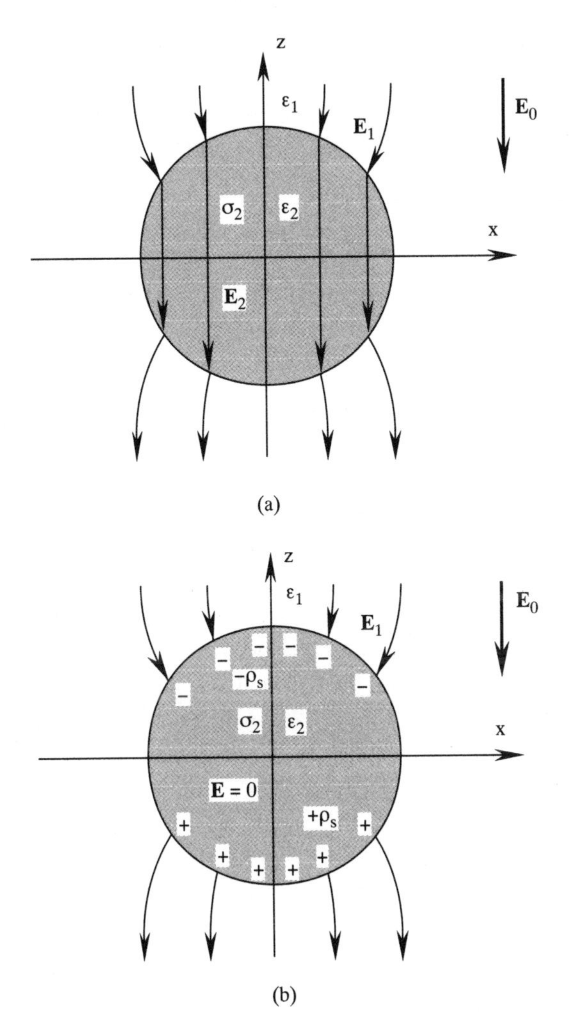

Fig. 8.10 *Semiconducting sphere in an external electric field. (a) Initially, no surface charge exists and the dielectric sphere solution applies. (b) After a very long time surface charge cancels the extenal field* ($\mathbf{J}_2 = 0$).

respond to the external field. At that time the solution for the dielectric sphere of Chapter 6 applies since it was derived assuming no surface charge. Outside the sphere:

$$\mathbf{E}_1 = -E_0\left[1 + 2\frac{\varepsilon_2 - \varepsilon_1}{\varepsilon_2 + 2\varepsilon_1}\left(\frac{a}{r}\right)^3\right]\cos(\theta)\mathbf{a}_r + E_0\left[1 - \frac{\varepsilon_2 - \varepsilon_1}{\varepsilon_2 + 2\varepsilon_1}\left(\frac{a}{r}\right)^3\right]\sin(\theta)\mathbf{a}_\theta \quad (8.74a)$$

where the solution at $t = 0^+$ is determined by the real part of the permittivity of the sphere, ε_2, and region 2 is inside the sphere. We are assuming an idealized semiconductor in which all losses are characterized by σ_2 and assume that ε_2 is real in this discussion. The electric field inside the sphere at $t = 0^+$ is then:

$$\mathbf{E}_2 = \frac{3\varepsilon_1}{2\varepsilon_1 + \varepsilon_2}\mathbf{E}_0 = \frac{3\varepsilon_1 E_0}{2\varepsilon_1 + \varepsilon_2}[-\cos(\theta)\mathbf{a}_r + \sin(\theta)\mathbf{a}_\theta]. \quad (8.74b)$$

The reader may quickly verify the surface charge is zero by applying the normal electric field boundary condition at $r = a$.

The free charge in the semiconductor will move in response to the internal field since $\mathbf{J} = \sigma\mathbf{E}$. Consequently, a surface charge will be induced arising from the available free charge in the semiconductor. For a step electric field, charge will move until, as t becomes large and steady state is reached, the internal electric field is zero. At that point equations (8.73) above will apply with surface charge given by equation (8.73c). So, for a static external field the ideal conductor solution applies exactly to the semiconductor case. By contrast, at the other end of the frequency spectrum, when ω becomes very large indeed (say, at optical frequencies), the internal free charge in the semiconductor will not be able to move in response to the external field. No net surface charge will be observed at any instant in time and the dielectric sphere result applies exactly, equations (8.74). At either end of the frequency spectrum the semiconducting inclusion is a very poor absorber, though local fields may be influenced by its presence.

At some intermediate frequency, which depends on carrier mobility (i.e. conductivity) and sphere dimensions, the time required to establish the surface charge will be of the same order as the electric field half-period, and the semiconducting inclusion will absorb maximally from the external field due to Joule heating $(\mathbf{E}\cdot\mathbf{J})$ within the inclusion. These Maxwell–Wagner absorption peaks may be observed over the RF range in some materials (usually at frequencies less than about 50 MHz or so). The sphere dimensions and conductivity determine the resonance time constant, τ.

Following the discussion in Chapter 4, we may estimate the surface charge for a harmonic external electric field by modifying the dielectric sphere result to include the semiconducting properties of the sphere: $\varepsilon_2^* = \varepsilon_2' - \mathrm{j}\sigma_2/\omega$. The external and internal fields are (respectively):

$$\mathbf{E}_1 = -E_0\left[1 + 2\frac{\varepsilon_2^* - \varepsilon_1}{\varepsilon_2^* + 2\varepsilon_1}\left(\frac{a}{r}\right)^3\right]\cos(\theta)\mathbf{a}_r + E_0\left[1 - \frac{\varepsilon_2^* - \varepsilon_1}{\varepsilon_2^* + 2\varepsilon_1}\left(\frac{a}{r}\right)^3\right]\sin(\theta)\mathbf{a}_\theta \quad (8.75a)$$

$$\mathbf{E}_2 = \frac{3\varepsilon_1}{2\varepsilon_1 + \varepsilon_2^*}\mathbf{E}_0 = \frac{3\varepsilon_1 E_0}{2\varepsilon_1 + \varepsilon_2^*}[-\cos(\theta)\mathbf{a}_r + \sin(\theta)\mathbf{a}_\theta]. \quad (8.75b)$$

From these two relations and the electric flux density boundary condition—$\rho_s = \varepsilon_1'\mathrm{E}_{1n} - \varepsilon_2'\mathrm{E}_{2n}$—it may be shown that:

$$\rho_s = \frac{\mathrm{j}\sigma/\omega}{\varepsilon_2^* + 2\varepsilon_1}3\varepsilon_1 E_0\cos(\theta). \quad (8.75c)$$

Note that the surface charge is not in phase with the electric field. It reduces to equation (8.73c) when either σ becomes very large or ω becomes very small. The surface charge is zero for $\sigma = 0$, as must be the case. This "low-pass" type function has a single pole at $\omega_p = \sigma/(\varepsilon_1 + \varepsilon_2)$, which may be interpreted as the roll off frequency for ρ_s.

Application to a suspension of metallic spheres

We may use these observations to obtain an estimate of the permittivity of a heterogeneous material containing small spherical conducting inclusions. We will consider the case in which an ideal lossless dielectric material ($\varepsilon_1^* = \varepsilon_1'$) has a conductor ε_2^* dispersed within it:

$$\varepsilon_2^* = \varepsilon_2' - \mathrm{j}\frac{\sigma_2}{\omega}. \quad (8.76)$$

An example of such a situation would be a plastic reinforced by dispersed carbon or metal particles, such as copper or aluminum powder. Due to the geometry of the particles, the simple Rayleigh formula yields the permittivity equation (since the assumptions and geometry used by Rayleigh apply):

$$\varepsilon_{\mathrm{m}}^{*} = \frac{\varepsilon_1'(2\varepsilon_1' + \varepsilon_2^*) + 2\Phi\varepsilon_1'(\varepsilon_2^* - \varepsilon_1')}{2\varepsilon_1' + \varepsilon_2^* + \Phi(\varepsilon_2^* - \varepsilon_1')}. \tag{8.77}$$

The complex permittivity consequently follows a Debye-type dispersion:

$$\varepsilon_{\mathrm{m}}^{*} = \varepsilon'(\infty) + \frac{\varepsilon'(0) - \varepsilon'(\infty)}{1 + \mathrm{j}\omega\tau}. \tag{8.78}$$

The values of the high frequency permittivity, $\varepsilon'(\infty)$, the amplitude of relaxation phenomena, $\varepsilon'(0) - \varepsilon'(\infty)$, and the relaxation time, τ, may be deduced from the above:

$$\varepsilon'(\infty) = \varepsilon_1' + 3\Phi\frac{\varepsilon_2' - \varepsilon_1'}{2\varepsilon_1' + \varepsilon_2' - \Phi(\varepsilon_2' - \varepsilon_1')} \tag{8.79}$$

$$\varepsilon'(0) = \varepsilon_1'\left(1 + \frac{3\Phi}{1 - \Phi}\right) \tag{8.80}$$

and

$$\tau = \frac{2\varepsilon_1' + \varepsilon_2' - \Phi(\varepsilon_2' - \varepsilon_1')}{\sigma_2(1 - \Phi)}. \tag{8.81}$$

These formulae can be used to indicate what happens in a practical heterogeneous material. They are, nevertheless, insufficient to describe the full behavior. Additionally, in some cases a "percolation" phenomenon appears in dispersed heterogeneous media when the dispersed volume fraction, Φ, exceeds 30%. That is, there exists a significant probability of finding a statistically continuous path in the dispersed medium from one inclusion boundary to another. Of course, these simple formulae cannot handle the complex case of high volume fractions.

8.5.7 Bergman relation for effective permittivity

The Bergman relation represents a very significant step forward in the general approach to determining the effective permittivity of isotropic composite materials. To date, the problem of calculating the effective permittivity of a heterogeneous material from the properties of its components does not have a general solution. Attempts to generate multipolar expansions of the fields around inclusions and their neighbors have not yielded sufficiently accurate estimates; even though the effective permittivity was obtained from successive terms which take into account higher order multiple interactions and though the precision could be increased as desired. Therefore, it seems unlikely that deterministic microscopic field analyses will yield convenient results without a considerable alteration in the methods used.

Since about 1980 statistical approaches based on probabilistic descriptions of the physical mixture in terms of a density function have shown the most promise. In brief, the probability density function describes the random location of the dispersed component inside the host material. The governing relationships and associated mathematical treatments lead to sets of bounds for the permittivity. The bounds are limiting values within which the effective permittivity of the mixture must reside; and the bounds take into account the geometry of the structure. In spite of the success, the bounds are not immediately applicable in engineering analyses.

However, in a connection with boundary value theory, Bergman did find the effective permittivity of any two-component composite mixture, $\varepsilon_{\mathrm{m}}^*$ in terms of a spatial density function, $g(x)$ in an integral form:

$$\frac{\varepsilon_{\mathrm{m}}^* - \varepsilon_1^*}{\varepsilon_1^*} = \int_0^1 \frac{g(x)\,\mathrm{d}x}{x + \varepsilon_1^*/(\varepsilon_2^* - \varepsilon_1^*)} \tag{8.82}$$

where material 2 is dispersed in medium 1 and the integration is over all possible positions—for reasons which will be made clear.

Origin of the Bergman relation

The origin of this relation is as follows. First, Bergman observed that $\varepsilon_{\mathrm{m}}^*$ is a homogeneous function of both variables ε_1^* and ε_2^* which means that $\varepsilon_{\mathrm{m}}^*/\varepsilon_1^*$ depends on $\varepsilon_2^*/\varepsilon_1^*$ rather than on ε_1^* and ε_2^* separately. Second, he constructed $f(s)$:

$$f(s) = \frac{\varepsilon_{\mathrm{m}}^* - \varepsilon_1^*}{\varepsilon_1^*} \qquad \text{where} \qquad s = \frac{\varepsilon_1^*}{\varepsilon_2^* - \varepsilon_1^*}. \tag{8.83}$$

It turns out that $f(s)$ is an analytic function with the particularly interesting property that $\mathrm{Im}\{\varepsilon_{\mathrm{m}}^*\} \neq 0$ if and only if $[\mathrm{Im}\{\varepsilon_1^*\} \neq 0$ or $\mathrm{Im}\{\varepsilon_2^*\} \neq 0]$—the mixture has loss only if at least one of the components is lossy. From that he deduced that $\mathrm{Im}\{f(s)\}/\,\mathrm{Im}\{s\} < 0$ everywhere and $f(s)$ has only simple real poles which are situated between 0 and 1. Consequently, a standard complex partial fraction expansion can be used to represent f:

$$f(s) = \sum_n \frac{B_n}{s_n - s} \tag{8.84a}$$

or by an integral form:

$$f(s) = \frac{A}{s} + \int_0^1 \frac{g(x)\,\mathrm{d}x}{s - x}. \tag{8.84b}$$

In this equation the residue corresponding to the zeroth pole, A, has been separately considered because of its significance: when $A \neq 0$ medium 2 is percolating (significant contact conduction pathways). Additionally, the probability density function describing the distribution of medium 2 in medium 1 has two restrictions. First, the upper limit must satisfy volume fraction (i.e. Φ) limits:

$$\int_0^1 g(x)\,\mathrm{d}x = \Phi - A \tag{8.85}$$

and second, the analysis assumes global isotropy (no overall particle orientation effects) so:

$$\int_0^1 x g(x)\,\mathrm{d}x = \frac{\Phi(1 - \Phi)}{3} \tag{8.86}$$

Application to real materials

The domain of application of this formula is obviously much wider than the empirical fits in the above equations. This relation is also more powerful because it separates the effects of the electrical properties from the geometrical distribution. As might be imagined, its use has been limited in practice by the lack of "good" $g(x)$ functions for describing most real mixtures.

However, it is interesting to note that when $g(x) = \Phi\delta\{x-(1-\Phi)/3\}$ — that is, for small particles — the Bergman relation reduces to the Rayleigh formula above. Other expressions for $g(x)$ have been proposed. Stroud, for example, used:

$$g(x) = Cx^{-b}(1-x)^{e} \qquad \text{where } 0 < b, e < 1 \tag{8.87}$$

to describe water-saturated rocks in the RF range. And, when the distribution function is of the form:

$$g(x) = \frac{2\Phi(1-\Phi)}{\pi}\sqrt{\frac{1-x}{x}} \qquad A = \Phi^2 \tag{8.88}$$

the Bergman relation reduces to the Looyenga formula.

Obtaining the density function from relaxation times

The Bergman relation is powerful because it separates the geometric features of the mixture from the electric properties of the dispersant. If, for example, the permittivity of a dielectric suspended in a dielectric is known, then the mixture permittivity can be calculated when any other material is suspended with the same distribution. We demonstrated (Roussy *et al.* 1992) that the geometric function of any isotropic mixture can be related to the distribution of relaxation times which describes the mixture frequency response when medium 2 is a metal with identical geometric dispersion to the dispersed dielectric case. By setting $x = 1/(1+u)$ in equation (8.84b) one obtains:

$$\frac{\varepsilon_{\mathrm{m}}^* - \varepsilon_1^*}{\varepsilon_1^*} = A\frac{\varepsilon_2^* - \varepsilon_1^*}{\varepsilon_1^*} + \int_0^\infty \frac{(\varepsilon_2^* - \varepsilon_1^*)g(1/(1+u))}{(1+u)(\varepsilon_2^* + u\varepsilon_1^*)}\,\mathrm{d}u. \tag{8.89}$$

When medium 2 consists of metallic particles ($\varepsilon_2^* = \sigma_2/\mathrm{j}\omega$) dispersed in a lossless dielectric ($\varepsilon_1^* = \varepsilon_1$, real) the resulting expression can be reduced, after some algebra, to:

$$\varepsilon_{\mathrm{m}}^* = \left\{1 - A - \left[\int_0^\infty \frac{g(1/(1+u))}{u(1+u)}\,\mathrm{d}u\right]\right\}\varepsilon_1 + A\frac{\sigma_2}{\mathrm{j}\omega} + \varepsilon_1\int_0^\infty \frac{g(1/(1+u))}{u(1+\mathrm{j}\omega(\sigma_2/\varepsilon_1)u)}\,\mathrm{d}u. \tag{8.90}$$

Three significant features of equation (8.90) are of interest. First, the first term is the high frequency limit of the permittivity for the mixture:

$$\varepsilon_{\mathrm{m}}^*(\infty) = \left\{1 - A - \left[\int_0^\infty \frac{g(1/(1+u))}{u(1+u)}\,\mathrm{d}u\right]\right\}\varepsilon_1. \tag{8.91}$$

The second term is the effective conductivity of the mixture:

$$\sigma_{\mathrm{eff}} = A\sigma_2 \leqslant \sigma_2 \qquad \text{so that} \qquad 0 \leqslant A \leqslant 1. \tag{8.92}$$

Finally, the third term is a relaxation term with reference time constant $\tau_0 = \varepsilon_1/\sigma_2$ or with reference to any time constant proportional to ε_1/σ_2; $\sigma_0 = \lambda\varepsilon_1/\sigma_2$, (see equation (8.16)). The distribution function of relaxation times is proportional to $g\{1/(1 + u\sigma/\varepsilon_1)\}$ so that:

$$g\left(\frac{1}{1 + u(\sigma/\lambda\varepsilon_1)}\right) = k_y G\left(\frac{u}{\tau_0}\right) \tag{8.93}$$

and the functions can be superimposed if both axes are scaled appropriately. The y-axis scale factor is k_y and the x-axis scale factor is λ.

Two assumptions must be applied: (1) the susceptibility describes an asymmetric Cole–Cole plot which satisfies Dissado and Hill requirements, and (2) $x = 1/(1 + u)$ so that a reduced geometric density, $d[1 - n, m, \lambda; x]$, function can be used:

$$d[1 - n, m, \lambda; x] \text{ is proportional to } G\left[1 - n, m; \frac{1 - x}{\lambda x}\right] \tag{8.94a}$$

and if $0 < x < 1/(1 + \lambda)$ then:

$$d[1 - n, m, l; x] = \sin(m\pi) \int_0^x \frac{\lambda^m \, du}{u^{1-m}(1 - u)^n[1 - (1 + \lambda)u]^{1-n+m}} \tag{8.94b}$$

while if $1/(1 + \lambda) < x < 1$ then:

$$d[1 - n, m, l; x] = \sin((1 - n)\pi) \int_x^1 \frac{\lambda^m \, du}{u^{1-m}(1 - u)^n[(1 + \lambda)u - 1]^{1-n+m}} \tag{8.95}$$

from which the mixture formula can be written as:

$$\varepsilon_m^* = (1 - A)\varepsilon_1^* + A\varepsilon_2^* + k_y(\varepsilon_2^* - \varepsilon_1^*)\varepsilon_1^* \int_0^1 \frac{d[1 - n, m, \lambda; x]}{\varepsilon_1^* + x(\varepsilon_2^* - \varepsilon_1^*)} \, dx. \tag{8.96}$$

The geometry of the mixture then depends on the quantities $A, k_y, \lambda, 1 - n$ and m. These quantities are linked together and to the volume fraction of material 2 in material 1 by equations (8.85) and (8.86). The calculation of the above mixture relations is not complicated and can be accomplished on a personal computer. Experimental data can be fitted using least squares techniques to determine the values of $1 - n, m$ and λ.

It has been also possible to extend the Bergman–Milton formula permitting the calculation of the effective permittivity of a two-component heterogeneous dielectric mixture to the case of a three-component mixture. If ε_1^* and ε_2^* are the complex permittivity and Φ_1 and Φ_2 are the volume fraction of the dispersed media:

$$\frac{\varepsilon_m^* - \varepsilon_3^*}{\varepsilon_3^*} = A_1\frac{\varepsilon_1^* - \varepsilon_3^*}{\varepsilon_3^*} + \int_0^1 \frac{\mu(r_1)\,dr_1}{r_1 - \dfrac{\varepsilon_3^*}{\varepsilon_1^* - \varepsilon_3^*}} + A_2\frac{\varepsilon_2^* - \varepsilon_3^*}{\varepsilon_3^*} + \int_0^1 \frac{v(r_2)\,dr_2}{r_2 - \dfrac{\varepsilon_3^*}{\varepsilon_2^* - \varepsilon_3^*}}$$
$$+ \int_0^1\int_0^1 \frac{K(r_1, r_2)\,dr_1\,dr_2}{\left(r_1 - \dfrac{\varepsilon_3^*}{\varepsilon_1^* - \varepsilon_3^*}\right)\left(r_2 - \dfrac{\varepsilon_3^*}{\varepsilon_2^* - \varepsilon_3^*}\right)} \tag{8.97}$$

with

$$\mu(r_1) = \int_0^1 \frac{K(r_1, r_2)\,\mathrm{d}r_2}{1 - r_2} \tag{8.98}$$

$$\nu(r_2) = \int_0^1 \frac{K(r_1, r_2)\,\mathrm{d}r_1}{1 - r_1}. \tag{8.99}$$

In these relations, $K(r_1, r_2)$ is a two-variable density function, which characterizes the geometrical dispersion of both dispersed media. The functions μ and v are the partial density function of Bergman–Milton theory. Their first moments measure the volume fraction of each component.

$$\int_0^1 \mu(r_1)\,\mathrm{d}r_1 = \Phi_1 - A_1 \tag{8.100}$$

$$\int_0^1 \mathrm{v}(r_2)\,\mathrm{d}r_2 = \Phi_2 - A_2. \tag{8.101}$$

The A_i are essentially positive quantities, which are involved only if material i percolates. If the distributions of the dispersed media are isotropic:

$$\int_0^1 r_1\mu(r_1)\,\mathrm{d}r_1 = \frac{\Phi_1(1 - \Phi_1)}{3} \tag{8.102}$$

and:

$$\int_0^1 r_2\mathrm{v}(r_2)\,\mathrm{d}r_2 = \frac{\Phi_2(1 - \Phi_2)}{3} \tag{8.103}$$

as in the two-media Bergman–Milton theory

8.5.8 Comments on mixture formulae

As a practical observation it may be seen that the above mixture formulae work for macroscopic composites. Although some of the earliest ones, by Böttcher and Maxwell, were originally derived for liquid mixtures, they may fail to adequately describe microscopic mixtures. A ready example is that neither pure water nor solid paper (bone dry) exhibit measurable loss at radio frequencies; nonetheless, wet paper has significant loss and is well heated in RF applicators. None of the mixture formulae allow interaction between the phases. Neither do they permit a non-zero imaginary part of $\varepsilon_{\mathrm{m}}^*$ when both ε_1^* and ε_2^* are real. Thus they cannot explain the loss observed in wet paper.

The mixture formulae treat only the geometric aspects of the problem. To explain why paper heats in an RF field we must reconsider the cluster theory of dielectrics. For wet paper the cluster is a section of solid on which some water molecules are fixed. The electrically active particles, the fluctuations to be considered, are the dipolar water molecules which we have treated in the liquid state; but their movement on the solid substrate is not the same as it would be in the liquid phase. The mechanical motion of the first layer is probably dominated by surface effects — the surface free energy — rather than by viscous dissipation, as in our previous analysis (Chapter 7). Consequently, the motion of second layer molecules is also different than in the liquid state owing to first layer effects. So, we suppose that the mobility of a large number of water molecules (a much larger fraction than in the bulk liquid state) is modified.

In the liquid state 27 MHz is much lower than the relaxation frequency; which means that the fluctuations we are concerned with are those between the dominant clusters, single water molecules. In wet paper each cluster includes large numbers of water molecules, so the relaxation frequency is expected to be much lower. We postulate that the relaxation frequency approaches closer to the RF regime and significant heating of the paper results. The frequency of the RF field will not be so much lower than the relaxation frequency that new cluster zones will appear, however.

8.6 SUMMARY

This chapter has presented the duality between dependence of permittivity on temperature and frequency. In that sense the permittivity describes the average mobility of electrically significant particles: electrons, ions, and dipolar molecules. We have summarized both the simple and not so simple improvements in the Debye theory which have been used to describe real materials. We have also seen that heterogeneous materials may be modeled in simple, more accurate, and more complicated ways for two-component systems. Although simple algebraic functions do not result, the calculations can be performed on a small computer. Finally, we have discussed the importance of interactions between host material and suspended phase, which is not described by mixture formulae.

REFERENCES

Barriol, J. (1957) *Les Moments Dipolaires* Gauthier-Villars, Paris.

Böttcher, C. J. F. and Bordewijk, P. (1978) *Theory of Electric Polarization* Elsevier, Amsterdam.

Digest of Literature on Dielectrics vol 1–29, NAS–NRC Washington DC, 1950–1969.

Dissedo, L. A., Nigmatullin, R. R. and Hill, R. M. *Dynamic Properties in condensed Matter, Advances in Chemical Physics*, (M.W. Evans, Ed.), Wiley, New York.

Fröhlich, H. (1955) *Theory of Dielectrics* Oxford University Press, Oxford.

Johnscher, A. K. (1983) *Dielectric Relaxation in Solids* Chelsea Dielectrics Press, London.

Morse, P. M. and Feshbach, H. (1968) *Methods of Theoretical Physics* Vols 1 and 2, Academic Press, New York.

Nelson, S. O. (1992) Estimation of permittivities of solids from measurements on pulverized or granular materials *Dielectric Properties of Heterogeneous Materials* (A. Priou, Ed.), *Progress in Electromagnetics Research*, Vol. 6 (J. A. Kong, Ed.), Elsevier, Amsterdam.

Roussy, G., Spaak, E. and Thiebaut, J. M. (1992) Universal equation for the effective complex permittivity of mixtures valid for dielectric–dielectric and dielectric–conductor mixtures *Phys. Rev. B*, **46**, 11452.

PART 3

Processing Aspects

In the first two parts of the book we were concerned with the effect of the material on the internal fields created by an externally applied source of high frequency energy. In the third part of the book we will consider the effect of the applied fields on the material. It will be clear that the two points of view are only separable in the simplest of cases. Much more frequently the changes in the material induced by the applied fields result in changes in the electrophysical properties which, in turn, affect the field distribution in the material. Consequently, the equations of thermodynamic state and the energy balance are inextricably linked to Maxwell's equations through the properties (both electrical and thermal) and a complete description of the interaction often involves iterative relaxation among the governing equations resulting in multiple determinations of the field distribution in a single model calculation. In some cases the interaction is such that there are regions of operation which are multiply-valued, and thus unstable heating (and thermal runaway) may be encountered. Chapter 9 presents the physical principles of thermal interactions between materials and fields from both analytical and experimental standpoints. Chapter 10 discusses examples of radio frequency and microwave interactions in homogeneous media, Chapter 11 contains examples in heterogeneous media and Chapter 12 discusses microwave catalysis in some detail.

9 Theoretical Models and Experimental Methods in High Power Density Electromagnetic Fields

We begin the discussion of the governing principles of material-field interaction in this chapter with a description of frequently used transient thermal models derived from the First Law of Thermodynamics. We then extend the framework with simple models for phase change under near-equilibrium conditions and follow with rate-limited processes including very rapid vaporization. Models of this type often require a large number of physical parameters to describe relatively simple field interactions. In many cases the required parameters are neither available nor determinable by independent means. In those cases an alternative global approach yields more satisfying results. Techniques for evaluating experimental results using the alternative strategy are developed in the concluding section of this chapter.

9.1 THERMAL AND THERMODYNAMIC MODELS

The term "model" refers to both analytical and numerical descriptions which respond to inputs and conditions in the same manner, or at least in a comparable manner, to the physical system under study. The attractive feature of this class of models is that each process is separately treated with its own set of controlling parameters. Consequently, in the model one is able to vary system characteristics and note the changes which result in system response — each process is controlled separately. This type of parametric study is essential to the intelligent design of a process. Many geometries and materials are too complex to be studied entirely with closed-form analytical solutions, so numerical methods are frequently used, as we have already seen in the solution of Maxwell's equations in Chapters 1 and 4. In this section we develop analytical and numerical approximations which may be used to study physical systems. Even though it may not be possible to create a model which predicts or simulates the results obtained in a particular set of experiments, the model will usually predict the trends with good accuracy. Individual features of a model may be turned on or off to assess their contribution to the observations. Also, the model may be followed with spatial and temporal detail completely unachievable by experiment. So while it is a mistake to expect too much from a model or to believe the results without substantial experimental verification, it is imperative to make use of models to fully understand the implications of a sequence of experimental observations.

9.1.1 Volumetric power dissipation

The dominant coupling between the electromagnetic field and the material is through power absorption by the imaginary part of the permittivity, ε''. While there are some materials which have significant coupling due to direct magnetic field interactions (i.e. materials with high values of complex magnetic permeability), they are few in number at high frequency. As was described in Chapter 4, even in magnetic field heating applications, in the vast majority of materials, the actual field effect is heating from an electric field induced in the material by the external magnetic field. Therefore, we will focus attention on the electric field, **E**; and for purposes of discussion the electrical conductivity effects will usually be included within ε'', as described in previous chapters. The power dissipated per unit volume, Q_{gen} (W/m^3), is thus given by:

$$Q_{\text{gen}} = \text{Re}\{-\nabla \cdot \mathbf{S}\} = \omega\varepsilon''|\mathbf{E}|^2 = \left(\varepsilon_0\chi''(\omega) + \frac{\sigma}{\omega}\right)\omega|\mathbf{E}|^2 \tag{9.1}$$

where **S** is the Poynting vector field (Chapter 1) and all **E**-field energy loss terms are included in ϵ''. This energy absorption term results in a source of localized internal heating without considering the precise nature of the molecular interactions, which are the origin of dielectric losses. The material can then undergo a transformation such as vaporization, reaction, fusion, or curing.

In any model of the phenomena under study, the internal heating appears as a contribution to the internal thermal energy of the material. We have seen the energy balance in a simple form in Chapter 1, which ignores changes in material properties. It remains to discuss how material changes and transformations affect the electric field.

Consider the example of a given volume of matter with a local electric field $\mathbf{E}(x, y, z, t)$, transient scalar temperature field $T(x, y, z, t)$, and $R(x, y, z, t)$ a conversion or transformation (Figure 9.1). The variable R can be, for example, evaporation, change of color, or a chemical reaction. All of the variables are functions of location and time. They are connected to one another through the system of governing equations. The components of the field, **E**, are solutions of the Maxwell equations. One of Maxwell's equations, Ampère's Law, introduces the complex dielectric permittivity ε^* as was seen in Chapter 1. Maxwell's equations are

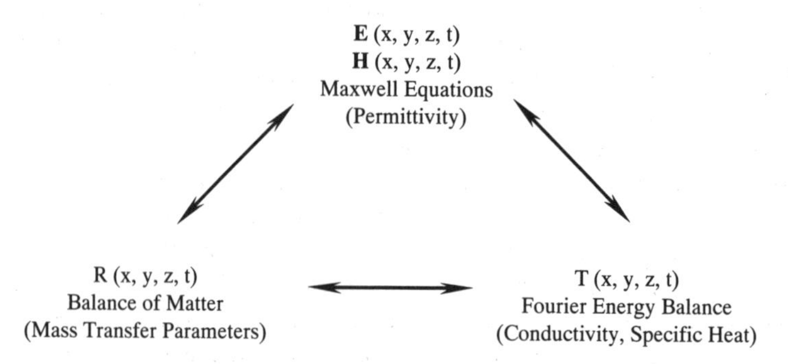

Fig. 9.1 *Triangular interaction among models.*

coupled to the energy balance in two ways: the electric field constitutes the power generation term (9.1) while temperature rise in the material may change the permittivity, and thus **E**. The transformation process, R, may be a thermal process, in which the transformation is induced entirely by the local temperature history, another kind of process in which the presence of the electric field alone initiates the transformation, or it may be a combination of the two. In any case, the temperature dependence of the electrical properties couples the equation sets, so a consideration of temperature response is the first order of business. We will present examples of all three types of transformation.

9.1.2 Transient thermal model with constant properties

In the simplest kind of transformation, temperature increase alone, the temperature in the bulk material is determined from the Fourier energy balance equation, in which we have included the local absorbed power term, Q_{gen}. The relationship applies for both conventional ($Q_{gen} = 0$) and electromagnetic heating, and is second order in the space variables and first order in time:

$$\rho c \frac{\partial T}{\partial t} - \nabla \cdot (k \nabla T) = Q_{gen}(x, y, z, t) \tag{9.2}$$

where ρ is the density (kg/m^3) and c the specific heat (J/kg K), so the first term on the left hand side is the rate of change of the internal thermal energy, and where k is the thermal conductivity (W/m K) and the second term represents Fourier conduction heat transfer. The traditional energy balance in equation (9.2) assumes that the heating time is long compared to the thermal relaxation time of atoms and molecules, the usual case in MW and RF heating. Very short pulses must include a relaxation term, which results in thermal wave solutions, and is beyond the scope of this book. In this formulation we have not assumed that the material is an isotropic heat conductor; if it is then the second term reduces to $k\nabla^2 T$, as we saw in Chapter 1.

The generation term on the right hand side includes the energy dissipated by the electric field and the energy of any local reaction, which can be either exo- or endo-thermic. In the conventional heating case the electric field term is zero and material heating is by boundary conditions and reaction heat alone. Possible thermal boundary conditions include:

(1) An isothermal (Dirichlet) boundary:

$$T_s = \text{constant} \tag{9.3}$$

where T_s is the surface temperature.

(2) An adiabatic (zero heat flux), or Neumann, boundary:

$$\left[\frac{\partial T}{\partial n}\right]_s = 0 \tag{9.4}$$

where n is the coordinate normal to the surface.

(3) A convection boundary (Figure 9.2a):

$$k\left[\frac{\partial T}{\partial n}\right]_s = h(T_s - T_\infty) \tag{9.5}$$

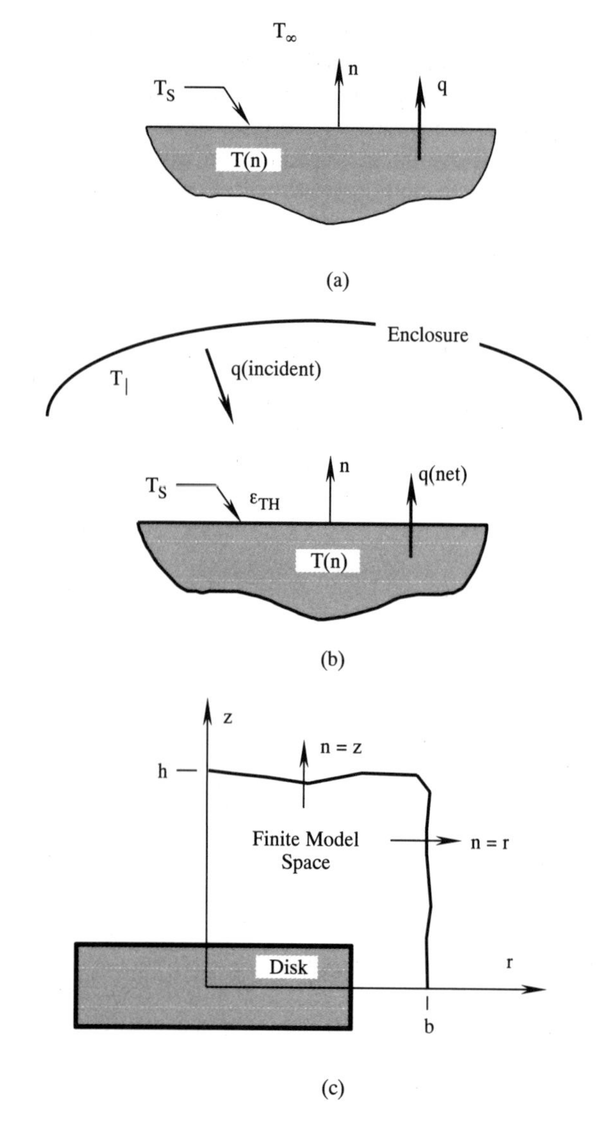

Fig. 9.2 *(a) Convection boundary. (b) Radiation boundary. (c) Mixed boundary condition at edges of model space.*

where h is the convection heat transfer coefficient (W/m^2 K), and T_∞ is the environment temperature (K).

(4) A radiation boundary (Figure 9.2b):

$$k\left[\frac{\partial T}{\partial n}\right]_{\mathrm{s}} = F\varepsilon_{\mathrm{TH}}\sigma_{\mathrm{B}}\left(T_{\mathrm{s}}^{4} - T_{\infty}^{4}\right) \tag{9.6}$$

where F is a factor describing the optical geometry of the object and its enclosure ($F = 1$ if the object is convex and the enclosure is isothermal at T_∞, complete and unobstructed),

ε_{TH} is the surface emissivity ($0 \leqslant \varepsilon_{TH} \leqslant 1$, a measure of its efficiency as a radiator), and σ_B is the Stefan–Boltzmann constant (5.67×10^{-8} W/m^2 K^4).

(5) A conduction heat transfer boundary between materials of differing thermal conductivity (similar to the normal **E**-field condition):

$$k_1 \left[\frac{\partial T_1}{\partial n}\right]_s = k_2 \left[\frac{\partial T_2}{\partial n}\right]_s . \tag{9.7}$$

(6) A mixed thermal boundary condition:

$$\left[\frac{\partial T}{\partial n}\right]_s + P(n)T_s = Q(n) \tag{9.8}$$

where $P(n)$ and $Q(n)$ are chosen so that the temperature field at the boundary follows a desired functional form.

Convection (9.5) and radiation (9.6) are both surface heat transfer phenomena and are sometimes combined into a single surface heat flux term. It is frequently convenient to express surface heat transfer in terms of a "thermal resistance" R_{th}, where the heat flux is related to the temperature difference in an analog of Ohm's law—heat flux corresponds to current and temperature difference to voltage difference. So, for one-dimensional (vector) conduction heat transfer, $\mathbf{q}$(W), in a thin slab Δx thick with temperature difference across the slab ΔT (not dependent on y or z):

$$\mathbf{q}_{cond} = -kA\frac{\partial T}{\partial n}\mathbf{a}_n \cong -kA\frac{\Delta T}{\Delta x}\mathbf{a}_x \tag{9.9}$$

where A is the area of the cross-section of heat flux (m^2) and the negative sign indicates heat flux from high to low temperature. The thermal resistance (K/W) associated with 1-D conduction is therefore:

$$R_{th}(\text{conduction}) = \frac{\Delta x}{kA}. \tag{9.10}$$

The convection boundary condition may also be rewritten in the same way:

$$R_{th}(\text{convection}) = \frac{1}{hA}. \tag{9.11}$$

In practice the resistance is often expressed for a unit surface area so that one often sees $R_{th} = 1/h$ (K/W m^2) for a convection boundary—the reader should be careful of the units used. The convection coefficient is very sensitive to the velocity of the external medium. Even in free convection (no external velocity) h varies due to surface orientation owing to induced convection. In free (natural) convection h typically ranges from about 5 to 25 W/m^2 K. A power law approximation for a natural-convection air interface based on an empirical fit is sometimes used:

$$q_{conv} = \mathcal{H}\,[T_s - T_\infty]^{5/4} \tag{9.12}$$

where $\mathcal{H}$ is a constant of proportionality, and the thermal resistance is thus not a constant. Of course, convection boundaries are, in fact, quite complex and the effective convection heat transfer coefficient depends critically on the flow field, especially on whether it is laminar or turbulent, even for natural convection. There are many methods for estimating h in differing fluids and flow conditions in heat transfer literature.

Using thermal resistance for a radiation boundary is a risky concept because of the highly nonlinear nature of the heat transfer. Nevertheless, the radiation boundary is sometimes approximated by a thermal resistance with an algebraic manipulation of T:

$$\left[T_s^4 - T_\infty^4\right] = [T_s - T_\infty]\left(T_s^3 + T_s^2 T_\infty + T_s T_\infty^2 + T_\infty^3\right) \tag{9.13}$$

and:

$$R_{th}(\text{radiation}) = \frac{1}{\varepsilon_{TH}\sigma_B A\left(T_s^3 + T_s^2 T_\infty + T_s T_\infty^2 + T_\infty^3\right)} \tag{9.14}$$

which is, of course, rather strongly nonlinear. In fact, the sensitivity of R_{th}(radiation) to temperature changes is -0.5%/K at 300K and varies reciprocally with T so that it is about -0.38%/K at 600K. The nonlinear nature of surface heat transfer processes is usually not a strong influence on the results obtained, providing the desired temperatures are not too high and heating times are not too long, because their contribution to the overall heat transfer is often negligible when rapid heating is studied. Ceramics sintering and steady state (constant temperature) treatment are notable exceptions to this generalization.

The mixed boundary of equation (9.8) is often used in numerical modeling studies to simulate a model field unlimited in extent—it is not particularly useful in closed-form analytical solutions. That is, suppose a transient numerical model of a finite disk in an infinite semi-solid (non-convecting) medium was desired, such as shown in Figure 9.2c. The bottom of the disk is a Dirichlet boundary, but Neumann (insulated) boundaries are inappropriate at large r and z since they would make the model space prohibitively large and no steady-state solution could be obtained. We may use a mixed boundary condition at both $r = b$ and $z = h$ to simulate the infinite medium. As an example, we may fit an exponential decrease to the surface temperature at the boundaries of the model space—set $P(n) = +a$ and $Q(n) = 0$ so that:

$$T(n) = T_s e^{-a_n} \tag{9.15}$$

where n is normal to the boundary and the value of T_s may estimated from neighboring locations (inside the model space) using this rule. This rule fits the disk case well since the analytical solution is exponential in the z-direction and Bessel functions (r-direction) tend toward exponentials at very large arguments. Other rules may be formulated as desired.

9.1.3 Inclusion of material transformation effects

A typical transformation, R, would in all likelihood satisfy a volumetric equation and a specific interface condition. These may be derived from the balance of matter during the transformation process. The corresponding relations for specific example cases will be described in other sections of this chapter.

Direct inclusion of transformation effects is mathematically complex since the systems of equations are coupled: the electric field which appears in the Maxwell equations is also in the Fourier equation, and in the transformation equation. Further, the characteristics of the material, such as permittivity, thermal conductivity, and mass transfer are also functions of the temperature. With the advent of numerical methods and convenient personal computers, the more complex interaction problems can be solved numerically. However, the most difficult part is to define the inter-dependencies of the characteristics of the matter with the temperature, and with the transformations. At the end of the mathematical process the numerical fit of the calculated values with the measured values describes the quality of the dependencies assumed. Even if it provides a good statistical fit, however, the model may not adequately describe the actual transformation process.

9.1.4 Near-equilibrium vaporization processes

One of the most common transformation processes is drying of wet materials. The essential process in drying is phase change from liquid to gas phase with removal of the vapor at or near the surface to prevent reflux condensation. The interactions among the electromagnetic field model, fluid flow (or mass diffusion) model and the medium thermodynamics can be quite complex. We will approach the overall question by considering individual components using simplified geometries in order to illustrate the interconnections. There are two points of view which may be applied to the phase change process: thermodynamic and kinetic. The thermodynamic approach considers the end point states of the material with little concern for the rate at or processes by which events occur. The kinetic approach is primarily concerned with the rate of transformation, which is, of course, determined by the thermodynamic state. The thermodynamic state is specified by a complete set of state variables: temperature, pressure, density (or specific volume) enthalpy and entropy.

This section will apply a thermodynamic model to describe vaporization under the assumption that the material never strays too far from equilibrium. The specific example used will be water vaporization, but the concepts apply to many other liquids, as well. When a liquid phase is in equilibrium with its vapor the mass rate of vaporization is equal to the mass rate of condensation. If the local pressure is lower than the saturation pressure at the liquid temperature then the vaporization rate exceeds the condensation rate. Near-equilibrium means that as equilibrium is disturbed by the addition of electromagnetic energy, the vaporization process is rapid enough to restore and maintain thermodynamic equilibrium over the time scale considered.

Thermodynamic properties of water

Most liquid phase materials are essentially incompressible at ordinary pressures, and we will make that assumption in this discussion. This essentially separates the pressure from the temperature and density in the state equation; thus the density is determined primarily by the temperature. Therefore, the thermodynamic state of the liquid will be determined by its temperature alone. We will use saturation values for the density, enthalpy and entropy at the liquid temperature which have been tabulated for water and other common liquids. Polynomial fits are easily derived from the tables for the saturation pressure of water, $\mathcal{P}_{sat}$ (kPa) (Figure 9.3):

$$\mathcal{P}_{sat} = -43.205 + 2.3887T - 3.6229 \times 10^{-2}T^2 + 2.4111 \times 10^{-4}T^3 + 6.6812 \times 10^{-8}T^4 + 1.6859 \times 10^{-9}T^5. \tag{9.16}$$

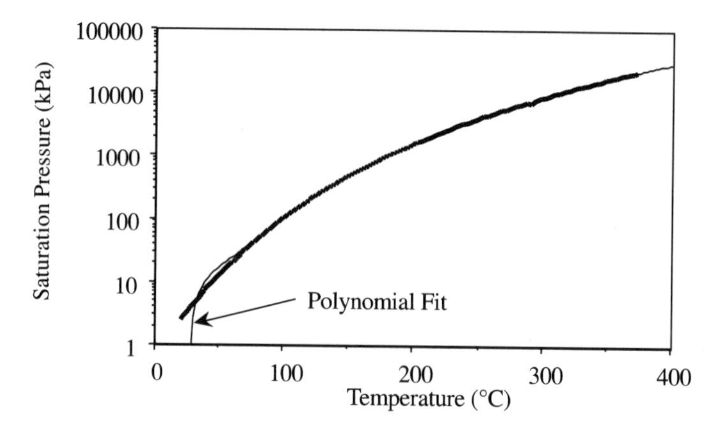

Fig. 9.3 *Saturation pressure vs. temperature.*

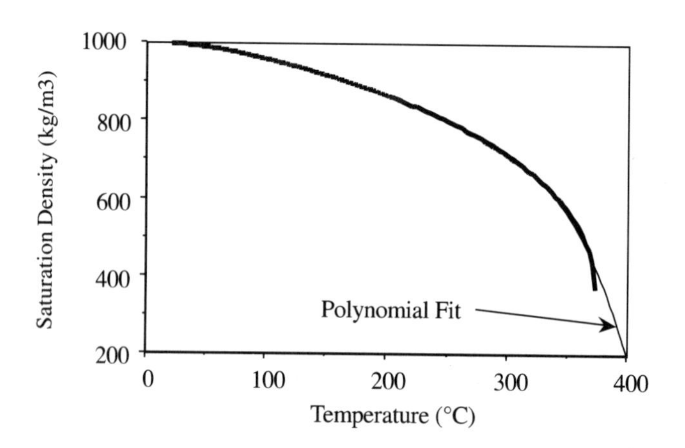

Fig. 9.4 *Saturation density vs. temperature.*

We will use the $\mathcal{P}$ to signify pressure so that it may be distinguished from electromagnetic power, italic P. The polynomial fit for the saturation density, $\rho_{\rm sat}$ (kg/m^3) (Figure 9.4) is:

$$\rho_{\rm sat} = 1025.8 - 1.4127T + 1.9122 \times 10^{-2}T^2 - 1.6499 \times 10^{-4}T^3 + 5.5939 \times 10^{-7}T^4 - 6.9179 \times 10^{-10}T^5 \quad (9.17)$$

for the liquid phase saturation enthalpy, $h_{\rm f}$ (J/kg) (Figure 9.5):

$$h_{\rm f}(T) = -3.4912 \times 10^4 + 6109.4T - 33.907T^2 + 0.25146T^3 - 8.1248 \times 10^{-4}T^4 + 9.8008 \times 10^{-7}T^5 \quad (9.18)$$

and for the liquid phase saturation entropy, $s_{\rm f}$ (J/kg K) (Figure 9.6)

$$s_{\rm f}(T) = -48.69 + 18.044T - 7.6485 \times 10^{-2}T^2 + 4.2569 \times 10^{-4}T^3 - 1.2829 \times 10^{-6}T^4 + 1.5047 \times 10^{-9}T^5 \quad (9.19)$$

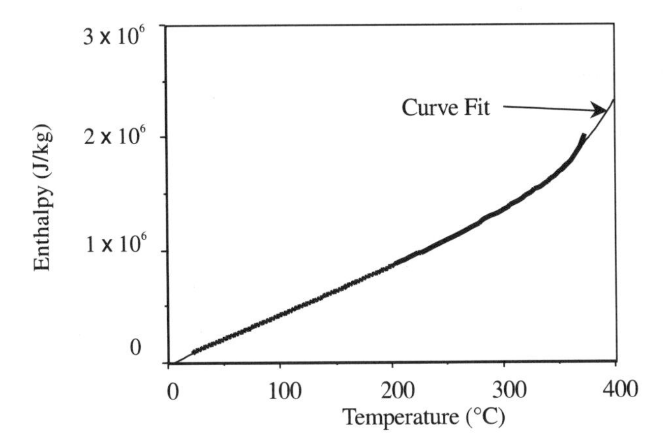

Fig. 9.5 *Saturation enthalpy vs. temperature.*

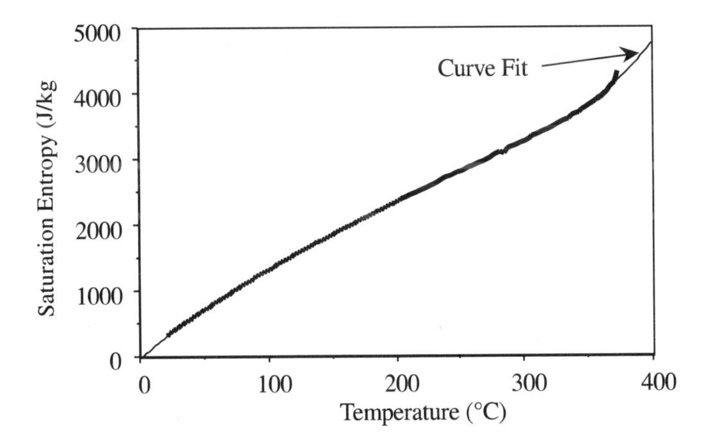

Fig. 9.6 *Saturation entropy vs. temperature.*

where the polynomial fit is over the liquid phase range from 20°C to the critical point, 375°C. Notice that the saturation pressure at 100°C is 1 atmosphere, 101.3 kPa. We will return to this point in future discussion.

Thermodynamic view of vaporization

The change in phase from liquid to gas states is easiest to visualize on a temperature-entropy (T–s) diagram, such as Figure 9.7. The major single-phase regions in the diagram are the solid zone (S, very thin), the liquid zone (F, also quite thin because of essential incompressibility of the liquid phase), and the superheated vapor zone (gaseous phase, G), as indicated. There are also two mixed phase zones—liquid + gas (FG) and solid + gas (SG)—which are separated by the triple point (the triple point in a pressure–volume plot is the projection of the triple point line in the T–s diagram onto P–V axes). Under the mixed-phase FG "steam dome" region liquid and gas phase water can exist in equilibrium (the relative mass fraction of gas is called the "quality") and constant pressure lines are parallel

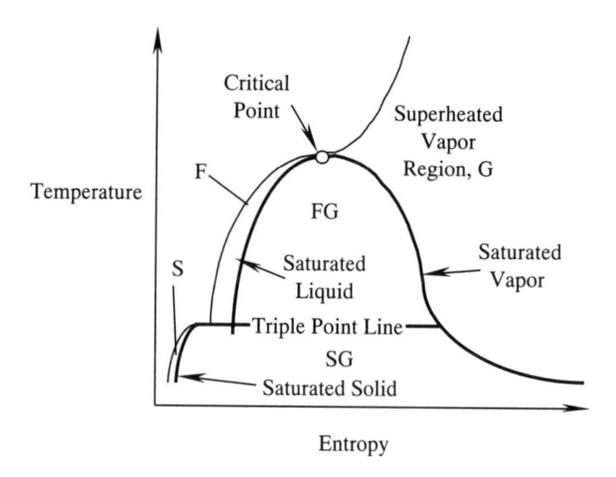

Fig. 9.7 *Typical temperature–entropy diagram showing the solidus region (S), liquidus region (F), the phase change regions (FG and SG), and superheated vapor region (G).*

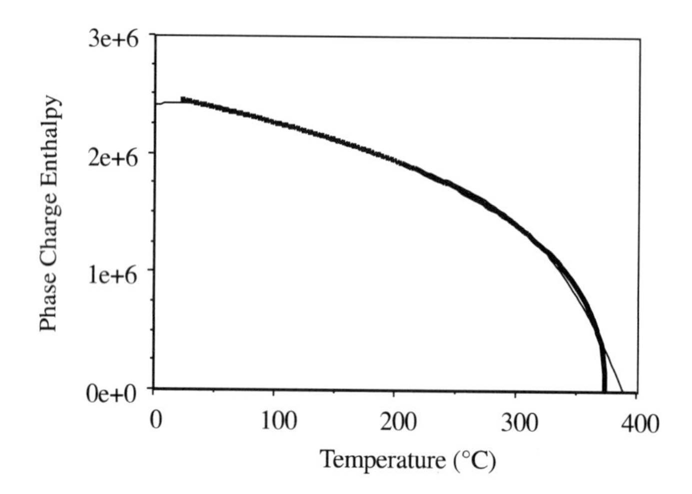

Fig. 9.8 *Phase change enthalpy, Δh_{fg}, vs. temperature.*

to constant temperature lines. In changing from liquid phase (f) to gas phase (g), the water undergoes a substantial increase in entropy ($\Delta s_{fg} = \Delta s_{gf}$) and enthalpy ($\Delta h_{fg} = \Delta h_{gf}$) where the sequence of subscript denotes the direction of phase change—fg is vaporization and gf condensation. The entropy increase is a measure of the increased randomness of the gas phase and the enthalpy increase is due to an increase in the pressure–volume product. The polynomial fit for the phase change enthalpy of water, Δh_{fg} (J/kg) (Figure 9.8) is:

$$\Delta h_{fg} = 2.402 \times 10^6 - 1975T + 54.3T^2 - 0.247T^3 - 4.1431 \times 10^{-4}T^4 \tag{9.20}$$

and for the phase change entropy, Δs_{fg} (J/kg K) (Figure 9.9) is:

$$\Delta s_{fg} = 8950 - 33.39T + 3.25 \times 10^{-2}T^2 - 1.49 \times 10^{-4}T^3 - 4.217 \times 10^{-7}T^4. \tag{9.21}$$

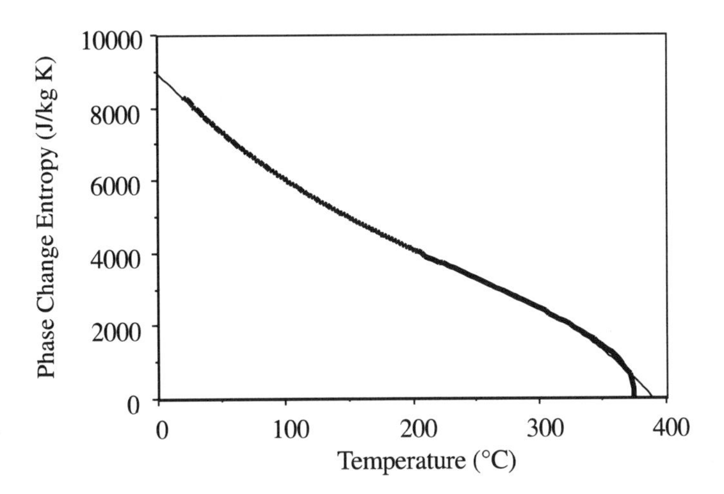

Fig. 9.9 *Phase change entropy, Δs_{fg}, vs. temperature.*

Vaporization of saturated water

At equilibrium the change of phase from saturated liquid state water to saturated gas phase occurs at constant pressure and temperature. The line at $P = 1$ atmosphere (101.3 kPa) and 100°C in Figure 9.10 illustrates such a phase change. A condensation process under saturation conditions occurs along the same constant pressure line but is accompanied by enthalpy and entropy decreases of the same magnitude. In the thermodynamic view of phase change the mass of liquid water vaporized, Δm (kg), at constant temperature and pressure is determined by the energy added to the water, ΔE (J):

$$\Delta m = \frac{\Delta E}{\Delta h_{fg}}. \tag{9.22}$$

So, for liquid water at the saturation temperature and pressure, the mass rate of vaporization at equilibrium is determined by the total power, P (W), added to the water from the RF or

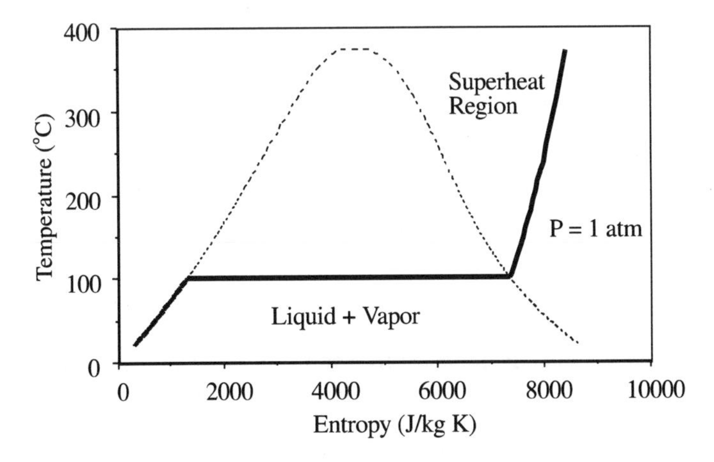

Fig. 9.10 *Constant pressure vaporization at 1 atmosphere (101.3 kPa).*

MW source:

$$\frac{\Delta m}{\Delta t} \rightarrow \frac{\partial m}{\partial t} = \frac{P}{\Delta h_{fg}(T)}. \tag{9.23}$$

The power, P, is the power in excess of that consumed by heat transfer or other processes in the volume.

Vaporization of water below saturation temperatures

Liquid water at temperatures below saturation will also vaporize from the surface. The difference between this phenomenon and vaporization under saturation conditions is that it is a surface and rate process phenomenon rather than a volumetric power deposition phenomenon. That is, the rate of water phase change is driven by the difference between the saturation pressure of the water, $\mathcal{P}_{sat}(T)$, and the partial pressure of the water vapor in the atmosphere, $\mathcal{P}_p(T_\infty)$. Stelling's formula is an empirical fit derived from data for solar ponds (free water surface):

$$\zeta = (A_s + B_s u)\left(\mathcal{P}_{sat}(T) - \mathcal{P}_p(T_\infty)\right) \tag{9.24}$$

where ζ is the rate of pond depth decrease (m/s), $A_s = 7.31 \times 10^{-11}$ (m/Pa s) and $B_s = 1.2 \times 10^{-11}$ (Pa^{-1}) and u is the velocity of the air 10 m above the water surface (i.e. the free stream air velocity outside of the boundary layer). The partial pressure of the water vapor in the atmosphere, $\mathcal{P}_p(T_\infty)$ may be found from the relative humidity, RH:

$$\mathcal{P}_p(T_\infty) = \text{RH}\mathcal{P}_{sat}(T_\infty). \tag{9.25}$$

To use this formulation in a model we apply the observation that ζA is the volume rate of vaporization (m^3/s) and $\rho\zeta A$ is the mass rate in a control volume (kg/s) where A is the surface area of the control volume. The heat of vaporization, Δh_{fg}, comes from the liquid phase and represents a heat loss from the surface control volume.

In this simple example of free surface vaporization the phase change process is "unhindered" in that it occurs at the surface and the vapor phase encounters no mass transfer resistance in migrating into the air above the liquid. In wet materials, such as wet paper or foam, the phase change process may be hindered by local surface free energy or other effects, creating an extra energy barrier which the vapor must surmount and resulting in slower evaporation rates than predicted by equation (9.24). In either case, surface vaporization cannot be neglected since it makes a very important contribution to surface heat transfer losses—for free water above about 45°C at 50% RH surface vaporization is about an order of magnitude more important than convection and radiation combined.

Inclusion of vaporization in the energy balance

The contribution of vaporization processes to the energy balance depends on the mass rate of vaporization, $\partial m/\partial t$, and on the temperature at which it occurs. We may add vaporization to the energy balance as a net decrease (energy sink) since the enthalpy of vaporization must come from the liquid phase water—this is how evaporation coolers work. Equation (9.2) then takes the form:

$$\frac{\Delta h_{fg}(T)}{\Delta v}\frac{\partial m}{\partial t} + \rho c \frac{\partial T}{\partial t} - k\nabla^2 T = Q_{gen} \tag{9.26}$$

where we have assumed an isotropic material and Δv is a small volume of the material in which vaporization occurs. In the bulk material (away from a surface) the effect of the addition of vapor volume to Δv is critical to the analysis of the process and to the execution of the energy balance. Numerical models of surface evaporation may also be structured this way (if surface losses are included), but doing so creates a moving boundary model.

Saturation vaporization example at constant pressure

Equation (9.26) is indeed a clumsy form for the energy balance under saturation vaporization since it changes the calculation from a point form differential equation into an integral form finite control volume (FCV) equation. We must apply some geometric or physical argument in order to recast the vaporization term. Perhaps the simplest approach is to recast m as ξ, the mass which changes phase *per unit volume of material* (kg/m^3):

$$\Delta h_{\text{fg}}(T)\frac{\partial \xi}{\partial t} + \rho c\frac{\partial T}{\partial t} - k\nabla^2 T = Q_{\text{gen}}. \tag{9.27}$$

In a numerical model of vaporization under saturation conditions, where excess Q_{gen} goes into vaporization in order to maintain equilibrium, one would set $\partial T/\partial t = 0$ and calculate $\partial \xi/\partial t$:

$$\frac{\partial \xi}{\partial t} = \frac{1}{\Delta h_{\text{fg}}(T)}\left[Q_{\text{gen}} + k\nabla^2 T\right]. \tag{9.28}$$

That is, when the saturation temperature has been reached we consider that all of the power generation term in excess of the conduction heat transfer (or other losses) goes into vaporization until all of the available water is exhausted, at which time the temperature of the matrix may rise or fall depending on ε'', ρ and c of the matrix, as depicted in Figure 9.11a. In the figure, we imagine a free water surface at atmospheric pressure with a constant and uniform volume rate of power deposition and adiabatic boundaries except for vaporization (no surface losses or internal heat transfer), and that the electric field in the material is constant. The temperature rises at a constant rate up to 100°C at which time all of Q_{gen} is consumed by vaporization until the material water is exhausted, and the matrix temperature is then determined by its properties alone.

If the dielectric losses in the wet material are due only to the water, then Q_{gen} will decrease as the material dries out. As a very simple example, assume that ε'' is linear with volume fraction of water, Φ:

$$\varepsilon''_{\text{material}} = \varepsilon''_{\text{water}}\Phi. \tag{9.29}$$

Note that this very crude approximation is applied so that we may inspect the nature of the interactions. It is necessarily much simpler than even the simplest mixture formula, the Rayleigh formula (8.65), so that some calculations may be performed in closed form. If a more realistic mixture formula was used numerical calculations would be necessary and the functional behavior would be less easily seen. In this hypothetical case, then, the generation term for a constant electric field would be:

$$Q_{\text{gen}} = \omega\varepsilon''_{\text{water}}\Phi|E|^2 = a\Phi \tag{9.30}$$

where a is a constant of proportionality and a very simple mixture relation has been assumed. In a real situation the electric field in the material would be affected by the loss of water, so assuming constant E is questionable. Nevertheless, for the simple example the mass rate of vaporization of water from equation (9.28) divided by the density of water gives the rate of decrease of water volume fraction:

$$\frac{\partial \Phi}{\partial t} = \frac{-1}{\rho_{\text{water}}} \frac{\partial \xi}{\partial t} = \frac{-a\Phi}{\Delta h_{\text{fg}}} = -b\Phi \tag{9.31}$$

which has a solution of:

$$\Phi(t) = \Phi_0 e^{-b(t-t_{100})} \tag{9.32}$$

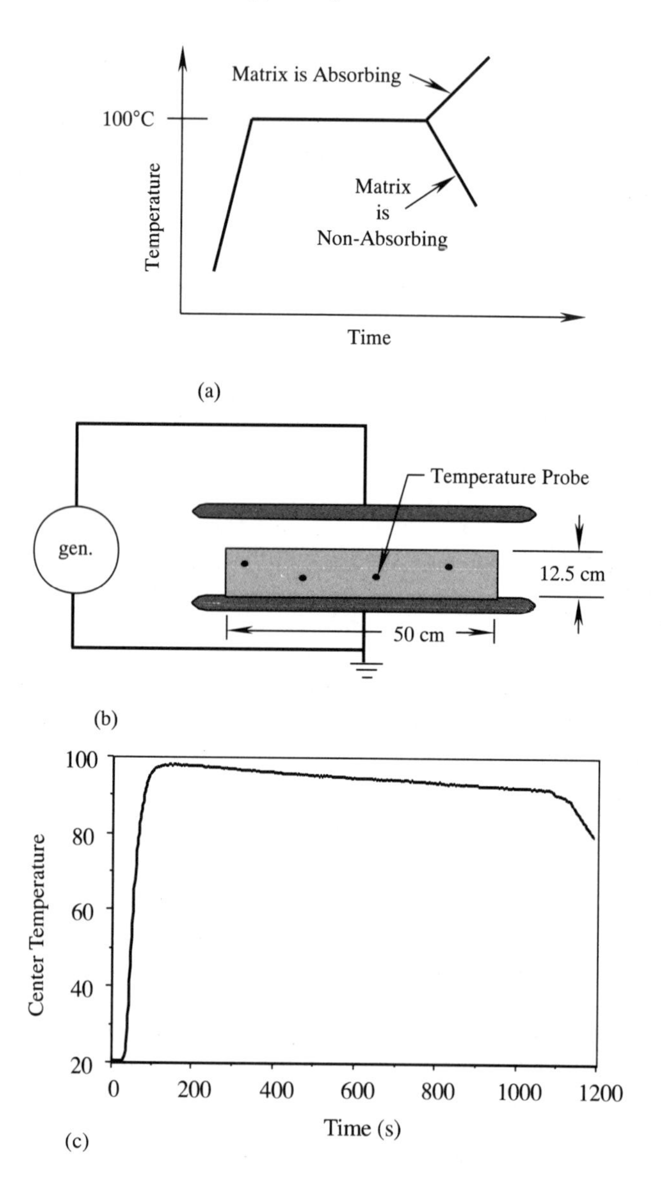

Fig. 9.11 *(a) Typical unhindered water vapor migration heating curve. (b) RF heating of wet urethane foam matrix. (c) Center temperature of large foam rubber block under RF heating. (d) Measured block mass for RF heating; water loss effects are clear.*

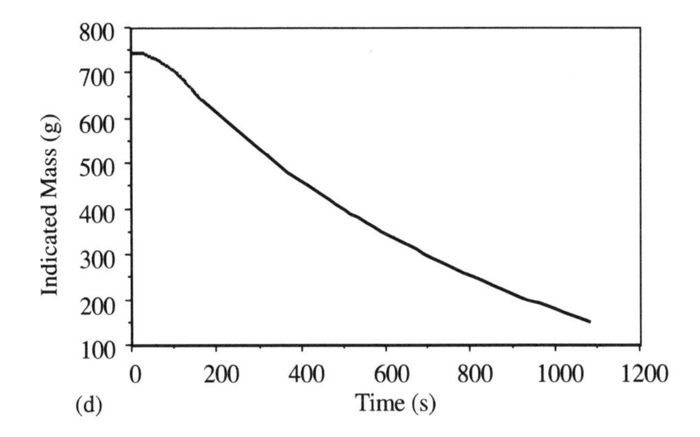

Fig. 9.11 *(continued)*

for an initial volume fraction of Φ_0 and where t_{100} is the time to reach 100°C. Note that Q_{gen} will also follow the same exponential decrease, as shown in Figure 9.11a.

Figure 9.11c is a temperature record from radio frequency heating of urethane foam (sketched in 9.11b) with an initial water content of 0.83 kg of tap water ($\sigma = 0.05$ S/m) suspended more or less uniformly in $31.81 \times 10^{-3}\text{m}^3$ of foam ($50 \times 50 \times 12.5\ \text{cm}^3$, mass 774 g). The moisture content was $M = 51.8\%$ on a wet weight basis—$\Phi_0 = 2.6\%$. The total RF power into the foam was $P = 1.02$ kW (about 32 kW/m^3) based on the constant drying rate period (100 to 1000 s). At 1000 s the RF power was turned off (total water removed 0.57 kg, about 87% of the initial water content, 0.74 kg, see Figure 9.11d). In the experiment fiber optic probes were placed at the corner and central locations indicated. Constant temperature vaporization occurred at an apparent temperature of about 95°C. The deviation from 100°C here is probably attributable to the low volume concentration of water and the influence of lower temperature adjacent foam parts in contact with the temperature probes. This simple material is not hygroscopic.

Saturation vaporization example at constant volume

In this example we will assume a different geometry—vaporization in a closed constant volume, V. The initial volume fraction of liquid phase water is Φ_0, as above, which we consider as uniformly dispersed in the volume in physical space, Figure 9.12a. Thermodynamically, but not electrically, this is equivalent to a liquid volume $V_f = \Phi V$ in a closed flask, as in Figure 9.12b; bear in mind that Q_{gen} is calculated per unit of V, not of ΦV. Initially, the pressure in the volume is one atmosphere. Also, we will assume, as above, that the electric field is uniform in space and of constant amplitude and that the only loss mechanism is in the liquid phase water, so that (9.29), (9.30) and (9.31) apply. Again, for purposes of analysis we neglect any heat transfer and vaporization below 100°, and assume that the initial temperature history is linear to 100°. Now, however, when the water begins to vaporize into the gas phase volume the vessel pressure rises. As the internal pressure rises the saturation temperature also rises (Figure 9.12c). By taking very small time steps we may consider that the vaporization process occurs along constant temperature phase change lines (Figure 9.10) at successively higher temperatures, and thus at higher saturation pressures.

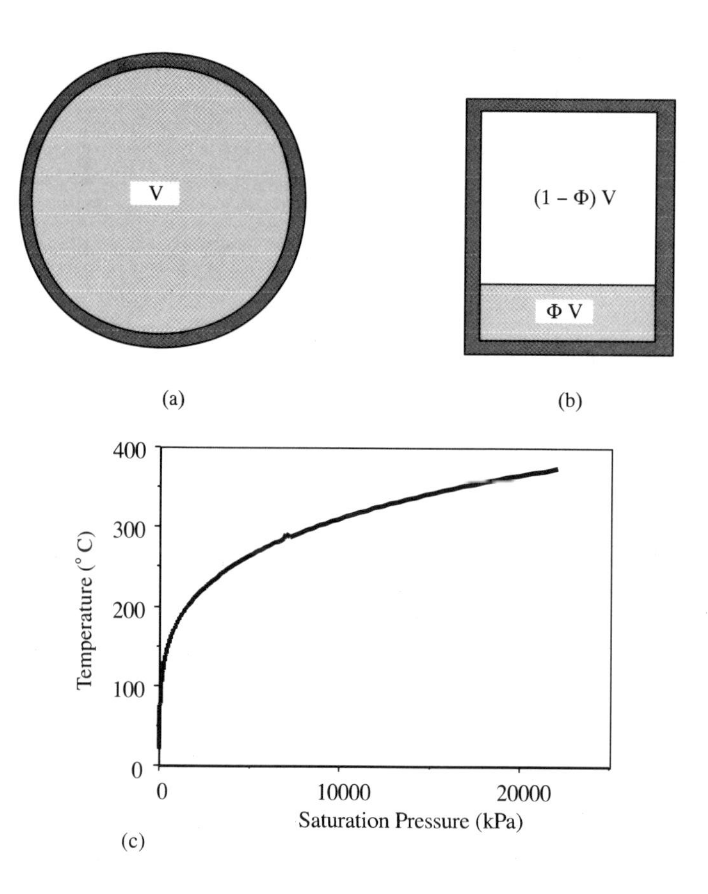

Fig. 9.12 *(a) Water uniformly dispersed in closed volume, V. (b) Closed flask water with the same volume fraction as in (a). (c) Saturation temperature vs. saturation pressure.*

The newly formed gas phase contributes to the mass in the vapor volume, $V_g = (1-\Phi)V$, with the state equation:

$$\rho_g = \frac{\mathcal{P}}{RT}. \tag{9.33}$$

Consequently, the rate of change in pressure due to vaporization is:

$$\frac{\partial \mathcal{P}}{\partial t} = RT\frac{\partial \rho_g}{\partial t} = \frac{\mathrm{RT}}{(1-\Phi)V}\frac{\partial m}{\partial t} \tag{9.34}$$

and the mass rate of liquid vaporization, $\partial m/\partial t$, comes from equation (9.26) (ignoring conduction heat transfer and temperature rise):

$$\frac{\partial m}{\partial t} = \frac{V}{\Delta h_{fg}}Q_{gen} = \frac{aV}{\Delta h_{fg}}\Phi. \tag{9.35}$$

Here we are using the finite volume form since it is a constant total volume situation and the constant a has the same significance as in the previous discussion.

At this point the thermodynamic analysis must be carried out in a finite time step fashion to continue. That is, we imagine that: (1) vaporization occurs at constant temperature, (2) the additional gas phase mass changes the pressure, and (3) the temperature of the liquid phase water increases to a new saturation temperature at the new pressure because of Q_{gen}, where (4) the cycle begins again. For stability, the time steps must be small enough that Q_{gen} is larger than the energy required to raise the liquid phase temperature to the new saturation point:

$$\Delta t < \Phi \rho_f c_f \frac{(T_{sat}(\text{new}) - T_{sat}(\text{old}))}{Q_{gen}}. \tag{9.36}$$

If this is satisfied then the rate of pressure rise may be estimated from recursive iteration among:

$$\frac{\Delta \mathcal{P}}{\Delta t} = \frac{RT}{(1-\Phi)V} \frac{aV}{\Delta h_{fg}} \Phi = \frac{aRT}{\Delta h_{fg}} \frac{\Phi}{(1-\Phi)} \tag{9.37}$$

and:

$$\frac{\Delta \Phi}{\Delta t} = \frac{-1}{\rho_f V} \frac{\Delta m}{\Delta t} \tag{9.38a}$$

and:

$$\frac{\Delta m}{\Delta t} = \frac{V}{\Delta h_{fg}} \left[Q_{gen} - \Phi \rho_f c_f \frac{\Delta T_{sat}}{\Delta t} \right] \tag{9.38b}$$

so that:

$$\frac{\Delta \Phi}{\Delta t} = \frac{-1}{\rho_f \Delta h_{fg}} \left[a - \rho_f c_f \frac{\Delta T_{sat}}{\Delta t} \right] \Phi = -b\Phi \tag{9.39}$$

and the new change in saturation temperature, ΔT_{sat}, is obtained from the previous change in saturation pressure. This equation set may be solved iteratively on a small computer for the assumed conditions.

For a small initial water volume the vaporization temperature steadily increases to a maximum value reached when the entire liquid phase water volume has been completely vaporized and, if the vapor is non-absorbing, interaction with the applied field ceases. For larger volumes the critical point will be reached at some point (375°C and 22048 kPa for water) where there is no distinction between liquid and vapor phases ($\Delta h_{fg} = \Delta s_{fg} = 0$) and phase change occurs with no power input. Above the critical point and outside of the saturation line (see Figure 9.10) the vapor is superheated. In the superheat region (all of the liquid must have been vaporized) if the vapor phase is still absorbing power from the electromagnetic field the temperature, pressure, entropy and enthalpy are dependent on the gas phase state equations.

Very few real materials can maintain vapor-tight isolated pockets at extreme pressures. However, the above discussion also gives a realistic view of the temperature–pressure history observed when the volume is "leaky". That is, in many cellular materials — fresh-cut wood, for example — the individual volumes (cells) will allow diffusion of vapor (and in many cases liquid water as well) as the pressure rises. Consequently, in wood heating one may observe a steady state vaporization temperature plateau in excess of 100°C as the internal pressure rises to a steady state value at which the rate of liquid and vapor flow equals the rate of vaporization. Figure 9.13a shows four sections of fresh-cut "live oak" log between RF plate electrodes. Four temperature probes (T1 to T4) were placed 3 cm deep to one end of the log pieces. Live oak is an especially dense hardwood found in the southern regions of the US. The frequency was 25 MHz, and the initial wood mass was 6 kg. One may see a rapid rise to a short plateau at 100°C in the temperature record (Figure 9.13b), as local free water vaporizes around all four probe sites, with subsequent rises to steady state temperatures in excess of 100°C ranging from about 120 to 170°C at different locations. The higher temperatures result from water captured within the wood structure which is vaporizing at higher pressures. We will use the leaky volume description again in our discussion of the behavior of zeolites under microwave irradiation (Chapter 10).

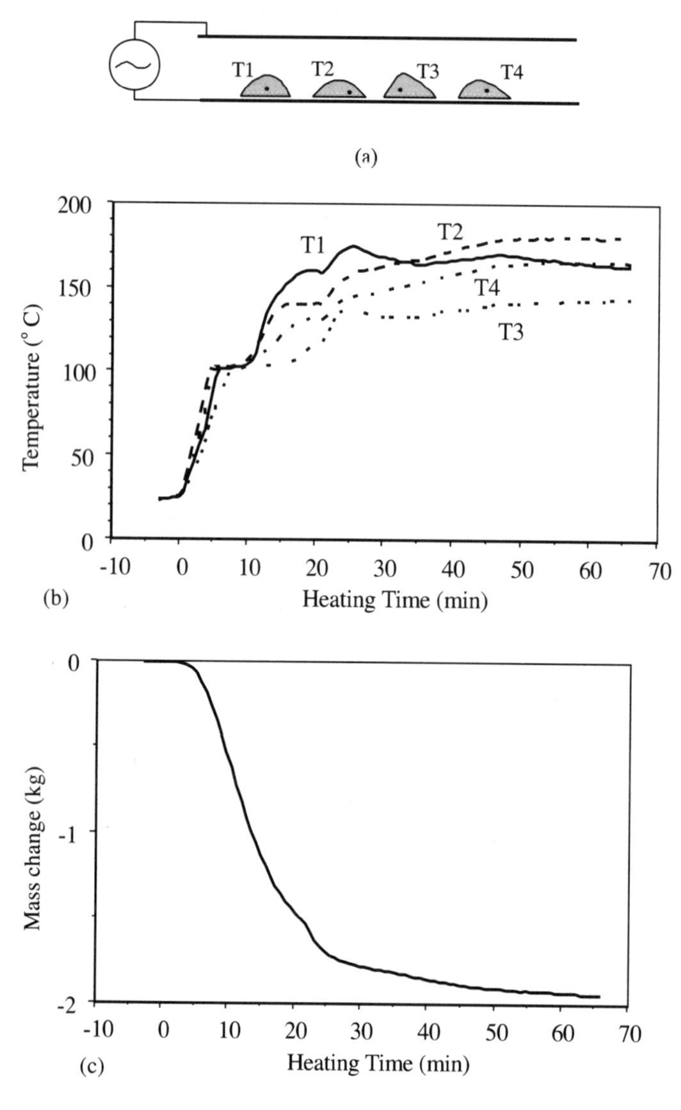

Fig. 9.13 *(a) Heating of oak log sections in RF field with locations of temperature probes T1 through T4. (b) Temperature records of oak log sections heated in RF electric field. (c) Oak log mass change for the same experiment, initial mass 6 kg.*

It is apparent that even simple models of vaporization lead to complex descriptions of drying behavior due primarily to the interaction among the governing equations.

Dehydration of zeolites: example of a complex system

As a further example, we will consider a simplified model of microwave dehydration of zeolites. In experimental studies a small diameter cylindrical volume of wet zeolite was subjected to an approximately constant electric field in WR340 waveguide ($a = 8.64$ cm) at a field maximum, as shown in Figure 9.14a. By assuming that **E** is constant, we can eliminate the solution of Maxwell's equations, as we have done in the previous discussion. The geometry of the zeolites is more complex than the previous examples, however. The zeolite crystals are formed into spheres about 1 mm in diameter and the space between the spheres is initially occupied by air. Spheres have a volumetric packing factor of about 70% so the air gaps are about 30% of the total zeolite bed volume. The interiors of the spheres contain a hollow cage framework formed by the zeolite matrix. At certain positions in the cage framework water is adsorbed onto the zeolite crystal. Experiments which will be described later show that the dehydration problem is complicated because there are at least three types of water in the system: (1) water vapor external to the spheres in the gaseous phase with density ρ_g and temperature T_g, (2) adsorbed water in the solid, which we will treat initially as a liquid phase, with apparent density ρ_a and temperature T_a, and (3) an intermediate phase of water not clearly identifiable as either a gas or a liquid which remains in the solid matrix (moving from cage to cage, for example) but is not fixed to the solid, with density ρ_i and temperature T_i.

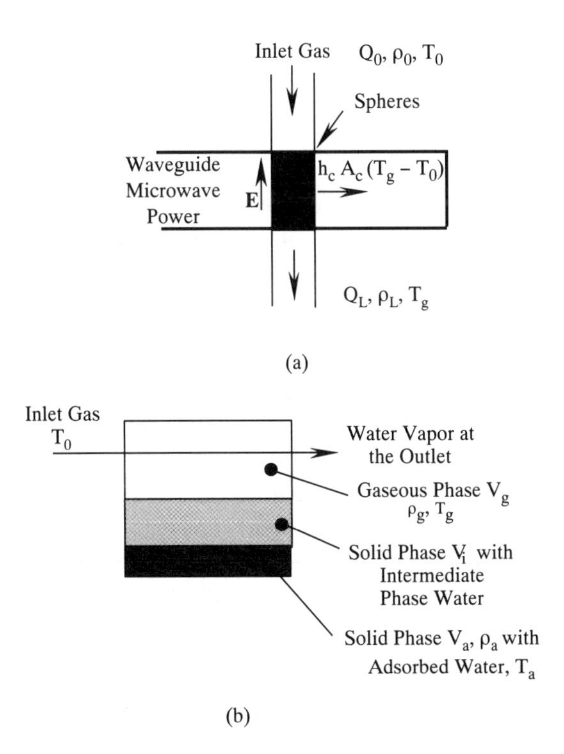

Fig. 9.14 *Thermodynamic model of zelolite drying. (a) Physical system. (b) Thermodynamic compartments.*

The associated volumes are V_g, V_a, and V_i—the "gas" represented by V_i has been "vaporized" but has not been transported out of the zeolite matrix into the space between the spheres. The phenomenological model for the system is shown in Figure 9.14b. Suppose that the three thermodynamic compartments each have homogeneous temperatures and compositions, which is reasonable because of the small size of the sample. The temperature of the solid crystal, T_m, is assumed equal to the internal gas phase, $T_m = T_i$, and the temperature T_0 of the inlet gas is equal to the environment temperature, $T_0 = T_\infty$. Q_0 is the volume rate of the inlet air flow (m^3/s), and Q_L the volume flowrate of the mixture of inlet gas and water vapor at the outlet with density ρ_L. T_g is measured at the outlet as well. In the model we imagine that Q_{gen} is dissipated in the adsorbed water phase and that the zeolite matrix is heated by conduction heat transfer from the adsorbed water, to obtain an elementary model, so that the zeolite does not affect the results.

We will use the double subscript notation as before so "ig" means a transition from intermediate to gas phase and "gi" the converse. Under these assumptions the mass balance for the gas phases is:

$$\rho_L Q_L = \rho_0 Q_0 + \frac{\partial m_{ig}}{\partial t} - \frac{\partial m_{gi}}{\partial t} \tag{9.40}$$

where ρ_L is the density of the mixed air (now at T_g) and water vapor at the outlet and m is, again, the mass (kg) which makes the transition. The water mass balance in the solid defines the "vaporization" process:

$$Q_w(t) = \frac{1}{\rho_i}\left(\frac{\partial m_{ai}}{\partial t} - \frac{\partial m_{ia}}{\partial t}\right) \tag{9.41}$$

where $Q_w(t)$ is the water volume flow rate between the adsorbed and intermediate phases (m^3/s), the defining relationship for the transport, about which we have not yet been specific. In order to preserve moderate pressures inside the cages it is necessary that the net volume flow be cleared into the external space, so this rate also describes the rate for equation (9.40).

The total energy balance involves the heating of the air between the inlet and the outlet, the mass of vaporized water, the heating of the intermediate phase inside the sample, and the temperature rise of the solid and of the water absorbed in the solid:

$$\begin{aligned} Q_0\rho_0 C_{p0}\int_0^{\Delta t}\left(T_g - T_0\right)\,dt + \int_0^{\Delta t}\Delta h_{fg}\frac{\partial m_{ai}}{\partial t}\,dt + V_m\rho_m c_{pm}\left(T_g - T_0\right) \\ + V_a\rho_a c_{pa}\left(T_{sat} - T_0\right) + V_i\rho_i c_{pi}\int_0^{\Delta t}\left(T_g - T_{sat}\right)dt = V_m Q_{gen}\Delta t - \text{losses} \end{aligned} \tag{9.42}$$

where the total heating time is Δt, T_g is time dependent within the integral (assumed to be the steady state value outside the integral) and "losses" represents heat transfer to the surroundings from the reactor vessel. In this relationship the change in internal energy of the gas phase is neglected because it is small compared with those concerning the solid phase and the absorbed water. For the external gas phase, the enthalpy balance at an instant in time is:

$$\left[Q_0\rho_0 c_{p0} + h_c A_c\right]\left(T_g - T_0\right) + Q_w\rho_i c_{pi}\left(T_g - T_i\right) = h_m A_m\left(T_m - T_g\right) \tag{9.43}$$

where the c subscript refers to the heat transfer from the entire zeolite bed and the m subscript to the coupling between the zeolite spheres and the gas stream.

One can partially interpret the experimental results by calculating the water conversion rate from the mass flow in the outlet gas:

$$(Q_w(t) = Q_L(t) - Q_0 \frac{T_g(t)}{T_0}. \tag{9.44}$$

Experimentally, we recorded ρ_L, $T_g(t)$, and $Q_{gen}(t)$—Q_{gen} varies as the water is eliminated. If all the physical and chemical characteristics of the fluid, of the solid, and of the heat and mass transfer were known, the equations could be solved by successive iteration in order to determine $T_m(t), m_a(t)$, and $m_i(t)$. The water transfer kinetics could then be evaluated. The thermodynamic relationship between the adsorbed water and the water of the intermediate phase could also be evaluated. This would specify the equilibrium adsorption/desorption of water on microwave irradiated zeolite. The suggested analysis is not simple to perform because of the lack of precise values of the parameters introduced in the model equations. Although the water mass flow in the outlet gas $Q_w(t)$ and other global variables were experimentally determined, it was not possible to determine all aspects of the system. It is important to remember that the system was extremely simplified since the difficulties of the Maxwell equations were eliminated—but it is still much too complex to construct a reliable model. We will explore an alternative approach in a later section of this chapter.

9.1.5 Rate process models of transformations

Thermally mediated transformations include bi-molecular and multi-molecular chemical reactions and also mono-molecular reactions. Examples of mono-molecular processes are the vaporization of water, the denaturization of tissue proteins and polymerization. The same type of governing equation based on an Arrhenius formulation can be used for all of these processes if first order reaction kinetics apply. Transformations may be exothermic or endothermic, but in either case the goal of the model is to predict the amount of material transformed so that the contribution to the energy balance may be estimated. This section presents the calculation strategy for including reaction heats in the energy balance and a model for rate-limited vaporization.

Reaction product formation

The transformation process is often conveniently described by a first order rate of formation equation, as we have used on previous occasions. In this formulation the rate of decrease in concentration of reactants is proportional to the remaining concentration of untransformed material, C (mole/m^3):

$$\frac{dC}{dt} = -uC \tag{9.45}$$

where the constant of proportionality, u(s^{-1}), is the reaction speed. The solution for the Bernoulli differential equation of (9.45), is:

$$C(\tau) = C(0)\exp\left(-\int_0^\tau u(t)dt\right) \tag{9.46}$$

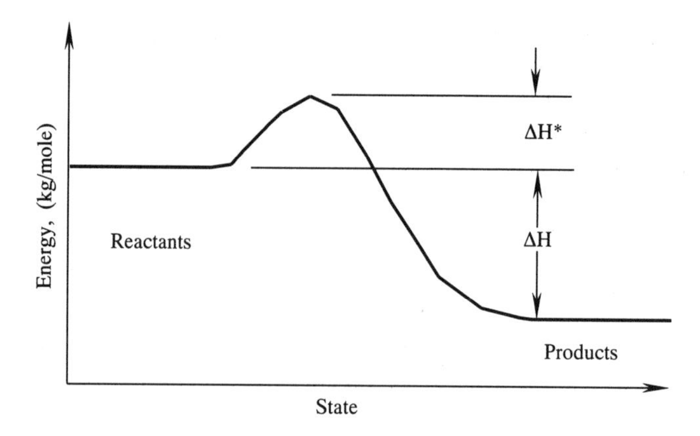

Fig. 9.15 *Energy–state diagram for rate process model.*

where we allow the reaction speed to be time dependent since the temperature is time dependent. The reaction speed for a first order thermal process is described by an Arrhenius relation:

$$u(t) = A\exp\left(-\frac{\Delta H^*}{RT(t)}\right) \tag{9.47}$$

where the time dependence in the reaction speed is contained in the temperature history, ΔH^* (J/mole) is an energy barrier which material must surmount to transform (Figure 9.15) from reactants to products, the pre-exponential "frequency factor", $A(\text{s}^{-1})$ is a measure of the collision rate between reactants, and R is the universal gas constant. The energy barrier should not be confused with the difference in enthalpy between the reactant and combined states, ΔH. In the universal reaction rate model, as we mentioned in Chapter 8, the frequency factor, A, can be expressed in terms of an entropy associated with the barrier, ΔS^* (J/mole K), so that:

$$u = \left(\frac{N_A h}{kT}\right)\exp\left(\frac{\Delta S^*}{R}\right)\exp\left(-\frac{\Delta H^*}{RT(t)}\right) \tag{9.48}$$

where N_A is Avogadro's number, h is Planck's constant (6.63×10^{-34} J s), and k is Boltzmann's constant (1.38×10^{-23} J/K). The reaction constants ΔH^* and ΔS^* are only calculable in the simplest of cases, so they must be determined by experiment.

The concentration of reactants at any point during the process may be calculated from:

$$\ln\left\{\frac{C(0)}{C(\tau)}\right\} = \int_0^{\tau}\left(\frac{N_A h}{kT(t)}\right)\exp\left(\frac{\Delta S^*}{R}\right)\exp\left(\frac{-\Delta H^*}{RT(t)}\right)\,dt. \tag{9.49}$$

Interestingly, the temperature dependence of the frequency factor, contained in the pre-exponential term, makes a negligible contribution to the calculation of $C(t)$ in most cases because of the very strong exponential nature of the other terms. Therefore, A can usually be treated as a constant and be brought outside of the integral with little loss in accuracy. Typical reactions in liquid phase systems have ΔH^* of the order of 10^5 J/mole with A between 10^{20} and 10^{90} s^{-1}. The highest numbers for A are associated with measurements of thermal damage accumulation in biological tissues and are not usually observed in typical chemical reactions.

The reaction heat is specified by the difference in state enthalpies, ΔH, which may be either positive (exothermic) or negative (endothermic). Reaction heat may be added to the energy balance (usually only in numerical models due to the computational complexity) having the same sense as the power generation term, Q_{gen}. First, the change in molar concentration during a Δt step is calculated from equation (9.49) where the integral is carried from t to $t+\Delta t$. Then the reaction heat is added to Q_{gen} to get the total heat source term for the energy balance:

$$Q_{\mathrm{source}} = Q_{\mathrm{gen}} + \Delta H(C(t+\Delta t) - C(t)). \tag{9.50}$$

Non-equilibrium vaporization processes

In our previous discussion the rate of vaporization was assumed large enough that equilibrium between the liquid and vapor phase was maintained and the rate of boiling was determined by the rate of energy addition in the bulk liquid. If, however, the rate of energy addition to the bulk liquid is higher than the rate at which vapor phase forms then the liquid phase temperature exceeds the saturation temperature at the system pressure — the liquid is super-saturated. The liquid and vapor phases are not in equilibrium and vapor formation is rate-limited. In that case we must apply a rate process model and describe the vaporization process kinetically rather than using equilibrium thermodynamics. The kinetic model is called nucleation theory and was first elucidated in the 1930s. Nucleation theory can be applied equally to the formation of vapor bubbles in liquid or the condensation of droplets or particles of solid from vapor.

In the kinetic model for boiling in bulk liquid phase (not at the surface) two events occur in cascade: (1) a bubble of critical radius must be formed from the bulk liquid called a nucleating center, and (2) the bubble then grows in size until equilibrium is approached. A bubble of critical size has a probability of growth greater than the probability of collapse. The critical radius depends on the surface tension of the water:

$$r_{\mathrm{crit}} = \frac{2\sigma_{\mathrm{l}}}{(\mathcal{P}_{\infty} - \mathcal{P})} \tag{9.51}$$

where σ_{l} is the liquid–vapor surface tension (N/m) at temperature $T(^{\circ}\mathrm{C})$, $\mathcal{P}_{\infty}$ is the saturation pressure at the liquid temperature and $\mathcal{P}$ is the pressure of the metastable liquid phase (less than $\mathcal{P}_{\infty}$). The surface tension of water is approximately linear with temperature:

$$\sigma_{\mathrm{l}}(T) = 7.642 \times 10^{-6} - 17.3 \times 10^{-9} T. \tag{9.52}$$

The rate of formation of nucleating centers follows the Arrhenius formulation which we have seen on several previous occasions. In this case, however, there are two energy barriers to surmount: (1) the surface energy of the bubble radius, and (2) Δh_{fg} since liquid phase molecules must make a transition to higher enthalpy vapor phase molecules. The rate of formation of bubbles of critical radius, $R_{\mathrm{n}}(t)$ (number/m^3 s) is:

$$R_{\mathrm{n}}(T) = Z_1 \exp\left(-\frac{\Delta h_{\mathrm{fg}}}{RT}\right)\sqrt{\frac{6\sigma_{\mathrm{l}}}{(3-b)\pi m}}\exp\left(\frac{-16\pi\sigma_{\mathrm{l}}^3}{3k(\mathcal{P}_{\infty}-\mathcal{P})^2 T}\right) \tag{9.53}$$

where Z_1 is the number of molecules per unit volume (for water):

$$Z_1 = \frac{\rho N_A}{0.0018} \tag{9.54a}$$

b is the super-saturation ratio:

$$b = \frac{(\mathcal{P}_\infty - \mathcal{P})}{\mathcal{P}_\infty} \tag{9.54b}$$

the phase change enthalpy is expressed on a per mole basis, and other variables are as previously described. A nucleation bubble is considered to have formed at the saturation pressure of the liquid phase. Once a nucleation bubble is formed it grows at a rate determined by its internal pressure:

$$\frac{dr}{dt} = \frac{c}{\rho_g}\left[\frac{\mathcal{P}_\infty}{\sqrt{2\pi R T_g}} - \frac{\mathcal{P}}{\sqrt{2\pi R T_f}}\right] \tag{9.55}$$

where r is the bubble radius (in excess of r_{crit} for a growing bubble), the density is now the gas phase density, ρ_g and T_f is the liquid temperature. The phase change energy for

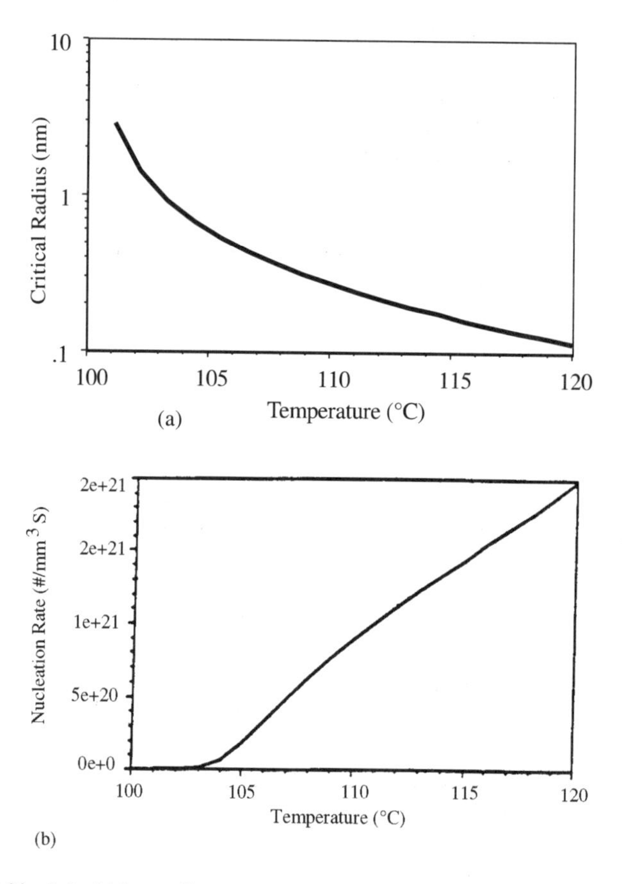

Fig. 9.16 *(a) Critical bubble radius vs. temperature. (b) Nucleation rate, $R_n(T)$, at 1 atmosphere.*

nucleation and growth is calculated using the same approach as in the near-equilibrium vaporization case, equation (9.26) or (9.27).

The rate of critical bubble formation is very large even for small degrees of supersaturation in the liquid. Calculated values for $R_n(T)$ are shown in Figure 9.16 assuming that the pressure is 1 atmosphere. As can be seen, the nucleation rate is so high that only a very few degrees of supersaturation is to be expected even over very short time intervals. So it will only be in the most rapid heating cases (on the order of nanoseconds) that appreciable superheating of liquid water is to be expected.

9.2 LINEAR SYSTEM MODELS

We have seen several examples of systems in which the number of parameters required to execute models based either on conservation of energy or rate process descriptions is prohibitive. To understand why the conservation of energy approach is ineffective, consider the analogy of inserting a nail into a piece of wood. If one applies a constant force to the nail the description of the results entails detailed analysis of the angle of attack, the structural mechanics of the wood, and frictional forces. For example, when a press is used to apply a constant force to a nail, if the applied force is larger than the resistive force of the wood the nail penetrates the wood with equilibrium or near-equilibrium between the active and reactive forces. The technique of conservation of energy can be applied; but the difficulty is to describe the frictional forces and the wood deformation mechanics adequately. An alternative method is to insert the nail with a hammer. From a mechanical point of view, one examines the total impulsive energy provided by the hammer and the net penetration of the nail which results—conservation of total momentum rather than force balance. The number of parameters required is much smaller than in the constant force case, and the model can be successfully applied.

We will now consider how the aforementioned difficulties with the detailed approach can be overcome. Our initial concern is to describe the influence of the electromagnetic field, which is the source of the transformation process, in a simpler way. For many cases we may formulate a linear model of the transforming system by applying an analogous dc circuit model, as in Figure 9.17. The resistance, R_m, represents the material under study. This R_m is connected to a generator with Thevenin equivalent voltage V_{TH} (dc), which represents the electric field, and has dc output impedance $Z_{TH} = R_{TH}$. The load voltage, current, and power are:

$$V_m = \frac{V_{TH} R_m}{R_m + R_{TH}} \tag{9.56}$$

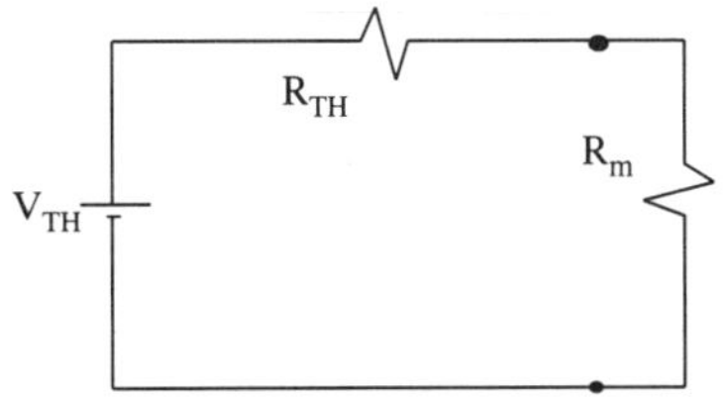

R_{TH}

R_m

V_{TH}

Fig. 9.17 *Simple series resistor combination.*

$$I_m = \frac{V_{TH}}{R_m + R_{TH}} \tag{9.57}$$

$$P_m = \frac{V_{TH}^2 R_m}{(R_m + R_{TH})^2}. \tag{9.58}$$

As the transformation of the matter takes place, the material equivalent resistance R_m changes. Thus, the voltage applied to the material is a function of the electromagnetic field of the generator (represented by V_{TH}), the source resistance of the generator, R_{TH}, and of the material, R_m. The current (I_m) and the absorbed power (P_m) are also functions of these three variables — as R_m changes, V_m, I_m, and P_m also vary. It is difficult to characterize R_m changes intrinsically; thus, it is necessary to maintain a constant V_{TH} in order to determine the quantities V_m, I_m, or P_m, and, of course, R_m. Analogously, to determine the change in permittivity of the material as it is transformed the electromagnetic field must be held constant. This is easily done with a servo-feedback loop, as illustrated in Figure 9.18. First, the voltage (E-field) applied to the material is measured and compared with a reference voltage (represented by the zener diode). The difference between the voltages is amplified and a variable resistance is changed according to the variations in R_m detected. This strategy can be easily adapted to a microwave experiment by placing the material in a waveguide and measuring the field near the sample. Using a computer, the equivalent field near the sample can be controlled by the movement of a short circuit behind the sample and by simultaneously controlling the amplitude of the microwaves produced by the generator, as illustrated in Figure 9.19.

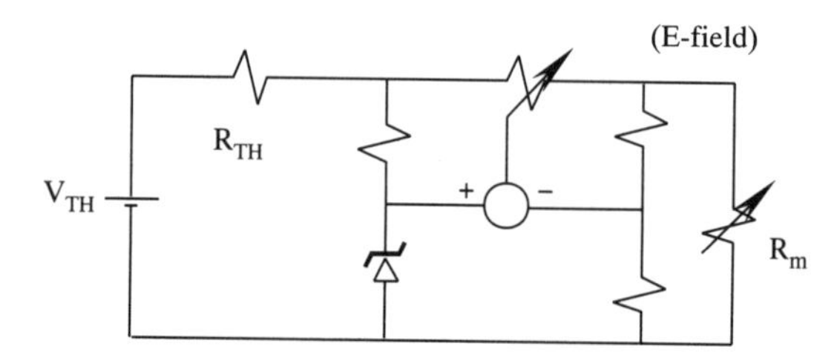

Fig. 9.18 *Servo-feedback loop control circuit.*

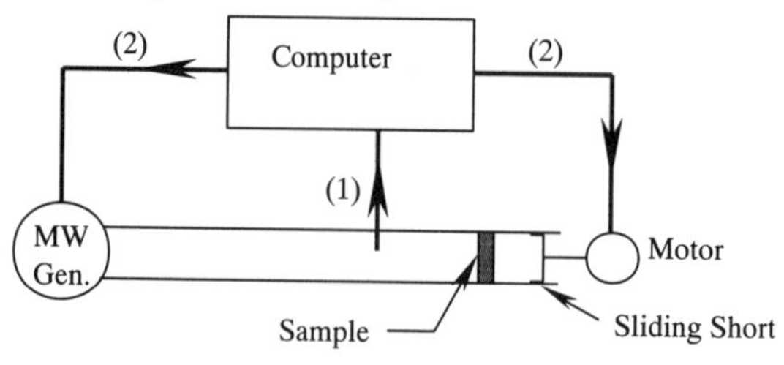

Fig. 9.19 *Waveguide set-up for controlling electric field in sample under MW irradiation.*

9.2.1 Transient response to a step change in field strength

We can, in principle, measure the effects of the electromagnetic field on the material with such a design. If we apply a constant field in a sequence of increasing steps, then we can measure and analyze the system response to this excitation to determine the response parameters: one response variable could be the temperature, another could be a change of color, etc. When using electromagnetic fields for dehydration, the measured response could be the amount of water vapor produced. Other measurable responses might include the actual absorbed power, permittivity, conductivity or the like. The response is measured as a function of time, and is analyzed as an increasing step function.

First, suppose that when the power is increased by an arbitrary factor, the response is shifted proportionally. In this case, the response is linear with the excitation and the dynamic behavior can be analyzed with the Laplace transform. The Laplace transform is defined by:

$$\mathcal{L}\{f(t)\} = F(s) = \int_0^\infty f(t)\mathrm{e}^{-st}\,\mathrm{d}t. \tag{9.59}$$

Table 9.1 lists several useful Laplace transform pairs.

The Laplace transform of the response can be compared with the known Laplace transform of the excitation — the ratio of the two yields a transfer function description

Table 9.1 *Short table of Laplace transforms.*

The Laplace transform of $f(t)$ is defined by $\mathcal{L}\{f(t)\} = \int_0^\infty \mathrm{e}^{-st} f(t)\,\mathrm{d}t = F(s)$.

$\mathcal{L}\{Af(t)\} = A\mathcal{L}\{f(t)\}$	$\mathcal{L}\{f(\alpha t)\} = \dfrac{1}{\alpha}F\left(\dfrac{s}{\alpha}\right)$
$\mathcal{L}\left\{\dfrac{\mathrm{d}f}{\mathrm{d}t}\right\} = sF(s) - f(0)$	$\mathcal{L}\left\{\int_0^1 f(t)\,\mathrm{d}t\right\} = \dfrac{1}{s}F(s)$
$\mathcal{L}\{t^n f(t)\} = (-1)^n \dfrac{\mathrm{d}^n}{\mathrm{d}p^n}(F)$	
$\mathcal{L}\left\{\int_0^{+\infty} f_1(r) f_2(t-r)\,\mathrm{d}r\right\} = F_1(s)F_2(s)$	
$\mathcal{L}\begin{Bmatrix} f(t-a) & t \geqslant 0 \\ 0 & t < a \end{Bmatrix} = \mathrm{e}^{-as}F(s)$	
$\mathcal{L}\{\mathrm{e}^{at} f(t)\} = F(s-a)$	$\mathcal{L}\{t^v\} = \dfrac{\Gamma(v+1)}{s^{v+1}}$ if $(v) > -1$
$\mathcal{L}\{\sin at\} = \dfrac{a}{s^2+a^2}$	$\mathcal{L}\{\cos at\} = \dfrac{p}{s^2+a^2}$
$\mathcal{L}\left\{\dfrac{1}{t}\sin at\right\} = \tan^{-1}\left(\dfrac{a}{s}\right)$	
$\mathcal{L}\{\mathrm{e}^{-pt}\sin(at+q)\} = \dfrac{a\cos\phi + (s+b)\sin\phi}{(s+b)^2+a^2}$	
$\mathcal{L}\{\bar{F}(\alpha;\beta;t)\} = \dfrac{1}{s}\bar{F}(\alpha;1;\gamma;1/s)$ (Confluent hypergeometric function)	
$\mathcal{L}\{t^v \mathrm{e}^{at}\} = \dfrac{\Gamma(v+1)}{(s-a)^{v+1}}$ Real for $v > -1$	$\mathcal{L}\{\delta(t-a)\} = \mathrm{e}^{-as}$
$\mathcal{L}u(t-a)\begin{Bmatrix} 0 & t < a \\ 1 & t > a \end{Bmatrix} = \dfrac{\mathrm{e}^{-as}}{s}$	$\mathcal{L}\left\{\sum_0^\infty \delta(t-na)\right\} = \dfrac{1}{1-\mathrm{e}^{as}}$
$\mathcal{L}\{\sinh\ at\} = \dfrac{a}{s^2-a^2}$	$\mathcal{L}\{\cosh\ at\} = \dfrac{s}{s^2-a^2}$

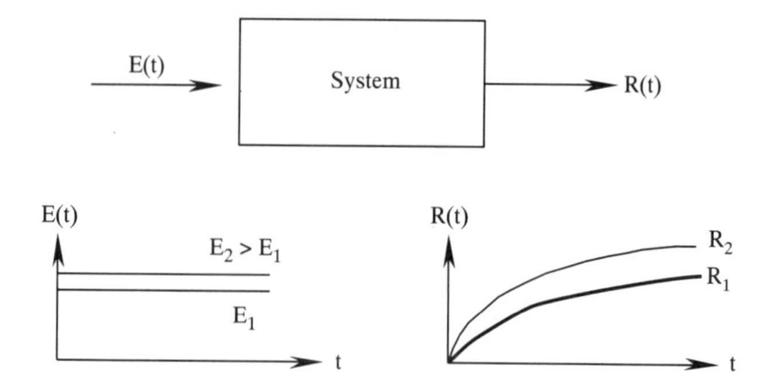

Fig. 9.20 *Linear system model for heated material.*

of the material, $F(s)$:

$$F(s) = \frac{\mathcal{L}\{R(t)\}}{\mathcal{L}\{E(t)\}} \tag{9.60}$$

where $E(t)$ is the excitation function and $R(t)$ is the material response (Figure 9.20). Using the transfer function, the response for any type of excitation can be predicted by forward transform of the excitation and reverse transform of the response.

An alternative and equivalent method is to convert the transfer function into a differential equation to which the response is a solution when the field is varying. For example, suppose that an arbitrary response, G, is an exponential function of time. The response is coupled to the electric field by a first order differential equation with time constant τ:

$$\frac{\mathrm{d}G}{\mathrm{d}t} + \frac{1}{\tau}G = \alpha E(t) \tag{9.61}$$

$$F(s) = \frac{1}{1 + \tau s}. \tag{9.62}$$

If the response is oscillating (under-damped), or not a single exponential, it can be modeled by a second order differential equation. When a linear response is observed in a material, one finds that the response is correlated to the field by a differential equation with coefficients which are constant. When the response is not linear with the electric field, the mathematical expressions become very difficult to manage indeed. However, there are mathematical techniques to overcome these difficulties. In all of the cases we have studied, the temperature response has never proved linear. So models based on temperature are not desirable for analyzing the particular responses (as we have just seen) using linear system formulations. However, in many cases other properties respond in a linear fashion to the electric field and this analytical approach can be used.

Two comments are in order at this point:

(1) Even if we do not have a physical interpretation for the transfer function, we can store it in a computer and by numerical methods predict the response of any type of excitation in time by calculating the inverse of the Laplace domain representation. This may be done numerically without having an algebraic equivalent form for the function in the time domain.

(2) We are primarily concerned with scalar responses: absorbed power, permittivities, temperature, etc. These scalars are independent of the field orientation (in isotropic materials). This means that if the field intensity is changed by a factor of −1, the response

will be the same. From this observation we can conclude that these responses actually ought to be functions of the square of the field (E) intensity. So, when we say that the response is linear with the field, it must be understood that we mean that it is linear with the square of the field intensity, the applied power.

9.2.2 Experimental apparatus to obtain the linear model

Although the instrumentation has been discussed in Chapter 5 we will briefly review the pertinent principles for the problem at hand with special attention to the problem of control. The experimental apparatus (Figure 9.21) consists of a small cylindrical sample which is placed in the center of a RG 112/u (WR340) waveguide, parallel to its small side wall. This arrangement allows access to the sample through its extremities in order to introduce the gaseous products which may be needed to assess the sample transformation; it also permits evacuation of the sample when necessary. If the sample diameter and electric permittivity are small it can be treated as a thin obstacle and its equivalent circuit may be reduced to a localized impedance (Chapter 3). The value of the admittance is a linear function of the complex permittivity:

$$Y = Y_0(\varepsilon^* - 1). \tag{9.63}$$

The coefficient Y_0 depends on the geometrical dimensions and on the wavelength. There are several restrictive conditions under which this relation is applicable. For larger values of the electric diameter of the sample — the electric diameter is $\sqrt{\varepsilon'}d/\lambda_0$ — the obstacle equivalent circuit must have series impedances added and its equivalent shunt admittance varies. The effect of permittivity on the measured reflection (Γ) and transmission (T) coefficients is shown in Figure 9.22. The curve in the figure shows the transition from Rayleigh to Mie scattering as the sample diameter increases (see Chapter 4). The linear segment of the curve is the operating region over which the electric field may be considered uniform in the sample. This is the region over which perturbation theory (equation (9.63)) is applicable.

The experimental method consists of compensating the phase change of the sample equivalent impedance by moving the short circuit placed behind the sample (Figure 9.21). Four probes S_1, S_2, S_3, S_4, are placed in the waveguide ahead of the sample, spaced $\lambda_g/8$ apart to monitor the field. Probe S_1 is located $\lambda_g/4$ from the sample centerline. At the beginning of the experiment before introduction of the sample, the sliding short circuit is

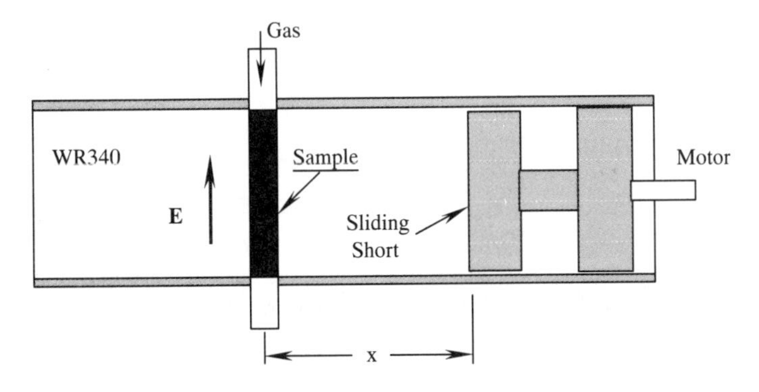

Fig. 9.21 *Waveguide applicator for* **E***-field control.*

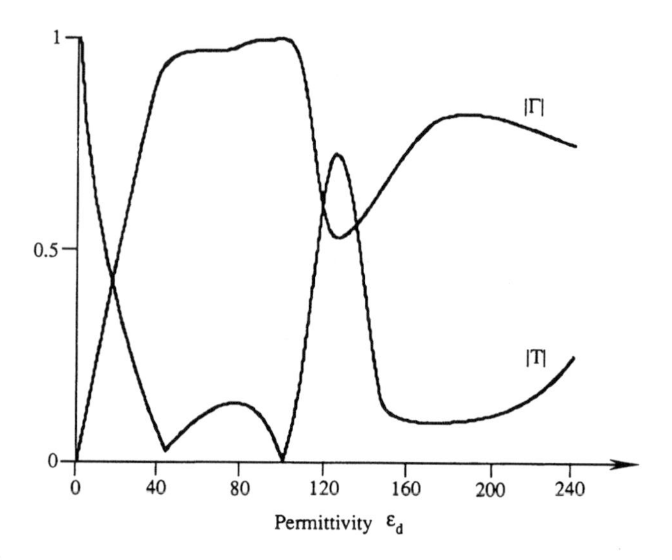

Fig. 9.22 *Amplitude of the reflection and transmission coefficients* Γ *and* T *as a function of the permittivity of a cylindrical dielectric of fixed diameter in a rectangular waveguide.*

positioned $\lambda_g/4$ behind the sample centerline—position x_0 (Figure 9.23a). A minimum of the field appears on the probe S_1, and a maximum on the probe S_2 ($\lambda_g/4$ from S_1). When the sample is introduced the short circuit piston is moved from x_0 to x_1 in order to maintain the minima and maxima at the same positions (Figure 9.23b). The capacitive nature of the dielectric specimen dictates that the plunger will be moved closer to the specimen when the material is inserted.

The standing wave ratio, VSWR=r, can be deduced from the relative amplitude of the probe signals. If the signals from the diode detectors are quadratic, as is typical, the standing wave ratio will be:

$$r = \sqrt{\frac{V_2}{V_1}}. \tag{9.64}$$

The incident power will be;

$$P_i = \tfrac{1}{2}k\left(\sqrt{V_2} + \sqrt{V_1}\right)^2 \tag{9.65}$$

and the absorbed power will be:

$$P_{abs} = P_i - P_R = k\sqrt{V_2 V_1} \tag{9.66}$$

The permittivity of the material will then theoretically be:

$$\varepsilon' = 1 + \frac{a\lambda^2}{\lambda_g \pi^2 d^2} \tan\left(\frac{2\pi}{\lambda_g}(x_1 - x_0)\right) \tag{9.67}$$

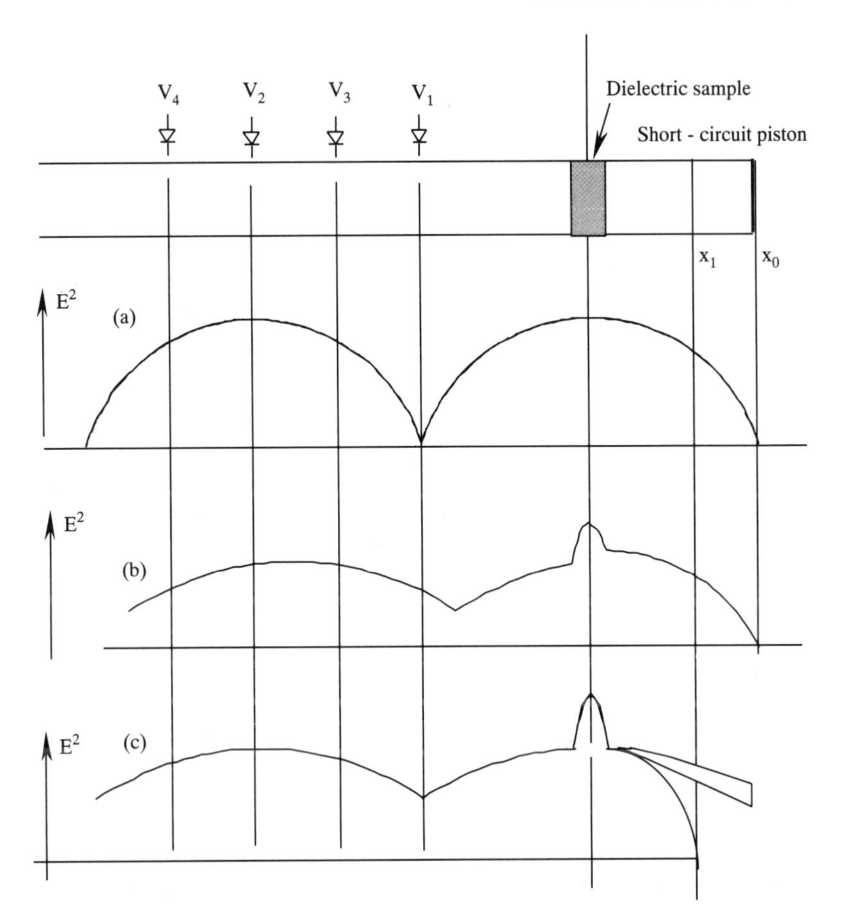

Fig. 9.23 *Standing waves in waveguide. (a) No sample, short circuit piston at x_0. (b) With the sample, short circuit piston at x_0. (c) With the sample, short circuit piston at x_1.*

$$\varepsilon'' = \frac{a\lambda^2}{\lambda_g \pi^2 d^2} \frac{1}{r} \tag{9.68}$$

where a is the large waveguide dimension (8.63 cm for WR340), λ is the wavelength in a vacuum (12.2 cm at 2450 MHz), λ_g is the wavelength in the guide (17.3 cm at 2450 MHz in RG 112/u), and d is the internal diameter of the sample (7 mm, for example).

Probes S_3 and S_4 have been inserted at a distance of $\lambda_g/8$ from S_2 to provide feedback for the automatic control of the moving short circuit. The value of $\phi = \arctan\{(V_4 - V_3)/(V_2 - V_1)\}$ yields the argument of the complex coefficient of the sample reflection with respect to the reference plane of probe S_1. If the standing wave is well positioned, probes S_3 and S_4 will have equal dc signals.

The displacement of the short circuit piston is controlled by a fine-thread screw which is moved by a stepping motor under computer control. The experiment is monitored by a computer which carries out three primary tasks: data acquisition, automatic positioning of the moving short circuit and analysis of experimental results.

9.2.3 Example: microwave heating of an inert product

We will first consider heating of an inert (non-reacting) product with a permittivity which is invariant with temperature; the system consists of granular pieces of solid product in a bed through which air can flow. The solid material has density ρ_m, specific heat c_{pm}, and total mass $(1-e)V\rho_m$; where e is the porosity of the bed expressed as a volume fraction and V is the bed volume. The mass flux of air through the sample is denoted by W_g (kg/s). The temperatures T_0, T_g and T_m are the input air temperature, the output gas temperature, and the solid material temperature, respectively. The energy balance for the solid mass is:

$$(l-e)V\rho_m c_{pm}\frac{dT_m}{dt} + hSV\left(T_m - T_g\right) = P_{abs}$$
$$= (1-e)V\omega\varepsilon''|\mathbf{E}_a|^2 \tag{9.69}$$

where h is the convection coefficient for the particles, and S is the particle surface area per unit volume of bed. The energy balance for the gas phase is:

$$W_g c_{pg}\left(T_g - T_0\right) = hSV\left(T_m - T_g\right). \tag{9.70}$$

If the electric field is applied in a stepwise increasing manner—steps of E_a^2, $2E_a^2$, $3E_a^2$, etc.—both the solid temperature, T_m and the outlet temperature, T_g, will increase as exponential curves from field application at $t = 0$:

$$T_g = T_0 + \frac{P_{abs}}{W_g c_{pg}}\left[1 - \exp\left(\frac{-BC}{A(B+C)}\right)t\right] \tag{9.71}$$

$$T_m = T_g + \frac{C}{B}\left(T_g - T_0\right) \tag{9.72}$$

where:

$$A = \rho_m c_{pm} \qquad B = \frac{hS}{1-e} \qquad C = \frac{W_g c_{pg}}{(1-e)V}. \tag{9.73}$$

Typical curves of this type are shown in Figure 9.24. The experiment in that figure was conducted on a sample of spherical alumina-silica particles (8 to 12 mesh) which was 7 mm in diameter. Power levels and air flow rates are given in the figure. The sample parameters can be determined from the recorded curves, including the effective convection heat transfer coefficient (as a function of W_g). We may also use the results to verify the applicability of conventionally used formulas like:

$$\frac{Sh}{W_g c_{pg}} = 0.91\left(\frac{W_g}{AS\mu}\right)^{-0.50}\left(\frac{c_{pg}\mu}{k}\right)^{-2/3} \tag{9.74}$$

where $W_g/AS\mu$ is the Reynolds number, μ is the viscosity, $c_{pg}\mu/k$ is the Prandtl number, and S is a characteristic area.

The curves in Figure 9.24 show that the steady state outlet temperature is given by:

$$T_g(\infty) = \left(T_0 + \frac{P_{abs}}{W_g c_{pg}}\right) \tag{9.75}$$

and is thus proportional to P_{abs} and to $1/W_g$ (see Figure 9.25).

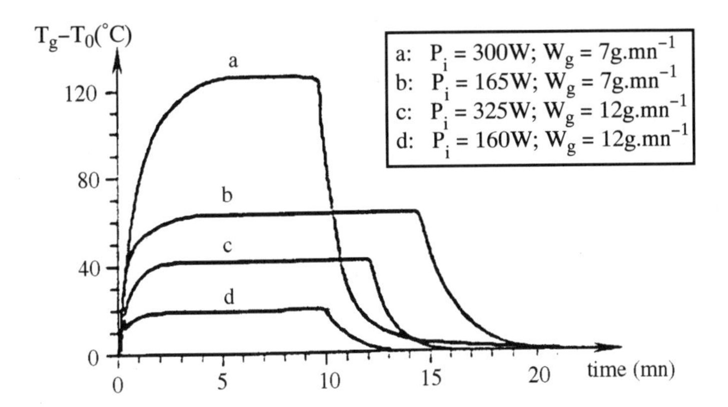

Fig. 9.24 *Temperatures, $T_g(t)$, for various absorbed powers, P_{abs} and gas flows, W_g.*

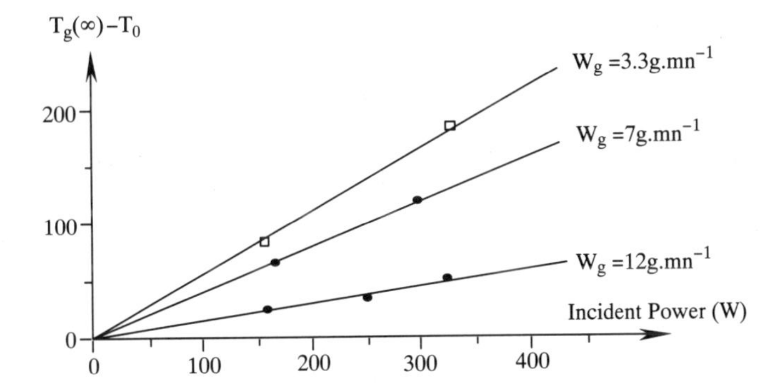

Fig. 9.25 *Steady state gas temperature, $T_g(\infty)$ vs P_{abs}.*

The experiment can be continued by determining T_m from equation (9.72) and then plotting ε'' vs. T_m to determine whether ε'' is dependent on temperature — note from (9.69) that ε'' is $P_{abs}/[k(1-e)VE_a^2]$. This is a simple method for step-by-step verification of the calorimetric hypothesis.

9.3 THERMAL RUNAWAY

9.3.1 Origins of thermal runaway

In some cases, the temperature does not change in a single exponential fashion with an increase in applied power. Up to a given value of the electric field, the curve does not increase as in Figure 9.24 (curve a in Figure 9.26); but above the threshold field an abrupt increase is encountered at higher temperatures (curve b in (9.26)). This type of curve cannot be understood as a variation of h or other transport parameter. To gain some understanding of this behavior, one must include the temperature dependence of the permittivity. For example, suppose that the imaginary part of the permittivity varies quadratically with temperature:

$$\varepsilon''(T) = \varepsilon''(T_0)\left[1 + \alpha\left(\frac{T - T_0}{T_0}\right) + \beta\left(\frac{T - T_0}{T_0}\right)^2\right] \tag{9.76}$$

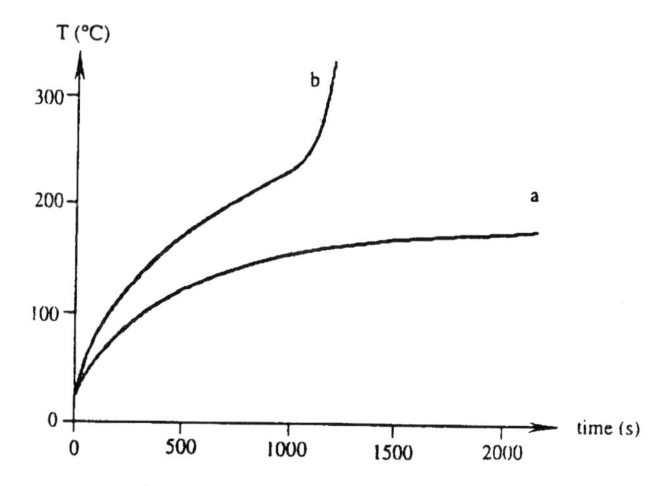

Fig. 9.26 *Normal response (curve a) and thermal runaway (curve b) under electromagnetic heating.*

where T_0 is the temperature at which $\varepsilon''(T_0)$ is measured and α and β are curve fit coefficients (without units). For purposes of insight we may initially consider a very small diameter sample in which radial thermal gradients are negligible (i.e. thermal conductivity k is very large). By this and other simplifying assumptions we may reduce the energy balance to:

$$(1-e)V\rho_{\mathrm{m}}c_{\mathrm{pm}}\frac{\mathrm{d}T}{\mathrm{d}t} + hSV(T-T_0) = P_{\mathrm{abs}} = \frac{\varepsilon'' P_{\mathrm{i}}}{\varepsilon_0}. \tag{9.77}$$

where P_{i} is the incident power (here we include geometric factors in with P_{i}). The differential equation under these assumptions is separable when the temperature dependence of ε'' is included on the right hand side, and may be recast into the form of equation (9.78) below:

$$\int_0^t \mathrm{d}\xi = t = \int_0^{(T-T_0)/T_0} \frac{\mathrm{d}u}{a_0\left[1+(\alpha-H)u+\beta u^2\right]} \tag{9.78}$$

where ξ is a dummy variable of integration (time), t is the duration of the heating, and the other parameters in equation (9.78) are:

$$u = \frac{T-T_0}{T_0} \qquad a_0 = \frac{\varepsilon''(T_0)P_{\mathrm{i}}}{(1-e)V\varepsilon_0 r_{\mathrm{m}} c_{\mathrm{pm}}}$$

$$H = \frac{\varepsilon_0 hSVT_0}{\varepsilon''(T_0)}.$$

This integral has a closed form solution which varies depending on the relative values of a_0, $(\alpha - H)$ and β, and on the sign of the discriminant of the denominator. When $(\alpha-H)^2 - 4\beta > 0$ the analytical solution involves a hyperbolic arctangent:

$$\frac{T-T_0}{T_0} = \frac{\sqrt{(\alpha-H)^2-4\beta}}{2\beta}\tanh\left\{\tanh^{-1}\left[\frac{\alpha-H}{\sqrt{(\alpha-H)^2-4\beta}}\right] + a_0\frac{\sqrt{(\alpha-H)^2-4\beta}}{2}t\right\} + \frac{H-\alpha}{2\beta} \tag{9.79}$$

which has no singular points and results in finite temperature at every instant — no thermal runaway. When the discriminant is negative the solution involves the tangent:

$$\frac{T - T_0}{T_0} = \frac{\sqrt{4\beta - (\alpha - H)^2}}{2\beta} \tan\left\{\arctan\left[\frac{\alpha - H}{\sqrt{4\beta - (\alpha - H)^2}}\right] + a_0 \frac{\sqrt{4\beta - (\alpha - H)^2}}{2} t\right\} + \frac{H - \alpha}{2\beta} \tag{9.80}$$

which has singular points when the argument is $n(\pi/2)$ and thermal runaway results at large t. The runaway temperature curves have the forms shown in curve b of Figure 9.26. The analytical result has several interesting features. As long as β is small the temperature approaches a finite steady state value; this system is inherently stable and the thermal behavior is denoted *Type I*. When β exceeds the value $(\alpha - H)^2/4$, the system becomes unstable, *Type II*. The temperature can increase without bound. The limiting curve is a hyperbola, the horizontal asymptote of which defines the critical temperature, T_c:

$$T_c - T_0 = T_0/\sqrt{\beta}. \tag{9.81}$$

The analytical model provides a good match to experimental results if the applied power (and thus rate of temperature rise) are not too large. In practice, real experimental results (curve b of Figure 9.26) do not follow the analytical result at high temperatures but have a much sharper "knee" and more precipitous rate of climb than predicted. This indicates that too many simplifications have been made in equation (9.76). Nevertheless, much can be learned about control strategies from the simple model.

9.3.2 Servo-control based on the simple model

The critical temperature, for example, is a characteristic of the material. So, we expect that the choices of the electric field strength and of the value of the convective losses in the system are independent and that one can compensate for the other. In fact, the two parameters play somewhat different roles. If the thermal losses are low, the system is unstable. To obtain stable behavior, it is necessary choose a sufficiently high value for H/P_i. But this condition alone is not sufficient to ensure stable heating. For some materials the temperature is unstable in all cases, whatever the conditions chosen, beyond the previously defined critical temperature. In other words, each material is characterized by a critical temperature which is the highest stable value which can be reached, whatever the microwave radiating and convection cooling conditions.

Temperature control of a system governed by equations (9.76) and (9.77) is simple. The electromagnetic incident power can be corrected as a function of the measured instantaneous temperature. A computer can be used to control $P_i(t)$ with a feedback loop as shown in Figure 9.27a. At regular time intervals the computer measures the temperature and resets the incident power according to:

$$P_i(t + \Delta t) = P_i(t) - a[T(t) - T(t - \Delta t)] + b[T_{set} - T(t)]. \tag{9.82}$$

The choice of the coefficients a and b is made to achieve the desired dynamic response and stabilize the servo-loop. Figure 9.27b shows the temperature response of the computer

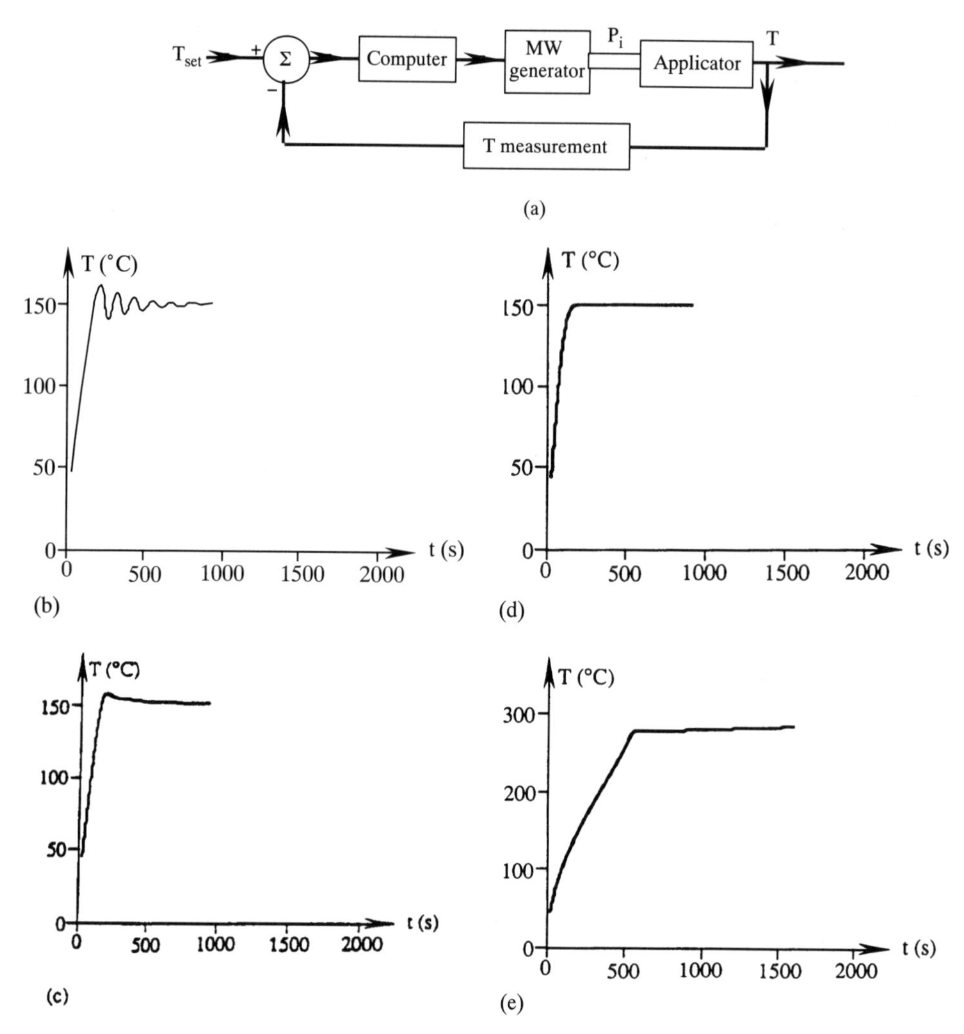

Fig. 9.27 *(a) Servo-control loop for temperature control. (b)–(e) Temperature response (see text).*

controlled loop. For a typical experiment the oscillation frequency of the loop is around 6×10^{-2} Hz when $a = 0.1$ and $b = 0.4$ W/°C—the sample is 9.78 g of EPDM rubber, $c_{pm} = 1.39 \times 10^3$ J/kg C, $\varepsilon''(T_0) = 0.42\epsilon_0, \alpha = 0, \beta = 8.6 \times 10^{-3}$, and $P_i = 35$ W (Figure 9.27c). The critical temperature of the material is 240°C. When $a = 0.8$ and $b = 0.1$ $W/^\circ$C the dynamic frequency of loop oscillation is reduced to 4.1×10^{-2} Hz; critical damping is obtained when $a = 2$ and $b = 0.01$ (Figures 9.27d and e).

The servo-loop is useful for controlling the temperatures above the critical temperature and also during rubber depolymerization, which occurs around 280°C. When approaching the depolymerization temperatures the loss factor of the rubber increases more rapidly than equation (9.76) permits. Nevertheless, the loop can retain control of the temperature during depolymerization when critical damping settings are used.

9.3.3 Thermal runaway in a material with finite thermal conductivity

The possibility of thermal runaway is one of the significant differences between classical convection and conduction heating and electromagnetic heating. There is much more to be discussed about this important topic. The origin of the unstable temperature rise is the dependence of the material permittivity on temperature. For the case studied previously the thermal conductivity was assumed large, thus the temperature within the material was considered approximately uniform. Suppose a material of finite thermal conductivity is heated using the same axisymmetric and axially uniform geometry. Then the instantaneous temperature distribution, $T(r, t)$, will be a function of the radius of the sample and a second order differential heat equation (in cylindrical coordinates) results:

$$\rho_{\mathrm{m}}c_{\mathrm{pm}}\frac{\partial T}{\partial t} - k\left(\frac{\partial^2 T}{\partial r^2} + \frac{1}{r}\frac{\partial T}{\partial r}\right) = \frac{\varepsilon''(T_0)P_{\mathrm{i}}}{\varepsilon_0 V}\left[1 + \alpha\frac{T - T_0}{T_0} + \beta\left(\frac{T - T_0}{T_0}\right)^2\right] \tag{9.83}$$

in the cylindrically homogeneous sample ($0 \leqslant r \leqslant a$) where a is the sample radius and V the sample volume. The initial condition and boundary conditions at the centerline and surface, $r = a$, (ignoring radiation) are given by:

$$\begin{aligned} &T(r, 0) = T_0 \\ &\text{at } r = 0 \qquad \frac{\partial T}{\partial r} = 0 \\ &\text{at } r = a \qquad k\frac{\partial T}{\partial r} + h(T - T_0) = 0. \end{aligned} \tag{9.84}$$

The temperature at all points can be calculated using numerical techniques. As a first step, we will study the stability condition with a finite difference model.

We divide the sample into a sequence of n concentric annular shells of material, each shell lying between $\{(i-1)a/n\} \leqslant r \leqslant \{ia/n\}$. The energy balance for each shell is equivalent to solving a system of finite difference equations. The average instantaneous temperature of each tubular shell is $T_{\mathrm{i}}(t)$. Equations (9.81) and (9.82) are rewritten for each shell to obtain a system of n equations:

$$\begin{aligned} n^2 K v_1 - n^2 K v_2 &= 1 + \alpha v_1 + \beta v_1^2 \\ \frac{-Kn^2}{3}v_1 + n^2 K v_2 - \frac{2K}{3}n^2 v_3 &= 1 + \alpha v_2 + \beta v_2^2 \\ \frac{-Kn^2(i-1)}{2i-1}v_{i-1} + Kn^2 v_i - \frac{Kn^2 i}{2i-1}v_{i+1} &= 1 + \alpha v_i + \beta v_i^2 \\ -K\frac{n-1}{2n-1}n^2 v_{n-1} + K\frac{n^2(n-1)}{2n-1}v_n + \frac{n^2 H}{2n-1}v_n &= 1 + \alpha v_n + \beta v_n^2 \end{aligned} \tag{9.85}$$

where the system of equations has been simplified by introducing reduced parameters:

$$K = \frac{2kT_0}{a^2\varepsilon_0'' P_i} \qquad H = \frac{2hT_0}{a\varepsilon_0'' P_i} \qquad v_i = \frac{T_i - T_0}{T_0}. \tag{9.86}$$

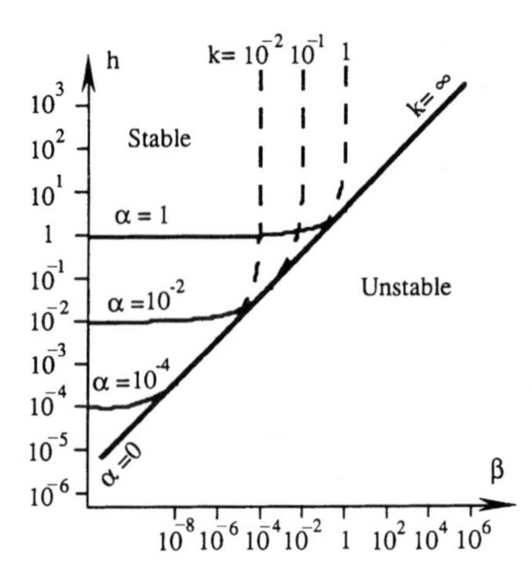

Fig. 9.28 *Stability thresholds for different thermal conductivities.*

The system of equations is nonlinear because of the $\beta v_i{}^2$ term. Nevertheless it can be solved by successive approximation. The stability conditions can be expressed using the dimensionless parameters. The solution is stable (no thermal runaway) if:

$$H > \alpha \qquad (H - \alpha)^2 > 4\beta \qquad \beta < 3K^2. \tag{9.87}$$

Figure 9.28 illustrates the effect of the stability limits established in (9.87). Note that runaway can always be prevented by using lower power settings and longer heating times. Note also that as the sample radius increases the influence of H on the stability is much reduced, as would be expected, except when the sample is of very high thermal conductivity. This model is useful for moderate heating where surface radiation is not too important. The radiation boundary condition can be added to make the model useful in analyzing ceramics sintering applications where the radiation losses are the most significant contribution; however, it is less likely that closed form expressions for the stability criterion can be obtained due to the highly nonlinear contribution of the nth shell.

9.3.4 Other special behavior of electromagnetic heating

Other heating situations have been studied. Kriegsmann pointed out that nonlinearities can induce bifurcation (hysteresis and other multi-valued effects) in a temperature history. He considered 1-D heat transfer in a slab infinite in the x and y directions irradiated by a uniform plane wave propagating in the z-direction, as we discussed in Chapter 1. The particularly interesting case is when the surface losses (convection and radiation) are small and changes in the material thermal conductivity are small compared with the temperature dependence of the permittivity. The imaginary part of the permittivity is modeled as exponentially dependent on temperature (Arrhenius form):

$$\varepsilon''(T) = \varepsilon''_A + \varepsilon''_B \exp\left(\frac{-\chi T_0}{T - T_0}\right). \tag{9.88}$$

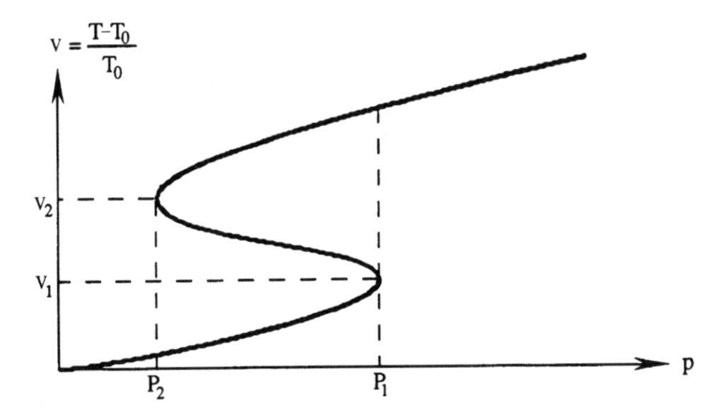

Fig. 9.29 *Thermal S-shaped curve in microwave heating.*

These hypotheses restrict the nonlinearity to the permittivity term. The differential equation describing the temperature history has one or three solutions depending on the initial condition, T_0, and on the normalized absorbed power, p:

$$p = \frac{\text{absorbed power}}{\text{convection losses}} = \frac{\varepsilon'' P}{hT_0} \tag{9.89}$$

as shown in Figure 9.29. The S-shaped response curve shows two critical power levels, P_1 and P_2 to which normalized temperatures v_1 and v_2 correspond, where $v_i = (T_i - T_0)/T_0$, as before.

Although the proposed theory has not yet been confirmed by experimental data, it shows that microwave heating might be expected to have special dynamic behavior with a jump in operating point from the lower branch to the upper one and back, with significant hysteresis when the power level is varied. Again, the effect of pre-heating or supplemental infra-red surface heating may be extremely important determinants of the dynamic behavior of electromagnetic heating. The primary lesson is that multiple operating points may be encountered when effective thermal insulation is used.

9.4 MICROWAVE HEATING WITH NONUNIFORM ELECTRIC FIELDS

The discussion in this chapter assumes uniform electric fields over the region of interest. This drastic simplification has made closed form analysis and discussion possible since the solutions are decoupled from Maxwell's equations. Experimentally, of course, it is nearly impossible to obtain truly uniform electric fields. Even when a small cylindrical reactor is introduced into a waveguide some field variations are expected due to the effect of the aperture created in the waveguide wall and the disturbance in the waveguide created by the boundary conditions at the surface of the substance under test. The fact that the aperture is essentially non-radiating is not sufficient to conclude that the electric field is uniform at the aperture: some fringing field is to be expected.

Consequently, the events associated with thermal runaway and other special behavior can occur locally and independently, as in an inhomogeneous field. We have seen the influence of geometry on the field distribution in homogeneous materials. We will discuss examples of the resulting material transformations in Chapter 10. Some of the effects of inhomogeneous fields will be reviewed in the discussion of heterogeneous materials (Chapter 11), in which the field is inhomogeneous because of the components of the material.

However, many experimental observations are difficult to interpret because of the large number of parameters which the effects depend upon. While the governing concepts are universally applicable, each case introduces particular interactions among the governing relations and must be analyzed separately.

REFERENCES

Bergman, D. J. (1978) Electrical Transport and Optical Properties of Inhomogeneous Media, AIP Conf. Proc. 40, (Garland, J. C. and Tanner, E. B. Eds), American Institute of Physics, New York.

Keenan J. H. and Keyes F. G. (1936). *Thermodynamic Properties of Steam tables* New York.

Kreigsman G. A. (1992) Thermal Runaway and its Control in Microwave Heated Ceramics, *Microwave Processing of Materials III* (Beatty, Sutton and Iskander, Eds.) *Proc. Materials Res. Soc.*, **269**, 257–264.

Incropera F. P. and Dewitt D. P. (1985) *Fundamentals of Heat and Mass Transfer* Wiley, New York.

Rosenhow, W. M. and Choi, H. Y. (1961) *Heat Mass and Momentum Transfer* Prentice-Hall, Englewood Cliffs, NJ.

Roussy, G. Bennani, A. and Thiebaut, J. M. (1987) Temperature runaway of microwave irradiated materials. *J. Appl. Phys.* **62** p1167.

Zemanski, M. W. (1957) *Heat and Thermodynamics* McGraw-Hill, New york.

10 Electromagnetic Processing of Homogeneous Materials at High Power Density

10.1 INTRODUCTION

In this chapter we will apply both thermodynamic models and the linear system characterization method (that is, step increases in field strength) to study electromagnetic processing in approximately homogeneous materials. The aims are two fold: (1) to illustrate the response of the material to the electromagnetic field from experiments and parametric studies in numerical and analytical models, and (2) to investigate the significance of changes in permittivity in the material undergoing transformation. The first section of the chapter considers drying. The results obtained are dependent, at the very least, on water content, temperature, mass transport properties and electric field. The second section considers curing and welding of composite materials wherein the results are dominated by heat transfer processes.

The drying and dehydration cases to be considered have been selected for their significance from the point of view of the industrial applications they cover. They have been placed in a sequence of concepts which, in the best knowledge of the authors, reveals the specificity of electromagnetic processing of materials. First we will consider drying of paper and gypsum. Then we present a more detailed look at the dehydration of zeolites, briefly introduced in the last chapter, a much more complex material. Finally, we will take an academic look at the special physics of the vaporization of polar liquids in electromagnetic fields — the goal being to develop a useful description of liquid to gas conversion through an interface, an irreversible process. Electromagnetic heating and many of the phase transport parameters (such as viscosity, surface tension, etc.) arise from molecular motion. We expect that the ordinary thermophysical processes due to molecular motion ought to be modified in the presence of an EM field.

The concluding section of this chapter deals with curing processes, in which phase change is not encountered. The specific examples described are in composite materials with some emphasis on graphite fiber composites.

10.2 PHYSICAL DESCRIPTIONS OF DRYING PROCESSES

It is usually very inconvenient to describe the moisture content of materials in terms of volume fractions, as we have done to this point. This is because nearly all materials swell as water is absorbed or shrink as it is desorbed. Consequently, the volume of the dry support structure is considerably less than the volume of the wet material. The relative volume shrinkage is mechanically different from material to material. We therefore use mass fractions to overcome this ambiguity. The moisture content of a material is described by a unitless fraction or percentage, either on a wet weight basis, M (wwb), or on a dry weight basis, X (dwb). There are slight differences in notation among authors regarding moisture content. Traditional drying literature (see for example, Keey (1978) or Strumillo and Kudra (1986)) typically uses X for dry weight moisture content whereas Metaxas and Meredith (1983) (and others) use M for dry weight moisture content. We adopt the notation of Keey and others in this text and use X for dry weight moisture content. The dry weight moisture content, X (in %), is:

$$X = 100\frac{m_w}{m_d} \tag{10.1}$$

where m_w (kg) is the mass of water in the wet sample and m_d is the dry mass of the solid matrix. The moisture content on a wet weight basis, M (%), is:

$$M = 100\frac{m_w}{m_w + m_d}. \tag{10.2}$$

The usual form is to use the dry weight expression, X, when describing material moisture content. X may be quickly found from M (decimal form):

$$X = \frac{M}{1 - M} \tag{10.3}$$

and the converse:

$$M = \frac{X}{1 + X}. \tag{10.4}$$

Moisture which is free to move in the solid matrix migrates by diffusion processes mitigated by capillary forces. Moisture may also be "bound" within the matrix by chemical attachment (for example, when Ca_2CO_3, Plaster of Paris, forms $Ca_2CO_3{\cdot}5H_2O$, gypsum), by relatively weak van der Waals attraction, or by other forces such as surface tension. Figure 10.1 illustrates the typical features of measured drying curves. The OA segment is the time required to heat the sample to temperatures at which significant drying occurs. When free water vaporizes at the surface, and is replaced by capillarity such that the vaporization conditions remain constant, the drying process is in the constant rate period and dX/dt is constant (the AB segment in Figure 10.1). This happens in many loose fibrous matrices and the mass and heat transfer convection boundary conditions are relatively simple to construct. Other porous matrices may have such large pore geometries that capillary action is ineffective. While the constant rate period may still be observed, the moisture front recedes into the matrix, the convection boundary conditions are difficult to express reliably (since it is a moving boundary), and the model has a changing geometry which complicates the calculation. In either case, as the free water is exhausted, the drying continues and more tightly restricted water is vaporized. The energy barriers which must be surmounted increase with decreasing moisture, so the drying rate decreases — this is the falling rate period (BC in Figure 10.1).

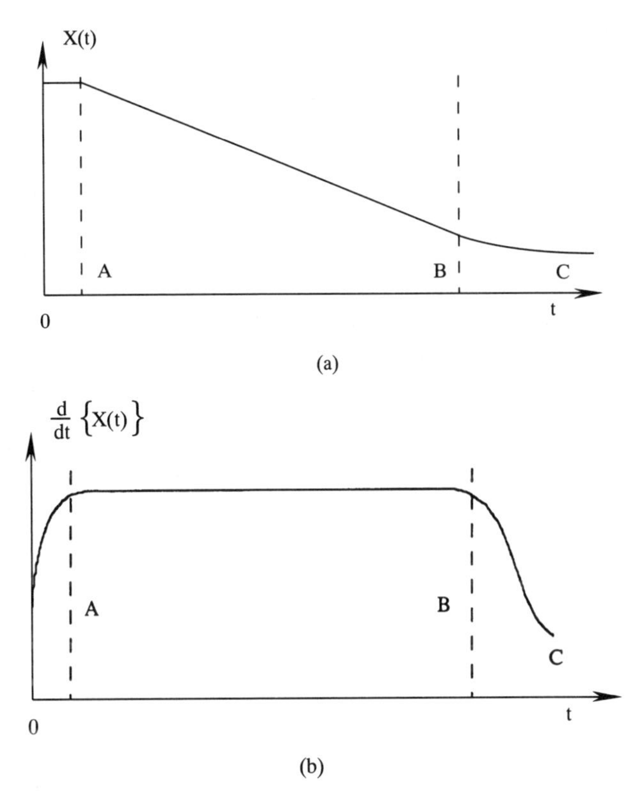

Fig. 10.1 *Typical drying curves: OA = set-up period, AB = constant rate period, and BC = falling rate period. (a) Moisture content. (b) Drying rate.*

Electromagnetic heating is directed toward the material water in both the RF and MW regimes: residual ion content gives effective absorption at radio frequencies in most materials, and both water and ions (if in sufficient concentration) absorb at microwave frequencies. The difference between conventional and EM heating is most clearly seen in the falling rate period. A material which must be dried to some point well within the falling rate period requires a very long conventional dryer. Electromagnetically induced drying is effective even at low moisture contents for many reasons, only some of which are presently well understood. So, electromagnetic heating may be used at some advantageous point (or at several points) during the falling rate period to reduce overall dryer volume and material residence time. From the standpoint of economics, the falling rate period is the most advantageous position for the EM equipment since the required power is much reduced (and thus the capital and operating costs are reduced) and EM heating has the most significant advantage over conventional methods there.

As a general design rule, conventional drying is most effective in the constant rate regime while electromagnetic heating is usually most sensibly applied in the falling rate regime. There are notable exceptions to this rule; for example, when a high value product is subject to surface damage at elevated temperatures. Electromagnetic heating has the unique effect of inverting the usual temperature profile. And it is sometimes advantageous, from a product quality viewpoint, to use cooler air at the surface to extract moisture—EM heating can often achieve this at high drying rates because the power is deposited volumetrically.

10.3 THERMODYNAMIC MODEL OF PAPER DRYING

The drying of cellulose is of major importance in the paper and cardboard industry. This process consumes significant amounts of energy and has been the subject of many studies. For many years, it has been recognized that radio frequency fields or microwaves could perform a useful function in paper drying, or in the leveling of moisture profiles across a wet sheet. This is not surprising because water is more receptive than most other materials to dielectric heating so that water removal is accelerated. However, in economic terms it must be recognized that all EM processes are capital intensive and that mechanical water removal processes, when feasible, are always more economical. Also, the standard drying method, steam heated drums (collanders), have a long history of successful application to thin webs or sheets. The potential advantage that EM heating offers is that it can be locally controlled (using multiple applicators) with a very short response time in order to compensate for uneven moisture distributions. Uneven moisture can be detected by localized permittivity measurements or by thermographic imaging downstream of a drying drum, as described in Chapter 5. The disadvantage, besides capital costs, is that the high velocity of paper through the drying system makes applicator design and the control algorithm critical. Consequently, the product must be a high value product in order to justify EM heating techniques. There are many instances in which the studies indicate significant technical advantages for EM heating, but the product is just of too low value to justify its use—gypsum drying is a prime example of this and will be discussed in a future section.

10.3.1 Drying of wet paper spheres

The following study concerns the microwave drying of spheres of wet paper irradiated with a constant electromagnetic field in a waveguide. We should note at the outset that we are not aware of any real process in which the drying of small paper spheres is the desired end point. We use this simplified experimental framework as an example to separate the physical processes so that we may thus clearly determine how the electric field removes the water molecules from the cellulose surfaces, and so that we may justify the conclusions derived from a theoretical model which describes the microwave dielectric heating process. This experiment submits to a simplified parametric analysis because the mass transport processes are quite simple in this material, especially in the thin web case.

The experimental apparatus is similar to that described previously (Chapter 9). A packed bed of spheres, 3 mm in diameter, is placed in the center of an RG 112/U (WR340) rectangular waveguide as shown in Figure 9.21. The cylindrical sample, 1 cm in diameter, is parallel to the TE_{10} electric field. A constant electric field strength is maintained by controlling the displacement of a movable short circuit behind the sample as a function of the position of the standing wave pattern induced by the dielectric sample. The water removed is evacuated by a downward draft of dry air which circulates through the bed. The flow rate, the temperature and the humidity of the gas are continuously measured at the output. All experiments were performed with "granucel long fiber" samples. The dry mass of the sample was about 0.80 g and the initial moisture content, X, was 20%. Microwaves were applied in a sequence of step increases starting at $t = 0$. The experimental variables were recorded.

The following notation is used: m_0 is the mass of the dry paper; the moisture content of the wet paper is X and $X(0)$ is the initial moisture content of the paper. The mass rate of drying, W_{m} (t) (kg/s) is:

$$W_{\mathrm{m}}(t) = -m_0 \frac{\mathrm{d}X}{\mathrm{d}t} \tag{10.5}$$

where X is in decimal rather than percent form. P_{i} and P_{abs} are the incident and the absorbed power (W), respectively, of the sample and the complex permittivity of the sample is $\varepsilon^* = (\varepsilon_{\mathrm{r}}' - \mathrm{j}\varepsilon_{\mathrm{r}}'')\varepsilon_0$.

Irradiation of the wet paper was performed at different values of incident power and for three different air flows. Typical drying rate curves are shown in Figure 10.2. Figures 10.3a and 10.3b show typical changes in the real and the imaginary parts, respectively, of the sample permittivity during the process determined according to the method described in Chapter 5. The outlet plane gas temperature record, $T_{\mathrm{g}}(t)$, is shown in Figure 10.4. The mass drying rate curves, $W_{\mathrm{m}}(t)$, were observed to be well described by single exponential curves:

$$W_{\mathrm{m}}(t) = A \exp\left(-\frac{t}{\alpha}\right). \tag{10.6}$$

The amplitude factor A (kg/s) and the time constant α (s) are obtained by numerical fits to the experimental data; both quantities vary with the experimental conditions. The drying rate, $1/\alpha$, depends linearly on the incident power, as shown in Figure 10.5. A complete analysis of the data also shows that the rate constant, $1/\alpha$, is the sum of two terms which

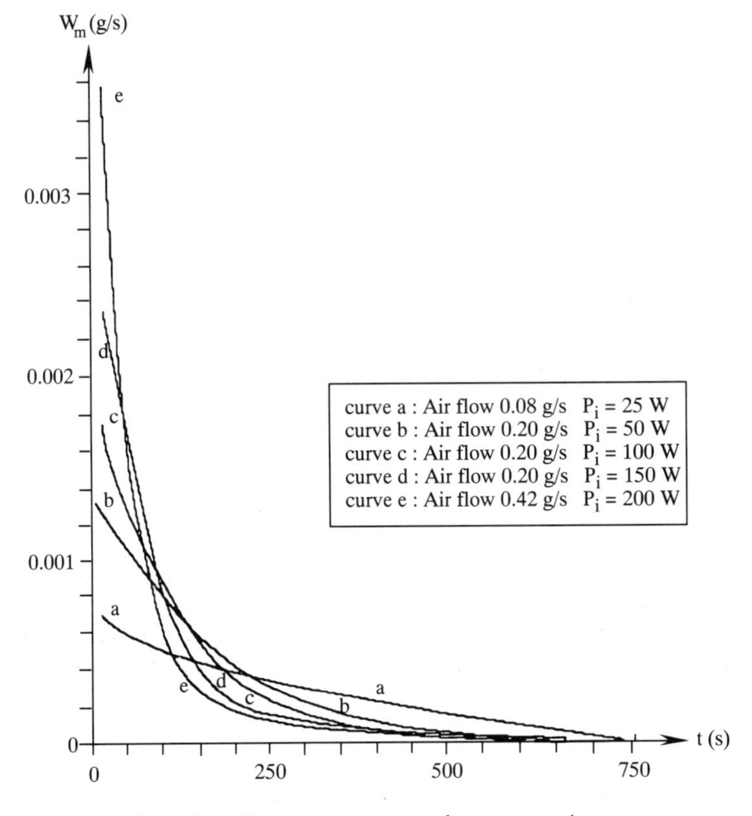

Fig. 10.2 *Drying rate curves for paper spheres.*

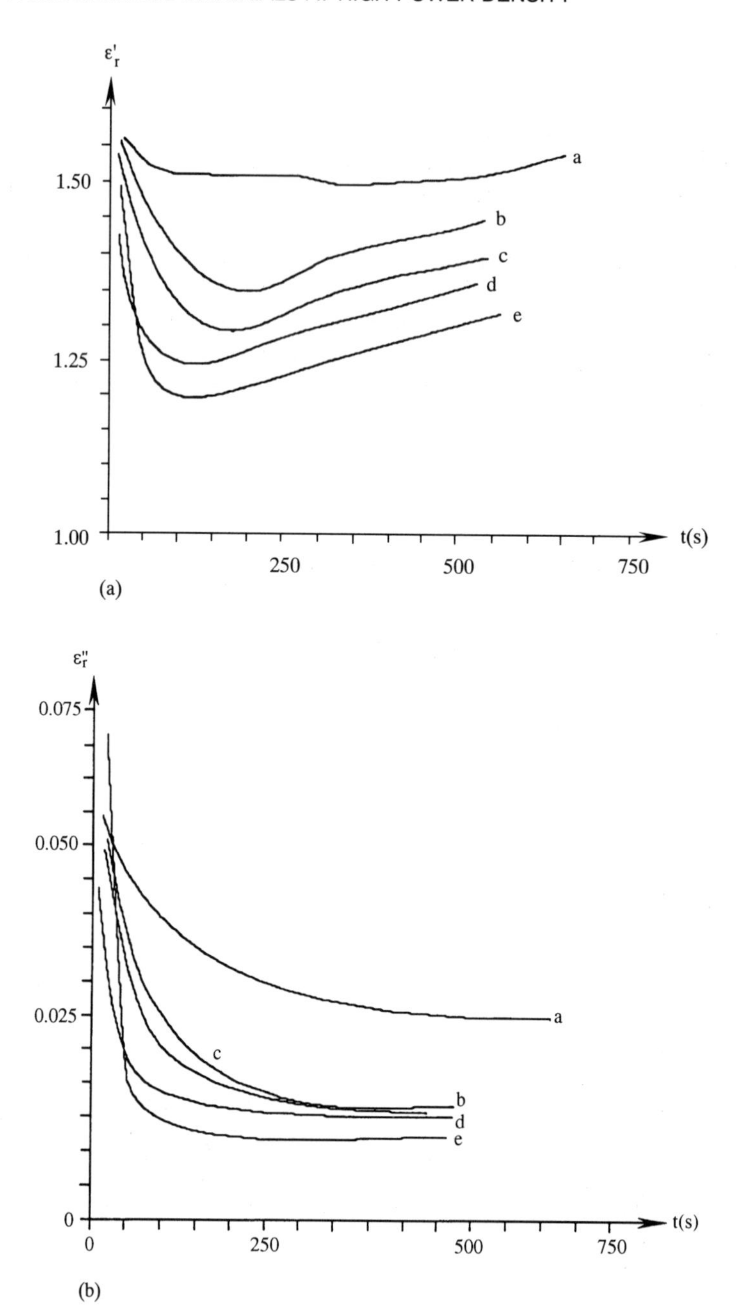

Fig. 10.3 *(a) Typical change of ϵ_r' and (b) Variation of ϵ_r'' during paper drying. The curve designations, a–e, have the same significance as in Figure 10.2.*

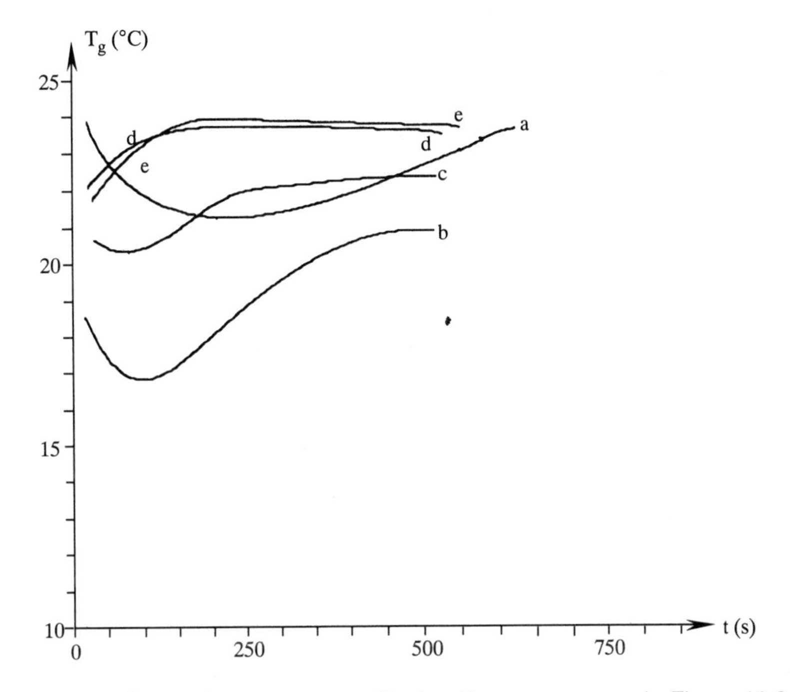

Fig. 10.4 *Gas outlet temperature. Designations a–e are as in Figure 10.2.*

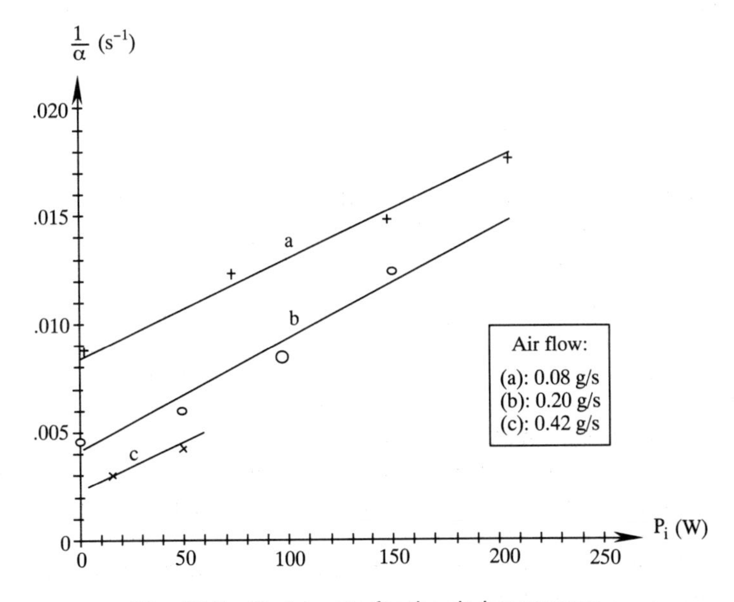

Fig. 10.5 *Drying rate for the drying process.*

independently describe the actions of the microwaves and the flow of dry air:

$$\frac{1}{\alpha} = \frac{1}{\alpha_0} + \xi P_{\mathrm{i}}. \tag{10.7}$$

Furthermore, at the end of the experiment, there is no residual water, so the amplitude constant comes from the initial drying rate:

$$A = \frac{m_0 X(0)}{\alpha}. \tag{10.8}$$

By integrating the water loss rate, the instantaneous moisture content in the paper is:

$$X(t) = X(0) \exp\left(-\frac{t}{\alpha}\right) \tag{10.9}$$

and the drying rate can be expressed by the following first order differential equation:

$$\frac{\mathrm{d}X}{\mathrm{d}t} = -\frac{X}{\alpha}. \tag{10.10}$$

These three equations represent a complete model for the paper drying process under microwave illumination. Note that equations (10.9) and (10.10) describe a falling rate period, so at the initial moisture content of 20% we are already in the falling rate region of the drying curve.

The gas flow temperature response is shown in Figure 10.4. The evolution of the gas temperature history is complex. When the microwave power is high, $T_{\mathrm{g}}(t)$ increases up to a maximum value which corresponds to the steady state heating of the dry paper. Under other conditions, when the incident power is low and mass flow of air is high, the temperature of the gas first decreases and then increases and stabilizes at some steady state temperature, as at high power.

Numerical analysis of the curves shows that they are well represented by the sum of two exponentials:

$$T_{\mathrm{g}}(t) = p\left[1 - \exp\left(-\frac{t}{u}\right)\right] + q\left[1 - \exp\left(-\frac{t}{v}\right)\right] + T_0 \tag{10.11}$$

where T_0 is the inlet air temperature; p, q, u and v are curve fit parameters. Although the fit is statistically valid it is not possible to correlate the values of the parameters p, q, u, and v with the experimental conditions (air flow and incident power).

In fact, the temperature of the outlet gas reflects that of the material, because of heat exchange between gas and paper spheres. If we could adequately model the heat transfer processes, we would be able to calculate an energy balance. In order to do this, we would introduce a term proportional to $\mathrm{d}X/\mathrm{d}t$ which would represent the enthalpy of desorption. Although the complete equation has not been specified in this analysis, the mathematical procedure holds. The temperature of the sample and that of the gas will be given by the energy balance equation. So, the gas temperature does not dominate or determine the drying process, as it does in conventional drying. On the contrary, it is a dependent variable since the power is actually dissipated in the liquid phase water. This point of view includes the microscopic effects of the interaction between the field and the dielectric materials. The

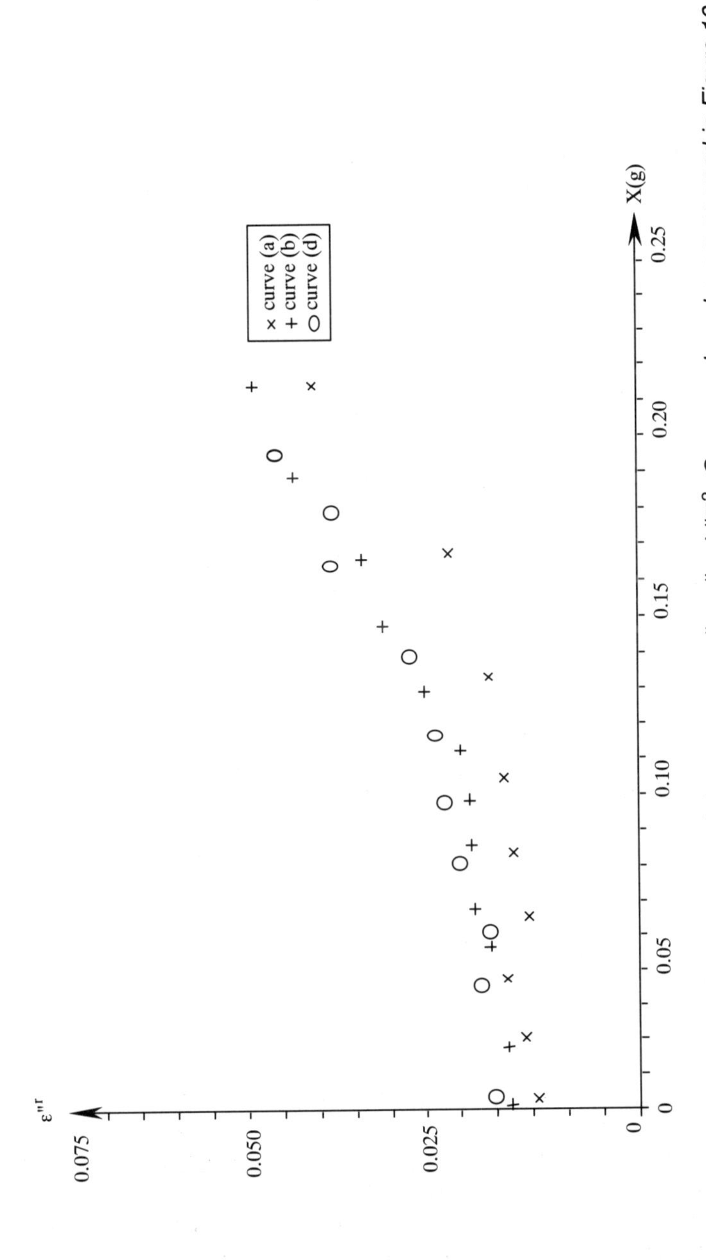

Fig. 10.6 *Imaginary part of relative permittivity of the wet paper,* $\epsilon''_{\mathrm{r}} = \epsilon''_{\mathrm{ri}} + k''X^2$. *Curves a, b and c are as used in Figure 10.5.*

MW field induces a torque on the molecules which are fixed on the surface and some energy is dissipated. The effect of the microwaves is a molecular electrochemical process described by equations (10.7) and (10.9). After this, the temperature of the macroscopic system is given by an energy balance, but the complications include power dissipation in the dry fibers.

Other responses of the system include property changes: the real and the imaginary part of the relative permittivity both vary with moisture content in the paper. The variation of $\varepsilon''(t)$ is not exponential with time; the permittivity is an approximately quadratic function of the instantaneous moisture content, as shown in Figure 10.6:

$$\varepsilon_{\mathrm{r}}'' = \varepsilon_{\mathrm{ri}}'' + k''X^2 \tag{10.12}$$

where $\varepsilon_{\mathrm{ri}}''$ represents the intrinsic loss of the dry paper fibers. As a consequence, the absorbed power per unit volume, Q_{gen}, turns out to be proportional to X^2:

$$Q_{\mathrm{gen}} = \frac{P_{\mathrm{abs}}}{V} = \eta(\varepsilon_{\mathrm{ri}}'' + k''X^2)P_{\mathrm{i}}. \tag{10.13}$$

In this experiment it is not possible to extract a clear interpretation of the response of ε'—first because the variation is small (20% is a low moisture content) and second because the geometry (volume) of the paper spheres is not constant. Also, since the electric field is maintained constant the value of ε' has little or no effect on the experimental results, and is not of particular importance. Other experiments (Section 10.5 of this chapter) reveal that ε' is proportional to X and is thus also an exponential function of time in this experiment.

In conclusion, we have proposed two fundamental equations to describe the falling rate region in this material, namely equations (10.7) and (10.10). The moisture content relations differ from a simple first order kinetics law because the rate constant does not depend on the reciprocal of absolute temperature:

$$\frac{1}{\alpha} \neq A \exp\left(-\frac{E}{RT}\right). \tag{10.14}$$

The origin of equations (10.7) and (10.10) is probably microscopic since they describe neither a thermal process nor a physical reaction between the phases. We will apply this semi-empirical model in the next discussion.

10.3.2 Laboratory scale web drying experiments

Further experiments confirm the usefulness of equations (10.7) and (10.10) for the drying of paper with microwaves by means of a larger scale study of the drying of a sheet of paper which is moving through a waveguide. The experiment simulates the situation in an industrial paper dryer; the apparatus is sketched in Figure 10.7.

A band of paper of width equal to the height of the waveguide (that is, b) is passed through two slots which have been cut in the narrow wall of the waveguide. The non-radiating slots do not significantly modify the electric field of the propagating TE_{10} mode, and the electric field is parallel to the surface of the paper. We are able to describe the drying process in terms of a constant electric field by again using the single-mode resonant microwave

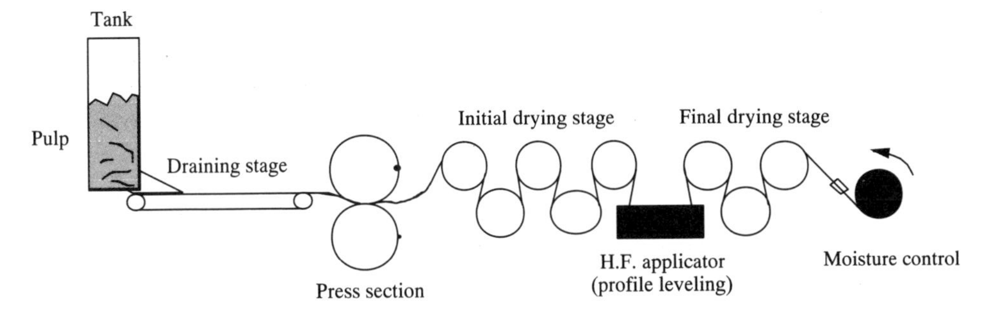

Fig. 10.7 *Typical configuration of a paper machine. (with H.F. leveling).*

applicator, as used in the small sphere experiment. This is because the distribution of the electromagnetic field is known when the cavity is tuned to resonance. Furthermore, when the moisture content of the paper varies, the corresponding change of the intensity of the electric field is corrected if the resonance of the cavity is restored or automatically controlled as shown Figure 10.8. The waveguide (RG 112/U, WR 340) of about 25 cm in length was made to oscillate in the TE_{103} mode at a frequency of 2.45 GHz. The paper moves through the waveguide on an axis perpendicular to that of the waveguide.

As in previous experiments, the incident and reflected power were measured in order to monitor the field at the resonance frequency by using a probe within the cavity. The mean permittivity of the paper during drying can be determined from these measurements. A servo-loop was used to adjust the cavity to maintain resonance during heating.

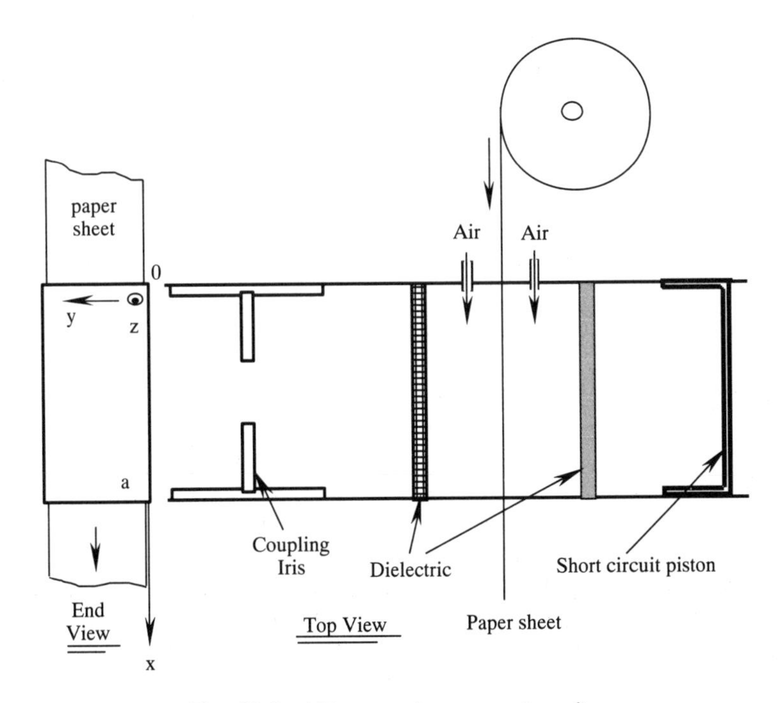

Fig. 10.8 *TE_{103} mode resonant cavity.*

Electric field distribution

When the applicator is at resonance, the electric field, which is instantaneously applied to the paper, varies sinusoidally from one side to the other in the waveguide (see Chapter 2 and Chapter 5). This distribution is stationary from side to side, $x = 0$ to $x = a$:

$$E(x) = E_0 \sin \frac{\pi x}{a}. \tag{10.15}$$

The paper moves in the x-direction and is irradiated by a variable distribution of electric field which is a function of time

$$E(t) = E_0 \sin \frac{\pi t}{\tau} \qquad \tau = \frac{a}{u} \tag{10.16}$$

where τ is the irradiation time (the residence time in the waveguide) and u is the linear speed of the web (m/s). The electric field distribution should be constant in the y-direction since we have a TE_{10} propagation mode.

The dry weight basic moisture constant of the paper, X, does not vary as y varies, but decreases inside the cavity with x from the input at $x = 0$ to $x = a$ when the paper leaves the cavity. Suppose also that the drying rate is a function of the instantaneous electric field distribution:

$$-\frac{\mathrm{d}X}{\mathrm{d}t} = (\zeta + \eta E^2)X. \tag{10.17}$$

The factors ζ and η are at this point unspecified parameters, but $\zeta = 1/\alpha_0$ from the previous discussion. Then, the moisture content of the paper at a distance $x(0 < x < a)$ can be obtained by solving the differential equation, with $x = ut$:

$$-\frac{1}{X}\frac{\mathrm{d}X}{\mathrm{d}x} = \left(\zeta + \eta E_0^2 \sin^2\left(\frac{\pi x}{a}\right)\right)\frac{\tau}{a} \tag{10.18}$$

from which we obtain:

$$X(x) = X(0) \exp\left[\frac{\eta\tau}{4\pi}E_0^2 \sin\left(\frac{2\pi \mathrm{x}}{a}\right) - \frac{\zeta\tau}{a}x - \frac{\eta\tau x}{2a}E_0^2\right]. \tag{10.19}$$

The paper entering the cavity with an initial moisture content $X(0)$ has an output moisture content of:

$$X(a) = X(0) \exp\left(-\zeta\tau - \frac{\eta\tau}{2}E_0^2\right). \tag{10.20}$$

This equation is valid if the final experimental moisture content of the paper is logarithmically dependent on the field power, E_0^2, and it is. Equivalently, the natural log, $\ln\{X(a)/X(0)\}$, should be a linear function of the applied power, as is shown in the experimental data of Figure 10.9.

The permittivity data can also be used to validate the empirical drying equation. The apparatus will yield a very sensitive measurement of the permittivity since the resonance conditions are much better defined for the thin slab of paper than for the paper spheres. However, we must first determine the permittivity of homogeneous moist paper in order to apply the results to the case where the paper is dried as it moves through the waveguide. We may do this by reducing the input power to a very low level and cutting off the airflow so that the paper does not dry when it passes through the cavity; then the moisture content

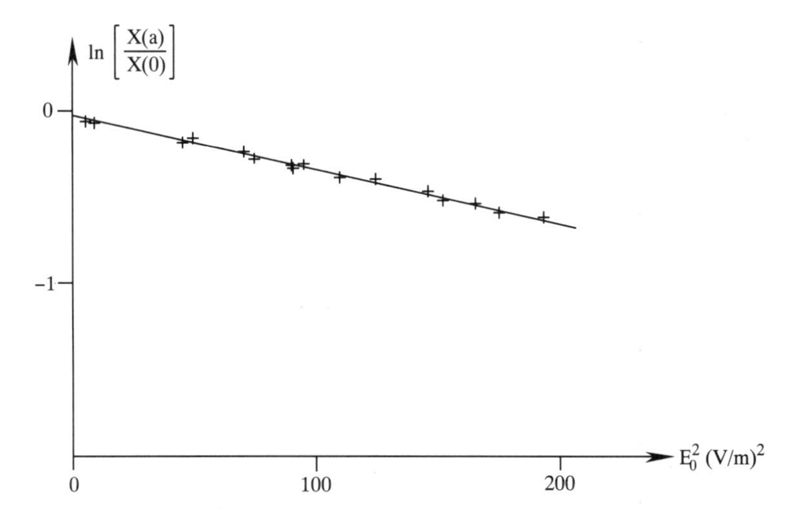

Fig. 10.9 *Output moisture content of the paper.*

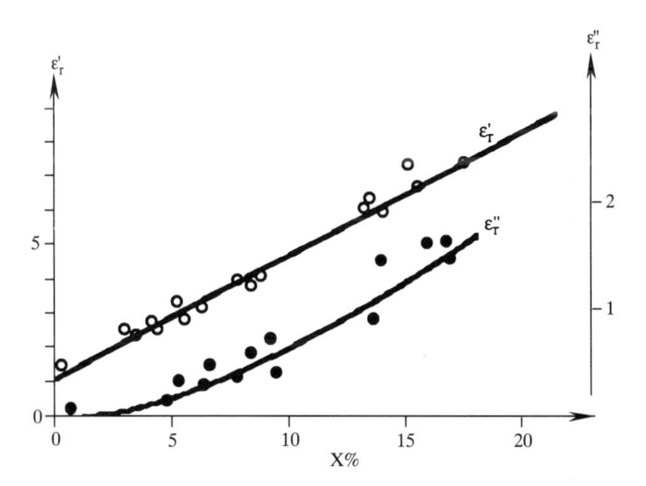

Fig. 10.10 *Complex permittivity of the wet paper.*

does not change — $X(a) = X(0)$. These experimental conditions create homogeneously wet paper, the moisture content and permittivity of which can be determined — the particular curve fit parameters derived were k_0', A' and A''. When wet paper is passed through the waveguide at high power, however, the material is not homogeneous since the moisture content depends on the x-position — each strip of paper makes a contribution to the apparent permittivity of the paper in the guide, $\langle \varepsilon^* \rangle$. The apparent permittivity at high power density is independently measurable. The low power permittivity results on homogeneous paper are presented in Figure 10.10 (with curve fits obtained similar to equation (10.11)). The real part of the permittivity is fitted by:

$$\varepsilon_r' = k_0' + A'X \tag{10.21}$$

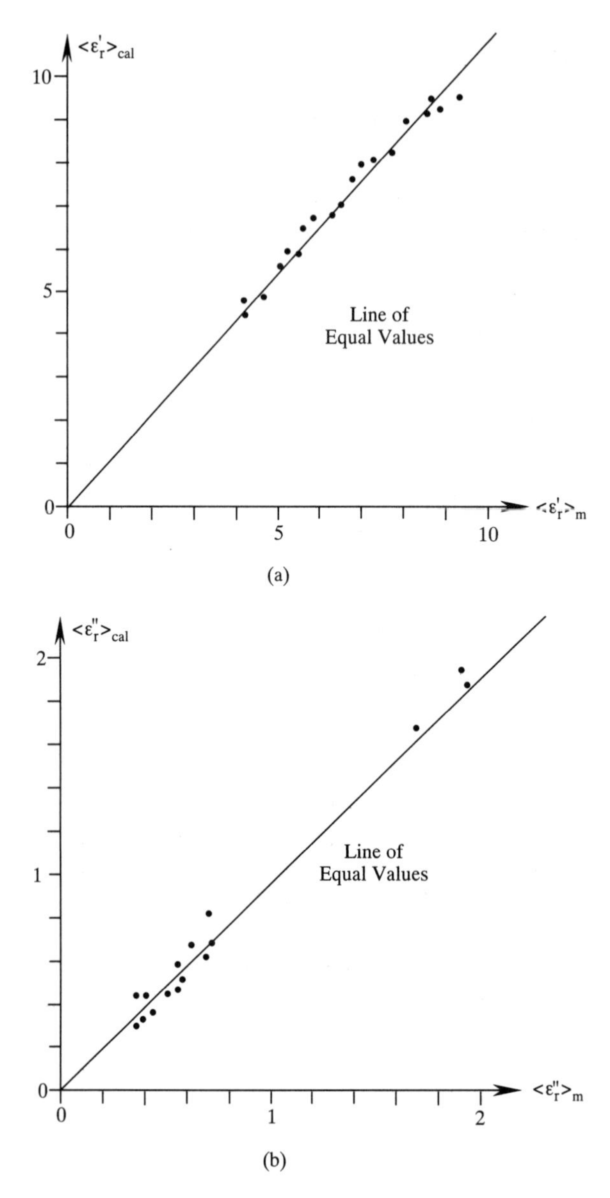

Fig. 10.11 *Comparison of measured and predicted values of (a) real and (b) imaginary parts of the permittivity.*

and the calculated spatial average value should be:

$$\langle \varepsilon_r' \rangle = k_0' + \frac{2A}{a} \int_0^a X(x) \sin^2 \left(\frac{\pi x}{a} \right) \, dx. \tag{10.22}$$

The imaginary part of the permittivity is fitted by:

$$\varepsilon_r'' = A'' X^2 \tag{10.23}$$

and the spatial average value at high power should be:

$$\langle \varepsilon_r'' \rangle = \frac{2A''}{a} \int_0^a X(x)^2 \sin^2\left(\frac{\pi x}{a}\right) dx \qquad (10.24)$$

where the curve fit coefficients are: $k_0' = 1.8$, $A' = 0.313$, and $A'' = 5.1 \times 10^{-3}$. In this data set the intrinsic loss of the paper fibers, ε_i'' from equation (10.11), is negligible compared to other terms and thus does not appear in the fit.

The comparison between predicted and measured permittivity is given in Figure 10.11. The overall agreement is excellent, indicating that the model works rather well in this case.

10.3.3 Conclusions

The experimental results clearly show that the simple model is valid for the initial conditions given. Nevertheless, it is worthwhile to comment on the difference between the model and the classical approach to paper drying. In Figure 10.12, the quantity $(X(a) - X(0))/\tau P_{\text{abs}}$ in kg of water produced per kWh, which are the usual industrial units, has been plotted as a function of the field intensity which irradiated the paper. The total rate of water evaporation is not given by the vaporization enthalpy alone:

$$\frac{X(a) - X(0)}{\tau P_{\text{abs}}} = \frac{1}{L} \qquad (10.25)$$

where $1/L = 0.625$ kg/kWh. In fact, the quantity is frequently less than $1/L$ in the figure. The evaporation of water under the influence of an electromagnetic field must be described by a more complex relation because the process occurs with spatially dependent material properties — and the response is therefore nonlinear. In the enthalpy balance, the enthalpy of the air and the temperature variation of the web of paper must be taken into account. The dispersion of experimental points in Figure 10.12 shows that the problem is not simple.

The model used here remains valid as long as the flow of air is high enough to avoid water condensation within the cavity. Some experiments were performed with a reduced

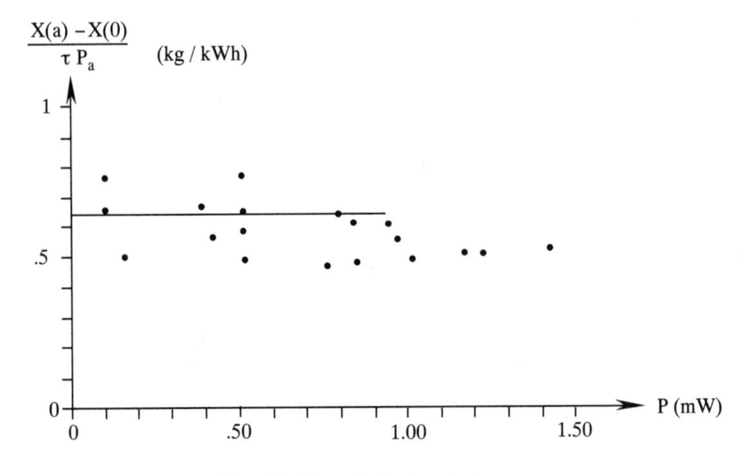

Fig. 10.12 *Enthalpy balance.*

flow of air, and the results were unsatisfactory since condensation of water was observed on the paper at the output and on the walls of the cavity. A high flow rate of cool air is required to evacuate the water. Also, as the material changes from non-hygroscopic (this case) to hygroscopic, this type of model loses validity. A more complete drying model will be described in the next section.

The experimental results show that microwave drying of paper can be described by a simple kinetic equation which, surprisingly, avoids the measurement of temperature. The validity of the resulting model has been demonstrated directly and by measuring the permittivity of the paper during drying. This model appears to be a simple method of predicting the efficiency of the process for thin non-hygroscopic materials. It opens the possibility for controlling the function of many on-line microwave or radio frequency dryers. However, even when the mass transfer processes are simple, as in this case, knowledge of the electromagnetic field distribution inside the applicator and during the process, when the fields may change, is required in order to use it.

10.4 MORE COMPLETE DRYING MODEL*

Drying processes in thick and/or hygroscopic materials are considerably more complex than for the previous case. Consequently, the model must be augmented to include the increased number of interactions. We now present a more complete model of drying processes. It will be seen that the complete model is much more difficult to apply due to the number of parameters which must be evaluated. The parametric model approach, as used for paper drying, is a preferred method in simple materials when it can be applied.

The three governing relations are:

(1) the mass balance:

$$\frac{\partial X}{\partial t} = a_{\mathrm{m}} \nabla^2 X + a_{\mathrm{m}} \beta_T \nabla^2 T + a_{\mathrm{m}} \beta_{\mathcal{P}} \nabla^2 \mathcal{P} \tag{10.26}$$

where a_{m} and β are generalized mass transport coefficients (unitless) and $\mathcal{P}$ is the pressure (Pa);

(2) the energy balance:

$$\frac{\partial T}{\partial t} = a_T \nabla^2 T + \frac{e_{\mathrm{v}}}{c_{\mathrm{p}}} \Delta h_{\mathrm{fg}} \frac{\partial X_f}{\partial \mathrm{t}} + \frac{Q_{\mathrm{gen}}}{\rho_{\mathrm{m}} c_{\mathrm{p}}} \tag{10.27}$$

where the subscript T refers to the thermal model, f refers to the free water, fg the phase change, e_{v} is the phase conversion factor (unitless), c_{p} is the constant pressure specific heat, and ρ_{m} is the density of the wet solid (a function of X);

(3) the pressure:

$$\frac{\partial \mathcal{P}}{\partial t} = a_{\mathcal{P}} \nabla^2 \mathcal{P} - \frac{e_{\mathrm{v}}}{c_{\mathrm{p}}} \Delta h_{\mathrm{fg}} \frac{\partial X_f}{\partial t}. \tag{10.28}$$

The reader may note that equation (10.27) is a modified form of equation (9.26) in which the mass rate of phase change has been recast in terms of the material moisture.

* This section was prepared with substantial assistance from Professors T. Bergman and P. Schmidt.

Complete models of this form require calculations and measurements of local temperatures and estimates of the mass transport coefficients, a and β The entire EM heating effect is included in the Q_{gen} term; and we note that this term often dominates the temperature profile. Also, we may easily see that the liquid heating process (under the influence of the EM field) may generate significant pressure gradients (even at low rates of evaporation) through the coupling between equations (10.28) and (10.26), which will result in pressure-induced migration of liquid phase water toward the surface. In some applications of EM heating, in hard woods and other cellular materials, pressure-induced migration can, in fact, be the primary "drying" process.

There are several other less obvious mass transfer effects in thick materials. First, internally vaporized water (or solvent) may recondense as it migrates due to local variations in saturation pressure (see Chapter 9) from temperature and moisture gradients. This is a so-called "reflux" process in which simultaneous condensation and vaporization occur. Moisture also affects the dielectric properties which, in turn, affect the heat generation term, Q_{gen}. Sometimes, the complex inter-connection of equations (10.26), (10.27) and (10.28) results in "moisture-leveling" in the material, an advantageous result which may be the primary motivation for using dielectric heating. Second, the direct coupling of EM energy with specific molecules (water or ions) may change the effective Δh_{fg} for adsorption in the material, and thus change the mass transport properties. We should note that the dielectric and thermodynamic properties are closely coupled in all materials: ε^* depends on both X and T, as does ρ, c, a and β. They must be considered as varying simultaneously during a drying process.

10.4.1 Nonhygroscopic materials

First, we return briefly to non-hygroscopic materials and enhance the discussion by including thick objects. The role of transport processes may be placed into perspective by considering limiting cases. The discussion of drying processes may be generalized by introducing the dimensionless moisture, X^* (Schmidt *et al.* 1992):

$$X^* = \frac{X(t)}{X(0)} \tag{10.29}$$

and dimensionless time, t^* (Schmidt *et al.* 1992):

$$t^* = \frac{Q_{gen}V}{m_0 \Delta h_{fg}} t. \tag{10.30}$$

We will see that it is now possible to compare results from many different experiments on the same basis. Also, use of these variables will help identify when the parametric model of the previous section is valid.

Drying regimes

The first limiting case, conventional drying, includes no volume heat generation term, and is thus described by $t^* = 0$. We may construct a "process map" (Figure 10.13) in which the limiting regimes are split into three straight line responses: (A) conventional drying ($t^* = 0$), (B) dielectrically-assisted drying response (equation (10.31)) in which a significant fraction of the EM power is consumed in raising the material temperature to the vaporization point ($T = T_{sat}$), and (3) an ideal dielectrically-dominated heating process in which all of the EM

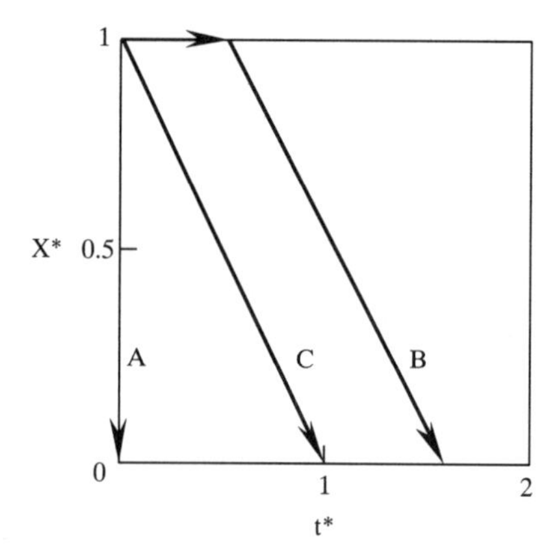

Fig. 10.13 *Process map for EM drying regimes showing conventional (A), dielectrically-assisted (B), and dielectrically-dominated (C) drying.*

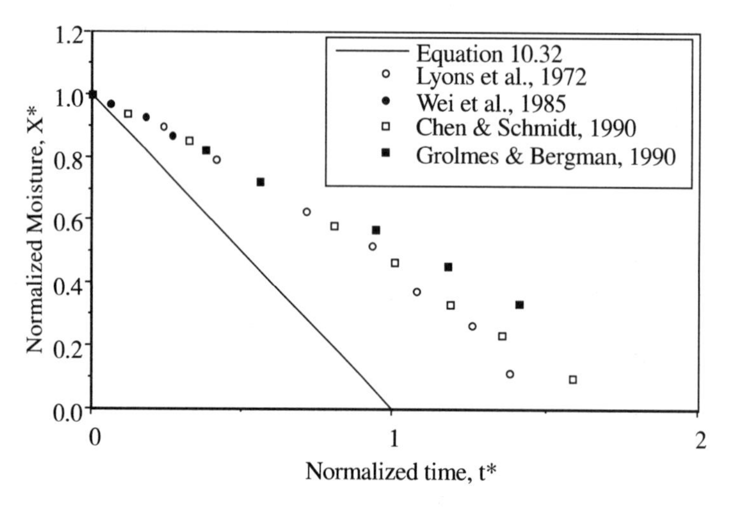

Fig. 10.14 *Measured normalized moisture content vs. normalizedtime from various studies compared to the bounding relation, equation (10.32) (Schmidt et al. 1992).*

Table 10.1 *Heating conditions in non-hygroscopic materials for plot in Figure 10.14.*

Reference	Material	MW power (W)	Environment temperature(°C)
Lyons *et al.* (1972)	cotton yarn	800	
Wei *et al.* (1985)	sandstone	60	25
Chen and Schmidt (1990)	alumina pellets	500	120
Grolmes and Bergman (1990)	glass beads	300	25

energy is used to evaporate liquid at constant temperature (equation (10.32)):

$$X^* = 1 + \frac{(\rho_m c_p)_{eff}[T_{sat} - T(0)]}{Q_{gen}} - t^* \qquad (10.31)$$

$$X^* = 1 - t^*. \qquad (10.32)$$

We expect real EM drying processes to fall between curves B (10.31) and C (10.32) in Figure 10.13. In fact, measured data in porous nonhygroscopic materials from many investigators (Figure 10.14) fall along a strikingly similar curve between B and C. The data plotted in Figure 10.14 were collected under the conditions in Table 10.1 below. In the cases shown the applicator was a large multimode cavity at 2.45 GHz. Even though conditions varied considerably for each of the experiments, the data essentially superpose in the figure. This indicates that proper normalization parameters have been used. We note that no consideration of transport processes has been included in the normalization parameters. So, for non-hygroscopic heating we may reasonably expect that transport processes might not significantly affect the drying rate and that parametric models (as above) will be quite effective. However, if internal pressures or local temperatures might exceed material damage thresholds, the full thermodynamic models are required.

Internal heat and mass transfer

The porous material is decomposed into solid matrix, liquid phase and vapor phase fractions, Figure 10.15. Both heat and mass transfer effects occur between the three phases, so inter- and intra-phase transport and conservation principles are applied separately. The resulting relations may be used for both EM and conventional drying. The individual phase energy relations may simply be added to obtain a global energy conservation relation if the assumption of local thermal equilibrium (LTE) among the phases, i.e. equal temperatures, is valid. While the LTE assumption is nearly always rigorously correct in conventional drying it is often invalid in EM heating cases since the power is usually deposited in one phase — ε'' is phase-specific in most applications, so drying rate is also locally phase-specific. Certainly, one should inspect the LTE assumption very carefully in EM heating models before applying it. We may, alternatively, compute a volume-averaged and phase-averaged energy (Figure 10.15) in order to assemble the global energy balance. This would then be coupled to similar mass conservation and transport relations.

The distinguishing feature of dielectrically-dominated EM drying is that sub-cooled drying may be obtained in which the environment temperature, T_∞ is much lower than the surface temperature, T_S, and the internal temperature profile is inverted with respect to conventional drying. In very porous materials inversion of the temperature profile leads to transport phenomena not observed in conventional drying. In general, the internal resistance to moisture migration and heat transfer is reduced in EM dominated drying. Reflux condensation processes, alluded to in the previous section, give rise to higher mass transfer rates than would be realized in conventional drying where the moisture front recedes into the material. The surface layer is kept moist, which enhances both surface heat and mass transfer. Effective thermal conductivities and mass diffusivities may be augmented by factors of about 2 and 10, respectively, over the dry solid phase material (Grolmes and Bergman 1990). Thus, the surface is maintained in a relatively isothermal and moist state in EM drying even in a cooled environment.

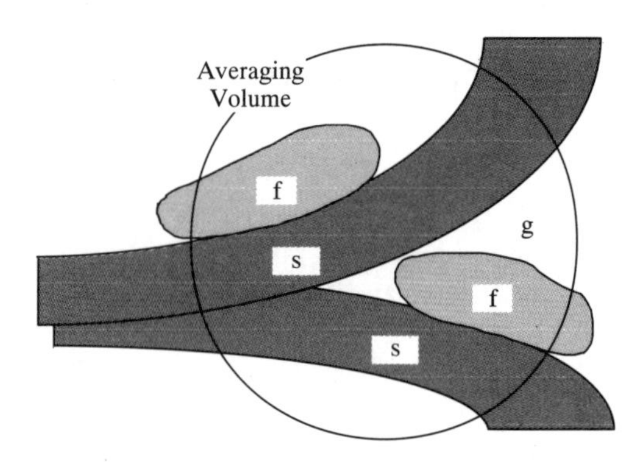

Fig. 10.15 *Schematic of non-hygroscopic solid matrix with solid (s), liquid (f), and vapor (g) regions and averaging volume.*

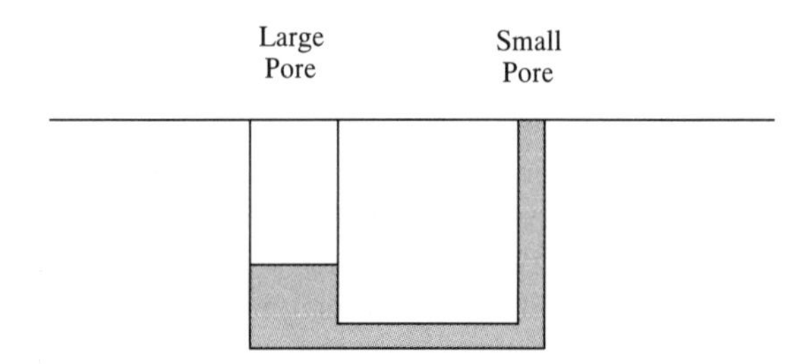

Fig. 10.16 *Two-pore capillary tube model of enhanced mass transfer in an inverted thermal gradient produced by EM heating.*

The companion case of a capillary porous material is similar in many ways. Mass transport is enhanced in a capillary material through the dependence of surface tension on temperature. The effect was described by Perkin (1990) where the inverted temperature profile enhances capillarity since most liquids have a positive slope for the dependence of surface tension on temperature. An example two-pore model of this effect is illustrated in Figure 10.16. The pores are imagined to communicate at some point below the surface. The material geometry results in a higher liquid level in a small pore relative to an adjacent large pore under isothermal conditions due to capillarity. The large pore will lose liquid preferentially relative to the small pore; thus the smaller pores will act to sustain the surface moisture.

Internal volumetric heating gives rise to high internal pressures due to thermal expansion in the liquid phase in parallel with vaporization processes. At high volumetric heating rates the gas phase is frequently discontinuous, evolving in the form of multiple gas bubbles of varying size at nucleation sites in the solid matrix (Lyons *et al.* (1972), Wei *et al.* (1985), Lefeuvre *et al.* (1978)). The high internal pressures which result push the liquid phase toward the surface, a process termed "microwave pumping" by Lefeuvre *et al.* Cellular materials — hard woods, for example — often experience this kind of heating and are particularly vulnerable to EM induced damage (stress cracking) at high power levels. The lower heating rates obtained in RF fields create less damage in cellular materials than microwaves, in addition to the deeper penetration of the fields, and are most often preferred for that reason.

External heat and mass transfer

The effect of dielectric heating on the convection boundary layer is not particularly pronounced, and the dominant surface heat transfer mechanisms are essentially unchanged in terms of physical principles. However, the evolution of larger amounts of vapor phase at the surface will usually significantly affect the values of the convection coefficients. This is because the evolved vapor phase may act as a gas injection in the boundary layer reducing the convection heat transfer coefficients, especially at the boundary layer–free stream interface. This will further decouple the drying rate from the environment conditions.

As a consequence of the above considerations, the common simplifying assumptions of local thermal equilibrium and a continuous gas phase are inappropriate when EM enhanced drying is considered. This motivates the parametric studies previously discussed since many of the required coefficients are difficult to determine or estimate individually, especially when they are temperature dependent.

10.4.2 Hygroscopic materials

In nonhygroscopic materials the free water is retained in the void space of the solid matrix. The equilibrium vapor pressure in a non-hygroscopic material is completely described by the Clapeyron equation since there are no other binding energies to be considered. We now consider the case of the bound liquid phase, in which the full (Clapeyron) vapor pressure is not realized due to the bond energies. There are several forms which the binding may take, so each form may exhibit different response to dielectric heating.

Mechanically bound moisture is found in interstitial spaces of a porous solid when the pore radii are less than about 0.1 mm. Surface tension is very high for water, and potentially for other liquids as well, and the high surface tension reduces the local vapor pressure. Binding energies are around 100 J/mol for water in these materials (Strumillo and Kudra 1986). Mechanically bound moisture is also characteristic of cellular structures such as wood, grains and other biological materials. In those materials the water is completely enclosed in a finite volume determined by the cell wall (plant) or membrane (animal) and elevation of internal temperatures causes both evolution of the vapor phase and liquid expansion. If the evolved vapor phase cannot diffuse readily through the cell walls the internal pressure will increase. Increasing internal pressure elevates the saturation temperature, so the vaporization rate will decrease. If no physical changes are encountered, the vapor evolution rate, temperature and pressure will reach steady state at a level determined by the mass diffusion resistance. We saw this effect in the wood heating experiment described in Chapter 9. The high internal pressures may result in damage to the solid matrix, and are usually to be avoided.

Physically bound moisture is adsorbed onto the solid matrix. The bonds are usually hydrogen bonds in the micropores or in the macropore space. We see this mechanism in adsorbents such as activated alumina and zeolites. Bond energies are typically of the order of about 3000 J/mol (Schmidt *et al.* 1992).

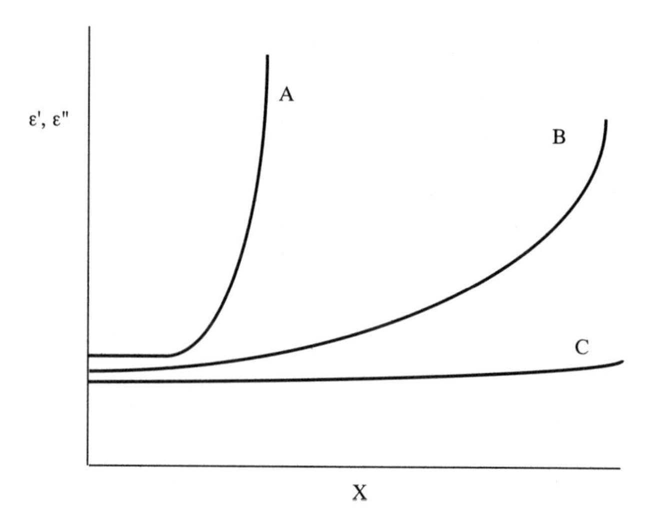

Fig. 10.17 *Variation in ϵ' and ϵ'' vs. X for a non-hygroscopic material (A), moderately hygroscopic (B) and highly hygroscopic material (C).*

Chemically bound moisture is found in hydrated materials such as gypsum. Actually, freshly made gypsum has both chemically bound and free water; and the goal in drying is to remove the free water without affecting the chemically bound water of hydration. Removal of the water of hydration will change the physical and chemical characteristics of the material; gypsum reduces to plaster of Paris if the bound water is removed.

The interaction between the bound moisture and an applied electromagnetic field differs significantly from free water in a non-hygroscopic material. The binding energies prevent rotational motion of polar molecules to varying degrees. Typically, we observe variations in ε' and ε'' similar to those sketched in Figure 10.17. In the Figure the non-hygroscopic material, sand for example (curve A), exhibits permittivities determined by the solid matrix at low moisture and by the water at high moisture. Highly hygroscopic materials, such as silica gel (curve C), show little change in permittivity as moisture increases, while the moderately hygroscopic material, say potato starch (curve B), falls between the limiting cases. We do not expect to be able to achieve moisture leveling in highly hygroscopic materials for this reason; non-hygroscopic materials are much better candidates.

The second way in which EM field interaction may be changed is that the EM field may couple directly to the bound water in the matrix. This has been observed in the zeolite matrix and will be discussed in a later section.

10.4.3 Model for dielectrically enhanced drying

We have briefly reviewed the physical processes underlying dielectrically assisted drying. The governing relations appear to be significantly entangled such that models may be quite difficult. However, several simplifications are useful and a tractable set of governing equations can be developed. The discussion follows the work of Chen and Pei (1989) and Whitaker (1985) and is based on the assumption of a homogeneous linear and isotropic material.

The point form of the liquid phase mass balance is:

$$\frac{\partial}{\partial t}\{\rho_w(1-\Phi_w)\} + \nabla \cdot \mathbf{J}_w = -m_{ev} \tag{10.33}$$

where $(1-\Phi_w)$ is the volume void fraction of liquid phase water, ρ_w the density, $\mathbf{J}_w$ the local mass flux (kg/m^2 s) and m_{ev} the evaporation rate (kg/m^3 s). The vapor transfer mass balance combines both the vapor phase and air volumes into the gas phase:

$$\frac{\partial}{\partial t}\{\rho_g(1-\Phi_g)\} + \nabla \cdot \mathbf{J}_g = m_{ev}. \tag{10.34}$$

The energy balance for all phases becomes:

$$\rho c_p \frac{\partial T}{\partial t} + (c_{pw}\mathbf{J}_w + c_{pg}\mathbf{J}_g) \cdot \nabla T = \nabla \cdot (k\nabla T) - m_{ev}\Delta h_{fg} + Q_{gen} \tag{10.35}$$

where ρ is the combined material density. The second term on the left hand side represents mass flux contributions to the enthalpy. The mass flux and vaporization enthalpies can be combined with the Fourier heat conduction (the first term on the right hand side) to obtain an effective thermal conductivity, k_{eff}. The resulting relation is considerably simplified:

$$\rho c_p \frac{\partial T}{\partial t} = \nabla \cdot (k_{eff}\nabla T) + Q_{gen}. \tag{10.36}$$

The mass flux terms contain the model intracies. Dominant mechanisms are capillary flow of free water, micropore diffusion of bound water and macropore vapor diffusion. The water flux, $\mathbf{J}_w$, is composed of capillary flow of free water, $\mathbf{J}_L$, and flow of bound water, $\mathbf{J}_b$. The forcing function for capillary flow is a gradient in either surface tension or pressure. Considering the material in a grossly macroscopic sense, the capillary pressure gradient may be related to the free water concentration gradient (Greenkorn 1981):

$$\begin{aligned}\mathbf{J}_L &= -\rho_w \frac{K_L}{\mu}(\nabla \mathcal{P}_g - \nabla \mathcal{P}_c - \rho_w \mathbf{g}) \\ &= -\rho_s D_L \nabla X_f - \rho_w \frac{K_L}{\mu}(\nabla \mathcal{P}_g - \rho_w \mathbf{g})\end{aligned} \tag{10.37}$$

where K is a mass permeability (m^2), $\mathcal{P}$ is pressure, $\mathbf{g}$ is gravitational acceleration, and D mass diffusivity (m^2/s). Bound water flux, $\mathbf{J}_b$, becomes important when free water approaches zero. There are several formulations for the forcing function of bound water flux. The most useful are the chemical potential formulation (Stannish et al. 1980) and the vapor pressure gradient because it is relatively easy to find the transport coefficients from measured drying data. A vapor pressure gradient gives:

$$\mathbf{J}_b = -\rho_w \frac{K_b}{\mu} \nabla \mathcal{P}_g. \tag{10.38}$$

The obvious and typical assumption is that the bound water is in equilibrium with its vapor phase. The bound mass fraction, X_b, and its associated relative humidity, ψ, at a given temperature are related by the sorption equilibrium isotherms:

$$\mathbf{J}_\mathrm{b} = -\rho_\mathrm{w}\frac{K_\mathrm{b}}{\mu}\mathcal{P}^*_\mathrm{g}\frac{\partial\Psi}{\partial X}\nabla X = -\rho_\mathrm{s}D_\mathrm{b}\nabla X \tag{10.39a}$$

where $\mathcal{P}^*$ is the partial pressure of the gas phase and the bound water diffusivity, D_b, has an Arrhenius dependence on temperature:

$$D_\mathrm{b} = D_\mathrm{b0}\exp\left(-\frac{E_\mathrm{d}}{RT}\right) \tag{10.39b}$$

with bond activation energy E_d.

The gas phase consists of vapor and air with associated fluxes $\mathbf{J}_\mathrm{v}$ and $\mathbf{J}_\mathrm{a}$ given by Darcy flow and diffusion combined:

$$\mathbf{J}_\mathrm{v} = -\frac{\mathcal{P}_\mathrm{v}M_\mathrm{w}}{RT}\frac{K_\mathrm{g}}{\mu_\mathrm{g}}\nabla\mathcal{P} - \frac{D_\mathrm{v}M_\mathrm{w}}{RT}\nabla\mathcal{P}_\mathrm{v} \tag{10.40a}$$

$$\mathbf{J}_\mathrm{a} = -\frac{(\mathcal{P}-\mathcal{P}_\mathrm{v})M}{RT}\frac{K_\mathrm{g}}{\mu_\mathrm{g}}\nabla\mathcal{P} - \frac{mD_\mathrm{v}M}{RT}\nabla(\mathcal{P}-\mathcal{P}_\mathrm{v}) \tag{10.40b}$$

where M is the molecular weight. Mass diffusivities and other parameters are determined by experimental studies for incorporation into the model

The governing equations may be solved using finite difference or finite element techniques (see Chapter 4). Boundary conditions are implemented in ways similar to those for EM fields. For wet material the normal component surface boundary condition is:

$$J_\mathrm{w} + J_\mathrm{v} = \frac{h_\mathrm{m}M_\mathrm{w}}{RT}[\mathcal{P}_\mathrm{v}(0) - \mathcal{P}_\mathrm{va}] = \phi_\mathrm{m} \tag{10.41a}$$

$$-J_\mathrm{h} = h[T_\mathrm{a} - T(0)] - \phi_\mathrm{m}\Delta h_\mathrm{fg} \tag{10.41b}$$

$$\mathcal{P}_\mathrm{g} = \mathcal{P}_\mathrm{a} \tag{10.41c}$$

where h_m is the convective mass transfer coefficient (h is convection heat transfer), ϕ_m is a drying mode coefficient, and the subscript a refers to the surrounding air. The receding moisture front case generates moving boundary conditions and the mass and heat balances must apply at the evaporation front:

$$T = T_\mathrm{S} \qquad X = X_\mathrm{mS} \qquad \text{and} \qquad \mathcal{P}_v = \mathcal{P}^*_v(T_\mathrm{S}) \tag{10.42a}$$

$$J_\mathrm{L} + J_\mathrm{v1} = J_\mathrm{b} + J_\mathrm{v2} \tag{10.42b}$$

$$J_\mathrm{h1} = J_\mathrm{h2} \tag{10.42c}$$

where the S subscript refers to the surface, X_mS is the maximum sorptive moisture content and the 1 and 2 subscripts refer to opposite sides of the evaporation boundary (1 is above, 2 below). The velocity of the moving evaporation front is estimated by setting the substantial derivative of the free water content to zero. The moving evaporation front model is handled by using a moving finite element grid (see Chen and Pei 1989).

10.5 ELECTROMAGNETIC DRYING OF GYPSUM

We have applied both RF and MW heating to the problem of drying freshly molded gypsum. We briefly described a finite element model constructed to calculate RF heating in finite gypsum slabs in Chapter 4. The goal in gypsum drying is to remove free water without changing the distribution of chemically bound water, as previously mentioned. Here we present temperature profiles from the model and experimental heating results. The stray field electrode configuration was used in the RF experiments, and a multimode cavity in the MW experiments.

10.5.1 Experiments with microwave heating

The fresh gypsum is a very effective absorber of microwaves at 2.45 GHz. Both the bound and free water interact with the applied field. In the multimode cavity experiments the preferential corner heating which is expected from the boundary conditions (see Chapter 4) is strong enough that the bound water is removed when field levels are high enough to obtain effective drying of free water in the bulk of the object. The result is that the corners of a rectangular shape are reduced to Plaster of Paris and crumble when handled. This is a most undesirable result in a drying application. Therefore, microwaves are not a particularly effective modality for gypsum drying because of excessive corner heating and shallow penetration depth.

10.5.2 Radio frequency heating of gypsum

It turns out that over the RF range the free water in the gypsum is an effective absorber primarily because of the residual ion content in the matrix. The bound water does not absorb from an RF field. Consequently, gypsum is an ideal candidate for RF heating, at least in a technical sense. As the free water migrates out of the matrix the material becomes non-absorbing and: (1) moisture is leveled, and (2) the heating drops to very near zero when the free water (alone) is exhausted. There are economic considerations as well, however, and on that scale RF heating is less attractive than conventional methods owing to the low cost of gypsum products. We will now consider RF heating of gypsum sheets.

Stationary rod electrode results

In Figure 10.18 we review the periodic stray field and staggered through-field electrode configurations. The arrays are periodic so that the model space (B to C in the figure) may be much smaller than the whole gypsum slab. The zero flux boundary conditions on planes of symmetry terminate the modeled section of gypsum. Figure 10.19a shows the round rod stray field electrode geometry of the first finite element solution. The electrical and thermal properties of the gypsum, paper covering and styrofoam insulators were included in the model as well as the air surroundings. The enclosure was grounded and the high voltage (left) and ground (right) rods were Dirichlet boundaries. The high voltage electrode was supported on a G-10 epoxy–glass fiber insulator and the ground electrode on a conducting plane (both support structures are included in the electrical model). The periodic nature of the electrode array was handled with Neumann boundaries at the electrode center lines. Predicted temperatures realized from heating for 40 seconds at 8 kV (rms) in an array

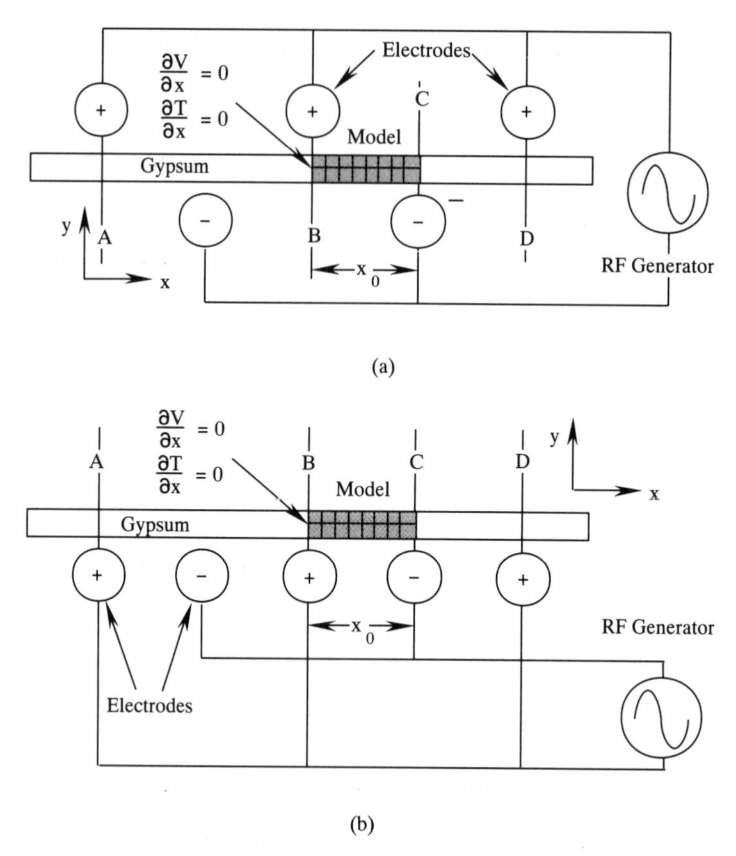

Fig. 10.18 *Rod electrode configurations. (a) Periodic (period = $2x_0$) staggered through-field electrodes showing model segment, electrical and thermal boundaries. Round electrodes are shown as an example. + indicates high voltage electrode, − ground. (b) Periodic stray field electrode configuration.*

with electrodes separated by 10.2 cm are shown in Figure 10.19b. The calculated maximum temperature (44°C) was 2°C less than that measured experimentally (for the same conditions) in dry gypsum using thermographic imaging. The maximum temperature was located within 0.5 mm of that observed experimentally.

Figure 10.20 shows the predicted temperature contours for a staggered through-field electrode array at a voltage of 6800 V (rms) (at a separation of 5.08 cm) and heating time of 50 s. The geometric distribution agrees well with thermal images of a heating experiment, although the predicted maximum temperature is 5°C higher in the model, compared to the matching experimental results, and occurred in the central region between the electrodes rather than under the electrodes, as was observed experimentally. This was the only case of disagreement between model calculations and experimental results in the study. We attribute most of the variance to uncertainty in electrode position.

In Figure 10.21 stray field rectangular rod array results are depicted for a comparable geometry. Here we see a maximum predicted temperature of 49°C (12°C above that measured) located within 3 mm of the maximum experimental maximum temperature, very near the center between the electrodes. The electrode potential was 9070 V (rms) with a rod

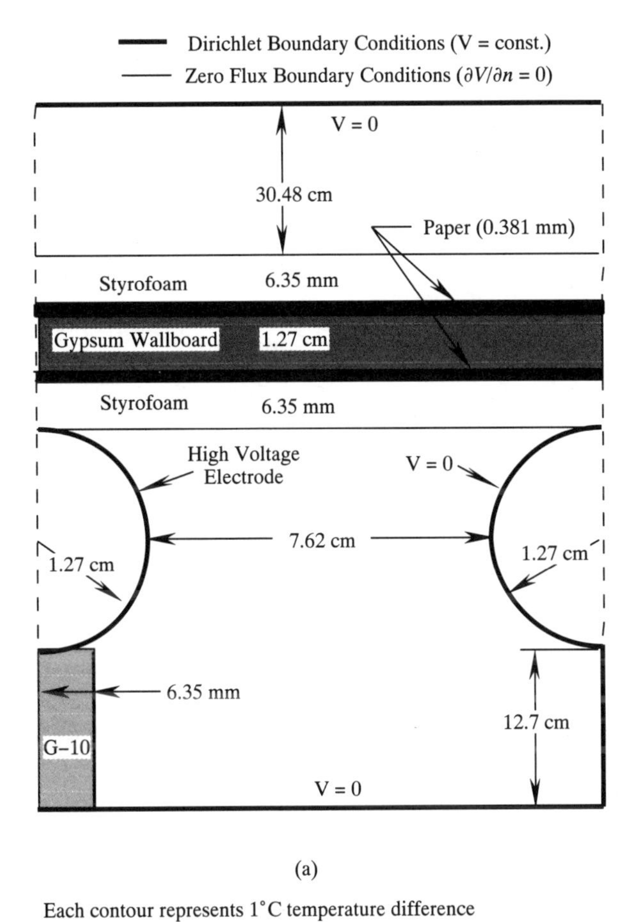

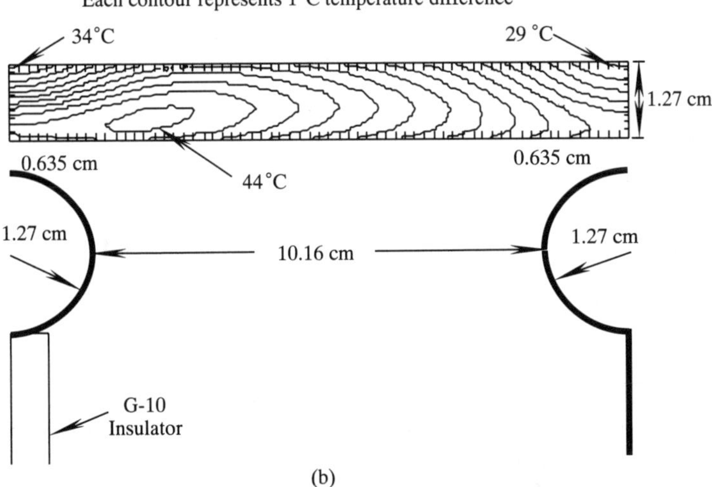

Fig. 10.19 *Stray field round rod electrode FEM model. (a) Model segment detailed geometry. (b) Predicted temperature distribution for 40 s of heating at an electrode voltage of 8 kV (rms). Each contour represents 1°C temperature difference.*

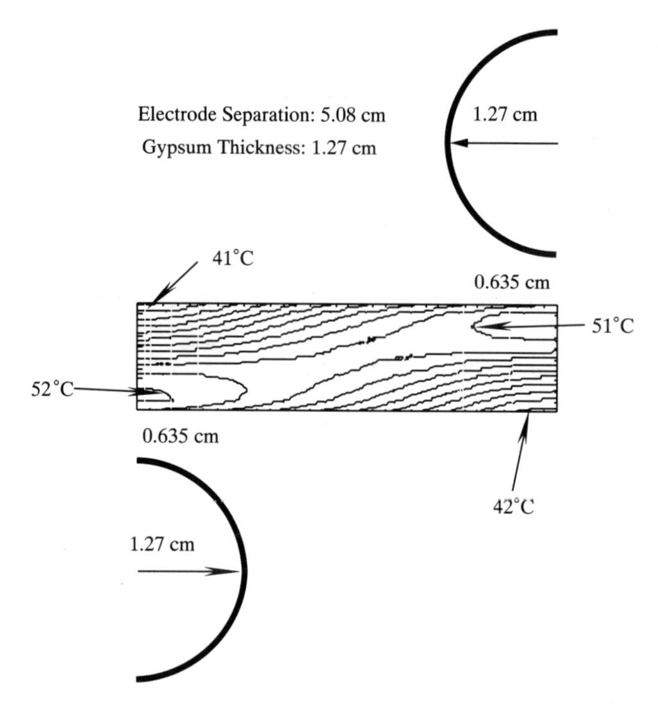

Fig. 10.20 *Isothermal contours in gypsum from staggered through-field rod electrodes after 50 s of heating, model results. Electrode voltage 6.8 kV (rms). Each contour represents a 1°C temperature difference.*

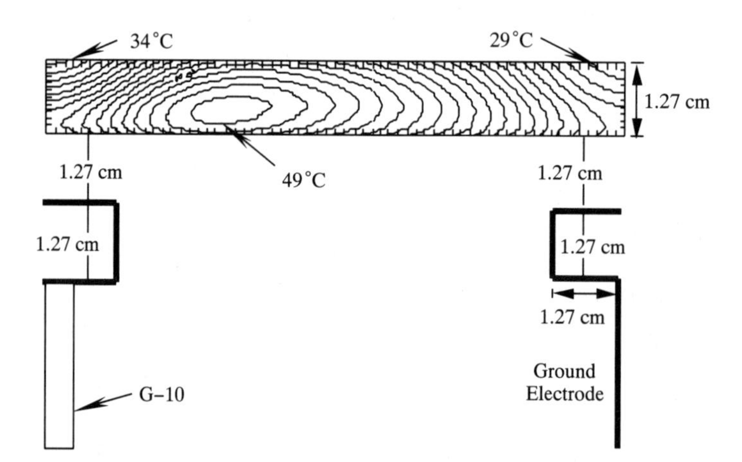

Fig. 10.21 *Rectangular rod stray field electrode model results for an electrode voltage of 9.07 kV (rms) and rod separation of 10.16 cm. Heating time was 50 s. Each contour represents a 1°C temperature difference.*

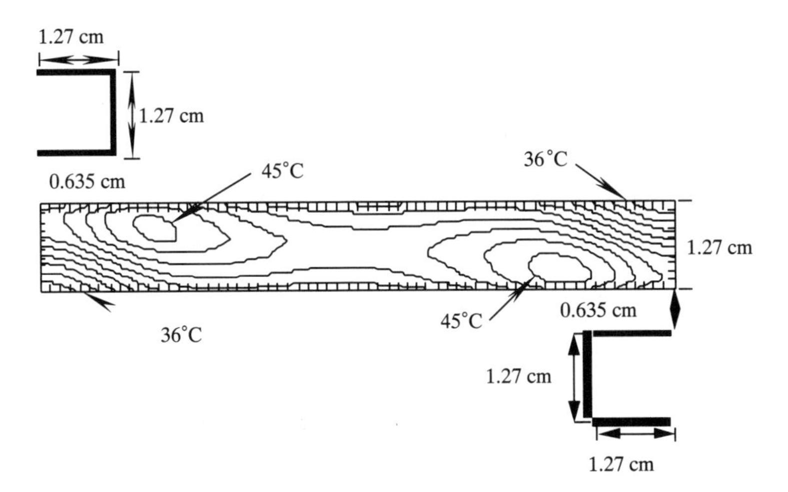

Fig. 10.22 *Rectangular rod staggered through-field electrode predicted temperature contours. Electrode voltage was 7.92 kV (rms), heating time 40 s and electrode separation 10.16 cm. Each contour represents a 1°C temperature difference.*

separation of 10.16 cm and the heating time was 50 s. Calculations for a staggered through-field array of the same rectangular rods are shown in Figure 10.22 (7920 V (rms) for 40 s). Note that the maximum power density occurs much closer to the electrode corners than for the comparable stray field array of Figure 10.21. The maximum predicted temperature (45°C) was 9°C below the measured temperature (54°C), but occurred within 1.8 mm of the same location.

Model and experimental results agree quite well in terms of spatial distribution even though there are uncertainties in the maximum values. The differences are most likely due to uncertainties in electrode voltage and electric permittivity. Since no trend may be seen in the differences, we conclude that they are due to randomizing influences rather than systematic error.

Moving-load results

The stationary model results may be used to study power density distributions, as we have just shown. A real drying application involves a moving gypsum slab load. Consequently, a model of gypsum wallboard moving through the round rod staggered-field array, shown in Figure 10.23, has been executed. The goal was to model the transient heating period to estimate the time required to establish a relatively uniform temperature distribution. A section of gypsum wallboard one electrode period in length was followed as it entered the rod array, passed through the repeating sections of electrodes, and finally exited the rod array. The temperature distribution in the gypsum wallboard at the four times marked t_1, t_2, t_3, and t_4 are shown in Figure 10.24. As in the stationary load cases, styrofoam was assumed present between the electrodes and the gypsum wallboard.

The velocity of the gypsum wallboard was set to 0.635 cm/s (0.25 inch/s). At this velocity, it took the gypsum wallboard section 16 s to traverse the distance between the electrode centers. For a thermal diffusivity of $\alpha = 1.93 \times 10^{-7}$ m^2/s (gypsum), electrode spatial period $x_0 = 0.2032$ m (8 inch), velocity $u = 0.006\,35$ m/s (0.25 inch/s), and

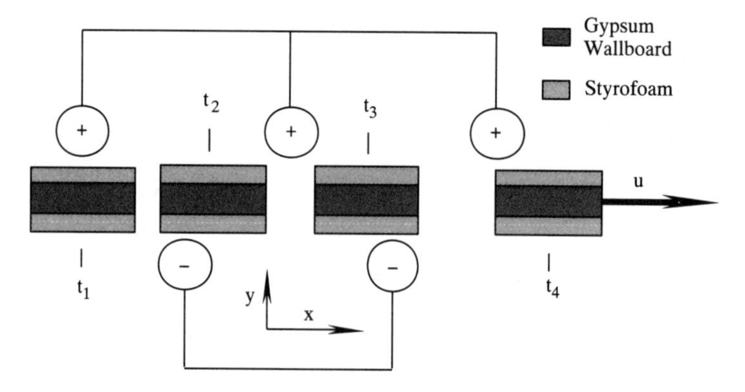

Fig. 10.23 *Moving-load model positions, staggered through-field electrode configuration.*

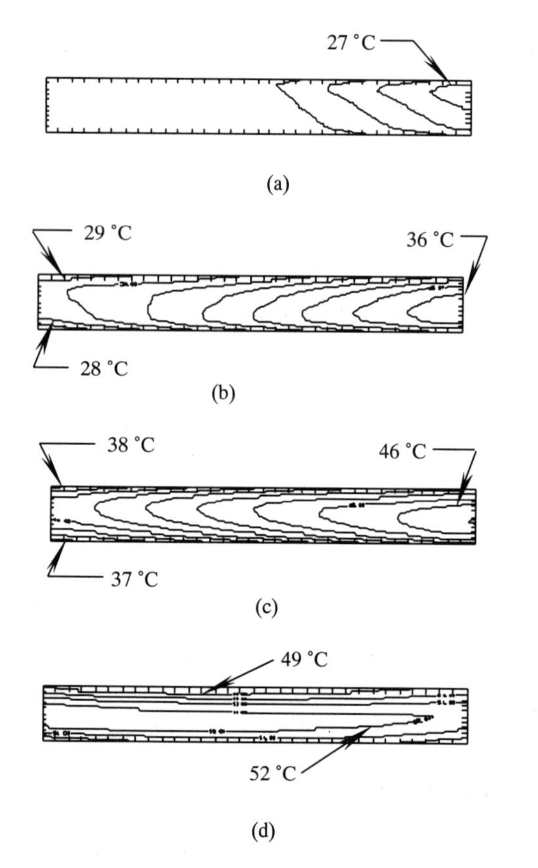

Fig. 10.24 *Temperature contouis in the moving-load model of Figure 10.23: (a) at time t_1, entering the first electrode set; (b) at time t_2; (c) at time t_3; (d) at time t_4. Each contour represents a 1°C temperature difference.*

gypsum half-thickness $L_c = 0.006\,35$ m (0.25 inch), the characteristic Fourier number (F_o) is calculated as

$$F_o = \frac{\alpha x_0}{u L_c^2} = 0.153. \qquad (10.43)$$

This value (< 1) indicates that heat conduction will be predominately in the vertical direction (direction of L_c) and that the temperature distribution in the load will be dominated by dielectric heating; however, some heat transfer in the horizontal direction will occur.

The transient nature of the heating can be seen in the plots of Figure 10.24. Only the right half of the wallboard was heated at time t_1 (Figure 10.24a) because no RF energy was applied to the gypsum until after it passed the center of the top electrode. Also, the top surface of the gypsum wallboard was heated more than the bottom surface since the top surface was slightly closer to the top electrode and experienced a higher power density. Figure 10.24b shows the temperature distribution in the gypsum wallboard at time t_2, in the second electrode section. As shown, the entire gypsum wallboard sample has been heated with the right portion of the gypsum wallboard warmer than the left. This is the sensible result as this portion was the first to enter the RF field and therefore it had been heated the longest. Also, the bottom half of the gypsum wallboard was warmer than the top half because: (1) the convection coefficient at the bottom surface of the gypsum wallboard was 2.5 while that on the top surface was 5.0 to compensate for the effect of air movement, and (2) the bottom half of the gypsum wallboard was slightly closer to the bottom electrode than the top half was to the top electrode. The same observations apply to Figure 10.24c. At this point the heat flux is primarily vertical. The final plot (10.24d) shows the gypsum temperature distribution at time t_4. In this position, the gypsum wallboard section was completely out of the electrode array and was not heated with RF energy. Also, the right half of the gypsum was cooler than the left half because the right half of the gypsum has been out of the electrode array the longest.

As shown in Figure 10.24d, heat transfer (i.e. the temperature gradient) was predominately in the vertical direction although some horizontal heat transfer was present. This result confirms the usefulness of the characteristic Fourier number in predicting the importance of heat transfer by conduction in dielectrically heated moving-load models.

Dryer simulation experiments

As mentioned in the introduction to this section, gypsum has the property that it stops absorbing RF energy when the free water is eliminated. We will now discuss the gypsum drying problem from an experimental perspective. These experimental results were obtained by J. Grolmes (unpublished data). The loss factor is very sensitive to moisture content (Figure 10.25a) and decreases from about 12 at 27 MHz and $X = 15\%$ to a negligible value as X approaches zero. In Figure 10.25b a gypsum slab has been exposed to a constant surface air temperature of 149°C (300°F) for a period of 68 min — a near simulation of a typical commercial convective gypsum dryer. When the free water is exhausted the gypsum temperature increases to near the air temperature since the moderating effect of the water has been eliminated. The product may thus be overheated to potentially damaging temperatures. In Figure 10.25c we have simulated a multi-zone radio frequency-assisted dryer in which the constant rate period drying is accomplished by convection (for time less than 31 min) with an air temperature of 130°C. At 31 min the radio frequency field was turned on and the air temperature dropped to 50°C. The transient heating experiment (Figure 10.25c) confirms the tendency of the gypsum to stop heating in an RF field when the free water is exhausted.

By dropping the surface air temperature the maximum gypsum temperature is reduced and product damage can be avoided. Drying in the falling rate period is accomplished by the RF field and the heated air is used to simply control surface heat transfer rates.

This application of RF drying is technically nearly ideal. The heating process is self-limiting and the gypsum temperatures can be kept low with no sacrifice in overall drying

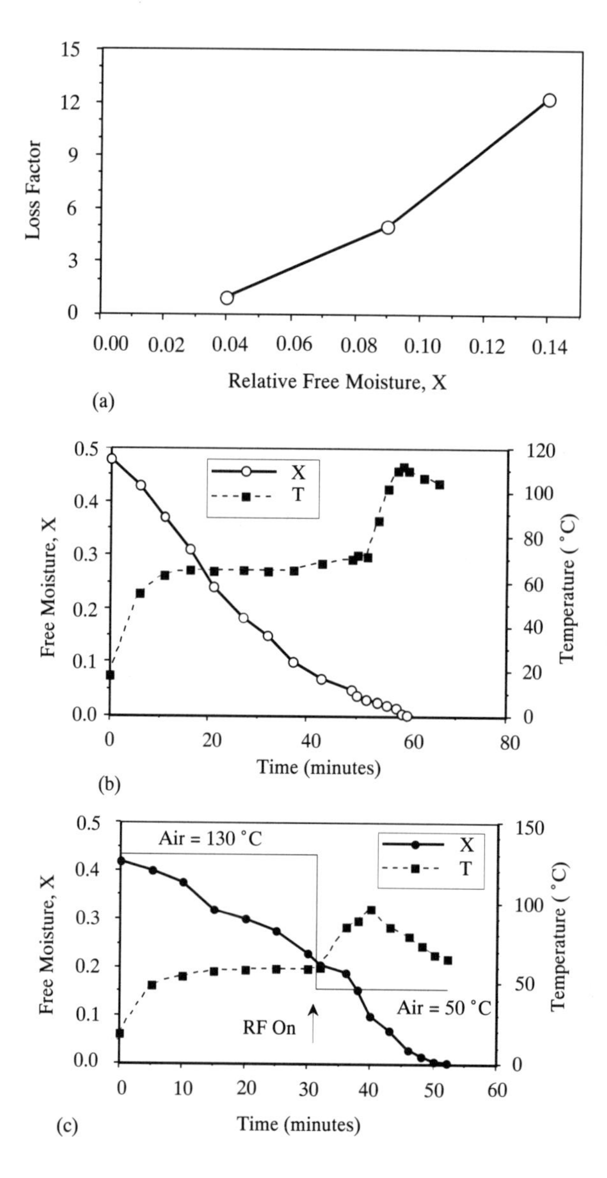

Fig. 10.25 *(a) Measured loss factor for wet gypsum wall board at 27 MHz as a function of relative free water content. (b) Gypsum drying experiment by forced convection heating, single zone dryer. The air temperature was 149°C. (c) Experimental simulation of multi-zone dryer with RF on and air temperature dropped from 130°C to 50°C at 31 minutes.*

rate. However, as previously mentioned, gypsum wall board is very inexpensive, and the cost of RF heating is prohibitive even if the RF section is limited to the end of the drying curve where a lower power (thus lower cost) system is required. So, while technically desirable, economic considerations make RF gypsum drying undesirable under present conditions.

10.6 DEHYDRATION OF ZEOLITES

There are many circumstances in industry where a liquid product must be dried. For example, automobile gasoline must contain no more than 3% water (by volume), otherwise the engine will not start when cold. Ethylene cannot be polymerized if it contains more than 0.1% water vapor. For drying materials, chemists often use adsorbents such as zeolites or silica gel. These drying products have a great affinity for water and are used to dry gases and liquids. In industrial use, a continuous flow drying system is used in which two drying reactors (columns) are used alternatively. While one is processing the product, the other is regenerated, as shown in Figure 10.26.

There are two very different drying problems associated with zeolite systems: (1) drying of greenware bricks (wet molded clay) prior to firing in the production of a zeolite catalyst element, and (2) regeneration of a saturated zeolite water-scavenging element in use. In the first case the relative water content is very high and there is a fairly high concentration of mobile ions in the greenware, so the depth of penetration of microwaves is quite shallow. Radio frequency heating is very suitable for this task because: (1) the zeolite structure is

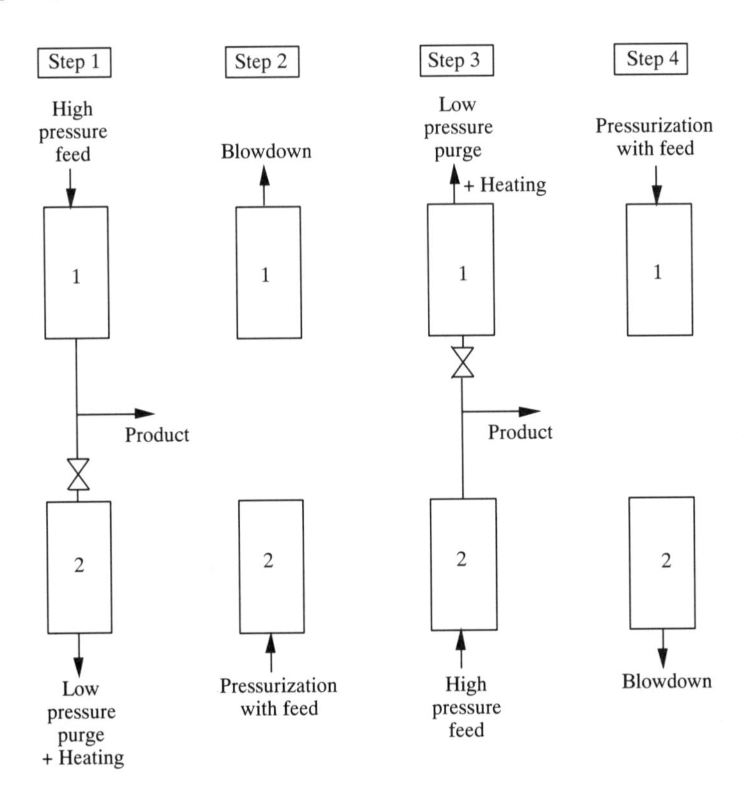

Fig. 10.26 *Alternate use and regeneration of zeolite water scavengers (two-bed cycle).*

an extremely high value product (about $1k for each element), (2) greenware drying is the production-limiting step, and (3) the water elimination rate can be finely and uniformly controlled in an RF field. In the second case, the fired zeolite element absorbs a relatively smaller amount of water from the liquid or gas stream and mobile ions are not abundant, so microwave heating is attractive as a regenerating technique. We will discuss both applications in this section.

10.6.1 Radio frequency drying of greenware

In the production of zeolite adsorbing elements a high moisture content "clay" material is molded or extruded to shape. A typical element is sketched in Figure 10.27a. The molded shape must be dried to extremely low moisture contents prior to firing in order to prevent rapid evolution of steam from cracking the element in the kiln. There is a trade-off between drying time of the greenware and passage geometry: a very effective adsorbing element must have a high surface area to volume ratio which is easier to achieve using small passage ways; however, a small passageway matrix is difficult to dry. It turns out that the drying of the greenware is often the limiting process in production. There are two effects which produce the drying bottleneck. First, air temperatures must not exceed about 28°C (82°F) so that the surface will not dry too quickly over the wet substrate and cause stress fractures. In fact, in the experiments which will be described it was necessary to coat the surfaces of the brick with mineral oil to prevent too-rapid drying in room air at 73% humidity. Second, the very small passageways saturate with evolved moisture very near the entrance ($x = 0$ in Figure 10.27b) due to the small passage cross-section. Consequently, high air velocities are required.

A typical brick is 40 cm (16 inches) long with square passageways either 2.5 or 4 mm on a side (Figure 10.27a). The bricks were 15.2 cm square (6 inches) on each side, as shown. A freshly extruded brick has a nominal moisture content of $X = 40\%$. The radio frequency impedance of a typical zeolite brick was measured using a network analyzer as shown in Figure 10.28a. By arranging the measuring electrodes to just fit the zeolite shape the measuring electric field is essentially uniform in the zeolite. The impedance of the wet brick is is shown in Figure 10.28b. The resistive component dominates the impedance at about 18Ω; thus, the fringing field is negligible in this measurement. The equivalent effective electrical conductivity for an intact wet brick (including the air cells) is thus in the neighborhood of 0.61 S/m.

Experimental analysis of the zeolite brick drying problem was conducted by J. Grolmes (unpublished data). In the experiments, greenware bricks were dried by convection and by an RF-dominated heating for comparison. Figure 10.29a is from a convection drying experiment in which room air at 23.3°C and 75% relative humidity was used in forced convection at a typical production velocity of 5.07 m/s (1000 ft/min). Approximately 7.2 hours were required to reduce the moisture from 38% to 8% (Figure 10.29a). Note that there is also an obvious transition from a constant drying rate to a falling rate period at about 15% moisture. The temperatures within the brick (T_1 to T_4) are shown in Figure 10.29b. Note that the temperatures differ only by a fraction of a degree among the probe locations during the entire experiment. The temperatures are initially several degrees below the 23.3°C room temperature owing to the heat of water vaporization. We may also see that the beginning of the falling rate period coincides with the rise in interior temperatures.

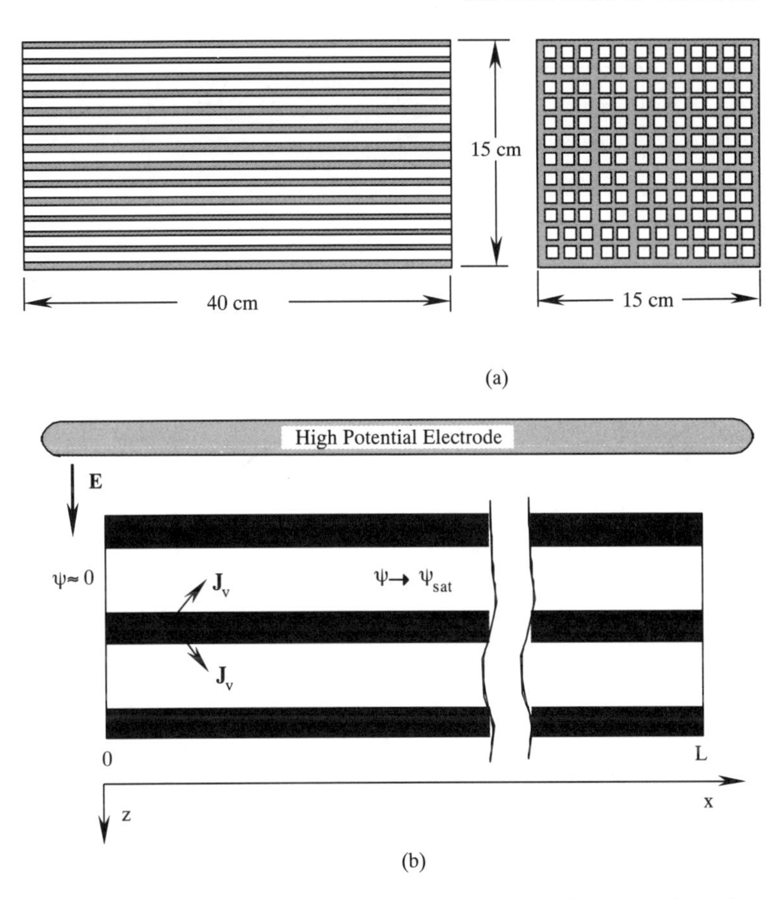

Fig. 10.27 *Zeolite greenware drying. (a) Catalyst geometry. (b) Saturation of air column in passageway.*

In a typical RF experiment the bricks were exposed to a plate voltage on the order of 7.5 kV (p-p) at 18 MHz with electrode spacing as shown in Figure 10.29c (approximately 2 inch air gap). The brick used in the RF experiment was about 7.6 cm thick (3 inches) to obtain effective coupling to the RF generator. Again, room air (22.6°C) was circulated through the pores at 5.07 m/s. The RF drying curve is shown in Figure 10.29d. The initial mass was 6.46 kg and 1.15 kg of water was removed during the experiment—drying from 40% to 15% in about 112 min. The average drying rate was 0.172 g/s for each brick.

The temperature record is shown in Figure 10.29e. The four optical probes—T_1 to T_4—were located so that T_1 and T_4 were at corners just under the upper surface (an electric field maximum, see Chapter 4) and probes T_2 and T_3 were near the center of the brick. The excess corner heating due to the electric field concentration can be seen near the end of the heating time in channels T_1 and T_4. While these temperatures would have caused damage at the initial moisture content, we did not observe any damage to the zeolite at the end of this test. Presumably, this is because the higher temperatures occur only when the local water content has been much reduced; and by the time the temperatures increase the local moisture is insufficient to cause cracking or surface checking.

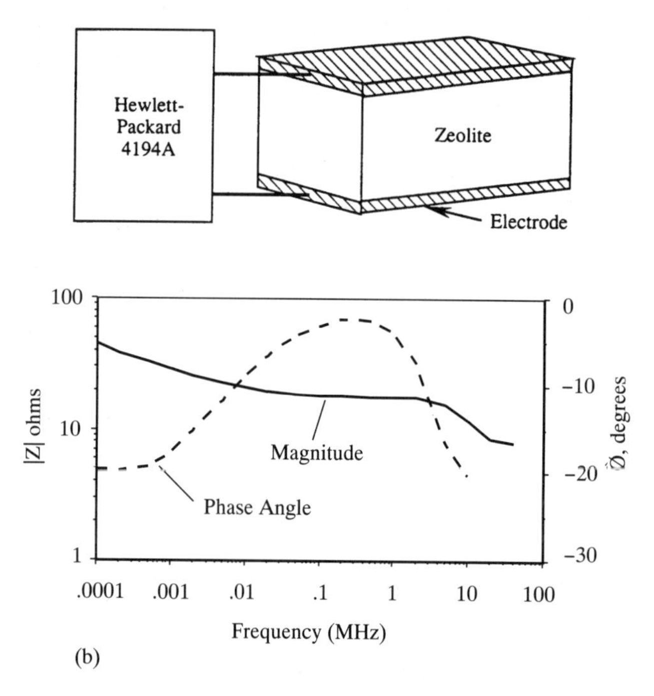

Fig. 10.28 *Determination of zeolite brick impedance. (a) Measurement system. (b) Measured impedance of zeolite greenware brick. The impedance is essentially resistive (at 18 Ω) from about 20 kHz to 5 MHz where external parasitic capacitances become important.*

The RF waveform was obtained from a generator (at about 18 MHz) controlled with a phase-proportional controller on the 60 Hz mains power. The controller was set for 50% output in this experiment. From the drying rate we estimate that 1.7±0.2 kW was coupled to the four-brick load, while from the plate current (0.3±0.02 A) and dc supply voltage (8 kV) of the generator we estimate that 2.4±0.2 kW was consumed in all of the elements connected to the tank circuit. The uncertainty in the estimation methods for RF power is apparent, though (in fact) both values could be acceptably accurate since some power is dissipated in the transmission lines, voltage measuring capacitor and tank circuit elements — the tank capacitor gets very warm during an experiment (up to about 90°C). On inspecting the RF drying curve we note that there is no falling rate period. The absence of a falling rate period characterizes all of the RF experiments, in contrast to all of the convection heating experiments. Also, the drying time has been reduced from about 8 hours to about 1.9 hours.

In the experiments the velocity of air through the matrix turned out to be a most important parameter. High velocities are required to clear the evolved water from the passage because in the small passages the cool air saturates within a few cm of the inlet if the velocity is low. Some improvement is obtained if the direction of airflow reverses periodically. The fired zeolite brick is also a very high value product, so the economics favor EM heating if production rates can be increased. Consequently, the zeolite greenware is, both technically and economically, an excellent candidate for radio frequency heating.

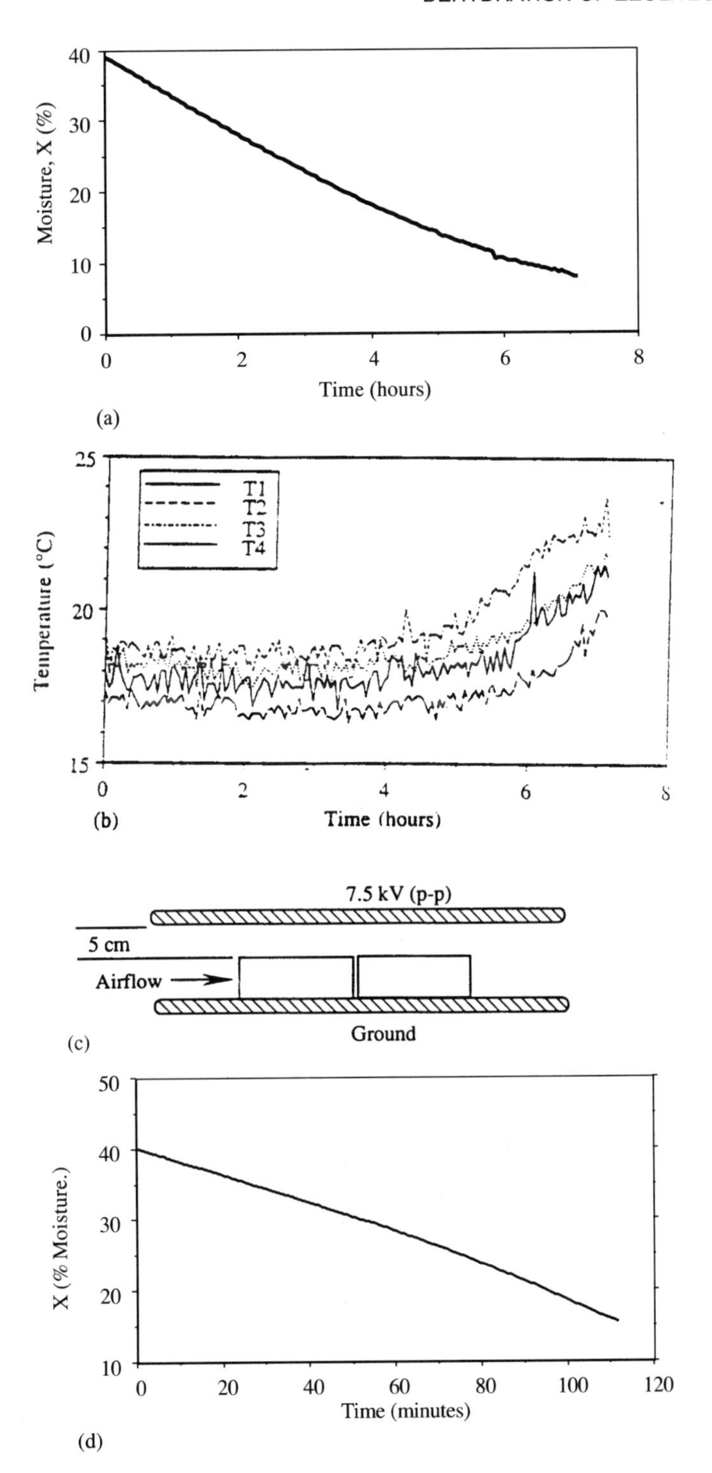

Fig. 10.29 *(continued)*

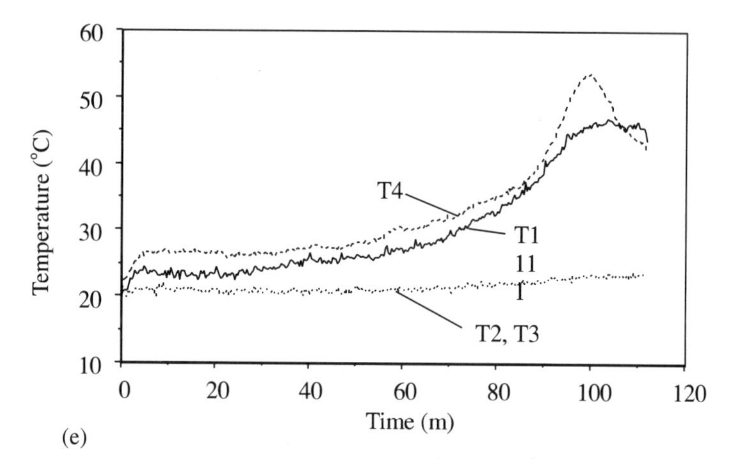

Fig. 10.29 *RF heating of zeolite greenware brick. (a) Drying curve for zeolite greenware, convection drying only. Air velocity 5.07 m/s (1000 fpm). (b) Temperatures for convection drying of zeolite greenware. Temperatures approach room temperature (23 °C) near end of drying during the falling rate period (see text). (c) Electrode geometry. (d) Drying curve for zeolite greenware, RF dominated convection drying. Air velocity 5.07 m/s (1000 fpm). (e) Temperatures for RF drying of zeolite greenware. Probes T1 and T4 were located in upper corners of the brick (at field concentrations) and probes T2 and T3 were located in the center of the brick.*

10.6.2 Microwave regeneration of saturated zeolite adsorber

In the last chapter we developed and inspected a linear model framework for the problem of microwave dehydration of zeolites for regeneration. The model contains a number of terms which are dependent on diffusion processes in the material. In this section we will take internal diffusion into account as a refinement of the model.

Consider the case of an adsorbent which can adsorb 10 % of its weight in water. If it must purify 10^3 kg of product containing 0.1% water in an hour, 1 kg of water must be retained. If a reactor has 100g of adsorbant the mass of water which can be retained is 10g. This means that the reactor is saturated 100 times in an hour; the cycle of adsorption and regeneration must occur every 36 seconds, which is not reasonable. Presently, a reactor can be regenerated in 24 hours. For this reason very large reactors are used (240 kg of zeolite is not uncommon).

In the case of a submarine for which the air must be purified in a confined volume, the concentration of CO_2 gas must be less than 1%. The amount of CO_2 produced by the crew is easily calculated and determines how much capacity the reactor must have. In a submarine, of course, the volume occupied by the reactor must be minimized, so a fast regeneration cycle would give a significant design advantage. In these two examples, it is obvious that the size of the reactor is very important. This parameter is directly related to the rapidity with which we can dehydrate (regenerate) the adsorbant.

Microwave processes for dehydration, desulfurization, and decarbonization have been studied for many years. The experimental design which we have used is based upon that discussed in Chapter 9 (see Figure 9.14) and the experiments were performed using the same technique. The sample tube situated in the RG-112/U (WR340) waveguide contains about 7 g of humid zeolite spheres, approximately 2 mm in diameter. The temperature of the material was measured as dry air flowed through the bed at inlet volume flowrate, Q_0 , at temperature T_0 and density ρ_0. The incident microwave power was increased stepwise, and the temperature and humidity of the outlet air was measured. We will attempt to interpret the results using the model equations from Chapter 9.

Experimental results

The first experimental result of microwave drying of zeolites shows that the gas outlet temperature, $T_g(t)$, (and thus the material temperature) is not a linear function of incident power since the curves are not of similar shape at low and high values of incident power (Figure 10.30). However, the water vaporization rate (as measured by the outlet plane humidity) is more linear in its dependence upon the incident power, indicating that bulk evacuation of water is a linear response (Figure 10.31). The rate of water production is a decreasing function of time, as might be expected, as was seen in the paper drying study. The difference between the rate of water loss in zeolites and paper drying occurs in the initial parts of the curves. In paper drying, the initial rate is at its highest value. In zeolite dehydration, the maximum rate is observed after a time delay. This can be simply interpreted by supposing that the water produced within the material will leave the system in the gas phase by a diffusion-induced delay such as appears in the first (increasing) part of the curve.

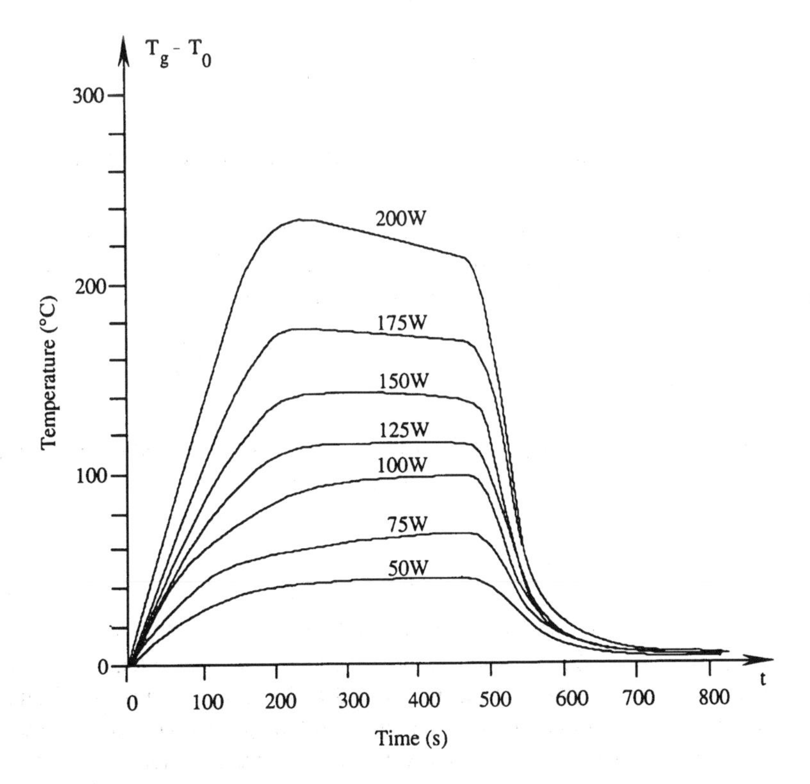

Fig. 10.30 *Outlet plane gas temperature during microwave regeneration of zeolite catalyst material at different power levels.*

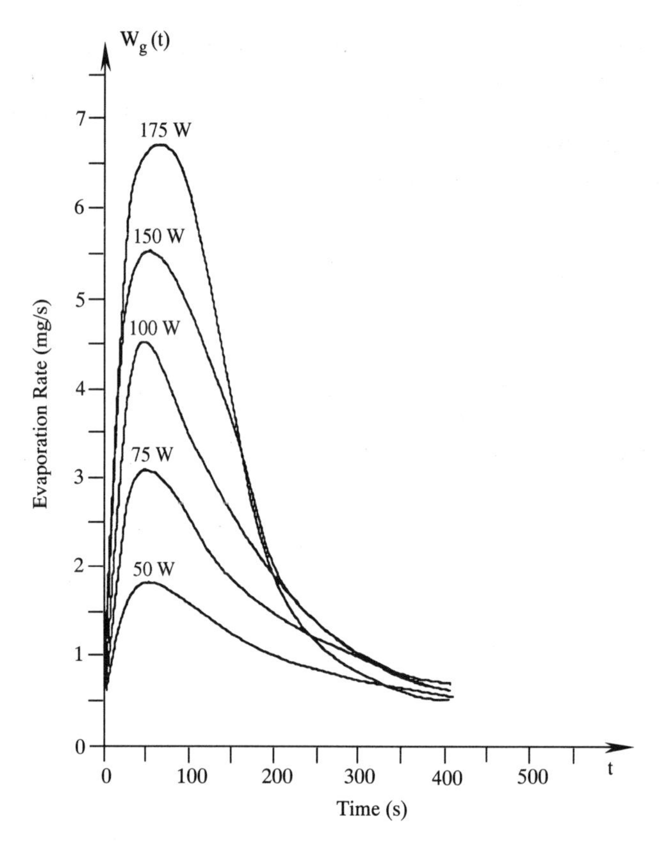

Fig. 10.31 *Evaporation rate for microwave dehydration of zeolite at different power levels.*

Within experimental error, the measured mass flow of water vapor, $W_g(t)$ (kg/s), appears to be successfully fitted by the combination of two exponentials. The gas moisture is related to $Q_w(t)$, the volume flowrate of gas phase water, by the density, so:

$$W_g(t) = \rho_g Q_w(t) = A\left(\exp\left(-\frac{t}{\alpha}\right) - \exp\left(-\frac{t}{\tau}\right)\right). \tag{10.44}$$

Therefore, the single exponential drying curve $A\{\exp(-t/\alpha)\}$ is delayed due to transport through the solid by the transfer function in the Laplace domain:

$$T(s) = \frac{1}{1 + s\tau}. \tag{10.45}$$

Other than characterizing its overall effect by this relation, the transport process cannot be clearly described from these experiments. The geometry of the zeolite cage is similar to a constant volume vessel with leaky walls, as mentioned within subsection 9.1.4 of Chapter 9. That is, we may (as a very simple picture) imagine that the space inside the zeolite cage is filled with saturated steam from the desorbed water at some pressure above the surrounding gas phase. Then, the delay in transporting desorbed water to the surrounding gas stream might be explained by a build-up of pressure required to expel, rather than resorb, the water vapor. Obviously, the definitive experiment has yet to be performed, but this explanation may serve as a starting point.

The production of free water from the moist zeolite has an exponential response. When we use the simple phenomenological model of dehydration, m_a (the mass of adsorbed H_2O) may be determined from a first order equation with experimentally derived parameters. In Chapter 9 we defined m_g as the mass of water in the gas phase (external to the zeolite spheres), with associated moisture fraction X_g ($X_g = m_g/m_o$ in each unit volume of outlet gas), and m_i as the mass of water in the intermediate state. The transfer of m_i will take place during a transport delay time, which we denote as τ.

If we suppose that the desorption rate of water inside the solid prior to transport is proportional to $\exp(-t/\alpha)$, the mass of adsorbed water, m_a, decreases with the same time constant. The algebraic form for the instantaneous mass of water adsorbed on the zeolite is:

$$m_a(t) = m_a(\infty) + (m_a(0) - m_a(\infty)) \exp\left(-\frac{t}{\alpha}\right). \tag{10.46}$$

The quantities $m_a(0)$ and $m_a(\infty)$ are the initial and final adsorbed water mass, respectively, which can now be directly evaluated from the experiment. Modeling the water transport as a dynamic first order process, the experimental mass flow of the evacuated water, $W_g(t)$ is:

$$\begin{aligned} W_g(t) = \rho_0 Q_0 M_g \frac{T_g}{T_0} &= \int_{x=0}^{x=t} \frac{dm_a}{dt} \exp\left[-\left(\frac{t-x}{\tau}\right)\right] dx \\ &= \frac{(m_a(0) - m_a(\infty))}{\tau - \alpha}\left(\exp\left(-\frac{t}{\tau}\right) - \exp\left(-\frac{t}{\alpha}\right)\right) \end{aligned} \tag{10.47}$$

where dm_a/dt comes from equation (10.46):

$$\frac{dm_a}{dt} = \frac{m_a(\infty) - m_a(0)}{\alpha} \exp\left(-\frac{t}{\alpha}\right). \tag{10.48}$$

Equation (10.47) gives the value of the coefficient, A, defined in equation (10.44) used for scaling the experimental curves. The values of τ and α are determined by the absorbed power, P_a, though it turns out that τ is relatively insensitive to P_a.

The interpretation of the response of water in the zeolite indicates that the process is comprised of two first order steps. First, water is desorbed from the zeolite, and, second, the free water is transported through the zeolite cage matrix. The number of molecules desorbed per unit time is proportional to the number of molecules which are presently adsorbed in the zeolite. The rate coefficient of the desorption kinetic process is $1/\alpha$. Figure 10.32 compares the experimental and theoretical (calculated) curves. The calculated values of $1/\alpha$, τ, and A are given in Table 10.2, from the numerical fits of the curves for $W_g(t) = \rho_g Q_w(t)$ at different values of the applied field. The equivalent time delay which characterizes the transport of water in the zeolite might be expected to vary somewhat with the applied power; but $1/\alpha$ is, in fact, the most sensitive parameter.

Interpretation of the permitivity measurements

Let us consider other aspects of the experiments. The instantaneous value of the real part of the permittivity, calculated from the admittance of the sample and the permittivity measurements, also support the validity of the first order model. In constructing the dehydration

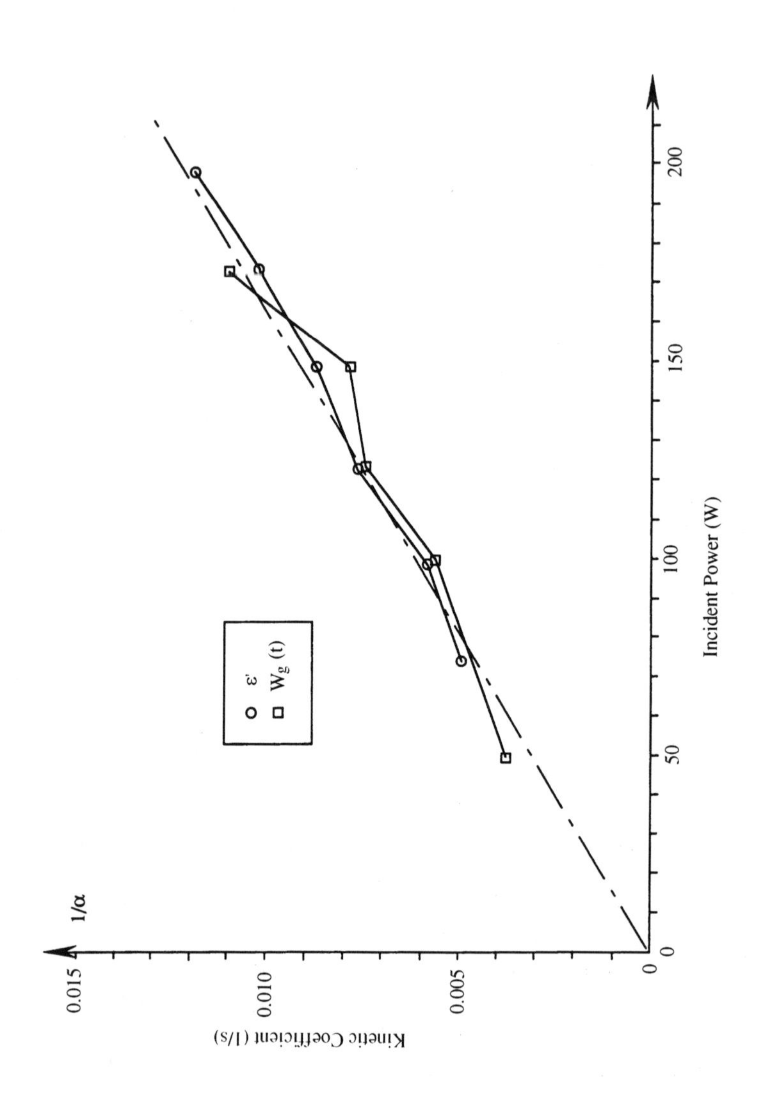

Fig. 10.32 *Kinetic transport coefficient, $1/\alpha$, vs. incident power measured in the zeolite material.*

model, it was supposed that two types of water were contained in the solid, denoted by m_a and m_i. The adsorbed water, m_a, behaves like liquid water. Its permittivity is high (about 60 ε_0) and must be taken into account in order to obtain the effective permittivity of the wet solid. On the other hand, the water for the two other phases can be neglected in the permittivity calculation. From the recorded data, we can determine whether or not the instantaneous complex permittivity of the system is a function of the number of molecules which are present in the zeolite and which type of water molecule(s) dominate the functional behavior. The answer lies in the applicability of the exponential relations between ε_r' (t) or ε_r'' (t) and m_a or m_i, that is:

$$\varepsilon_r'(t) = \varepsilon_r'(\infty) + (\varepsilon_r'(0) - \varepsilon_r'(\infty)) \exp\left(-\frac{t}{\alpha}\right) \tag{10.49a}$$

$$\varepsilon_r''(t) = \varepsilon_r''(\infty) + (\varepsilon_r''(0) - \varepsilon_r''(\infty)) \exp\left(-\frac{t}{\alpha}\right). \tag{10.49b}$$

We have used the technique previously discussed to determine the real and imaginary parts of the zeolite complex permittivity during dehydration, shown in Figures (10.33) and (10.34), respectively. The first relation (10.49a) only approximately fits the experimental data; it does not represent the measured curves of Figure 10.33 particularly well. This curve represents the change of ε_r' during microwave zeolite dehydration. It must be noted that the time constants determined either from the $W_g(t)$ or from the $\varepsilon_r'(t)$ curves is the same and is proportional to the square of the electric field strength, as is shown in Figure 10.32, since $1/\alpha$ is presented as a function of the incident power. The fit shows that the coefficient $1/\alpha$ of the exponential factor is, further, proportional to $(m_a(0) - m_a(t))$. So, the real part of the permittivity is proportional to the mass of adsorbed water, m_a:

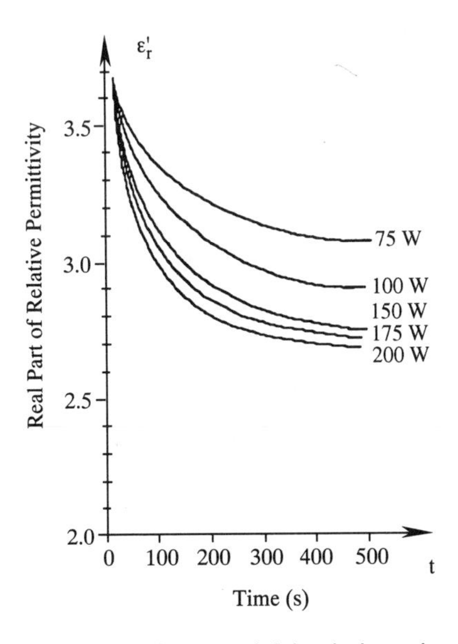

Fig. 10.33 *Measured real part of relative permittivity during microwave dehydration of zeolite material for different incident powers.*

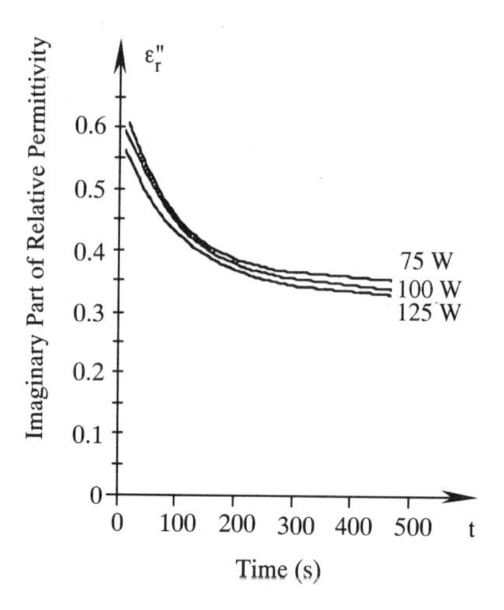

Fig. 10.34 *Measured imaginary part of relative permittivity during microwave dehydration of zeolite material for different incident powers.*

$$\varepsilon_r' = \varepsilon_{rs}' + k' m_a \tag{10.50}$$

where ε_{rs}' is the real part of the relative permittivity ($\varepsilon_{rs}' = 2.3$) of the dry zeolite solid. Figure 10.35, shows the measured real relative permittivity, ε_r', vs. m_a, from which equation (10.50) was derived.

Equation (10.49b) does not give a good description of the experimental values of $\varepsilon_r''(t)$, given Figure 10.34. However, if ε_r'' is plotted versus m_a, it is well represented by a quadratic function:

$$\varepsilon_r'' = \varepsilon_{rs}'' + k'' m_a^2 \tag{10.51}$$

where ε_{rs}'' is the imaginary part of the relative permittivity of the dry solid zeolite and k'' is a fit coefficient. These interpretations of the variations of ε' and ε'' give two independent experimental definitions for the instantaneous adsorbed water m_a during the dehydration. This illustrates once again that permittivity measurements allow one to follow transformations within the system at high power.

Intermediate conclusions

This model of microwave dehydration does not consider heating of the zeolite by the microwaves. It was shown, instead, that microwave dehydration is the result of the electric field which irradiates the adsorbed water molecules. The rate of internal desorption varies as the square of the applied electric field strength and is also proportional to the instantaneous mass of water which is locally present following a first order kinetic process. Thus, transport of the desorbed molecules is correctly represented by a single delay time, τ, which measures the mean time that the free water remains in the porous solid.

The relation:

$$m_i = \tau Q_0 \rho_g(t) \frac{T_g}{T_0} \tag{10.52}$$

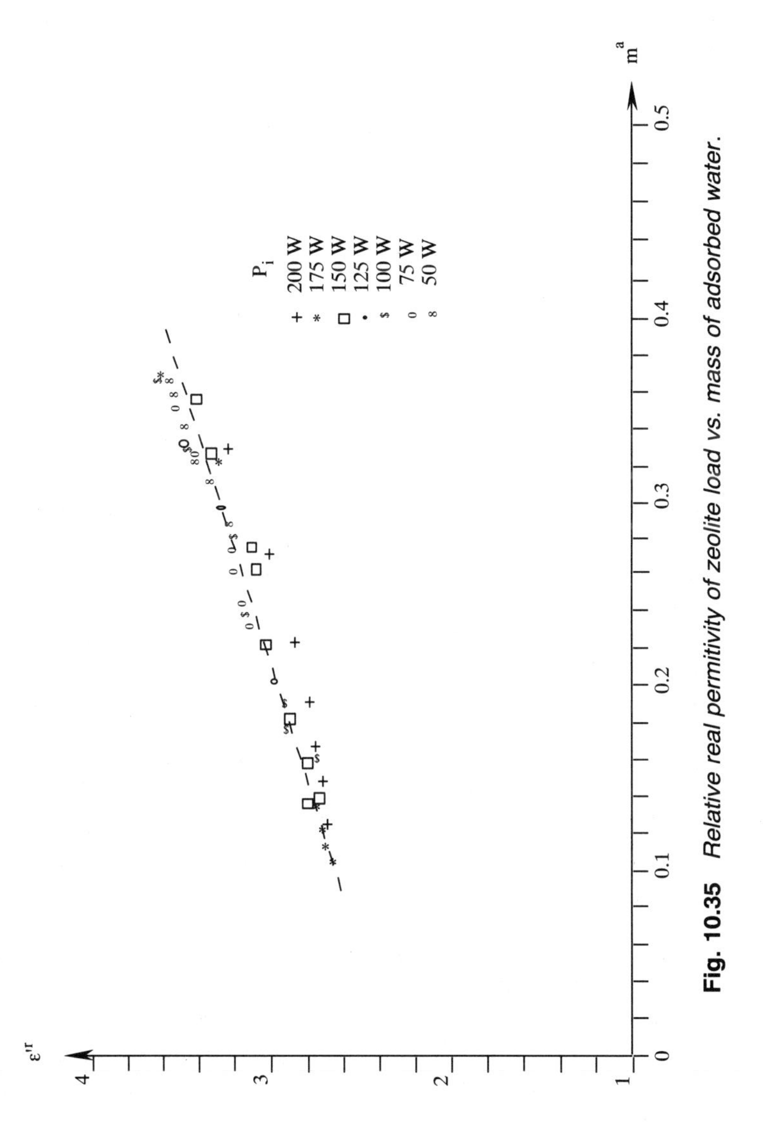

Fig. 10.35 *Relative real permittivity of zeolite load vs. mass of adsorbed water.*

permits the estimation of the instantaneous mass of water in the internal vapor phase which is circulating in the pores. High values of intermediate phase density, up to 70 kg/m^3, may be obtained under the given experimental conditions. Consequentially, it is reasonable to assume that the mobile water is neither a real gas nor a real liquid, but is better described as an intermediate state in which the molecules interact considerably with each other or with the solid. This is in accordance with the meaning of equation (10.52), and can be interpreted as follows: the instantaneous flow of water between the internal and the external gaseous phases is proportional to the difference in densities (i.e. the pressure difference across the zeolite cage):

$$W_\mathrm{i}(t) = W_\mathrm{i}(\rho_\mathrm{i}(t) - \rho_\mathrm{g}(t)) = c_\mathrm{i}\frac{m_\mathrm{i}(t)}{2\tau} \tag{10.53}$$

where W_i is the net mass flow of intermediate phase water and c_i is a constant of proportionality.

In the description of the system it has not been necessary to introduce the temperature of the solid in the relationships. The outlet gas temperature (which is nearly the same as the solid temperature) is used to correctly evaluate the instantaneous mass flow of water which is evacuated. Evidently, this description of the system is a coarse approximation. The solid zeolite is indeed heated by the irradiating field, even if it does not contain water. The dielectric losses of the solid can be explained by the oscillations of the mobile ions within the zeolite under the influence of the microwave field. In this description, the variations of many parameters with the surrounding temperature and with the hydration coverage inside the solid have been ignored since they are essentailly indeterminant. For example, it is known that the specific heat values are usually functions of the temperature. Furthermore, the quantity $\mathrm{d}m_\mathrm{a}/\mathrm{d}t$, defined as the instantaneous dehydration rate cannot have a single value throughout the sample when the temperature of the outlet gas varies from ambient (at the inlet plane, top of the waveguide) to as much as 150°C (at the outlet plane). The temperature dependence must be such that the relationship describes the equilibrium adsorption–desorption state. These observations explain why no simple interpretation of the temperature curves could be given up to this point.

Nevertheless, in spite of these limitations the agreement between the experimental results and the proposed relationships is adequate to justify the model, and to predict the behavior of any similar system. We remember that microwave dehydration is directly obtained from the square of the electric field so the desorption rate of water molecules varies as the square of the applied field strength and the transport of water inside the solid is correctly represented by a first order law.

Returning to the consideration of the significance of permittivity; we have shown that the real part of the permittivity is a linear function of the adsorbed water, and that the imaginary part is quadraticly dependent on adsorbed water. The electric field seems to have a stronger influence on ε^* than the temperature. The measured permittivity was the effective permittivity of the entire bed including the zeolite matrix, spherical particle shape (about 2 mm diameter) and air volume.

With the aid of a mixture formula, Böttcher or Bergman Hanai, (see Chapter 8) it is possible to estimate ε^* of the wet solid zeolite and to again compare the dependence of ε^* on the adsorbed water (mass m_a and volume fraction ξ). Using this approach lead to a similar interpretation of the observations with both mixture formulae: the permittivity of water adsorbed on the zeolite surface was $\varepsilon^* = (27 - \mathrm{j}8)\varepsilon_0$ F/m at 2.45 GHz. It is well known and thoroughly documented that the permittivity of distilled liquid phase water at this frequency is $\varepsilon^* = (78 - j10)\varepsilon_0$. We argue that adsorbed water is probably in a thin film on the zeolite surface and is less free to move than bulk liquid phase molecules owing to surface energy effects. This hypothesis is in accordance with what we already know about the degree of fixation of water molecules on the solid zeolite from thermodynamic studies.

It is clear, then, why the standard thermodynamic models of electromagnetic heating introduced in the first part of Chapter 9 — in which the primary goal was to calculate the transient temperature field — are difficult, if not impossible, to apply. The dehydration process must be modeled by itself and for itself and not through, or as a result of, the temperature variation. When starting with the energy balance (equation (9.1)) the model must reveal the ε^* dependence, the appropriate mixture law, the values of the permittivity and the volume fraction of the adsorbed water. There are just too many parameters to determine from this sort of model. For example, it is not clear at all that the adsorbed water boils at a fixed temperature determined by the saturation pressure (since it has more energy barriers to surmount than bulk quantities of liquid phase water); however, that view might, in fact, constitute a very accurate model if we could determine the transient interstitial pressure within the zeolite cage, but we cannot. The specific heat and vaporization enthalpy of the adsorbed water are not easily estimated. The main limitation in the energy balance approach to this problem is that there are too many sensitive parameters which must be determined by independent experiment to conveniently apply it in this case. In fact, we can often learn what we need to know about the material behavior from a semi-empirical approach and avoid the problems asociated with a detailed thermodynamic model.

10.7 MICROWAVE EVAPORATION OF POLAR LIQUIDS

In spite of the difficulties in assembling accurate thermodynamic models of complex processes, there are many effects which can be explained by considering the evaporation of polar liquids in electromagnetic fields in some theoretical detail. Consider a system such as that represented in Figure 10.36. The relation for the evaporation of the liquid is:

$$\Delta h_{fg} \frac{dm}{dt} = P_a \tag{10.54}$$

where P_a is the net absorbed power, dm/dt is the mass rate of evaporation, and Δh_{fg} is the latent heat of vaporization (J/kg). Equation (10.54) is easily obtained from the first law of thermodynamics.

The first law of thermodynamics describes the equilibrium state of the system and the second law leads to certain restrictions on the rate of evaporation, dm/dt. The first ambiguity in the situation at hand is that we must consider a non-equilibrium situation, i.e. that dm/dt has a net non-zero value. This is why we were careful to assume "near equilibrium" vaporization in our earlier discussion. However, it is interesting to note that, as we saw in the discussion of non-equilibrium phase change in Chapter 9, dm/dt can in fact be quite high without disturbing equilibrium very much.

The problem at hand, however, is to inspect the influence of an electromagnetic field at the liquid–vapor interface. It is of concern to know if the evaporation relation is as simple as it is usually presented, and whether it is sufficient to include only the absorbed power of the applied field, or whether other effects must be considered.

The experiment is as follows: consider a mass of liquid which, under reduced pressure, transfers from a liquid column to a cold bath (Figure 10.36). The cold bath is at a lower

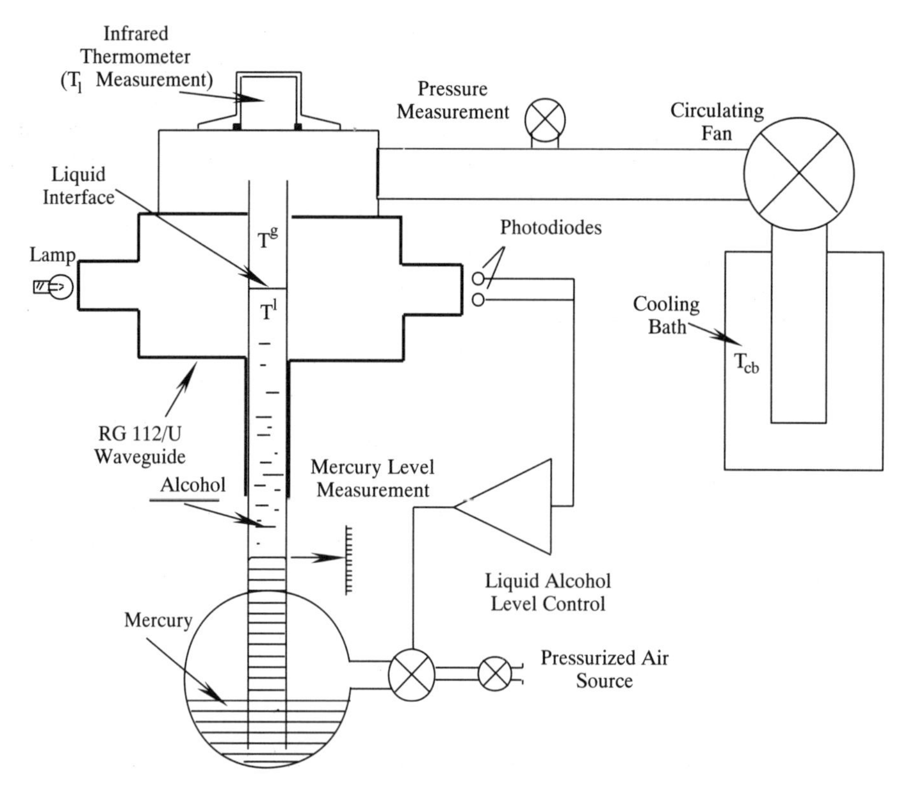

Fig. 10.36 *Apparatus for liquid evaporation studies.*

temperature, T_{cb}, than the liquid column, T_{l}. When $T_{\text{cb}} = T_{\text{l}}$, (thermal equilibrium) there is no net evaporation or transfer of liquid. The temperature of the gas above the surface is also equal to T_{cb} and T_{l}, and the system (gases plus liquids) is in equilibrium. When T_{cb} is lowered, T_{g} decreases and as the liquid evaporates, T_{l} decreases, and the transfer of mass can be observed by measuring how fast the liquid level changes. We use the height of the mercury column to do this while the level of the liquid alcohol is maintained constant by the feedback control circuit. In the case where $T_{\text{cb}} < T_0$, then $T_{\text{g}} > T_{\text{l}}$ and the liquid cools faster than the gas. The difference between T_{l} and T_{g} acts against the evaporation of liquid, so evaporation is an irreversible transformation. As soon as the rate of evaporation increases, the temperature difference between gas and liquid increases such that the evaporation stops. At this limit, the surface of the liquid may freeze. The permanent rate at which the liquid evaporates is a function of T_{l}, T_{g}, and the vapor pressure $\mathcal{P}_{\text{sat}}(T_{\text{l}})$ of the liquid, which is directly connected to T_{cb}.

The governing physical relationship:

$$\overset{\circ}{m} = f(T_{\text{L}}, T_{\text{g}}, \mathcal{P}_{\text{sat}}) \tag{10.55}$$

is difficult to define since thermodynamic events at the interface between the phases consist of irreversible processes. Remember that here we are describing vaporization at the interface rather than nucleation of gas phase bubbles in the liquid. The incremental production of entropy, δS, is the sum of terms for each of the processes, which are the products of forces and conjugated fluxes (J_k).

$$\delta S = \Sigma J_k \delta \Phi_k. \tag{10.56}$$

The Φ_k are the forces driving the corresponding fluxes. At the liquid–gas phase interface the forces include discontinuities in the temperature and in the chemical potential. The corresponding fluxes are the heat and mass flux.

Suppose that the irreversible process can be described as a linear process; that is, the fluxes are linearly dependent upon the corresponding forces:

$$J_k = \sum_{i=1}^{i=n} L_{ki} \delta \Phi_i \tag{10.57}$$

and the phenomena coefficients, L_{ki}, are symmetric over the indices (the Onsager principle):

$$L_{ki} = L_{ik} \tag{10.58}$$

From this relation one can obtain the production of entropy as a linear function of the temperature discontinuity and the chemical potential difference:

$$\delta S = g^* \left(\frac{1}{T_g} - \frac{1}{T_L} \right) + \overset{\circ}{m} \left[\left(\frac{u}{T} \right)_g - \left(\frac{u}{T} \right)_L \right] \tag{10.59}$$

where g^* is the heat flux, and u the chemical potential. The fluxes of heat and mass are also linearly dependent upon the same variables:

$$\overset{\circ *}{g} = L_{11} \left(\frac{1}{T_g} - \frac{1}{T_L} \right) - L_{12} \left[\left(\frac{u}{T} \right)_g - \left(\frac{u}{T} \right)_L \right] \tag{10.60}$$

$$\overset{\circ}{m} = L_{21} \left(\frac{1}{T_g} - \frac{1}{T_L} \right) - L_{22} \left[\left(\frac{u}{T} \right)_g - \left(\frac{u}{T} \right)_L \right]. \tag{10.61}$$

To represent the chemical potential, the mass flux and liquid heat flux are:

$$\overset{\circ}{m} = A \log \left[\frac{\mathcal{P}_{sat}(T_{cb})}{\mathcal{P}_{sat}(T_L)} \right] + B \left(\frac{1}{T_g} - \frac{1}{T_L} \right) \tag{10.62}$$

$$g_L = C \log \left[\frac{\mathcal{P}_{sat}(T_{cb})}{\mathcal{P}_{sat}(T_L)} \right] + D \left(\frac{1}{T_g} - \frac{1}{T_L} \right) \tag{10.63}$$

with the following relationships between the coefficients:

$$-A/r = L_{22} \tag{10.64}$$

$$B = L_{21} - L_{22} h_l \tag{10.65}$$

$$-C/r = L_{12} - L_{22} h_g \tag{10.66}$$

$$D = L_{11} - L_{21} h_l - L_{12} h_g + L_{22} h_g h_l. \tag{10.67}$$

The production of vapor phase in an evaporative system is the sum of two contributions: the first is a function of the temperature difference between T_{cb} and T_l, and the other

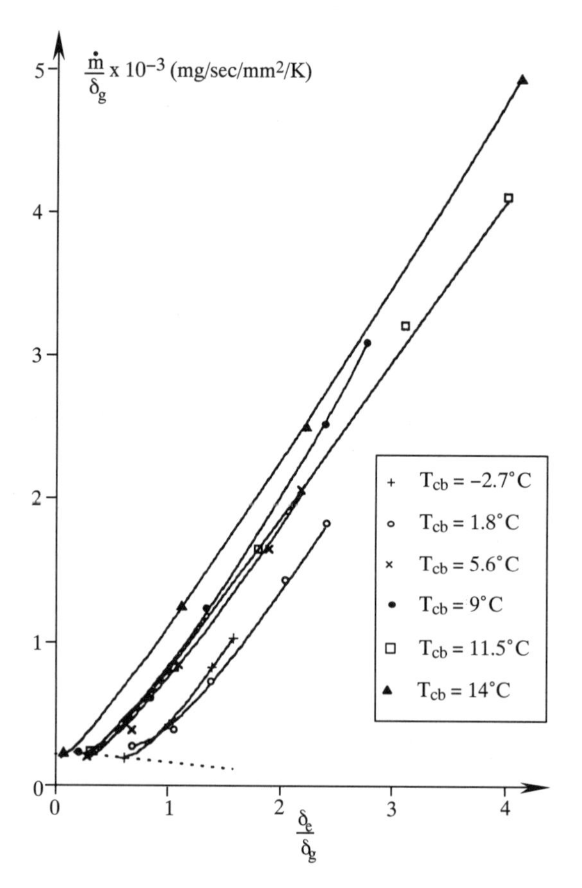

Fig. 10.37 *Microwave enhancement of evaporation at the liquid interface.*

contribution is the reaction of the system, T_{cb} and T_{g}. It should be noted that the production of vapor is a linear combination of both quantities:

$$\delta_{\text{L}} = \frac{1}{T_{\text{cb}}} - \frac{1}{T_{\text{L}}} \tag{10.68}$$

$$\delta_{\text{g}} = \frac{1}{T_{\text{cb}}} - \frac{1}{T_{\text{g}}} \tag{10.69}$$

$$\overset{\circ}{m} = A\delta_{\text{L}} + B\delta_{\text{g}}. \tag{10.70}$$

The influence of an applied electromagnetic field upon these quantities is the next question. The experiment can be done by placing the liquid–vapor interface within a waveguide. The level of the meniscus is maintained automatically in order to compensate for evaporation by pushing liquid upward in a column using air pressure, as shown in Figure 10.36. An optical system detects the meniscus level in order to control the external air pressure. The incident and reflected power levels are measured; and, as before, a moving short circuit is manipulated in order to maintain a constant electric field in the liquid column.

The temperatures of the liquid and of the cooling bath are measured with the aid of thermocouples (outside the waveguide).

With this design we can show whether or not the thermodynamic relations given previously are valid by calculating the coefficients A and B from the measurements of T_{cb}, T_l and T_g. Figure 10.37 shows the results. The quantity $\overset{\circ}{m}/\delta_g$ is presented as a function of δ_l/δ_g. The dashed line corresponds to the zero-field MW experiments. When an external EM field is applied the result of the experiment is that the thermodynamic relation (10.70) is no longer valid. The discrepancy increases as the field strength increases. It is surprising that thermodynamics alone does not give a correct interpretation of the process. The problem is that the electric field provides an additional force impinging upon the surface of the liquid, with a corresponding mass flux. The additional force has a discontinuity which introduces a new term in the production of entropy, which has not been accounted for in the entropy equation which excludes the electromagnetic field. So this relation cannot be used when there is a new source of entropy.

The electromagnetic field provides a generalized supplementary force because a polar molecule near the surface of the liquid experiences a field which is discontinuous. A field normal to the interface has a gradient through the thickness of the fluid–vapor interface since the permittivities of the liquid and vapor phases differ, and liquids always have higher permittivities than their vapor phases. It is usually a very large gradient because the thickness of the interface, in the thermodynamic sense, is extremely small. In an electric field gradient, a polar molecule (dipole moment, $\mathbf{m}$) is subjected to a force couple, $\mathbf{F}_1$ and $\mathbf{F}_2$, which has a torque and a net force ($\mathbf{F}_1 - \mathbf{F}_2 > 0$) since the electric field depends on location. This acts to accelerate molecules from the liquid into the gas (Figure 10.38a) for the polarity shown. Of course, on the other half-cycle of the time-varying electric field the force acts in the opposite direction (Figure 10.38b). It is entirely likely that the reverse process includes additional energy barriers, however, so one might expect that the condensation process would have less mass flux than the vaporization process. This is, at this point, a hypothetical description of the process and by no means represents a satisfying explanation for the observed vaporization enhancement of Figure 10.37.

Obtaining new relations for the average rate of evaporation as a function of the intensity of the field is quite difficult, and has not yet been done. To date we have only made two attempts at data fitting at the Université de Nancy. The first was to try to add a supplementary term proportional to the square of the field to equation (10.70). That result was unsatisfactory, but a fit was nevertheless obtained with a nonlinear mixed term:

$$\overset{\circ}{m} = A\delta_L + B\delta_g + c\pi + A_2\left(\exp - \frac{\tau}{\alpha}\right)\delta_L. \tag{10.71}$$

This new nonlinear term describes mass and temperature coupling to provide an improved fit.

The second technique, which was also successful, was to imagine that the phase transformation flux is enhanced by a driving electromagnetic "pressure" term. The value of the imagined pressure term is determined to obtain the fit. By fitting data numerically we found that the pressure required is a function only of δ_l (see Figure 10.39). Although this approach cannot be justified on a theoretical basis, it does provide a method of estimating the enhancement of evaporation due to the electromagnetic field by introducing a single pressure term dependent on the local temperature. So, we expect that a procedure which utilizes a local pressure in combination with temperature might be a useful model of vaporization enhancement.

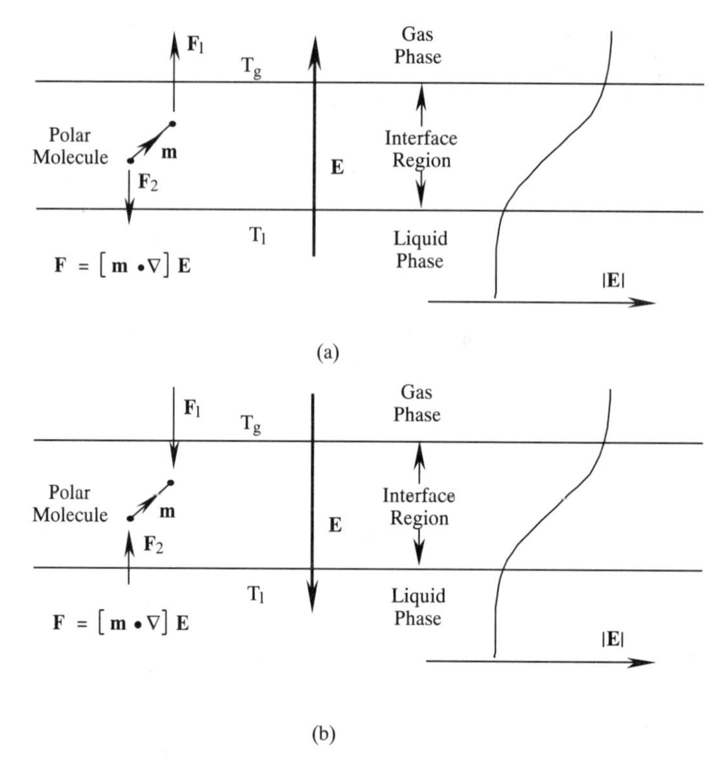

Fig. 10.38 *Detailed view of the liquid–gas phase interface region showing gradient in electric field magnitude due to transition in electric permittivity. (a) Positive half-cycle. (b) Negative half-cycle.*

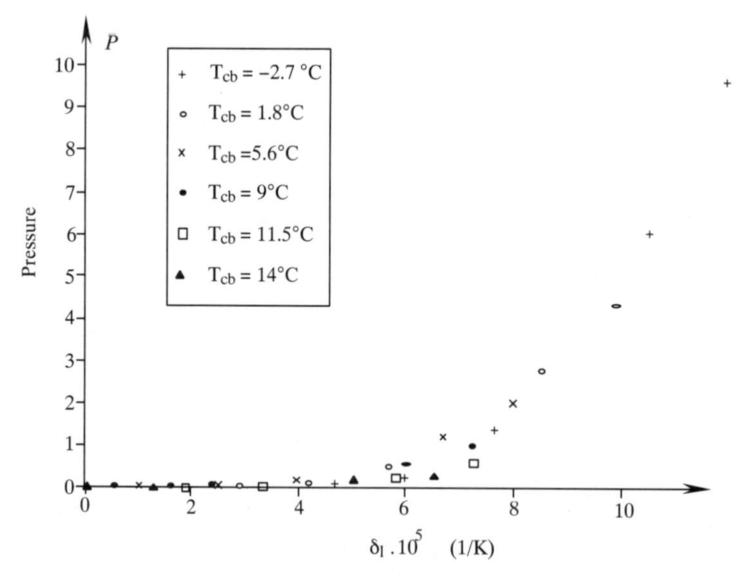

Fig. 10.39 *Representative curve of pressure as a function of δ_1, necessary to fit microwave vaporization measurements.*

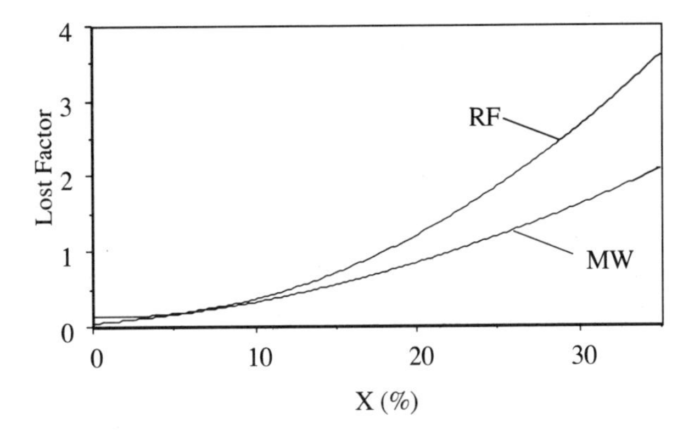

Fig. 10.40 *Variation of loss factor with moisture for paper board at RF and MW frequency.*

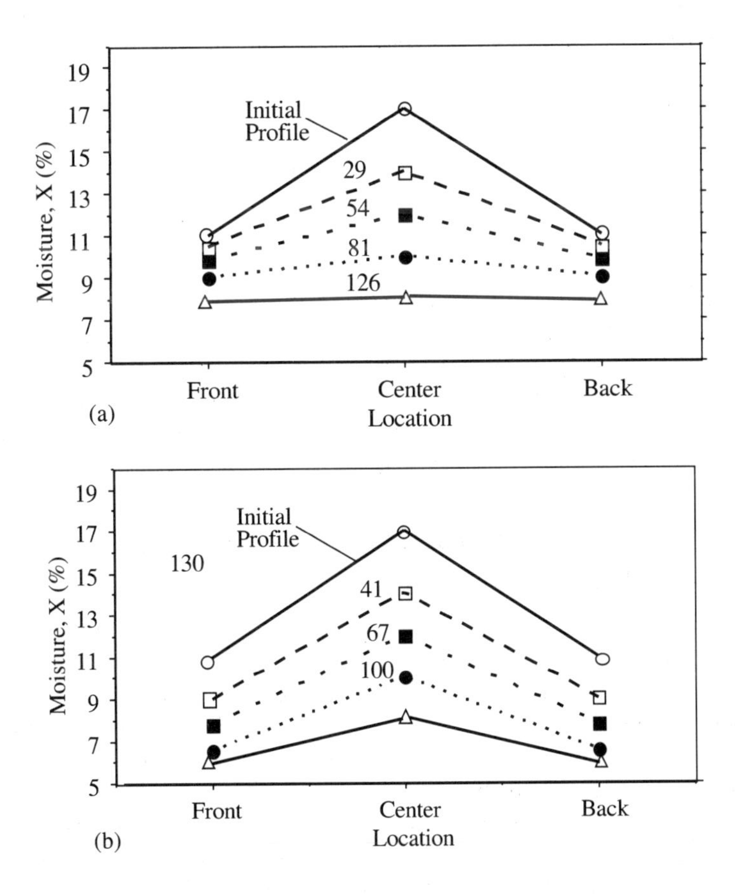

Fig. 10.41 *(a) Moisture leveling in a large radio frequency paper dryer (about 1.5 tons per hour throughput). The initial moisture profile is indicated and the numbers are absorbed RF power in KW. (b) Moisture leveling in a large microwave paper dryer (about 1.5 tons per hour throughput). The initial moisture profile is indicated and the numbers are absorbed MW power in kW.*

10.8 MICROWAVE AND RADIO FREQUENCY MOISTURE LEVELING

In 1974 Jones and Lawton (Electricity Council Research Centre, Capenhurst, UK) and sometime later le Maître (Centre Technique du Papier, Grenoble, France) observed that the relative imaginary permittivity (loss factor, ε_r'') of moist paper or fiber board increases as a nonlinear function of moisture and behaves differently at microwave and radio frequencies. The degree of curvature is more pronounced at RF (27.12 MHz) than at MW frequency (2.45 GHz), as is shown in Figure 10.40. Because the wetter portion of the web has a higher loss factor its drying rate will be higher in an electromagnetic field.

Jones and Lawton experimentally compared the efficiency of MW and RF fields for paper drying in full scale industrial driers (order of 1.5 ton/hour) using standard temperatures and air flow velocities. At both frequencies the electric field was parallel to the plane of the paper. The high frequency fields preferentially dried the wet regions since the substrate is not significantly lossy. The combination of properties is such that moisture leveling is obtained without overheating. The material response is similar to that observed in gypsum for many of the same reasons. The results are summarized in Figure 10.41. A representative incoming moisture profile is shown where the variation in X is between 11% at the edges and 17% in the center of the sheet. The center was subjected to varying RF (Figure 10.41a) or MW (Figure 10.41b) power levels with outlet plane moisture profiles indicated on the respective figures. For this application RF energy is much more effective in obtaining uniform moisture than is MW energy. There is an economic barrier which must be considered, however, created by the very high rates of production in a standard paper process. The capital costs of an RF installation are significant since the water elimination rate (in terms of mass per unit time) must be very large indeed. Also, the operating point for the RF unit is somewhat dependent on the physical characteristics of the particular paper in production. Taking into account these considerations, there is an industrial RF-assisted paper dryer presently being used in production in the south of France.

10.9 INDUCTION HEATING OF GRAPHITE FIBER EPOXY COMPOSITES

The heating of epoxy or thermoplastic composite materials with electromagnetic fields is a topic of current interest. The two major classes of composite materials are thermoset (epoxy) and thermoplastic composites, so-named because of the binding material used. Glass fiber or synthetic fiber composites are typically nearly electrically uniform materials since the permittivities of the epoxy or plastic binders are very similar to that of the fiber material. Usually both are relatively low loss, so radio frequency heating is less practical than microwave heating. Large diameter tubular structures of these types can be illuminated from within by leaky waveguide applicators while smaller diameter tubes can be externally irradiated in a multimode cavity or rectangular or cylindrical waveguide, depending on which mode is desired. The primary problem is temperature feedback since the epoxy curing process is exothermic.

The problems presented by graphite fiber composites are another matter entirely. Graphite fiber–thermoset epoxy resin composites and thermoplastic composite materials are frequently used in aerospace and related industries. Electromagnetic heating may have some advantages over the present autoclave techniques, especially for large objects. However, the loss factor of graphite composites at microwave frequencies (915 MHz and 2.45 GHz) is so high that the depth of penetration is no more than about 0.1 mm. Arcing may be encountered at high heating rates due to strong **E**-fields and inability to control the field polarization at these frequencies. Also, since the curing process is exothermic and the surface is at elevated temperature, thermal runaway may be encountered in the interior spaces and result in over-cured (that is, thermally damaged) material. RF may be readily used to heat graphite fiber composites, however.

10.9.1 RF heating strategies

Because the electrical conductivity of the graphite fibers is so high, **E**-field coupling is not practical — arcing often occurs and the coupling efficiency is poor even in the best of cases. The graphite fiber composites are excellent candidates for induction heating. Coupling the energy through a magnetic field has the advantage that the electric field is low everywhere, virtually eliminating arcs. Also, at these lower frequencies there are instances where the field polarization can be matched to the geometric boundary conditions of several important structural shapes in an advantageous way, as we mentioned in Chapter 4.

Induction heating mechanism

We have investigated the heating obtained at radio frequency (nominal 27 MHz) in magnetic induction fields in both experiments and numerical models. Inductive coupling has the advantage that arcing is rare even at high heating rates for typical cylindrical geometries. A possible benefit of this approach is that the heating time and the time at temperature may be reduced because heat is deposited volumetrically rather than being limited by surface coverings. Transient temperatures were recorded during heating by optical fiber probes and heating distribution was measured, where possible, using thermographic imaging. Finite difference electrical and thermal model results illustrate the dominance of conduction heat transfer in these materials.

The thermoset epoxy composite problem which will be discussed here is the heating of an arbitrarily long cylinder (inner radius a and outer radius b) in a "long" solenoid coil, Figure 10.42a. The coil is long in an RF sense if its length, L, is sufficiently greater than its outer diameter that we may assume an approximately uniform axial magnetic field, $\mathbf{H} \cong H_z \mathbf{a}_z$ under the coil. The coil may be moved coaxially along the cylinder to obtain zone heating, and thus need not be as long as the cylinder itself to be useful in curing the cylinder. We have already discussed induction heating in the cylindrical geometry in a long solenoid coil; see Chapter 4, Section 4.3. Briefly, the axial magnetic field induces a circumferential electric field which is parallel to the cylinder surface. The electric field coexists with induced current density according to Ohm's law. The induced currents act to oppose the external magnetic field, so the presence of the cylinder results in reduced field strengths compared to the unloaded coil. The radial dependence of the induced electric field is described by Bessel functions, as has been discussed (see Chapter 4). The fields of Figures 4.23 and 4.24 were used in a thermal model of the heating process.

Thermal energy balance

Temperature distributions within the cylinder are induced by the applied RF heating and moderated by heat transfer from the cylinder to the mandrel and bleeder cloth–vacuum bag covering (see Figure 10.42b). The mandrel is the substrate on which the cylinder is wound

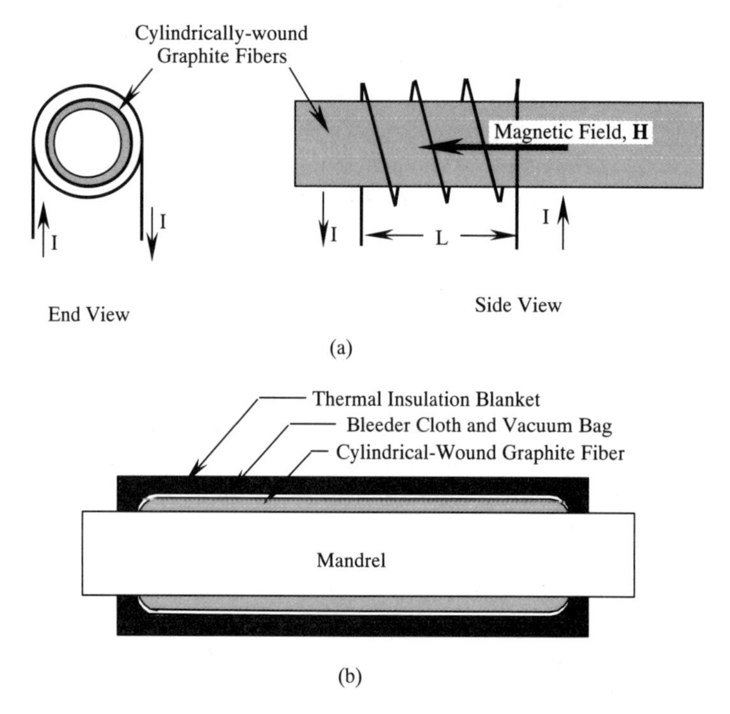

Fig. 10.42 *Induction heating of an epoxy-graphite cylinder. (a) Induction coil geometry. (b) Thermal layers comprising the complete cylinder.*

and establishes the inner radius, a. The bleeder cloth absorbs epoxy resin which exudes from the cylinder surface as the composite is heated and its viscosity decreases. The vacuum bag is used to remove vapors released during curing of the epoxy. Finally, an external insulation blanket of fiber glass was used to control heat losses to the room temperature surroundings. The numerical model includes radiation and convection at the insulation surface. The first law energy balance is thus:

$$\rho c \frac{\partial T}{\partial t} = \nabla \cdot (k \nabla T) + Q_{\text{gen}}(r) + Q_{\text{cure}} + Q_{\text{surface losses}}. \tag{10.72}$$

Above the cure temperature of the epoxy binder, additional power generation comes from the reaction enthalpy, Q_{cure}. Surface losses occur at the outer layer of the insulation.

Including reaction heat effects

Adding a curing process to the thermal model, whether endothermic or exothermic, requires additional knowledge of the kinetic coefficients, A and E, for the curing reaction. The remaining concentration of uncured epoxy at the end of a Δt step, i.e. at time $t_0 + \Delta t$, may be determined from the Arrhenius rate relation (see Chapter 9):

$$\ln\left[\frac{C(t_0)}{C(t_0 + \Delta t)}\right] = \int_{t_0}^{t_0+\Delta t} A \exp\left(-\frac{E}{RT}\right) \mathrm{d}t. \tag{10.73}$$

The mass of uncured epoxy which transforms, Δm (kg/m^3), is then determined from the concentration change:

$$\Delta m = M[C(t_0 + \Delta t) - C(t_0)] \tag{10.74a}$$

where M is the molecular weight, and Δm is subsequently used to calculate the heat liberated during the time step, Q_{cure}:

$$Q_{cure} = \Delta h_c M[C(t_0 + \Delta t) - C(t_0)] \tag{10.74b}$$

where Δh_c is the curing reaction enthalpy (> 0 for an exothermic reaction). This heat contributes to the energy balance of equation (10.72).

10.9.2 Comparison of experimental and numerical model results

A one-dimensional finite difference model of 45 total nodes was created for the cylinder heating problem — 21 nodes for the polypropylene mandrel (20.5Δr node spaces since the center line may not have a node), 15 nodes (14Δr spaces) for the cylinder and 10 node spaces for the insulation, soft rubber (1 node space) plus fiber glass (9 node spaces). The total power deposited in each cylinder node space was determined by trapezoidal integration of the Q_{gen} term. Thermal properties for the polypropylene mandrel composite cylinder and insulation blanket are shown in Table 10.3. The forward finite difference model was executed with a time step of 0.01 s to satisfy the standard FDM stability criterion — the maximum value of Δt which could be used was imposed by the composite cylinder properties. Curing heat was not included in the model since the kinetic coefficients were not available for the epoxy used. Shorter time steps are required for higher thermal diffusivity materials and smaller node separations, as we discussed in Chapter 4.

Model studies were performed on a simulated hollow cylinder of hand-wrapped pre-preg cloth (Hercules Magnamite AW193PW/3501-6) with length 15 cm, and radii $a = 1.27$ cm and $b = 2.47$ cm for comparison with experiment and with convection heating alone. The electrical conductivity of the cylinder was not determined independently. However, published values range between 12 000 and 36 000 S/m at 40 MHz for the plain weave cloth (Lind *et al.* 1990). We used $\sigma = 10\,000$ S/m in the initial calculations. It is very likely that the effective conductivity of our hand-wrapped specimen is even lower than the published values for the cloth alone, so an additional calculation for $\sigma = 1000$ S/m was also performed.

A short discussion regarding the absorbed power is warranted since considerable uncertainty accrues from several sources. Even though the coil current was measured with a Moebius loop (see Chapter 5) we cannot precisely determine the applied magnetic field strength. More importantly, the effective electrical conductivity of the cylinder is very uncertain. Consequently, the model heat generation term had to be scaled to match model predictions with experimental observations. In order to match the model heating rate with the experiment at 10 000 S/m, an external field of 302 A/m would be required, while at

Table 10.3 *Thermal properties of model materials.*

	Conductivity (W/m K)	Specific Heat (J/kg K)	Density (kg/m^3)
Mandrel	0.138	1930	900
Composite	1.1	935	1400
Insulation	0.038	835	32
Soft Rubber	0.13	2010	1100

1000 S/m effective conductivity an external applied field of 168 A/m would be required. We believe that the lower match field strength corresponding to the lower electrical conductivity is closer to the actual situation in this experiment. More power is coupled to the cylinder load at a lower effective electrical conductivity for reasons discussed in Chapter 4 (see Figure 4.25). The 15 cm long cylinder was placed in a 3.5 turn solenoid coil with inner radius 5 cm at 27 MHz. The coil current in the experiment was measured to be 1.89 A rms from the Moebius loop sensor. This would correspond to an external applied field strength of $H_0 = 47$ A/m, so it is very likely that the actual effective electrical conductivity of the composite cloth is less than 1000 S/m.

Figure 10.43 summarizes the numerical model results. In (10.43a) there is a comparison between heating rates predicted at an optical temperature point sensor location in the middle of the composite cylinder and the experimental results, from which the two assumed magnetic fields were derived. The uncertainty in electrical conductivity is of less importance than might at first be thought since the temperature profiles (Figure 10.43b) at 168 A/m (i.e. for $\sigma = 1000$ S/m) are virtually identical to those of the higher electrical conductivity case at 302 A/m ($\sigma = 10\,000$ S/m). So, the power density profile makes little difference for heating times used in this experiment—total absorbed power dominates the results. Also,

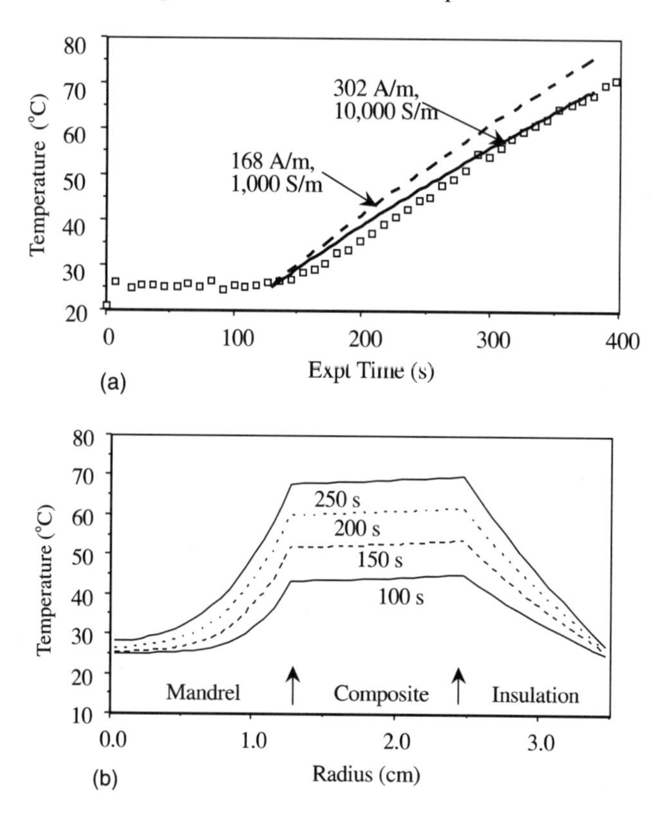

Fig. 10.43 *(a) Simulated heating record for temperature probe on mandrel surface at conductivities of 1000 and 10 000 S/m compared to measurements. The measuring probe was placed at mid-thickness in the cylinder wall. (b) Simulation results in small cylinder at times between 100 and 250 s for a magnetic field at 302 A/m.*

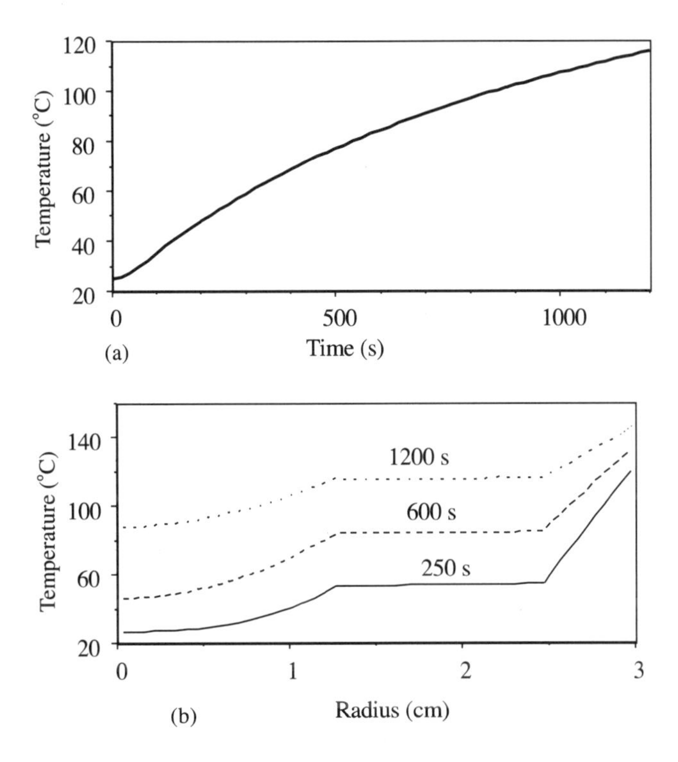

Fig. 10.44 *(a) Simulation of convection heating temperature record for heating the same cylinder as that modeled in Figure 10.41(a) at the probe location. (b) Simulated temperature profiles for convection heating of composite cylinder.*

at all model times the thermal gradient within the cylinder is negligible compared to that in the surrounding materials. This is due to the high thermal conductivity of the composite relative to the other materials.

For comparison, a convection heating model result for the same cylinder is shown in Figure 10.44. The surrounding temperature was fixed at 160°C in the convection calculations, just above the cure temperature in this material. In this model, the insulation was reduced to include only 0.5 cm of vacuum bag and bleeder cloth, as would be used in a convection curing oven, to make a fair comparison. Thermal properties for soft rubber were used to give a conservative estimate of heating rate (i.e. minimal insulation thermal delay). Even so, a much longer heating time is required to reach desired temperatures (600 s as opposed to 250 s to reach 70°C). As expected, the rate of temperature rise at the probe site (Figure 10.44a) decreases markedly as the cylinder approaches the surrounding temperature. The inverted spatial temperature profiles in the insulation (Figure 10.44b) are also apparent.

10.10 SUMMARY

This chapter has illustrated some of the difficulties of modeling transformations driven by an electromagnetic field. The two parts of the complex permittivity have different dominant interactions: the real part determines the electric field (for moderately lossy materials) and the imaginary part is the forcing function for the energy balance. Plainly, the complex permittivity is an electrical state variable (in the thermodynamic sense) of the material.

The value of interest, however, is not necessarily the value measured at low power, which we considered when discussing cluster theory in Chapter 8. When the material is undergoing a transformation — this is by definition a non-equilibrium process in time and/or in space — the complex permittivity must be determined by special experiments such as we have described in this chapter. For some materials, such as the zeolites, the complex permittivity depends only on the adsorbed water molecules and not on the molecules in the zeolite structure. Other complex systems will probably exhibit different behavior. In some cases the transformation may not change the value of the permittivity in a measurable way — as in the thermoset composite example where the permittivity is dominated by the graphite fibers, which do not transform. These transformations may be treated as quasi-equilibrium processes (in an electrical sense if not thermodynamically). In all cases, whether the transformation significantly affects the complex permittivity or not, the methods we have proposed can be used to study it.

REFERENCES

Chen, P. and Schmidt, P. S. (1990) An integral model for drying of hygroscopic and nonhygroscopic materials with dielectric heating *Drying Technology*, **8** 907–930.

Chen, P. and Pei, D. C. T. (1989) A mathematical model of drying processes *Int. J. Heat and Mass Transfer*, **32**, 297–310.

Greenkorn, R. A. (1981) Steady flow through porous media, *AIChE J.*, **27**, 529–545.

Grolmes, J. L. and Bergman, T. L. (1990) Dielectrically-assisted drying of a nonhygroscopic porous material *Drying Technology*, **8**, 953–975.

Jones, P. L. and Lawton, J. (1974) Comparison of microwave and radio frequency drying of paper and board *J. of Microwave Power*, **9**, 109–115.

Keey, R. B. (1978) *Introduction to Industrial Drying Operations* Pergamon Press, Oxford.

Lefeuvre, P. S., Mangin, B. and Rezvan, Y. (1978) Industrial materials drying by microwave and hot air *Proc. IMPI Microwave Power Symposium*, pp 65–67.

Lind, A. C., Fry, C. G., and Sotak, C. H. (1990) *J. Appl. Phys.*, **68**, 3518–3528.

Lyons, D. W., Hatcher, J. D. and Sunderland, J. E. (1972) Drying of a porous medium with internal heat generation, *Int. J. Heat and Mass Transfer*, **9**, 897–905.

Metaxas, A. C. and Meredith, R. J. (1983) *Industrial Microwave Heating* Peter Perigrinus, London.

Pearce, J. A. and Faulkner, C. M. (1992) RF Processing of Thermoset and Theromoplastic Composites *Microwave Processing of Materials III* (Beatty *et al.* Eds.) *Proc. Materials Res. Soc.*, **269**, 397–408.

Perkin, R. M. (1990) Simplified modeling for the drying of a nonhygroscopic capillary porous body using a combination of dielectric and convective heating *Drying Technology*, **8**, 931–951.

Roussy, G., Zoulalian, A., Charreyre, M. and Thiebaut, J. M. (1988) How microwaves dehydrate zeolites *J. Phys. Chem.* **88**, 5702.

Schmidt, P. S., Bergman, T. L. and Pearce, J. A. (1992) Heat and mass transfer considerations in dielectrically enhanced drying *Drying '92* (A. S. Majumdar, Ed.) Elsevier, Amsterdam, pp 137–160.

Stanish, M. A., Schajer, G. S. and Kaythan, F. (1986) A mathematical mode of drying for hygroscopic porous media *AIChE J.*, **32** (8), 1301–1311.

Strumillo, C. and Kudra, T. (1986) *Drying: Principles, Application and Design* Gordon and Breach, New York.

Thiebaut, J. M., Colin, P. and Roussy, G. (1983) Microwave enhancement of the evaporation of a polar liquid, I and II *J. Thermal Analysis*, **28** 37 and 49.

Wei, C. K., Davis, H. T., Davis, E. A. and Gordon, J. (1985) Heat and mass transfer in water-laden sandstone: microwave heating, *AICHE J.*, **31**, 842–848.

Whitaker, S. (1985) Moisture transport mechanisms during the drying of granular porous media *Drying'85* (A. S. Majumdar, Ed.), Hemisphere, New York, pp 21–32.

11 Electromagnetic Processing of Heterogeneous Materials at High Power Density

11.1 INTRODUCTION

In the previous chapter, drying of wet paper and dehydration of wet zeolites were studied. Although they are, in fact, heterogeneous materials, we have treated them as approximately homogeneous. Even so, it was seen that the absorbed power is a complicated function of the electric field. This is because the permittivity of heterogeneous materials depends in a complicated way on the constitution of the material and its microscopic and macroscopic organization. For example, in wet paper the real part of the permittivity is a linear function of the humidity whereas the imaginary part of the permittivity is a quadratic function. In other cases, where the heterogeneity is more pronounced, such simple dependence does not occur. Heterogeneous materials are nevertheless important and are very often found in industrial applications. This chapter will discuss three applications of EM heating: (1) the separation of ores by floatation, (2) neopentane pyrolysis, and (3) microwave acid dissolution of solids.

First, we consider a mixture of the minerals fluorite, barytite and silica. In a crushing process this mixture of minerals is pulverized and dispersed in water. A collector, such as high molecular weight oil, is used to obtain the desired mineral by fixation to the surface of one of the ore minerals, for example fluorite, making the surface of the mineral hydrophobic. Air bubbled through the liquid with the aid of a fritted disc allows oxygen to fix on the surface of the hydrophobic particles, which then float to the surface as a concentrated foam. The minerals are thus separated. This is one of the most common industrial processes for the separation of low concentration minerals from ores. In the laboratory, the process can be studied with a Hallimond tube, as shown in Figure 11.1. At the industrial level, a large vessel is used, and the foam is collected from the surface of the fluids, as in Figure 11.2. It will be shown here that the collection of minerals by floatation is a function of the pH, the temperature, the composition of the fluids (concentration of the collector), and is also a function of an applied electromagnetic field. The ore floatation example illustrates the thermal dependencies in such a system.

11.2 THERMAL AND ELECTROMAGNETIC ASPECTS

When an RF field is applied to the floatation bath, the first assumption was that the water was not heated. This is reasonable because a 27 MHz RF field was used in the demonstration

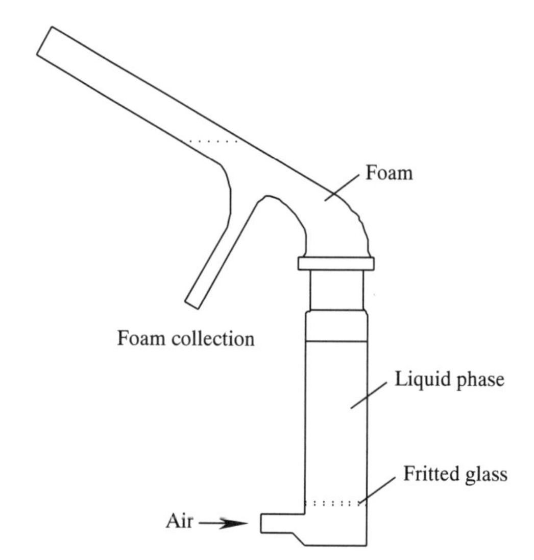

Fig. 11.1 *Hallimond tube.*

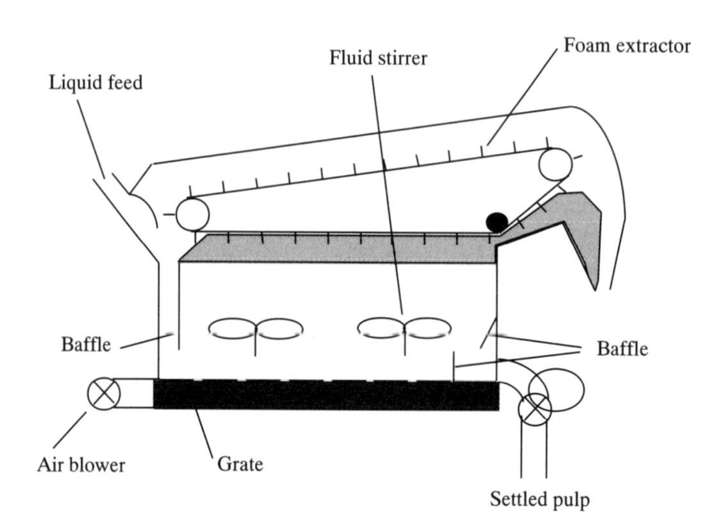

Fig. 11.2 *Flotation tank.*

experiments, the ion concentration is low in the bath, and the absorption of electromagnetic energy by water is only significant at much higher frequencies. The ore particles are sufficiently conductive to be preferentially heated in this field.

Consider a given volume of material in which the dispersion of particles of material 1 into 2 is homogeneous. Let us define an average temperature for each of the materials: T_1 for the particles and T_2 for the liquid. Thermal exchange between the two phases is given by:

$$m_1 c_{p_1} \frac{dT_1}{dt} + h(T_1 - T_2) = P_a(t) \qquad (11.1)$$

where m is the mass, c_p the specific heat, h is the coefficient of heat transfer, between particles and liquid; and P_a is the absorbed power in the mass. This relation assumes that only material 1 is heated. Material 1 (the particles) exchanges heat with material 2, and the heat transfer coefficient h is a function of the surface area of material 1. The second relation is:

$$m_2 c_{p_2} \frac{dT_2}{dt} + h(T_2 - T_1) = 0. \tag{11.2}$$

By combining the two equations and then eliminating T_2, we obtain a second order ordinary differential equation with constant coefficients for calculating T_1:

$$m_2 c_{p_2} m_1 c_{p_1} \frac{d^2 T_1}{dt^2} + h(m_1 c_{p_1} + m_2 c_{p_2}) \frac{dT_1}{dt} = hP_a(t) - \frac{m_2 c_{p_2}\, dP_a(t)}{dt}. \tag{11.3}$$

The second term is a function of the absorbed power from the applied field. For a step absorbed power, this equation leads to:

$$T_1 = T_0 + \frac{P}{h}\left(\frac{m_2 c_{p_2}}{m_1 c_{p_1} + m_2 c_{p_2}}\right)^2 \left[1 - \exp\left(\frac{-h(m_1 c_{p_1} + m_2 c_{p_2})t}{m_1 c_{p_1} m_2 c_{p_2}}\right)\right] + \frac{Pt}{m_1 c_{p_1} + m_2 c_{p_2}}. \tag{11.4}$$

T_2 can then be determined by equation (11.1) which leads to

$$T_2 = T_1 - \frac{P m_2 c_{p_2}}{(m_1 c_{p_1} + m_2 c_{p_2})h}\left[1 - \exp\left(\frac{-h(m_1 c_{p_1} + m_2 c_{p_2})t}{m_1 c_{p_1} m_2 c_{p_2}}\right)\right]. \tag{11.5}$$

The curves of $T_1(t)$ and $T_2(t)$ are given in Figure 11.3. It can be seen that the difference between T_1 and T_2 approaches a steady state at longer times. This is a consequence of the

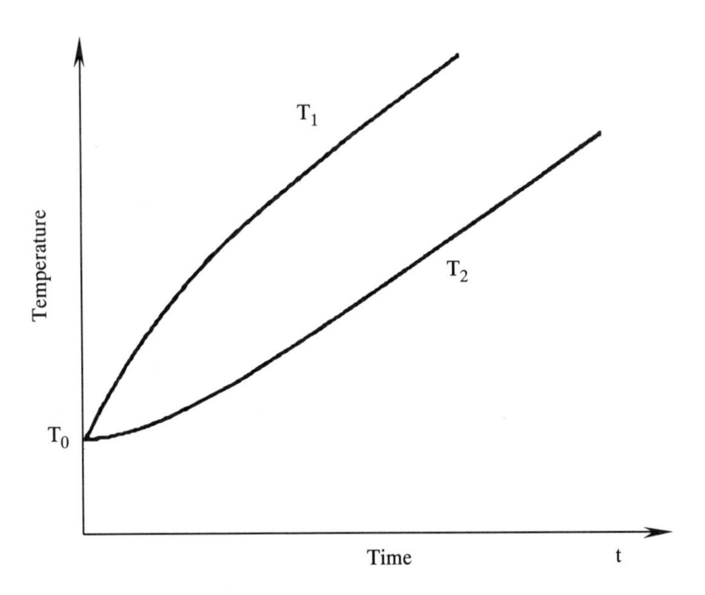

Fig. 11.3 *Temperatures of the two mixture components neglecting heat transfer with the surroundings.*

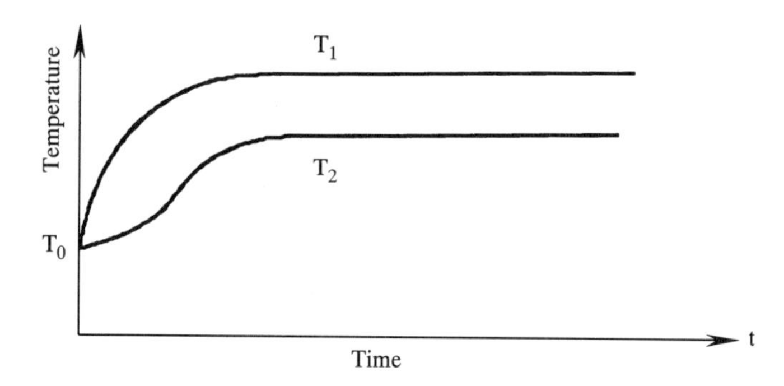

Fig. 11.4 *Temperatures of the two components when heat transfer with the surroundings is significant.*

fact that T_1 is induced by the field which heats material 1 alone. This temperature difference is a function of the specific heat, the mass, and the heat transfer between the materials.

This is a very simple model of the physical situation since it was assumed that the system was adiabatic — that heat transfer with the surroundings was negligible. In real systems, the liquid will exchange heat with its surroundings. Equation (11.2) must have an additional term, $H(T_2 - T_0)$, which describes the amount of heat transferred to the surroundings:

$$m_2 c_{p_2} \frac{dT_2}{dt} + h(T_2 - T_1) + H(T_2 - T_0) = 0. \tag{11.6}$$

In this formulation, H is a global heat transfer coefficient. The structure of the mathematical system is not modified. The solution of the equations is obtained as before. Figure 11.4 shows the time dependence of the temperature for a non-adiabatic system. There is still a steady state temperature difference between T_1 and T_2 after a certain time.

In simple cases such as this, where an absorbing material dispersed in a dielectric is heated by a microwave or RF field, the dielectric heating maintains a permanent temperature difference between the absorbing material and the suspending fluid. This temperature difference can be verified easily using known values of mass, specific heat and the measured heat transfer. For example, when small ore particles are considered, a temperature gradient of 40°C can be easily obtained with an absorbed power of 1kW per kg of material. This is equivalent to power densities provided by a microwave oven. This situation is advantageous for industrial applications because most of the energy introduced by means of the electromagnetic field selectively heats the particles and not the liquid phase. This situation is favorable for inducing a chemical reaction on the surface of the particles.

Before continuing the discussion of these aspects, consider a mixture of an electromagnetic absorbent material dispersed in a liquid which is also likely to be absorbent: permittivities ε_1^* and ε_2^*. We have discussed how the equivalent permittivity of the mixture can be estimated from the particle sizes and volumetric proportions of material 1 dispersed in material 2. The previous chapter has shown that, though the theories for dispersed media were not entirely adequate, correct results can be obtained with some adjustment. Knowledge of the permittivity of the mixture is not sufficient for the calculation of the portion of the incident energy absorbed by each component. The intensity of the electric field, localized for each material, must be known. Suppose that the intensity ($|E|^2$) is constant with spatial average magnitude $\langle E_1 \rangle$ for material 1 and $\langle E_2 \rangle$ for material 2, and that the local passive and active energy in both media are conserved, as in a homogeneous medium. The proportion of the spatial average electric field experienced by each material can be obtained from:

$$\varepsilon'\langle E^2\rangle = \varepsilon_1'(1-\Phi)\langle E_1^2\rangle + \varepsilon_2'\Phi\langle E_2^2\rangle \tag{11.7}$$

$$\varepsilon''\langle E^2\rangle = \varepsilon_1''(1-\Phi)\langle E_1^2\rangle + \varepsilon_2''\Phi\langle E_2^2\rangle \tag{11.8}$$

where $\langle E_1^2\rangle$, $\langle E_2^2\rangle$ and $\langle E^2\rangle$ are the electric field intensities in material 1, material 2, and the total material, respectively. This simple model leads to two important results. First, the proportion of absorbed power by the desired product in the dispersed medium can be estimated. Second, the temperature of each product can be estimated. The advantage of RF heating is that it establishes a temperature difference between the bath and the desired mineral. The model considers the influences of all parameters engaged in the system, such as the dispersion, the quantity of material 1 dispersed in material 2, and their permittivities.

As a consequence of the Bergman–Milton theory, which expresses the effective complex permittivity ε^* of a two-phase mixture, with an integral of a density function

$$\varepsilon^* = (1-A)\varepsilon_1^* + A\varepsilon_2^* + \varepsilon_1\int_0^1 \frac{f(x)\,\mathrm{d}x}{x + \varepsilon_1^*/(\varepsilon_2^* - \varepsilon_1^*)} \tag{11.9}$$

it is possible to evaluate the local electric field intensity inside any heterogeneous mixture. After some algebraic manipulation:

$$\frac{\langle E_1^2\rangle}{\langle E\rangle} = \frac{1}{(1-\Phi)}\left[1 - A + |\varepsilon_2^* - \varepsilon_1^*|^2\int_0^1 \frac{xf(x)\,\mathrm{d}x}{|(1-x)\varepsilon_1^* + x\varepsilon_2^*|^2} - \varepsilon_1^*\int_0^1 \frac{f(x)\mathrm{d}x}{|(1-x)\varepsilon_1^* + x\varepsilon_2^*|^2}\right] \tag{11.10}$$

and

$$\frac{\langle E_2^2\rangle}{\langle E\rangle} = \frac{1}{\Phi}\left[1 + |\varepsilon_1^*|^2\int_0^1 \frac{f(x)\,\mathrm{d}x}{|(1-x)\varepsilon_1^* + x\varepsilon_2^*|^2}\right]. \tag{11.11}$$

(In these relations, $|u^*|$ denote the magnitude of the complex number u^*.)

The same solutions apply to the case of textile dying, where an advantage is to be realized by heating the dye particles preferentially with respect to the bath. Although dispersed media theories are not especially precise, they are sufficient for delineating this problem and estimating the value of properties which require consideration for industrial application.

11.3 PHYSICAL-CHEMICAL ASPECTS OF HETEROGENEOUS MATERIALS

We have seen that a heterogeneous system irradiated by an electromagnetic field will result in a heterogeneous distribution of temperature. Previous sections described the methods for modeling this temperature distribution. The systems which will be examined in this section will illustrate that the previous model is inadequate to describe the physical-chemical aspects of the transformation of heterogeneous matter by an electromagnetic field. Processes at the interfaces of the heterogeneous medium must be more precisely described. Three different systems will be discussed.

11.3.1 Floatation under electromagnetic irradiation

All floatation processes involve, principally, the electrochemical reaction at the interface between the ore particles and the solution in which they are suspended. The irradiation of this system by an electromagnetic field gives rise to heterogeneously dissipated energy, and is very useful when the energy is preferentially dissipated near the interface to selectively modify the floatation process. The experimental results discussed here cannot be described either in terms of thermal gradients or non-thermal effects.

Consider an ore powder containing fluorite dispersed in water with an oil collecting solution added. The concentration of the collector in the solution was varied between 0 and 10^{-4} mol l^{-1}. The mixture was placed in an irradiation cell, a Hallimond tube as in Figure 11.1. Figure 11.5 shows the RF curved electrode capacitor used to expose the mixture. An intense, high frequency field was applied during the fixation of the collector on the fluorite particles. Two types of anionic collectors were used: sodium oleate (SO) and sodium dodecylsulfonate (SDDS). Air was bubbled into the cell from the base creating a foam which contained the fluorite. The foam was collected from the top of the solution. Identical floatation experiments were performed using classical heating methods at the same temperature or at higher temperatures. The results are summarized in Figures 11.6 and 11.7.

In the case of floatation of fluorite by sodium oleate, the classical heating shifts the floatation curve toward the lower concentrations of collector (curve labeled 2 in Figure 11.6) by an improvement of collector fixation. The high frequency irradiation has the same effect but is more efficient. For an equal energy input, where the mean temperature is the same for both types of heating, a higher recovery was obtained (curve 3) because the temperature of the ore particles is higher.

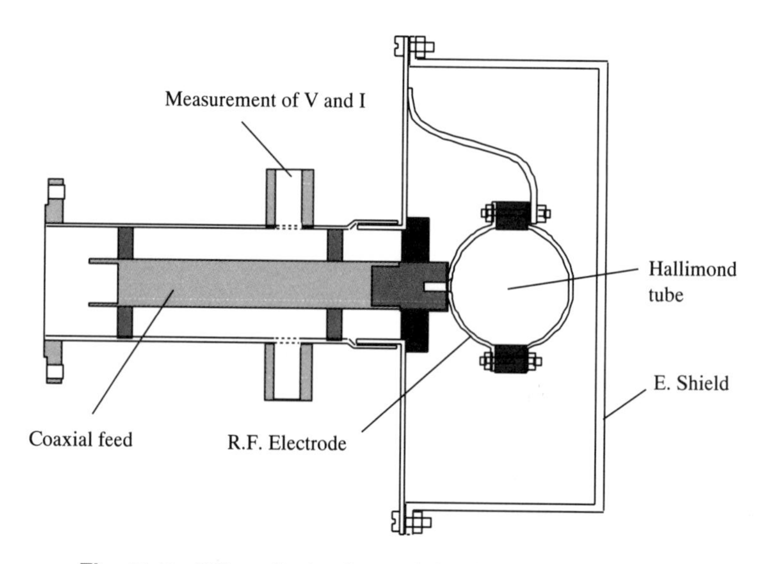

Fig. 11.5 *RF applicator for studying floatation processes.*

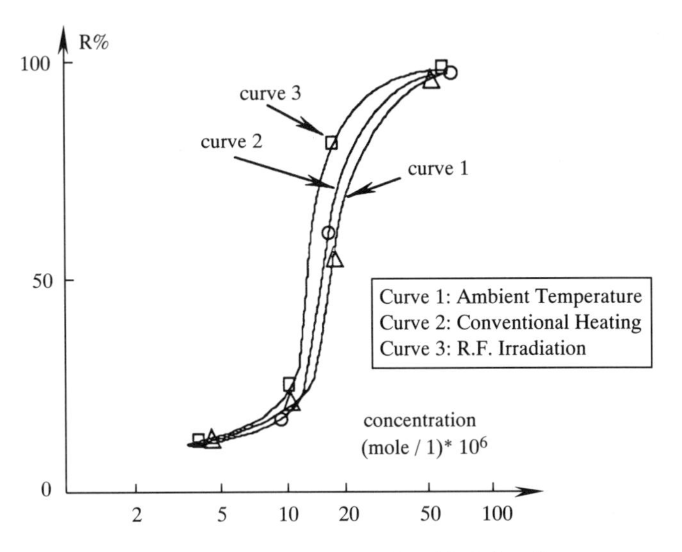

Fig. 11.6 *Fluorite recovery, R, with SO collector.*

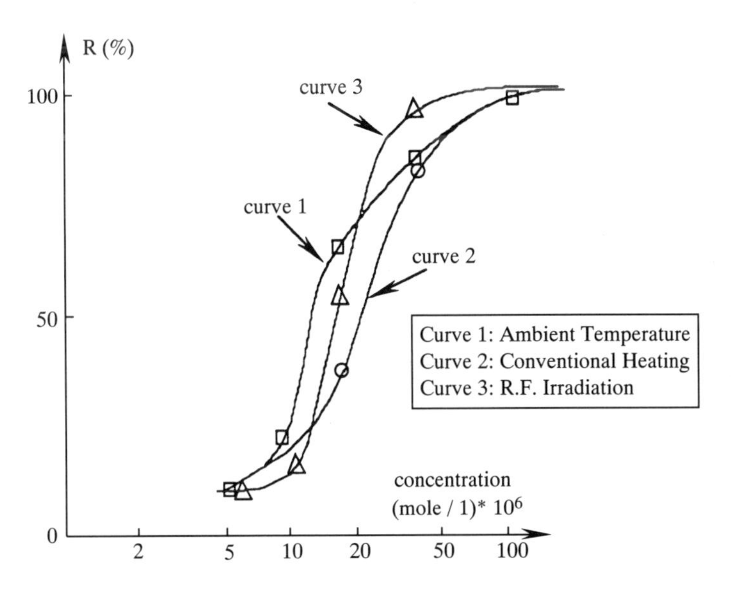

Fig. 11.7 *Fluorite recovery, R, with SDDS collector.*

In the case of floatation by sodium dodecylsulfonate, the effect of classical heating is to destabilize collector fixation. In Figure 11.7 curve 2 is shifted toward higher concentrations of collector in contrast to curve 1, which was obtained at ambient temperature. Curve 3, which corresponds to the high frequency radiation, has a change in form such that at low concentration, the recovery is lower, and at higher concentrations the recovery is enhanced and is larger than the other two methods. Again, the two curves which represent the heated floatation experiments were both given the same total amount of energy. These effects can be explained if the fixation of SDDS is considered as a two-step process.

In the first case, in which a simple chemical reaction describes the process offixation of collector upon the ore, the collector molecules react at the surface as an ordinary endothermic reaction. Both types of heating enhance the fixation of collector onto the surface of the particle.

In the SDDS collector case, there is a two-step mechanism. First, molecules of SDDS are physically adsorbed at the surface of a particle; second they become chemisorbed after some activation:

$$CaF_2 + SDDS \rightleftarrows (CaF_2, SDDS)^*_{\text{physical adsorption}}$$

$$(CaF_2, SDDS)^* \rightarrow CaF_2(CaF_2, SDDS)_{\text{chemisorption.}}$$

When this system is classically heated the first equilibrium reaction is shifted to the left. Then there are fewer adsorbed molecules; thus the fixation of collector is reduced, and floatation is reduced. That is the reason for the shift of curve 2 to the right of the ambient temperature curve.

When the collector concentration is lower than 10^{-5} mol/l, the effect of the electromagnetic field is to prevent the collector molecules from being fixed onto the ore. When the collector concentration is greater than 10^{-5} mol/l, the field cannot pull away all of the molecules which collide with the surface. Once a collector molecule is physically adsorbed onto the surface of an ore particle, it reacts rapidly because the second reaction is enhanced by the temperature gradient provided by the electromagnetic field. When that happens the recovery is enhanced.

Using this approach floatation can be controlled by RF irradiation; electromagnetic control of the process is very different from heating control. Thus we see that irradiation with an electromagnetic field need not be equivalent to conventional heating techniques.

11.3.2 Pyrolysis of neopentane on 13X–Na–zeolites

The second system of results to be discussed concerns the pyrolysis of neopentane. Neopentane is a spherical molecule which is difficult to break down, or "crack", by heat. Pyrolysis of neopentane occurs classically at 500°C, yielding equal amounts of methane and isobutene. The principal equation is:

$$\text{neo}\,.\,C_5H_{12} \xrightarrow{\Delta} CH_4 + \text{i-}\ C_4H_8.$$

In a classical process, this reaction has a complex mechanism which involves free-radical formation:

Initiation:

$$\text{neo}\,.\,C_5H_{12} \xrightarrow{k_1} (CH_3)_3C^{\cdot} + CH_3^{\cdot}$$
$$(CH_3)_3C^{\cdot} \rightarrow \text{i-}C_4H_8 + H^{\cdot}$$
$$H^{\cdot} + \text{neo}\,.\,C_5H_{12} \rightarrow H_2 + \text{neo}\,.\,C_5H_{11}^{\cdot}$$

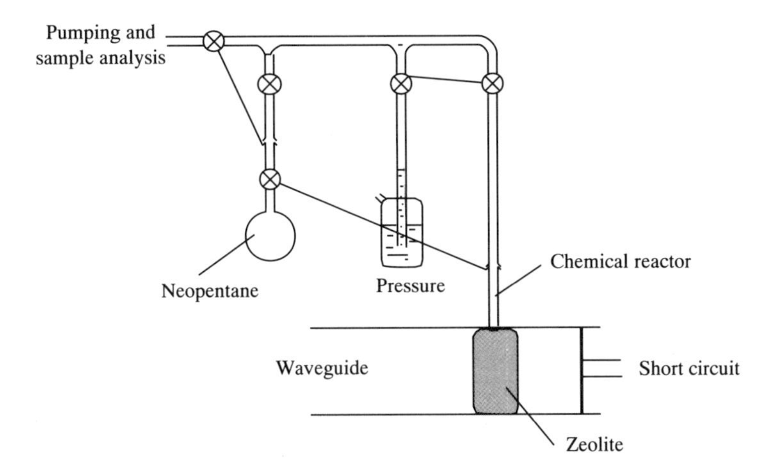

Fig. 11.8 *Neopentane pyrolysis apparatus.*

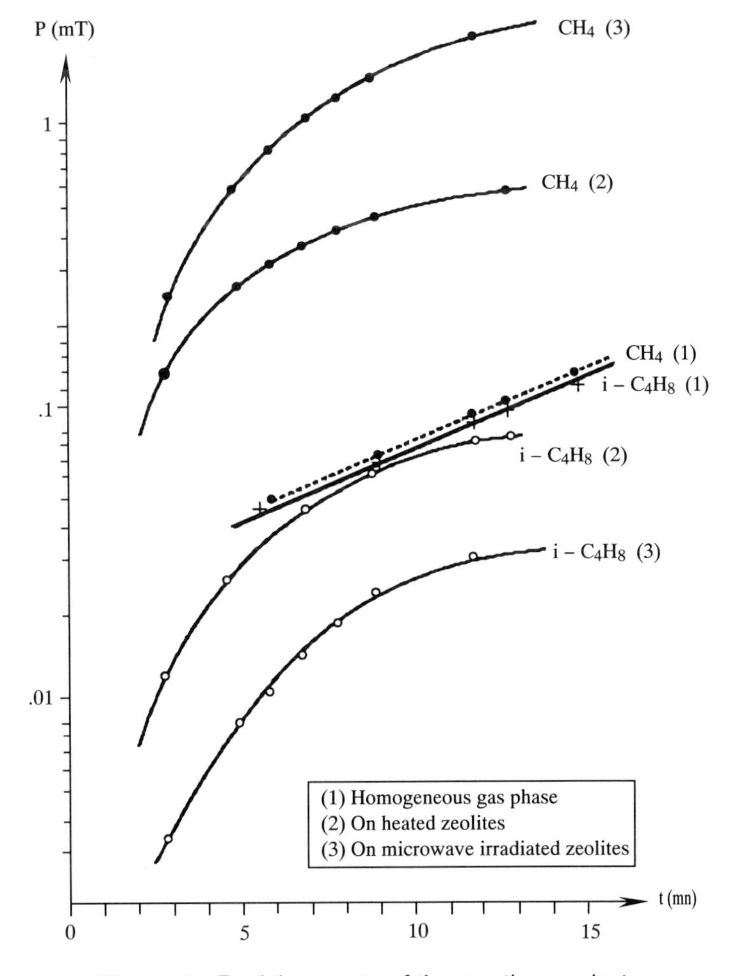

Fig. 11.9 *Partial pressure of the reaction products.*

Propagation:

$$\text{neo}\,.\,C_5H_{11}^{\cdot} \xrightarrow{k_2} \text{i-}\,C_4H_8 + CH_3^{\cdot}$$
$$CH_3^{\cdot} + \text{neo}\,.\,C_5H_{12} \xrightarrow{k_3} CH_4 + \text{neo}\,.\,C_5H_{11}^{\cdot}$$

Termination:

$$2CH_3^{\cdot} \rightarrow C_2H_6$$

It has been shown that neopentane pyrolysis is self-inhibited by isobutene and can also be inhibited by the addition of small amounts of other substances such as propene, toluene, butene, and cyclopentadiene. The reaction can be accelerated by the addition of hydrogenating substances such as HCl, H_2S, CH_3SH, and H_2.

The mechanisms of the reaction in the heterogeneous phase are more complicated due to surface phenomena and adsorption, as is usually the case with heterogeneous reactions. Experiments comparing the amounts of methane and isobutene obtained from the reaction on zeolites using classical heating and electromagnetic heating were performed. A vessel containing zeolites was placed in a waveguide, using the sliding short microwave applicator described previously (see Figure 11.8). The temperature of the zeolites was controlled by the continuous measurement of their dielectric properties. Using a servo-loop, the irradiation was manipulated to maintain a constant temperature of 500°C. A known amount of neopentane was introduced into the cell after the zeolite temperature was stabilized. After a given reaction time the products were analyzed by chromatographic methods in order to determine the progress of the reaction as well as the stoichiometric ratios of products.

The results are presented in Figure 11.9 where the pressure of each product is shown as a function of the reaction time. The curves labeled (1) represent classical pyrolysis without

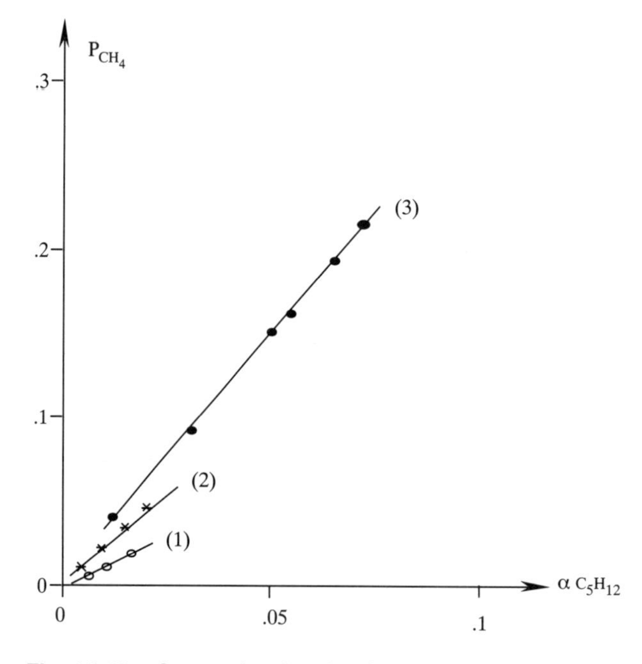

Fig. 11.10 *Conversion fraction for neopentane pyrolysis.*

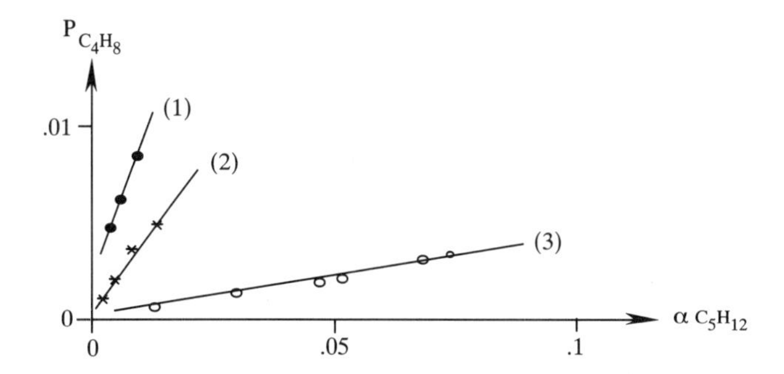

Fig. 11.11 *Product formed vs. neopentane.*

catalysis, in which the amounts of CH_4 and i-C_4H_8 are the same. The curves labeled (2) are for the reactor placed in a classically heated oven, and (3) represent the results obtained with microwave irradiated zeolites. It can be seen that the production of methane is enhanced by MW irradiation of the zeolites over the other methods.

The same results can be presented in another manner. We define the mole fraction of neopentane which has reacted as αC_5H_{12}, and plot the results versus α, as presented in Figures 11.10 and 11.11. We observed that the partial pressure of CH_4 and i-C_4H_8 are linear with α, but the slopes are different. So, the pyrolysis reactions are kinetically different. These results are explained by assuming that the Na ions which are in the solid are excited by the electromagnetic field such that cracking of isobutene occurs more easily. Conventional heating does not provide this extra excitation of the Na ions, and thus the process is slower. In conclusion, special results using irradiated catalysis can be obtained.

We will discuss microwave enhanced catalysis further in Chapter 12.

11.4 MICROWAVE ACID DISSOLUTION OF SOLIDS

Microwave ovens have been used to prepare samples in laboratories for many years. Acid dissolution of solids was first obtained in specially shaped open reflux vessels made of teflon or glass and heated on a carousel inside the oven (Figure 11.12).

Faster reaction rates result in high temperatures and high pressures in closed Teflon vessels with pressure relief valves. Kingston and Jassie have reported many curves to illustrate the histories of temperature and pressure increases. For example, a nitric acid bath (without a sample to dissolve) reaches a temperature of 176°C and a pressure of 5 atm in five minutes at an oven power of 580 W (Figure 11.13). These high levels exceed the normal nitric acid boiling point (120°C at 1 atm). The pressure history during dissolution of a sample gives additional information from other gases, such as CO_2, NO_2, SH_2, produced.

Continuous flow sample and reagent-injection MW devices with digitally controlled temperature and pressure have also been developed.

From the theoretical point of view the acceleration of the reaction occurs because of the increased temperature and pressure developed in the reaction vessels which causes superheating of the acid. Superheating is the major effect; however, in some special cases liquid–solid reaction rates are limited by the production of a surface layer of protective oxide. The oxide layer prevents the further reaction between the solid and liquid. This layer often consists of ions and polar molecules which the electric field is able to destabilize so that we observe that the electric field enhances the reaction rate.

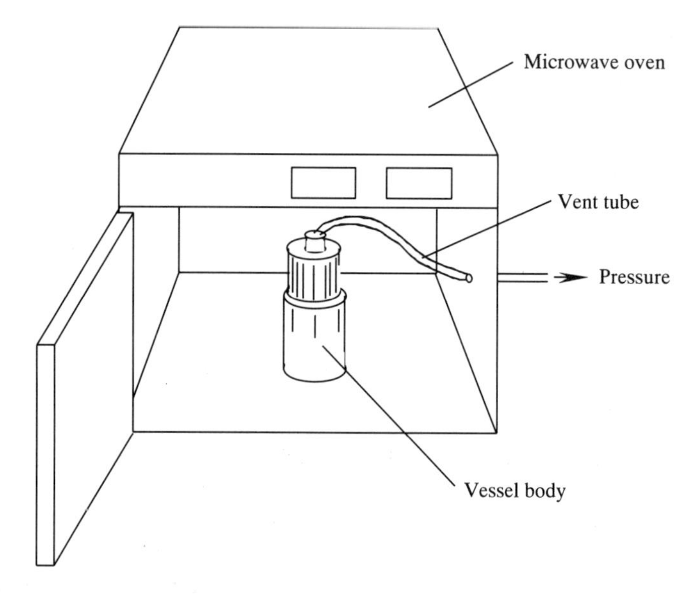

Fig. 11.12 *Apparatus for microwave dissolution of acids.*

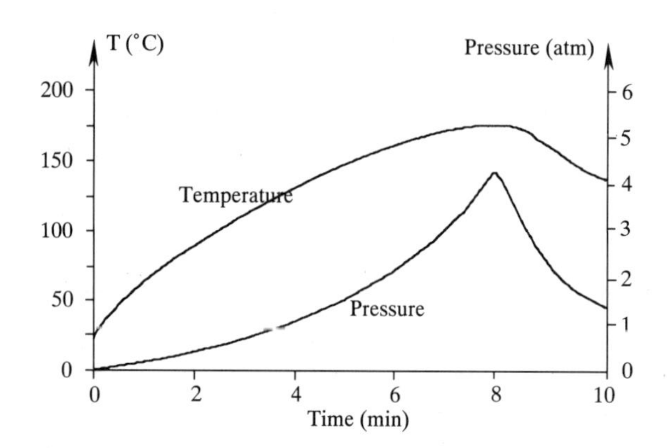

Fig. 11.13 *Temperature and pressure profile of 5 ml of nitric acid at 144 W MW power.*

11.5 SUMMARY

In this chapter, a basic model for calculating the temperature distribution in a heterogeneous material irradiated by an electromagnetic field — microwave field or high frequency RF field — has been presented. It has been shown that a permanent gradient is located at the interface between two components of a mixture. Although this thermal model is very simple, it allows us to understand many experimental results which have been reported in the literature regarding physical–chemical systems.

In the future, our knowledge of systems such as those for which limited understanding has been obtained in this presentation will be improved. The model we have presented will also be improved. Notably, much progress has recently been made in the computation of electric fields in cavities and applicators. In the near future, it will be possible to calculate the distribution of the temperature during the processing of a composite ceramic or food material and thus enable us to be more precise about the physical effects of microwave and RF heating.

REFERENCES

Further information can be found in the following texts:

Beatty, R. L., Sutton, W. H. and Iskander, M. F. (1992) (Eds.) *Microwave Processing of Materials III, Proc. Materials Res. Soc.*, **269**.

Decareau, R. V. (1983) *Microwaves in the Food Processing Industry* Academic Press, New York.

Kingston, H. M. and Jassie, L. B. (1988) *Introduction to the Microwave Sample Preparation* American Chemical Society, Washington, DC.

Snyder, W. B., Sutton, W. H., Iskander, M. F. and Johnson, D. L. (1990) (Eds.) *Microwave Processing of Materials II, Proc. Materials Res. Soc.*, **169**.

Sutton, W. H., Brooks, M. H. and Chabinsky, I. J. (1988) (Eds.) *Microwave processing of materials I, Proc. Materials Res. Soc.* **124**.

12 Microwave-enhanced Catalysis

12.1 INTRODUCTION

For many years microwaves have been used — or have been reported to have been used — for increasing the rate of catalytic reactions and for enhancing the yield of particular reaction products. We will sum up the situation by considering three illustrative examples in some detail.

The first deals with the isomerization of hexane. It will illustrate how the applied electric field can physically modify a catalyst, and thus modify the catalytic reaction. The particular example considered will be a catalyst made of metallic platinum (Pt) particles dispersed in alumina (Al_2O_3), a nearly lossless dielectric structure. In this catalyst the modification is of a physical rather than chemical nature: we will describe the effect of microwaves on the nature of the catalyst in terms of a new method of catalyst preparation.

The second example is that of partial oxidation of methane (CH_4) into ethane (C_2H_6) or ethylene (C_2H_4). In this case the catalyst is an oxide of crystalline samarium and lithium ($SmLiO_2$) doped with both calcium oxide (CaO) and magnesium oxide (MgO). Here the MW field excites the catalytic site, a labile oxygen atom, on which the first step of the CH_4 transformation occurs. Consequently, the first reaction step is enhanced relative to the other steps and the global reaction is thus "tuned".

The third example concerns the acidic properties of the support (e.g. the Al_2O_3) of a metallic catalyst. Here the microwave field excites ions originating in the intermediate species of the acid catalyzed mechanism. The advantageous effects of microwave irradiation are noticeably limited to certain types of reactions; also, certain catalysts and experimental conditions are required in order to enhance the selectivity resulting from microwave irradiation.

12.2 HEXANE ISOMERIZATION OVER A Pt/Al_2O_3 CATALYST

Isomerization reactions are important in chemistry because, as one example, they allow the transformation of organic molecules for which the carbon (C) atom skeleton is linear into molecules for which the skeleton is branched. A specific example is the attachment of a methyl group (CH_3) to the second atom of a five-atom carbon chain — this molecule, known as 2-methylpentane, is more interesting than n-pentane because it has a higher octane index. Of course, the isomerization reaction producing 2-methylpentane does not occur alone, but is complex, as it yields several other products during intermediate steps which are also hexane molecules with different carbon skeletons. The intermediate products may transform into each other. The reaction enhancement problem is to modify the product statistics in favor of the desired transformation.

12.2.1 The Pt/Al_2O_3 catalyst

In the laboratory the individual reactions can be studied by strictly limiting the experimental conditions — temperature, mass rate of reactants, and the rate of reaction — so that multiple reactions are avoided. The usual catalyst for isomerization is a platinum catalyst made by dispersing metal Pt particles, about 3 nm (30 Å) in diameter, over an inert support matrix, typically alumina.

The catalytic reaction occurs typically around 300°C. The hexane is initially diluted in a reducing atmosphere of hydrogen (H_2) gas. The hydrogen gas prevents oxidation of hydrocarbon molecules which would result in combustion of the hexane and other reaction products. The H_2 gas also induces hydrocracking, which splits the long molecule into several smaller hydrocarbon molecules. If the initial molecule is n-hexane and the first C–C bond is broken (de-methylization) a molecule each of methane and pentane is created (Figure 12.1). If the second bond of the carbon skeleton is broken an ethane and a butane molecule are obtained. Breaking the third bond on the carbon skeleton (deethylization) results in two molecules of propane. The results become even more complicated unless one avoids successive cracking of the products. Therefore, in order to obtain an easily interpreted result the reaction must be limited to slow rates.

Generally, cracking is not the desired result — we are mostly interested in isomerization, which has been studied for some time. Two basic mechanisms of hexane isomerization within the catalyst must be distinguished: (1) the bond-shift mechanism, and (2) the cyclic mechanism. The bond-shift mechanism is illustrated in Figure 12.2. From 2-methylpentane we get 3-methylpentane when bond A or B shifts to the other carbon atom (Figure 12.2a or 12.2b). We may likewise get n-hexane when bond B shifts to the extremity of the molecule (Figure 12.2c). The cyclic mechanism in hexane isomerization assumes the existence of an intermediate cyclopentane on the surface of the catalyst which is in equilibrium with four forms of hexane: (1) n-hexane, (2) 2-methylpentane,(3) 3-methylpentane, and (4) methyl-cyclopentane (Figure 12.3). The four forms are obtained by opening one of the C–C single bonds. So, each of these types of hexane can be obtained from one of the other forms.

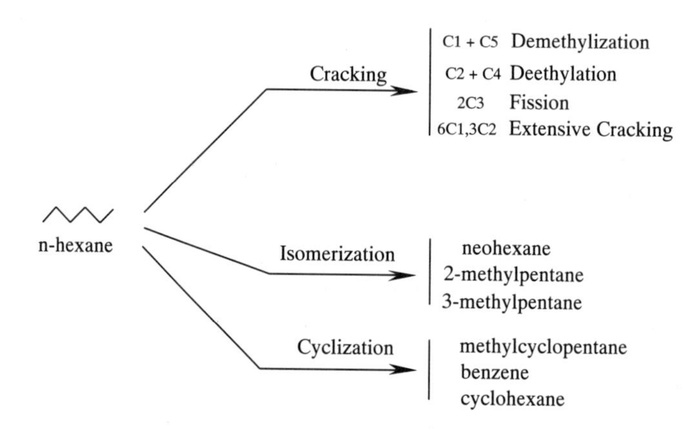

Fig. 12.1 *Possible n-hexane reactions.*

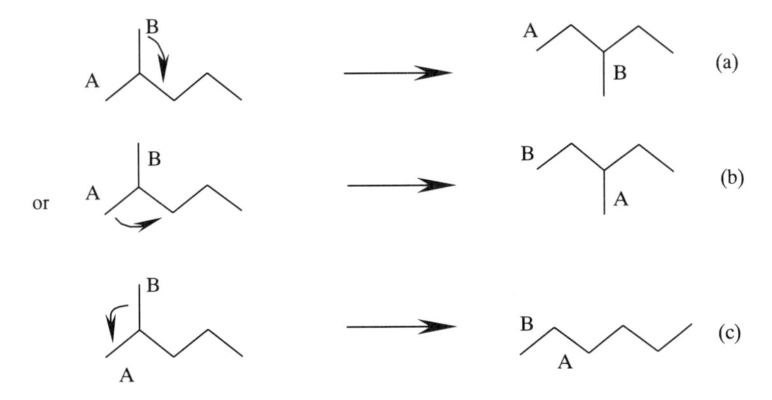

Fig. 12.2 *Bond-shift mechanism for hexane.*

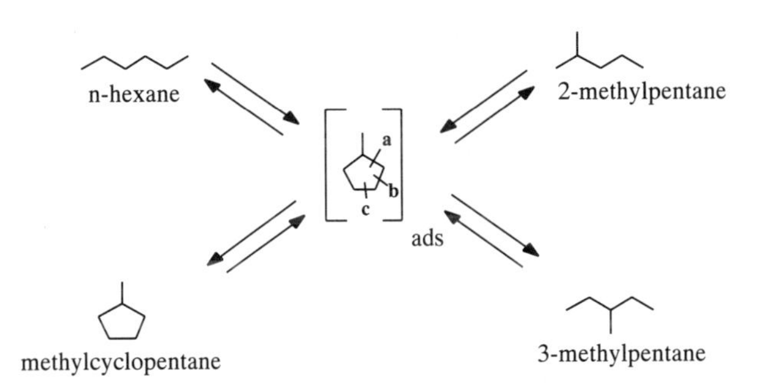

Fig. 12.3 *Cyclic mechanism for hexane.*

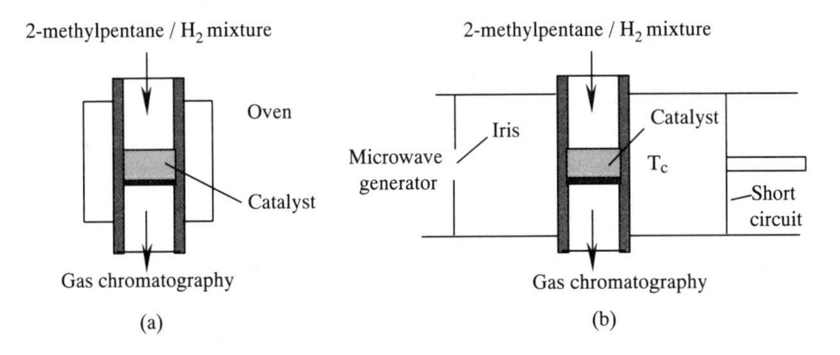

Fig. 12.4 *Catalyst heating experiment. (a) Conventional oven. (b) Microwave cavity.*

It is clear, then, that the work of the chemist will consist of measuring the distribution of the different molecules that the reaction yields using, say, gas chromatography results (and any other analytical technique results) to describe how the bonds are broken and how the molecules are isomerized. We will compare the results obtained in a conventional oven to those in a microwave system under the same (or nearly the same) conditions. In both experiments the temperature of the catalyst was maintained constant and a mixture of hydrogen and 2-methylpentane gas flowed through it, as in Figure 12.4. The output stream was analyzed using gas chromatography and the quantity of each product was determined. For a constant mass flow rate, $\partial m/\partial t$ (kg/s), the reaction time is the residence time within the catalyst, τ (s):

$$\tau = \frac{L}{u} = \frac{AL}{(\partial V/\partial t)} = \frac{\rho AL}{(\partial m/\partial t)} \tag{12.1}$$

where L is the length of the catalyst section (m), u is the velocity of gas flow (m/s), A is the cross-sectional area of the flow stream (m^2), $\partial V/\partial t$ is the volume flow rate of gas (m^3/s), and ρ the gas mixture density (kg/m^3). A fully developed turbulent velocity profile has been assumed with negligible boundary layer thickness.

The temperature in the conventional oven apparatus was regulated by a commercial PID (proportional-integral-differential) feedback control system. In the microwave experiments (at 2.45 GHz) the catalyst was suspended in a WR340 waveguide, as shown in Figure 12.4, and irradiated in the TE_{10} mode. The microwave application system and temperature estimation method are the same as those described in Chapter 5. Briefly, the catalyst bed was placed at an **E**-field maximum in the resonant cavity created by a sliding short circuit and waveguide iris. Feedback control was used to maintain tuning and to control generator output power to obtain constant catalyst temperature, as measured by the thermocouple.

The presentation of results warrants some discussion. It at first appears that a simple plot of reaction product quantity vs. temperature will provide an effective comparison. This is straightforward in the conventional oven but not in the microwave system. First, in the MW system there is the question of the meaning of the thermocouple reading. Even though the thermocouple reading has been corrected as described (Chapter 5) there remains some uncertainty in the estimate. Second, even if the MW field did not affect the thermocouple, the chances are that significant thermal gradients exist within the catalyst due to electromagnetic boundary conditions. So, it is not clear that the experimental conditions were completely identical. However, when plotted vs temperature the resultant curves from the two types of experiments have a similar functional form. The conclusion is that too much depends on rather sensitive measurements of temperature to trust these plots alone — remember that the reaction product formed depends approximately exponentially on temperature and linearly on residence time within the catalyst. In fact, comparison on a temperature basis alone does not yield the most useful information, and so is not pertinent to this discussion.

Among the more interesting physical characteristics of the reaction is the conversion, α (%). Suppose that a reaction converts the reactant, A, into a single product, B. For a fixed residence time within the catalyst of τ, α is the ratio of the number of molecules of A which have reacted to form product, B, to the initial number of molecules of reactant, A. The conversion typically follows an Arrhenius function increasing exponentially with temperature, for other conditions constant, according to:

$$\alpha = \frac{\text{number of molecules of A which react}}{\text{the initial number of molecules of A}}$$

$$\alpha = \alpha_0 \exp\left(-\frac{E}{RT}\right) \tag{12.2}$$

where α_0 is a constant of proportionality, E is the activation energy (J/mole), R the universal gas constant (J/mole K) and T the absolute temperature (K). So, it is most worthwhile to know whether the activation energy, E, is the same for conventional and microwave catalysis. The activation energy is obtained from the slope of the experimental results — on a plot of log{α} vs. $1/T$ — and is not very sensitive to the actual temperature values. In fact, an offset error (alone) in temperature measurement will result in a shift in the line rather than a change in its slope; so repeatable offset errors in temperature measurement would still yield acceptably accurate estimates of the activation energy. If the conventional and microwave curves are parallel then (even though there is some uncertainty in the temperature) the techniques would have the same activation energy and thus be indiscernable from each other. However, the trap created by temperature uncertainty has not been completely avoided in the Arrhenius plot because the resulting curves do not sufficiently reveal the underlying mechanisms. To get a clearer picture we need to look at the reaction products.

As an aside, the comparisons based on the conversion only make sense for multiple product reactions when they occur in parallel in single steps. That is, for two possible products B_1 and B_2 we can discuss α_1 and α_2, the respective conversions, independently and $\alpha_{tot} = \alpha_1 + \alpha_2$. However, this approach cannot uniquely describe reactions which contain several series steps and intermediate products. For those reactions we define the selectivity, S (%), for the jth product as:

$$S_j = \frac{\text{molecules of A converted to } B_j}{\text{molecules of A which have reacted}}. \qquad (12.3)$$

In this framework the yield for product B_j is the selectivity multiplied by α, where α is defined by equation (12.2). The most rational method for presenting the results is to plot the quantity of product B_j formed as a function of α. Then, if the points from the conventional and microwave experiments fall along a single line we will be able to say that the reaction occurs in the same way under both conditions and the two methods are equivalent: that is, that the microwave field effects are completely described as thermal effects. If so, and if the log{α} vs. $1/T$ plot shows a difference, then it is clear that errors were made in temperature measurement. The advantage of the yield vs. α plot is that the interpretation of all results will be more coherent, since the selectivities of each product should give the same curve for both heating methods in the absence of nonthermal effects.

12.2.2 Hydrogenolysis of methylcyclopentane

The catalytic hydrogenolysis of methylcyclopentane over a 0.2% Pt/Al_2O_3 catalyst will be presented as a clear illustration of the technique just described. Methylcyclopentane is one of the four molecules in equilibrium with intermediate cyclopentane (Figure 12.3) in the previous description of the cyclic mechanism. The study of the hydrogenolysis of methylcyclopentane tells how the bonds of the adsorbed species, a, b, or c, are opened. If they are opened statistically, then the ratios of the three products — n-hexane, 2-methylpentane and 3-methylpentane — are, respectively: 2, 2 and 1. The results of a verification experiment are displayed by plotting the respective ratios of numbers of molecules formed as a function of α : r_1 = 2-methylpentane/3-methylpentane and r_2 = 3-methylpentane/n-hexane. Both ratios are independent of α (Figure 12.5) and have the same value whether microwave (MW) or conventional heating (CH) is used: $r_1 \cong 2$ and $r_2 \cong 0.6$ — the small deviations at very low values of α are attributable to the difficulty of accurately measuring small quantities of the respective products. This indicates that the bonds are opened statistically in both types of heating. This is a reasonable result since the catalyst consists of small well-dispersed metallic Pt particles. Figure 12.6 is an Arrhenius plot of the same data which shows that both

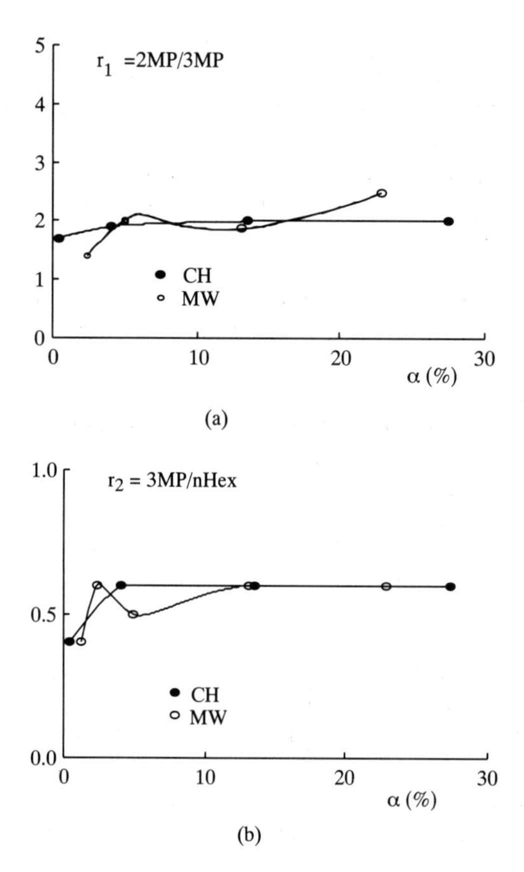

Fig. 12.5 *Hydrogenolysis of methylcyclopentane with conventional heating (CH) and microwave irradiation (MW).*

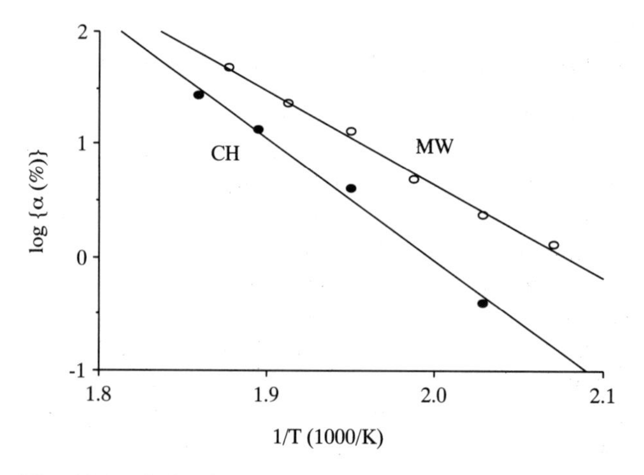

Fig. 12.6 *Arrhenius plot of cyclopentane hydrogenolysis.*

heating curves are linear on $\log\{\alpha\}$ vs. $1/T$ axes and the slopes are asymptotically the same within the accuracy of the experiment (remember that the $1/T$ axis is hyperbolic and the uncertainty bands differ in size at different locations for the same temperature uncertainty). Therefore, we may conclude that the activation energies can be considered equal (at about 45 kcal/mol). The shift between the two lines would correspond to an underestimation of the temperature in the microwave experiments of about 20°C, if that interpretation were applied to the data.

12.2.3 Isomerization of 2-methylpentane

Next, we will consider the isomerization of 2-methylpentane over the same 0.2% Pt/Al_2O_3 catalyst. In each case the conventional heating experiment was performed prior to the microwave experiment. As was done in Figure 12.5 we plot the isomeric selectivities, S_i, as a function of α; the results are shown in Figure 12.7. The total selectivity in isomerization is the ratio of the sum of all isomer molecules obtained to the number of original molecules which have reacted. For the microwave experiments S_i is around 80%, while in conventional heating S_i is only about 60%. In addition, the isomeric selectivity is not a function of temperature for either heating modality; thus, temperature is not the controlling variable and need not be discussed. The two heating methods do have different activation energies, as can be seen in Figure 12.8. Therefore, microwave irradiation enhances the isomeric selectivity of this reaction by a nonthermal mechanism. Furthermore, in the companion Figures, 12.9 and 12.10, microwave irradiation also enhances the isomeric selectivity of 3-methylpentane on the same catalyst. The selectivity is again independent of α and of temperature since the data points were obtained at different temperatures. Also, the Arrhenius plot of Figure 12.10 shows comparable differences in activation energy.

Two possible explanations occur which we must choose between: (1) the mechanism of the reaction might depend directly on the electric field under microwave irradiation, or (2) the catalyst might be physically modified by exposure to the microwave field. The second explanation appears to be the explanation of choice, but the first hypothesis is worthy of discussion as well.

To approach the question of specific mechanism, a special series of hexane isomerization experiments was performed using ^{13}C isotope-labeled molecules. The labeled molecules

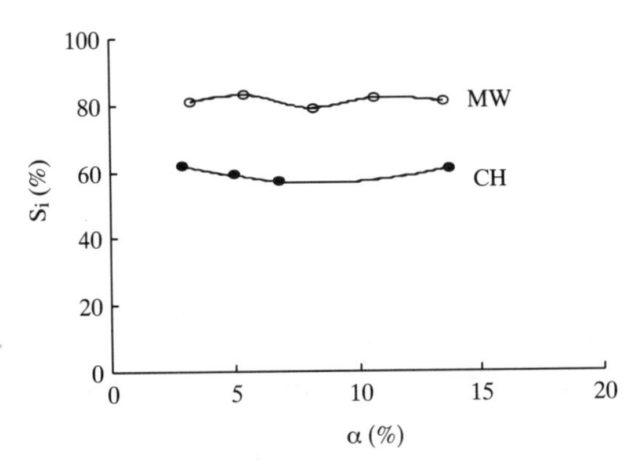

Fig. 12.7 *Selectivity of 2-methylpentane isomerization.*

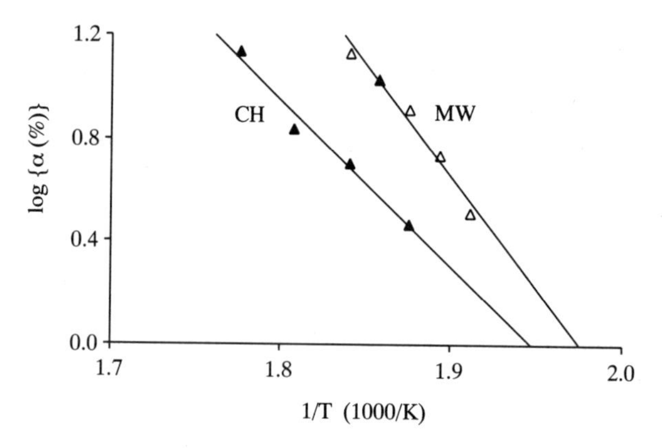

Fig. 12.8 *Arrhenius plot of 2-methylpentane isomerization.*

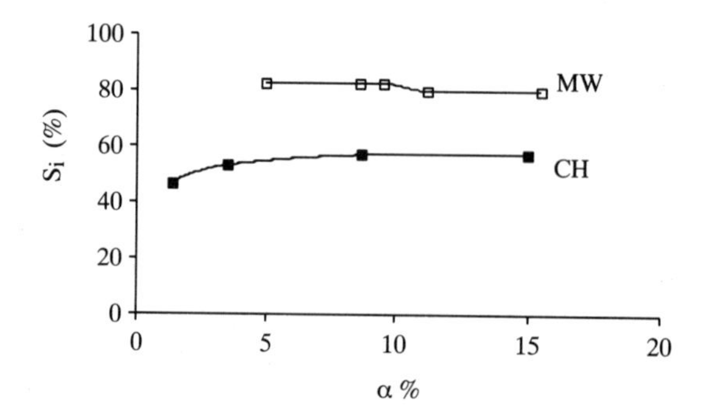

Fig. 12.9 *Selectivity of 3-methylpentane isomerization.*

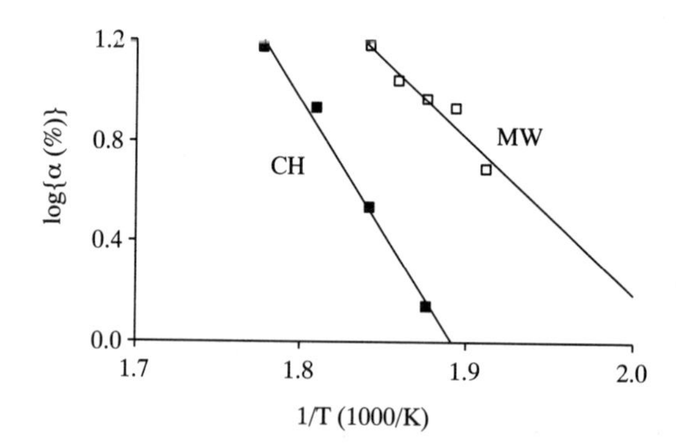

Fig. 12.10 *Arrhenius plot of 3-methylpentane isomerization.*

permit the identification of specific molecules in the product stream using mass spectroscopy. The experiments are tedious but clearly show that: (1) the mechanism is the same under microwave irradiation as in conventional heating, (2) under microwave irradiation the two chemical mechanisms which result in isomerization occur concurrently, as in conventional heating, and (3) the cyclic mechanism (Figure 12.3) is enhanced relative to the cracking process; however, the bond-shift mechanism is unaffected. Therefore, either the number of sites on which the cyclic mechanism occurs is increased or the sites are made more efficient. At this point we cannot distinguish between a number density increase and site modification effect due to the electric field.

A definitive experiment was performed in which a single catalyst was used in a conventional reaction, then in a microwave-irradiated reaction and then finally used again in a conventional reaction. The initial conventional heating reaction resulted in a selectivity of 60%, as in the original series. The selectivity in the microwave experiment was 90%, as expected. However, in the final conventional experiment the selectivity was 85%, similar to the microwave irradiation case. This is direct proof that the catalyst was physically modified (irreversibly) by irradiation with microwaves. So, microwave irradiation provides a new tool for the preparation of catalysts.

To see why the measured effects occur we need to look at the catalyst structure. The catalysts are made by impregnating a quantity of alumina with chloroplatinic acid in solution. The concentration of platinum is calculated to obtain the desired dispersion of platinum particles. The alumina is then dried and calcined under an air flux which oxidizes the platinum chloride and the Cl_2 is removed. Then the oxide is reduced by flowing hydrogen gas at high temperature. This method gives small particles or layers of platinum metal dispersed uniformly in the porous alumina matrix. The size and distribution of the particles is estimated by electron microscopy. Small particles are obtained from low concentrations of chloroplatinic acid. For example, a catalyst with 10% Pt usually has particles with mean diameters in the range of about 10 nm (100 Å) diameter.

The catalyst isomeric selectivity depends on the mean particle diameter. The cyclic mechanism percentage is a function of the relative number of particles with diameters less than 1 nm (10 Å). Microwave irradiation does change the particle size; however, it causes the particles to grow from about 10 Å to about 30 Å. But the growth of particles does not explain the enhancement of catalyst selectivity. Other experiments suggest that the reduction of platinum oxide is more effective under microwave irradiation than by conventional means, which may be a more satisfying explanation.

We have treated the question of the optimal treatment protocol for microwave irradiation of catalyst material — what irradiation parameters should be used at which step in catalyst preparation — in some detail in the literature (Thiebaut *et al.* 1993). Briefly, it turns out that when microwave irradiation is applied during the reduction step, the oxidation step (chloride transformation into oxide) is almost unnecessary. Therefore, the most important step in the selectivity of the catalyst is the reduction of platinum, and that step is made more effectively with microwaves.

In conclusion, microwaves allow the preparation of a new metallic catalyst in the sense that some of the catalyst properties are changed by MW irradiation in ways which are difficult to obtain with conventional heating. The effect of the electric field can be explained as a more efficient reduction of the platinum during preparation. This observation opens the possibility of finishing the preparation of the catalyst, or of regenerating the catalyst, *in situ* during use with microwaves.

12.3 METHANE ACTIVATION BY MICROWAVES

The activation of methane is a different type of microwave interaction. It has long been a goal of scientists to convert the abundant natural gas methane (CH_4) into higher forms of hydrocarbon molecules for use in the petroleum and chemical industry. The most elusive transformation has been the conversion of C_1 methane into the C_2 compounds ethane and ethylene because that reaction is more endothermic than other transformations. Many solutions have been attempted; one of them has been the partial oxidation of methane over a special catalyst to inhibit production of CO and CO_2 during the transformation. The possible reactions are:

$$2CH_4 + O_2 \rightarrow C_2H_4 + 2H_2O$$

$$2CH_4 + \tfrac{1}{2}O_2 \rightarrow C_2H_6 + H_2O$$

$$CH_4 + \tfrac{3}{2}O_2 \rightarrow CO + 2H_2O$$

$$CH_4 + 2O_2 \rightarrow CO_2 + 2H_2O$$

The most efficient catalysts for accomplishing the desired transformation are oxides such as MgO, CaO, La_2O_3, Sm_2O_3, Li_2O and the like. The problem is far from being solved. The best results obtained so far have a C_2 selectivity of only 60% and a conversion of $\alpha = 5\%$ for a total yield of only 3% (where α is the ratio of the number of CH_4 molecules which have reacted to the initial number of CH_4 molecules). The first step of the reaction occurs on an oxygen atom at the surface of the catalyst (investigators disagree on whether the actual ions are O^- or O^{2-}), which extracts a hydrogen atom from a methane molecule to create a methane radical, $CH_3^{\cdot}$. Then, two methane radicals combine in the gas phase to form C_2H_6 and further lead to C_2H_4 (see Figure 12.11).

On the catalyst surface the hydrogen atom combines with a hydroxyl (OH) group to form a water molecule. The surface oxygen atom is then replaced by oxygen atoms from the surrounding gas as they are consumed. This approximate explanation is sufficient for describing the action of microwaves in this process. A more complete explanation would include a description of the differences between the catalysts and the formation of CO and CO_2, but this is not necessary. Suffice it to say that the catalytic reaction may be represented by a triangular five-reaction schema, as shown in Figure 12.11, without separating the mono-oxide and dioxide of carbon, CO and CO_2, but treating them as a single species, CO_x.

Just as in the hexane isomerization problem, we seek to enhance the selectivity of C_2 products: that is, the quantity of CH_4 molecules which have converted into C_2 products over the quantity of CH_4 molecules which have engaged in the reaction. The experiments were performed in one of two reactors: (1) a reactor identical to that of Figure 12.4, as in the hexane isomerization study, or (2) a modified reactor, as shown in Figure 12.12, again in the TE_{10} mode but with a different polarization in the catalyst. This experiment has a different field polarization, but there is no difference in the results obtained. A mixture of CH_4 and O_2 gas, diluted with helium to prevent an explosion, flowed through the catalyst. When steady state conditions were reached (the temperature stabilized) a sample of outlet gas was analyzed by gas chromatography. The same experiment was performed in identical

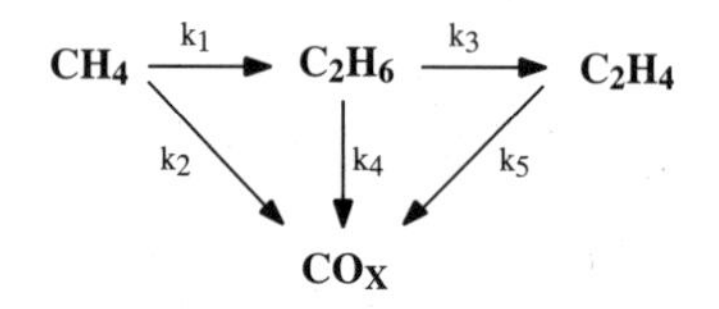

Fig. 12.11 *Methane five-reaction schema.*

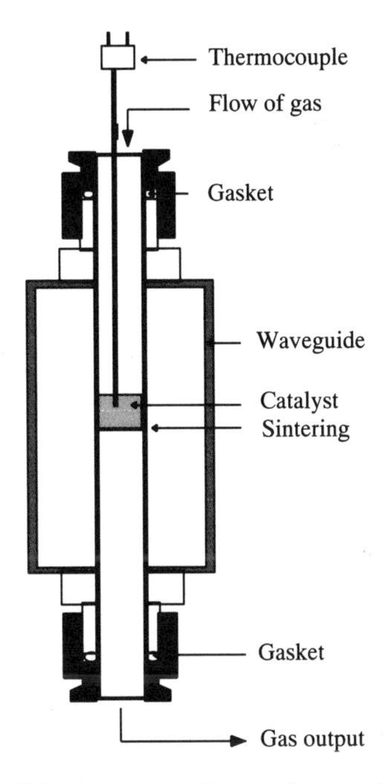

Fig. 12.12 *Microwave applicator for catalyst irradiation.*

reactors in a microwave waveguide and a conventional oven. The action of the microwaves is clearly illustrated by the results.

The presentation of yield or conversion as a function of temperature, again, is not as revealing as the selectivity as a function of conversion — in this case, we plot S_2 vs. α in Figure 12.13. For many catalysts (our example is $La_2Zr_2O_7$, Figure 12.13a) the curves superpose, indicating no essential difference between microwaves and conventional heating. For others, (Figure 12.13b) the microwave S_2 is lower than that of conventional heating; such catalysts are of no interest in microwave processing. The best results were obtained with a $SmLiO_2$ catalyst doped with CaO and MgO (Figure 12.13d). For this catalyst the selectivity under microwave irradiation has much higher S_2 values than conventional heating, especially at low conversion. The difference comes from a fundamentally different operating principle in the catalyst.

The partial oxidation of CH_4 is more difficult to analyze than hexane isomerization because the quantity and type of information available is limited. A more coherent interpretation of the results can be obtained from a plot of the ratio of the number of C_2 molecules to the number of CO_x as a function of the ratio of the number of C_2H_6 molecules to the number of C_2H_4 molecules for this catalyst (Figure 12.14). This plot illustrates, once more, that there is clearly a fundamental difference in mechanism under microwave irradiation.

The kinetics of each of the five reaction paths in Figure 12.11 hold the key to interpreting the results. We define the separate reaction velocities, k_1 through k_5 as shown in Figure 12.11. The work is tedious because the energy of activation must be determined for both processes for each intermediate reaction separately. That is, the initial gas was modified by adding varying concentrations of C_2H_6 and O_2 and also of C_2H_4 and O_2. The only simplifying assumption used in the study was that the order of all reactions was the

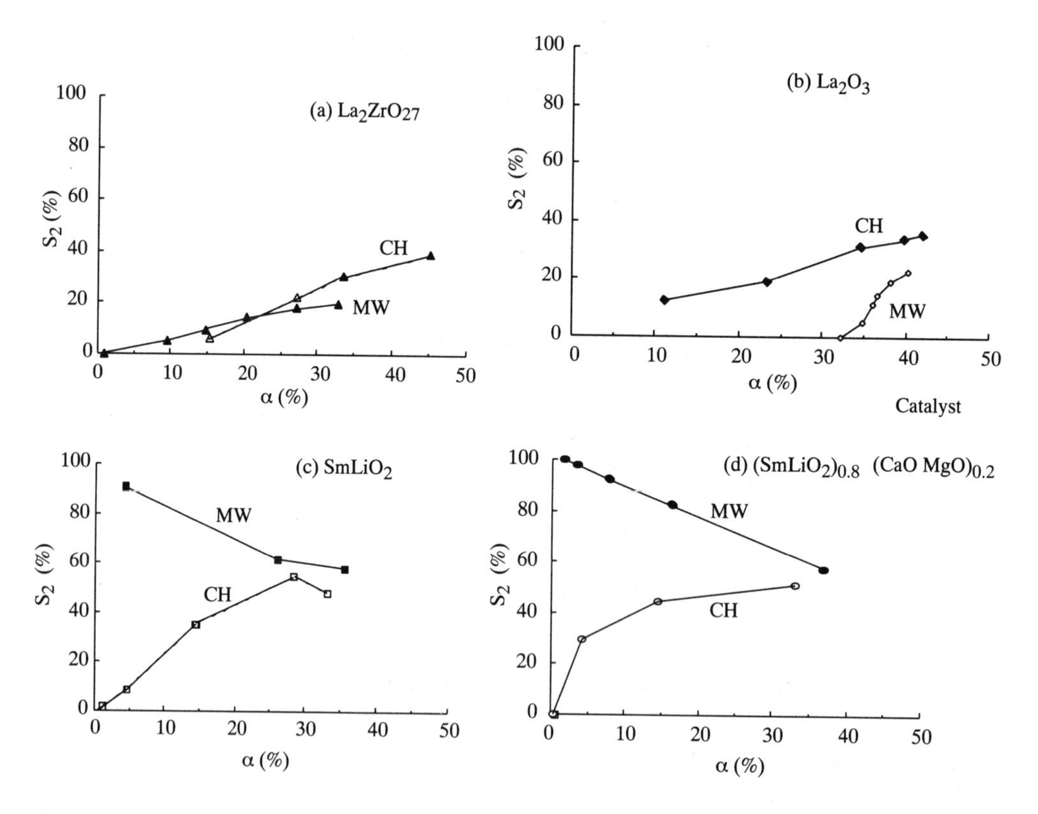

Fig. 12.13 *Selectivity plots for various catalysts. (a) $La_2Zr_2O_7$ (b) La_2O_3 (c) $SmLiO_2$ (d) $(SmLiO_2)_{0.8}$ (CaO, MgO)$_{0.2}$.*

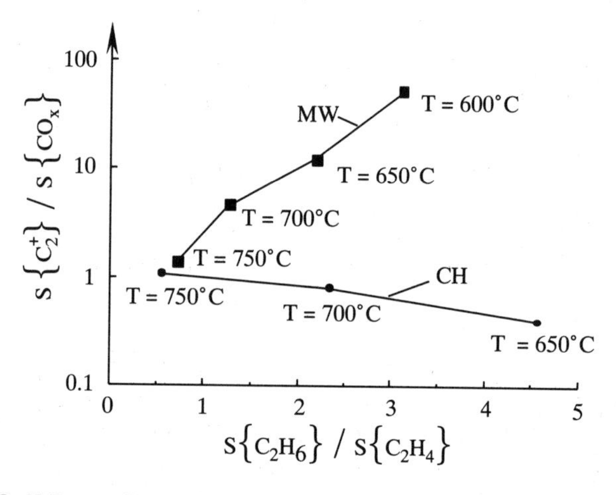

Fig. 12.14 *C_2/CO_x vs. C_2H_6/C_2H_4 for $(SmLiO_2)_{0.8}$ (CaO MgO)$_{0.2}$ catalyst reaction.*

same, as has been published elsewhere, and that the order did not change between the MW and CH experiments.

In spite of the necessary approximations, the rate of each reaction was determined by fitting a large amount of data to obtain k_i and E_i:

$$k_i = b_i \exp\left\{-\frac{E_i}{RT}\right\} \tag{12.4}$$

where k is the reaction velocity, b is a constant of proportionality, and E is the activation energy. Of course, under microwave irradiation precise interpretation of the meaning of "temperature" is somewhat difficult, as we have already discussed. We have chosen to use the catalyst temperature as the reference for the experiment. A plot of the ratio of k_i (MW)/k_i(CH) as a function of the measured catalyst temperature T for this reaction (Figure 12.15) reveals the following:

(1) The reactions labeled $i = 4$ and 5 in Figure 12.15a, through which the products C_2H_6 and C_2H_4 are oxidized, are much slower under microwave irradiation than in conventional heating. Both k_4 and k_5 are slower by factors of 10 or 1000.

(2) On the other hand, the initial selectivity, which is indicated by the ratio k_1/k_2, is nearly constant in microwave experiments while it varies significantly in conventional heating (Figure 12.15b).

We can interpret these results by noting that the degradation reactions ($i = 4$ and 5) occur in the gas phase and we might expect them to be slow if the gas temperature is less than the temperature of the solid catalyst. In a conventional heating system the gas and catalyst are at the same temperature. In the microwave experiments the power is deposited within the catalyst which is cooled by the gas flow and thermal conduction to the surroundings. So, if the catalyst is not thick compared to the thermal equilibration distance (a function of catalyst pore geometry and average gas velocity) the gas is always at a lower temperature than the solid catalyst. In a thick catalyst bed no difference between MW and CH was measurable. For experiments in which a shallow bed of catalyst was used the residence time of the gas was short, so that the gas was much cooler than the catalyst, and reactions $i = 4$ and 5 were slow. This explanation applies for this particular type of catalyst, but not in general, since Figure 12.12 tells us that the selectivity improvement realized depends upon the catalyst used.

We do not have a clear explanation of why the k_1/k_2 ratio changes, however. A plausible hypothesis can be derived from a comparison of the results obtained in different catalysts. The specific catalyst used was $(SmLiO_2)_{0.8}$ $(CaO, MgO)_{0.2}$ which gives the best catalysis in methane oxidation and has some dielectric loss. In general, crystallized oxides have low dielectric loss because the displacement of an atom is difficult, at best, and often not possible at all. $SmLiO_2$ has a small loss factor and a complex structure. The Sm atom has seven oxygen atoms around it and the Li atom is surrounded by five oxygen atoms; the distances between the atoms have been determined by X-ray crystallography. The separations in the unit cell show that there is one oxygen atom which is far from both Sm and Li and is relatively free to move about in the crystal. So, the electric fields can displace the weakly bound oxygen atom better because each oxygen atom has two lone pairs. When the crystal is doped with CaO and MgO, to form $(SmLiO_2)_{0.8}$ $(CaO, MgO)_{0.2}$, some Sm atom sites are occupied by Ca atoms and some of the Li sites are occupied by Mg. Generally, Sm is trivalent, Ca and Mg are bivalent, and Li is single-valent so that the electronic defect in the

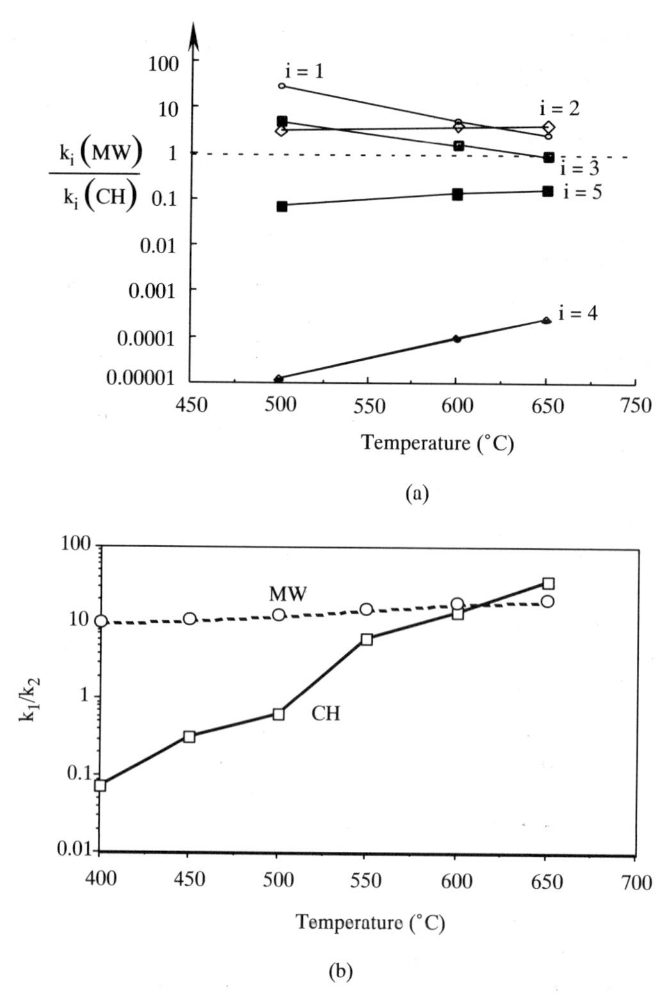

Fig. 12.15 *(a) Ratios of the kinetic rates for the five methane reactions under microwave irradiation and in conventional heating. (b) Reaction velocity ratio, k_1/k_2, for microwave and conventional heating.*

solid is enlarged, especially at the surface of the crystal. The dielectric loss in this catalyst is consequently significantly increased by the doping.

The increased loss factor favors the formation of CH_3 radicals because they are produced at "O_2" sites, and because these *specific sites* are preferentially excited by the microwave field. So, the observed enhancement of C_2 selectivity is, in fact, a "thermal" phenomena in that it can be completely explained by temperature gradients within the crystal,—i.e. by locally excited catalytic sites. In summary, we may explain this observed effect by assuming that under microwave irradiation the temperature of the reaction sites is higher than the mean temperature of the solid crystal. This example illustrates how microscopic studies of the catalytic system must be included in an explanation of microwave effects.

12.4 2-METHYL-2-PENTENE ACIDIC CATALYTIC REACTION OVER MIXED $AL_2O_3 \cdot WO_3$

In heterogeneous catalysis, the support structure of the catalyst plays an active role. At first, solids having large surface to mass ratio (m^2/kg) for gas adsorption permit generally larger specific activity values (activity/kg) when active metal is dispersed within them. The most common support materials for metal catalysts are simple or mixed oxides such as Al_2O_3, SiO_2, $Al_2O_3 \cdot MgO$ or $Al_2O_3 \cdot WO_3$. They are mostly inert but, at higher temperatures, above about 400°C, they participate in many catalytic reactions: alcohol dehydration, skeletal transposition of hydrocarbons, cracking, and alkylation or also de-alkylation. Most oxides are acidic and metal impregnated oxides are known to be bifunctional. For example, the isomerization of n-paraffin into isoparaffin does not occur on the metal catalyst (as we saw with hexane isomerization) but rather occurs on a metal-impregnated acid support. First, dehydrogenation occurs on the metal sites, forming an olefin molecule; then, the olefin molecule is isomerized on the acid sites into an iso-olefin, which is finally hydrogenated on the metal sites into an iso-paraffin. This simple explanation justifies the importance of considering microwave influences on the acid catalyst.

By analogy to solution chemistry, the catalytic activity is postulated to be acidic if the solid is able to combine with a base. An acid solid may be of the Brönsted type — in which it can donate a proton H^+ to unsaturated hydrocarbon — or of the Lewis type, in which it acts as an electron acceptor removing a hybrid ion from a hydrocarbon, as follows.

Brönsted:

$$R - CH = CH_2 + H^+ \quad \begin{array}{c} OH \\ | \quad O \\ {}^-Al \quad\;\; \\ | \quad O \\ O \end{array} \rightarrow (R - \overset{+}{C}H - CH_3)^- \quad \begin{array}{c} OH \\ | \quad O \\ Al \quad\;\; \\ | \quad O \\ O \end{array} \qquad (12.5)$$

Lewis:

$$R - CH_2 - CH_3 + Al - \begin{array}{c} O \\ O \\ O \end{array} \rightarrow (R - \overset{+}{C}H - CH_3) \ldots \bar{H} - \begin{array}{c} O \\ | \\ Al - O \\ | \\ O \end{array} \qquad (12.6)$$

The acid 2-methyl-2-pentene reaction forms intermediate carbo-cations in several ways. Investigators have proposed multiple-step mechanisms such as:

(1) Initial ion formation: in which a proton is added to the adsorbed olefin molecule:

$$\longrightarrow \left(\quad \right)_{ads} \xrightarrow{+H^+} \; + \qquad (12.7)$$

(2) Then, the ion undergoes a variety of shifts, as in a chain reaction:

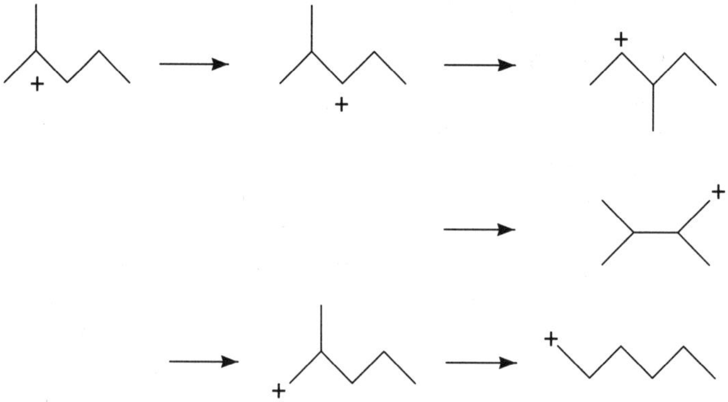

(12.8)

(3) Alternatively, the carbenium ion might be deactivated and react with another molecule to reverse the process:

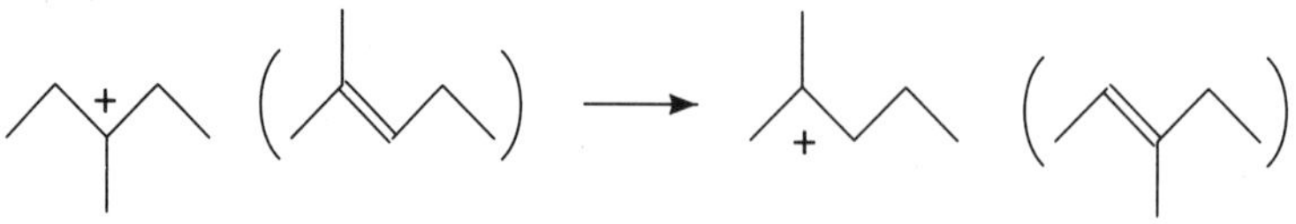

(12.9)

(4) Finally, the carbenium ion is deactivated and loses a proton to yield an adsorbed olefin molecule:

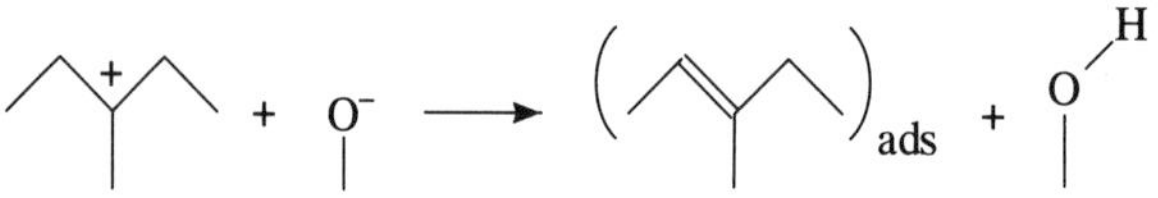

$$+ \ O^- \longrightarrow (\ )_{ads} + O{-}H \qquad (12.10)$$

By combining these elementary mechanisms, the 2-methyl-2-pentene acid reaction can be explained with the general schema of Figure 12.16, in which the carbenium ion is formed first and then transformed into four others. Three skeletal isomerization processes may then occur before the carbenium ions produce olefins.

The overall reaction has been studied on a Al_2O_3/WO_3 mixed support in a classical oven and a microwave waveguide. The catalyst support was prepared by depositing 5% by mass WO_3 on a wet Al_2O_3 solid. Then, the solid was reduced at 500°C in a controlled flow of hydrogen gas. The microwave experiments were performed as previously described. The experimental conditions (quantity of catalyst, pressure, flows of hydrocarbon and hydrogen) were maintained as constant as possible in both experiments. Following standard techniques, the products were analyzed and identified by gas pressure chromatography of the reactor output stream and after the reactor products were fully hydrogenated over a special Pt/Adams catalyst. The results are summarized in Table 12.1 for three typical temperature values.

The relative quantities of the products, given in %, are dependent variables of the reaction. In some cases, it was not possible to separate spectral lines on the chromatograph, so those products are reported together (for example 2-methyl-pentane and 2-methyl-2-pentene in the fifth column). As an aside, the results show that the reaction is not a pure acid reaction

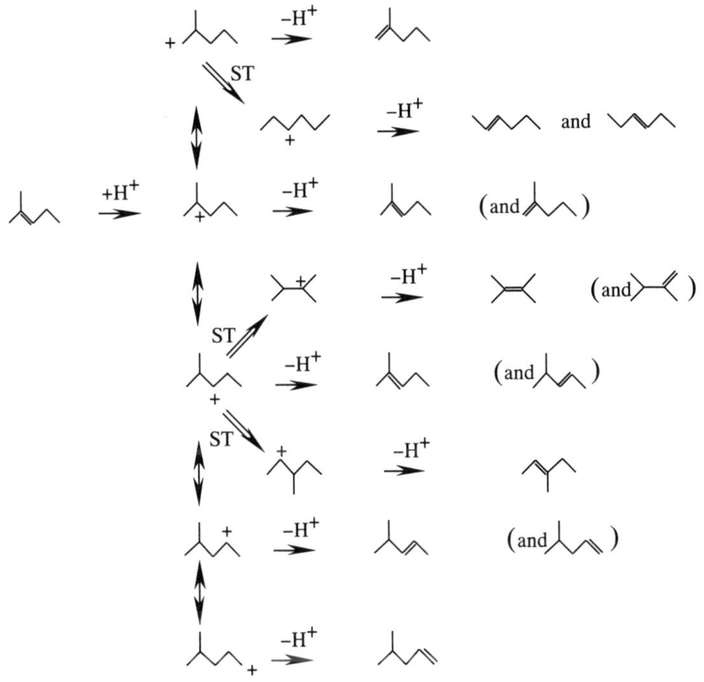

Fig. 12.16 *Acid reaction of 2-methyl-2-pentene.*

Table 12.1 *Results of 2-methyl-2-pentene catalytic acidic reaction, indicating relative quantities of reaction products (in %).*

	T (°C)									Σ cracking		Pt Adams		
	300	3.14	7.15	88.74	0.28	0.29	0.18	0.22	—	—	—	99.32	0.51	0.18
CH	350	7.27	11.95	78.08	0.71	1.11	0.24	1.31	—	0.04	0.29	98.0	1.42	0.24
	400	10.02	19.73	57.01	2.00	4.35	0.72	4.45	0.06	0.46	1.21	89.46	8.80	0.72
	275	0.94	0.06	96.8	—	1.88	0.22	0.02	0.01	0.08	0.01	97.98	2.00	0.22
MW	325	2.17	0.60	90.4	0.06	5.57	0.39	0.2	0.03	0.20	0.31	93.21	5.77	0.39
	375	5.42	3.88	76.20	0.41	8.96	0.93	1.63	0.15	0.90	1.16	87.21	10.59	0.93

because some 2-methyl-pentane is produced in both types of heating. The production of 2-methyl-pentane by direct hydrogenation of the initial molecule is probably due to metallic sites (impurities or tungsten metal which has been not oxidized during the preparation of the support). Nevertheless, the results show differences in the behavior of the catalyst when it is classically heated and when it is microwave irradiated. Although the conversion, α, of the reaction cannot be directly evaluated by the quantity of 2-methyl-2-pentene obtained in the output gas (because that product is produced reversibly during the reaction), the activity can be evaluated by the amount of 2-methyl-1-pentene (values of which are given in the third column of Table 12.1.). That variable has approximately the same temperature dependence

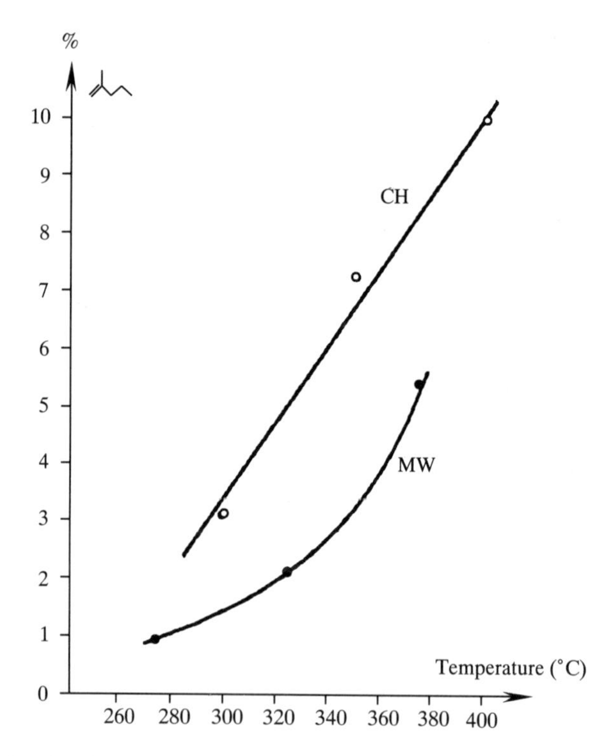

Fig. 12.17 *α estimation from the dependence of 2-methyl-1-pentene vs. temperature in both heating modes.*

under both heating methods (see Figure 12.17), so it represents a good first order estimation of the conversion α.

A large difference appears between MW and CH results when skeletal transposition (ST) is measured, either directly with the n-hexene and the 2,3 -dimethyl-2-butene (as in Figure 12.18) or by 3-methyl-pentane (after Pt/Adams hydrogenation) as in Figure 12.19. All sets of curves are presented as a function of 2-methyl-1-pentene and show the same enhancement of acidic isomerization due to microwave irradiation. A second difference is in the ratio of 2-methyl-2-pentene to 2-methyl-1-pentene. The ratio decreases with the temperature in conventional heating (and is about 2), but increases with the temperature when the support is microwave irradiated, and is less than 1 (Figure 12.20).

Before we open the discussion on the effects of the field in that case, we must observe that the pure alumina (Woelm), from which we prepared the mixed support, has no catalytic activity (no skeletal transposition, no acidity, and no hydrogenation). Also, the microwave field effect in acidity is not permanent, as was the metallic catalyst isomerisation of hexane. When a mixed support is reused in a conventional oven, after it has been irradiated in a microwave waveguide (even for a long time), its activity is the same, within the experimental error, as it was before irradiation with microwaves. Therefore, no permanent transformation of the catalyst occurs.

The modification of the activity of acid support we observed can then only be explained by the carbenium ions, which are intermediate species of the acid mechanisms. The microwave field enhances the skeletal transpositions (ST) in one direction, (indicated by double arrows

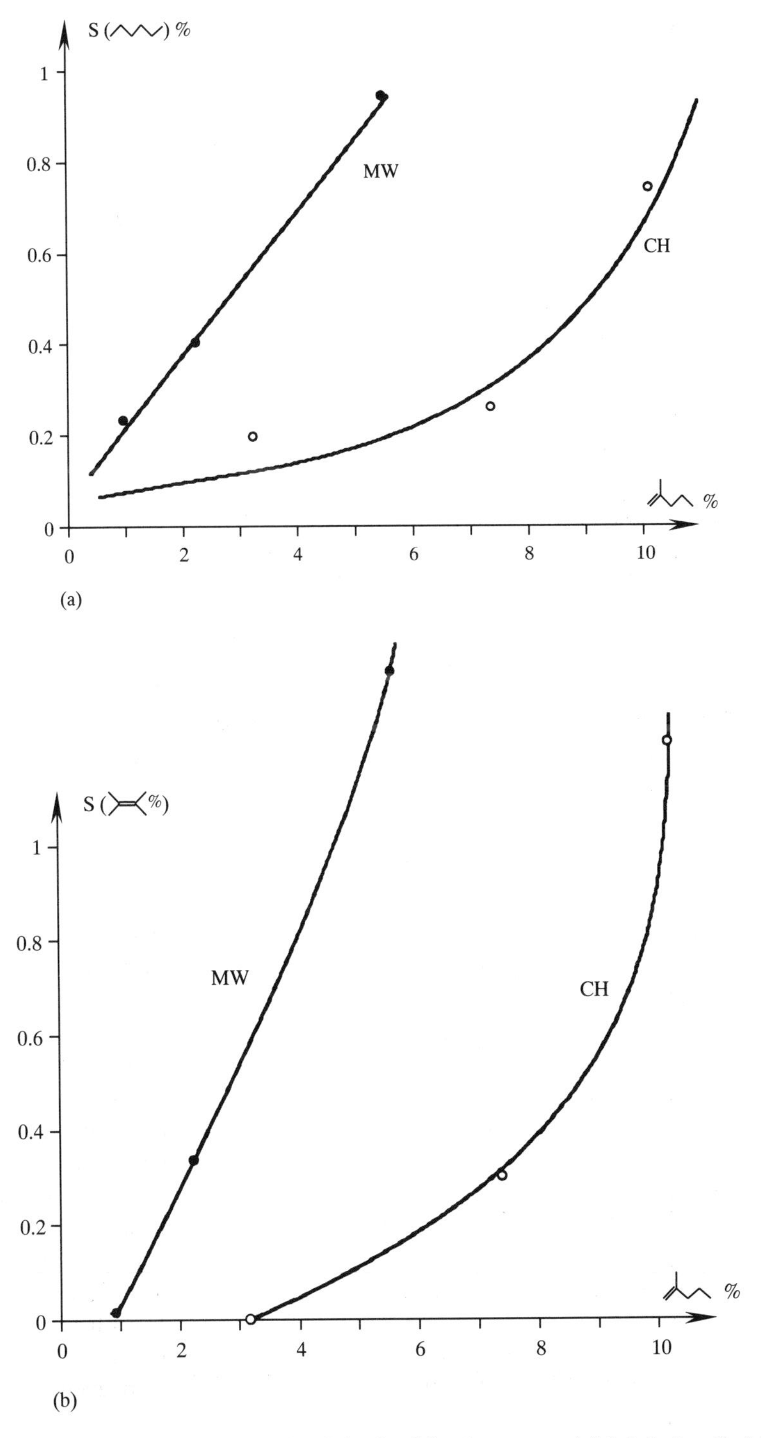

Fig. 12.18 *Skeletal transposition selectivity for (a) n-hexeneand (b) 2,3-dimethyl-2-butene.*

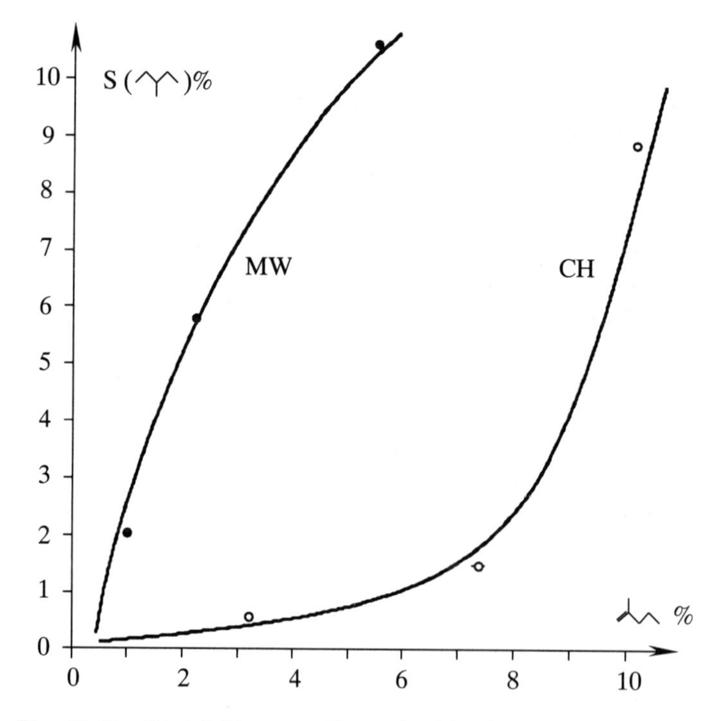

Fig. 12.19 *Skeletal transposition selectivity for 3-methyl-pentane.*

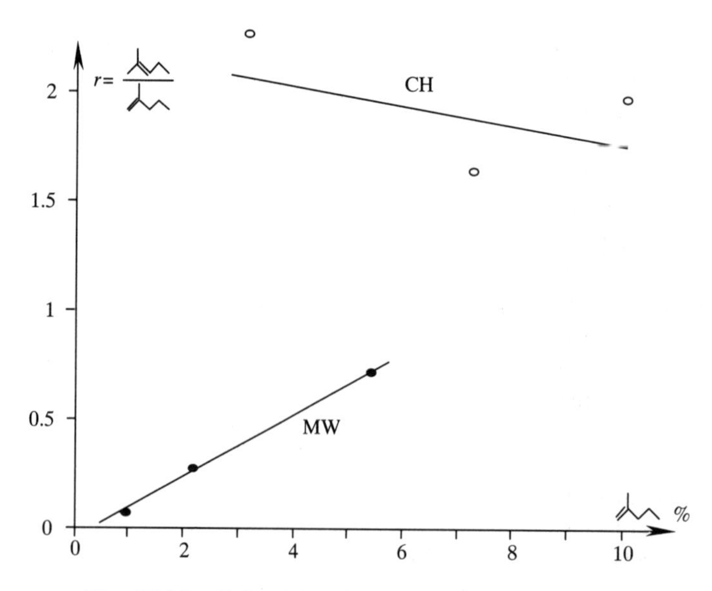

Fig. 12.20 *Selectivity of isomers in acidic reaction.*

in Figure 12.16), probably in preference to the charge shift pathways. This type of effect is easily understood as an electric field effect, rather than a thermal effect, since the carbenium ions are subjected to electric field forces. The detailed mechanisms have not yet been explained, but they have been confirmed by studying the corresponding reaction of 2-methyl-1-pentene on the same support, which gives similar results and conclusions.

12.5 COMMENTS ON METHODS OF STUDYING MICROWAVE-ENHANCED CATALYSIS

The case studies yield several useful observations on experimental methods for studying microwave enhanced catalytic reactions. It has been demonstrated that catalysis in a wave-guide at 2.45 GHz can be compared to a conventional oven under the same stationary physical conditions (composition and flow of gas and catalyst temperature). Also, the geometry of the MW catalyst reactor is nearly identical to that presently used, a distinct design advantage when changing from conventional to MW energy sources.

By ensuring that good thermal isolation of the reactor is maintained, and by using a small sample of catalyst heated for a long time, we have confidence that the temperature of the catalyst is nearly uniform (in spite of the non-uniform field) owing to heat transfer phenomena. So, the comparison of MW and conventional heating is significant. However, the art of the presentation is important. We can make useful estimates of catalyst temperature even though the thermocouple does accumulate error in the MW field. But, direct plots of output vs. temperature yield little useful information and obscure the significant differences between the processes.

It is much better to analyze the results in terms of a set of system responses, and to separate the variables into classes of response, including known information about the reaction of interest. A response variable is any measurement of the quantity of any product obtained, or is the measurement of the consumption of any initial product. Conversion and elementary selectivities, as defined in equations (12.2) and (12.3), are also valuable response variables for system analysis. Of course, the results may conveniently be presented by plotting any of the responses as a function of any other of them. Then, if MW and CH data coincide for all response variable pairs we can be sure that the catalyst behavior is the same under both types of heating.

On the other hand, if MW and CH data follow different curves in at least one plot we can be sure that the operating principle is different under MW irradiation. The plot of interest is the one which provides the best view of the differences between MW and CH energy sources. Two classes of responses result: those which show a difference and those which do not. The separation of the response variables into classes indicates where the MW and CH responses diverge.

Please note that this method does not work for simple single-step reactions such as decomposition of molecule A into two parts A' and A". Indeed, in that case the mass balance gives the relation between α and S' and also between α and S''. So, the three responses are not independent variables and belong to the same class. This situation is observed in many cases — for example, the full oxidation of CH_4 over perovskites into CO_2. In the simple reaction case the discussion can only concern reaction rate, and any observed enhancement comes from thermal effects.

12.6 SUMMARY

The three examples which have been presented illustrate that microwaves influence some catalytic reactions. The diversity of possible effects has been investigated, but the list is probably far from complete. To date we have identified four possible effects of the electric field: (1) microthermal treatment can permanently modify the catalyst, (2) heterogeneous phases (solid and gas) may give inter-phase temperature gradients, (3) the electric field can act directly on reaction sites (O_2 sites in the first step of activation of methane), and (4) the field can act on intermediate species, such as carbenium ions. We see that macroscopic and microscopic effects are coupled. Before engineers can take them into account in the design of a catalyst of significant volume we need to be more precise about the relationship between them.

REFERENCES

Further information can be found in the following texts:

Anderson, J. R. and Boudart M. (1984) *Catalysis Science and Technology* vol 2–6. Springer-Verlag, Berlin.

Imelik, B., Naccache C., Coudurier G., Ben Taarit, Y. and Vedrine J. C. (1985) *Catalysis by Acids and Bases* Elsevier, Amsterdam.

Satterfield, C. N. (1990) *Heterogeneous Catalysis in Industrial Practice. (2nd. Ed.)* McGraw-Hill.

Seyfried, L., Garin, F., Maire, G., Thiebaut, J. M. and Roussy, G. (1994) Microwave electromagnetic field effects on reforming catalysts *J. Catalysis*, **148**, 281–287.

Thiebaut, J. M., Roussy, G., Medjram, M. S., Garin, F., Seyfried, L. and Maire, G. (1993) Durable changes of the catalytic properties of alumina-supported platinum induced by microwave irradiation *Catalysis Letters*, **21**, 133–138.

and in the patents:

Roussy, G., and Maire, G. (1698) Procédé catalytique microonde de réformage des hydrocarbures EP-0519 824A1.

Roussy, G., Marchand, C., Thiebaut, J. M., Souiri, M., Kiennemann, A., Petit, C., Maire, G. Procédé catalytique d'oxydation ménagée du méthane par microondes pour la synthèse de l'ethane et de l'ethylène et catalyseurs mis en oeuvre dans le procédé EdF.F.92. 11676-US 08/131, 428. Filed 4 October 1993.

Index

Index compiled by Geoffrey C. Jones